Lecture Notes in Computer Science   14240

Founding Editors

Gerhard Goos
Juris Hartmanis

The series Lecture Notes in Computer Science (LNCS), including its subseries Lecture Notes in Artificial Intelligence (LNAI) and Lecture Notes in Bioinformatics (LNBI), has established itself as a medium for the publication of new developments in computer science and information technology research, teaching, and education.

LNCS enjoys close cooperation with the computer science R & D community, the series counts many renowned academics among its volume editors and paper authors, and collaborates with prestigious societies. Its mission is to serve this international community by providing an invaluable service, mainly focused on the publication of conference and workshop proceedings and postproceedings. LNCS commenced publication in 1973.

Franco Maria Nardini · Nadia Pisanti ·
Rossano Venturini

Editors

String Processing and Information Retrieval

30th International Symposium, SPIRE 2023
Pisa, Italy, September 26–28, 2023
Proceedings

 Springer

Editors
Franco Maria Nardini (iD)
ISTI-CNR
Pisa, Italy

Nadia Pisanti (iD)
University of Pisa
Pisa, Italy

Rossano Venturini (iD)
University of Pisa
Pisa, Italy

ISSN 0302-9743   ISSN 1611-3349 (electronic)
Lecture Notes in Computer Science
ISBN 978-3-031-43979-7   ISBN 978-3-031-43980-3 (eBook)
https://doi.org/10.1007/978-3-031-43980-3

Preface

The 30th International Symposium on String Processing and Information Retrieval (SPIRE) was held on September 26–28, 2023, in Pisa (Italy), followed by the 18th Workshop on Compression, Text, and Algorithms (WCTA) held on September 29, 2023.

SPIRE started in 1993 as the South American Workshop on String Processing. It was held in Latin America until 2000. Then, SPIRE moved to Europe, and from then on, it has been held in Australia, Japan, the UK, Spain, Italy, Finland, Portugal, Israel, Brazil, Chile, Colombia, Mexico, Argentina, Bolivia, Peru, the USA, and France. SPIRE continues the long and well-established tradition of encouraging high-quality research at the broad nexus of string processing, information retrieval, and computational biology.

This volume contains the accepted papers presented at SPIRE 2023. SPIRE 2023 received a total of 47 submissions. Each submission received at least three reviews. After an intensive discussion phase, the Scientific Program Committee accepted 31 papers. We thank all the authors for their valuable contributions and presentations at the conference and the Program Committee members for their valuable work during the review and discussion phases. We thank Springer for publishing the proceedings of SPIRE 2023 in the LNCS series and ACM SIGIR for sponsoring the conference.

The scientific program of SPIRE 2023 includes invited talks by three eminent researchers in the field: Sebastian Bruch (Pinecone, USA), Inge Li Gørtz (Technical University of Denmark, Denmark), and Jakub Radoszewski (University of Warsaw, Poland).

SPIRE 2023 had a Best Paper Award, sponsored by Springer. The award was announced during the conference.

Finally, we thank the Local Organizing Committee members for making the conference successful.

August 2023

Franco Maria Nardini
Nadia Pisanti
Rossano Venturini

Organization

Program Committee Chairs

Franco Maria Nardini	ISTI-CNR, Pisa, Italy
Nadia Pisanti	University of Pisa, Italy
Rossano Venturini	University of Pisa, Italy

Program Committee

Diego Arroyuelo	Universidad Técnica Federico Santa María, Chile
Jasmijn Baaijens	TU Delft, The Netherlands
Golnaz Badkobeh	Goldsmiths, University of London, UK
Ricardo Baeza-Yates	Northeastern University, USA, OptIA, Chile, and Universitat Pompeu Fabra, Spain
Giulia Bernardini	Università di Trieste, Italy
Paola Bonizzoni	Università degli Studi di Milano - Bicocca, Italy
Panagiotis Charalampopoulos	Birkbeck, University of London, UK
Giorgio Maria Di Nunzio	Università di Padova, Italy
Gabriele Fici	Università di Palermo, Italy
Travis Gagic	Dalhousic University, Canada
Pawel Gawrychowski	University of Wroclaw, Poland
Filippo Geraci	Institute for Informatics and Telematics of C.N.R., Italy
Daniel Gibney	Georgia Institute of Technology, USA
Shunsuke Inenaga	Kyushu University, Japan
Dominik Kempa	Stony Brook University, USA
Tomasz Kociumaka	Max Planck Institute for Informatics, Germany
Dominik Köppl	Tokyo Medical and Dental University, Japan
Susana Ladra	University of A Coruña, Spain
Thierry Lecroq	University of Rouen Normandy, France
Zsuzsanna Liptak	Università degli Studi di Verona, Italy
Felipe A. Louza	Universidade Federal de Uberlândia, Brazil
Sean MacAvaney	University of Glasgow, UK
Joel Mackenzie	University of Queensland, Australia
Cinzia Pizzi	Università di Padova, Italy
Giovanna Rosone	Università di Pisa, Italy
Kunihiko Sadakane	University of Tokyo, Japan

Blerina Sinaimeri	LUISS Guido Carli University, Italy
Jouni Sirén	University of California, Santa Cruz, USA
Hélène Touzet	CNRS, CRIStAL, Lille, France
Oren Weimann	University of Haifa, Israel
Wiktor Zuba	CWI, The Netherlands

Additional Reviewers

Abedin, Paniz
Baier, Uwe
Bannai, Hideo
Barton, Carl
Brown, Nathaniel
Chubet, Oliver
David, Julien
Della Vedova, Gianluca
Dondi, Riccardo
Fariña, Antonio
Fuentes, Jose
Gabory, Estéban
Guerrini, Veronica
I, Tomohiro
Jin, Ce
Kammer, Frank
Lefebvre, Arnaud
Limasset, Antoine

Lukasiewicz, Aleksander
Masillo, Francesco
Mendivelso, Juan
Nakashima, Yuto
Nogler, Jakob
Park, Kunsoo
Parmigiani, Luca
Pirola, Yuri
Prezza, Nicola
Radoszewski, Jakub
Rizzi, Raffaella
Sommer, Frank
Steiner, Teresa Anna
Telles, Guilherme
van Bemmelen, Jasper
Vinciguerra, Giorgio
Walen, Tomasz

Abstracts of Invited Talks

Information Retrieval Needs More Theoreticians

Sebastian Bruch 🆔

Pinecone, New York, NY, USA
sbruch@acm.org

Abstract. The juxtaposition of "theory" and "Information Retrieval" (IR) often invites much eyebrow-raising and chin-stroking. That is because a trove of exciting applied work in IR—particularly since the dawn of deep learning—along with a constant drift of the literature towards empirical studies and heuristics, tempt one to conclude that "experimental" is a more apt adjective to describe much of the discipline. The reality, however, is more nuanced: experimental explorations rest on foundational algorithmic and data structure research that is routinely overlooked.

Embarking on a quick journey through time helps unearth the theoretical underpinnings of IR. Historical examples abound, from *compactly* indexing massive amounts of text with compressed inverted indexes; to efficient set intersection algorithms and mechanisms for the dynamic pruning of inverted lists for fast top-k retrieval; to compressed representations of decision forests and their efficient inference for learning-to-rank. One can effortlessly draw a direct line from these developments to a range of applications such as search and recommendation systems that have flourished at massive scales.

As the IR landscape evolves and the field becomes more intertwined with deep learning, *vectors* and *matrices* replace *text* as units of information. That, in turn, gives way to fascinating new algorithmic problems for which the existing tools are not only insufficient but are also wrong. In fact, a strategy of avoiding new problems and resorting instead to the existing arsenal of data structures and algorithms holds us back and restrains us from looking beyond the lamppost for new ideas.

That is the argument I will make in this keynote: For IR research to remain innovative, more theoreticians must go back to the drawing board, reexamine and possibly reinvent much of what is at its core. To illustrate this point, I will give concrete examples of some of today's open problems. These include developing (non-linear) sketching algorithms for vectors and matrices for the purposes of fast retrieval at scale; building compact indexes; designing approximate top-k retrieval by maximal inner product, possibly subject to additional constraints; and generalizing retrieval from inner product to matrix norms. I hope that this talk will convince us of the need for more theoretical work and entice this particular community to explore new challenges in IR.

Regular Expression Matching

Inge Li Gørtz ⓘ

Technical University of Denmark, Denmark
inge@dtu.dk

A regular expression specifies a set of strings formed by single characters combined with concatenation, union, and Kleene star operators. Given a regular expression R and a string Q, the regular expression matching problem is to decide if Q matches any of the strings specified by R. Regular expressions are a fundamental concept in formal languages and regular expression matching is a basic primitive for searching and processing data. A standard textbook solution [Thompson, CACM 1968] constructs and simulates a nondeterministic finite automaton, leading to an $O(nm)$ time algorithm, where n is the length of Q and m is the length of R. Despite considerable research efforts only polylogarithmic improvements of this bound are known. Recently, conditional lower bounds provided evidence for this lack of progress when Backurs and Indyk [FOCS 2016] proved that, assuming the strong exponential time hypothesis (SETH), regular expression matching cannot be solved in $O\big((nm)^{1-\epsilon}\big)$, for any constant $\epsilon > 0$. Hence, the complexity of regular expression matching is essentially settled in terms of n and m.

In this talk we will give an overview of the main standard techniques for regular expression matching, and show a new approach that goes beyond worst-case analysis in n and m. We introduce a *density* parameter, Δ, that captures the amount of nondeterminism in the NFA simulation on Q. The density is at most $nm+1$ but can be significantly smaller. Our main result is a new algorithm that solves regular expression matching in time.

$$O\left(\Delta \mathrm{loglog}\frac{nm}{\Delta} + n + m\right)$$

Recent Results on the Longest Common Substring Problem

Jakub Radoszewski [ORCID]

Institute of Informatics, University of Warsaw, Poland
jrad@mimuw.edu.pl

Abstract. In the longest common substring (LCS) problem, we are given two strings S and T, each of length at most n, and are to find a longest string occurring as a fragment of both S and T. An $O(n)$-time solution for this classic problem, assuming that S and T are over a constant-sized alphabet, dates back to Weiner's seminal paper on suffix trees (SWAT 1973). This solution can be extended to strings over an integer alphabet using Farach's suffix tree construction algorithm (FOCS 1997) and to computing the LCS of many strings of total length n (Hui, CPM 1992).

Recently it was shown that, under the word RAM model of computation, the LCS of two strings over a constant-sized alphabet can be computed even faster. More precisely, Charalampopoulos, Kociumaka, Pissis, and Radoszewski (ESA 2021) showed that assuming the alphabet size is σ and that the strings are represented in a packed form of size $O(n \log \sigma / \log n)$, the LCS of S and T can be computed in $O(n \log \sigma / \sqrt{\log n})$ time.

In the last 10 years, several further results on the LCS problem and its variants have been presented. These include computing the LCS of two strings using only $O(s)$ space, where $s < n$, maintaining the LCS of two dynamic strings, internal queries for the LCS of two fragments of a given text, approximate variants of the LCS in which one searches for the longest fragment of string S that resembles a fragment of string T (the two fragments may be required to have a small Hamming or edit distance), and computing the LCS with the assumption that one of the strings is given in a compressed form (a less efficient algorithm assuming that both S and T are compressed was known earlier).

In the talk I will present a survey of known results on the LCS problem, with an emphasis on the result from ESA 2021 and the underlying technique by Charalampopoulos et al. (CPM 2018), and list open problems related to the LCS computation.

Contents

Longest Common Prefix Arrays
for Succinct k-Spectra

Jarno N. Alanko[ID], Elena Biagi[✉][ID], and Simon J. Puglisi[ID]

Helsinki Institute for Information Technology (HIIT), Department of Computer
Science, University of Helsinki, Helsinki, Finland
{jarno.alanko,elena.biagi,simon.puglisi}@helsinki.fi

Abstract. The k-spectrum of a string is the set of all distinct substrings
of length k occurring in the string. K-spectra have many applications
in bioinformatics including pseudoalignment and genome assembly. The
Spectral Burrows-Wheeler Transform (SBWT) has been recently introduced as an algorithmic tool to efficiently represent and query these
objects. The longest common prefix (LCP) array for a k-spectrum is
an array of length n that stores the length of the longest common prefix of adjacent k-mers as they occur in lexicographical order. The LCP
array has at least two important applications, namely to accelerate pseudoalignment algorithms using the SBWT and to allow simulation of
variable-order de Bruijn graphs within the SBWT framework. In this
paper we explore algorithms to compute the LCP array efficiently from
the SBWT representation of the k-spectrum. Starting with a straightforward $O(nk)$ time algorithm, we describe algorithms that are efficient
in both theory and practice. We show that the LCP array can be computed in optimal $O(n)$ time, where n is the length of the SBWT of the
spectrum. In practical genomics scenarios, we show that this theoretically optimal algorithm is indeed practical, but is often outperformed
on smaller values of k by an asymptotically suboptimal algorithm that
interacts better with the CPU cache. Our algorithms share some features
with both classical Burrows-Wheeler inversion algorithms and LCP array
construction algorithms for suffix arrays. Our C++ implementations of
these algorithms are available at https://github.com/jnalanko/kmer-lcs.

Keywords: longest common prefix · LCP · longest common suffix ·
k-mer · string algorithms · compressed data structures · de Bruijn
graph · Burrows-Wheeler transform · BWT

1 Introduction

The k-*spectrum* of a string S is the set of substrings of a given length k that occur
in S. Indexing k-spectra has become an important topic in bioinformatics, perhaps most notably in the form of de Bruijn graphs, which are a long-standing

---

Supported in part by the Academy of Finland via grants 339070 and 351150.

F. M. Nardini et al. (Eds.): SPIRE 2023, LNCS 14240, pp. 1–13, 2023.
https://doi.org/10.1007/978-3-031-43980-3_1

tool for genome assembly [6] and more recently for pangenomics [3,8,11]. In metagenomics, k-spectra are used as a concise approximation of the sequence content of the sample, allowing rapid similarity estimation between data collected from sequencing runs [10,12]. In most current genomics applications k is in the range from 20 to 100.

Recently, the Spectral Burrows-Wheeler transform (SBWT) [2] has been introduced as an efficient way to losslessly encode and query k-spectra. In particular, the SBWT encodes the k-mers of the spectrum in *colexicographical* order. Combining the SBWT with entropy-compressed bitvectors leads to a data structure that encodes the spectrum in little more than 2 bits per k-mer [1,2]. Remarkably, while in this form, it is also possible to answer lookup queries on the spectrum rapidly, in fact in $O(k)$ time.

The SBWT allows a lookup query for a given k-mer to be reduced to at most k *right-extension queries*. The input to a right-extension query is a letter c and an interval $[i, j]$ in the colexicographic ordering of the k-mers of the spectrum such that all k-mers in the interval share a suffix X. The query returns the interval $[i', j']$ that contains all the k-mers that have Xc as a suffix (or an empty interval if none do).

Our focus in this paper is on augmenting the SBWT with a data structure called the longest common suffix (LCS) array that stores the lengths of the longest common suffixes of adjacent k-mers in colexicographical order (we give a precise definition below[1]). The *LCS* array allows us to support so-called *left contraction* queries: given an interval $[i, j]$ in the colexicographical ordering of the k-mers of the spectrum containing all the k-mers that share a suffix X of length $k' \in (0, k]$ and a contraction point $t < k'$, a left contraction returns the interval $[i', j']$ containing all the k-mers having $X[t..k']$ as a suffix. Left contractions have at least two interesting applications, namely the implementation of variable-order de Bruijn graphs [5] and streaming k-mer queries [2]. We avoid further treatment of these applications here, and refer the reader to [2,5] for details. Our focus instead is on the efficient construction of the *LCS* array of a k-spectrum given its SBWT, which is also an interesting problem in its own right.

We are aware of little prior work on efficient *LCS* array construction for k-spectra. A naive approach is to expand the entire contents of the spectrum from the SBWT and scan it in colexicographical order. This requires $O(nk)$ time and $O(nk \log \sigma)$ bits of space, where σ is the size of the alphabet of the k-mers in the spectrum. Bowe et al. [5] make use of the *LCS* array for a k-spectra for variable order de Bruijn graphs, but do not address construction. Very recently, Conte et al. [7] introduced LCS arrays for Wheeler graphs[2], but describe

---

[1] We remark here that the LCS array of a colexicographically-ordered spectrum is equivalent to the longest common prefix (LCP) array of the lexicographically-ordered spectrum, and the algorithms we describe in this paper to compute the LCS array are trivially adapted to compute the LCP array.

[2] Wheeler graphs are a class of graphs including de Bruijn graphs, that admit a generalization of the Burrows-Wheeler transform. The SBWT can be seen as a special case of the Wheeler graph indexing framework.

no construction algorithm, mentioning only in passing that a polynomial-time algorithm is possible. Prophyle, due to Salikov et al. [13] uses k-LCP information for sliding window queries on the BWT and describes an $O(nk)$ construction algorithm, where n is the size of the full BWT, which can be orders of magnitude larger than the SBWT for repetitive datasets.

Contribution. We describe three different algorithms for computing the *LCS* array of a k-spectrum from its SBWT. The first of these essentially decodes the k-mers of the k-spectrum in colexicographical order starting from their rightmost symbols in k rounds, keeping track of when the suffixes become distinct using just $n(1 + \log \sigma)$ bits of side information (significantly less than the naive method mentioned above) and taking $O(nk)$ time overall. Our second approach is similar, but exploits the small DNA alphabet to decode multiple symbols per round with the effect of reducing computation and, importantly, CPU cache misses. Its running time is $O(cn + (k - c)n/c)$ overall with $O(nc \log \sigma + \sigma^c \log n)$ bits of extra space, where $c \le k$ is a parameter controlling a space-time tradeoff. Our final algorithm runs in time linear in n — independent of k — and while it is shaded by the second algorithm on smaller k, it becomes dominant as k grows.

The remainder of this article is organized as follows. The next section sets notation and basic definitions. Sections 3–5 then describe the three above-mentioned *LCS* array construction algorithms in turn. Section 6 presents an experimental analysis of their performance in the context of a real pangenomic indexing task. Conclusions, reflections, and avenues for future work are then offered.

2   Preliminaries

Throughout we consider a *string* $S = S[1..n] = S[1]S[2]\ldots S[n]$ on an integer alphabet Σ of σ symbols. In this article we are mostly interested in strings on the DNA alphabet, i.e. when $\Sigma = \{A, C, G, T\}$. The *colexicographic order* of two strings is the same as the lexicographic order of their reverse strings. The *substring* of S that starts at position i and ends at position j, $j \ge i$, denoted $S[i..j]$, is the string $S[i]S[i+1]\ldots S[j]$. If $i > j$, then $S[i..j]$ is the empty string ε. A suffix of S is a substring with ending position $j = n$, and a prefix is a substring with starting position $i = 1$. We use the term k-mer to refer to a (sub)string of length k. The following two basic definitions relate to k-spectra.

Definition 1. *(k-spectrum). The k-spectrum of a string T, denoted with $S_k(T)$, is the set of all distinct k-mers of the string T.*

Definition 2. *(k-prefix set). The k-prefix set of a string T is defined as the left-padded set of prefixes $P_k(T) = \{\$^{k-i}T[1..i] \mid i = 0,\ldots,k-1\}$, where $\$$ is a special character not found in the alphabet, that is smaller than all characters of the alphabet.*

The k-spectrum of a *set* of strings $T_1, \ldots T_m$, denoted with $S_k(T_1, \ldots T_m)$, is defined as the union of the k-spectra of the individual strings. For example, consider the strings AGGTAAA and ACAGGTAGGAAAGGAAAGT. The 4-spectrum is the set {GAAA, TAAA, GGAA, GTAA, AGGA, GGTA, AAAG, ACAG, GTAG, AAGG, CAGG, TAGG, AAGT, AGGT}. Likewise, the k-prefix set $P_k(T_1, \ldots T_m)$ is the union of the k-prefix sets of the individual strings. In this case, the 4-prefix set is {\$\$\$\$, \$\$\$A, \$\$AG \$AGG, \$\$AC, \$ACA}.

Definition 3. *(k-source set). The k-source set $R_k(K)$ of a set of k-mers K is the set $R_k(K) = \{x \in K \mid \nexists y \in K \text{ such that } y[2..k] = x[1..k-1]\}$*

In our running example, the 4-source set of the 4-spectrum has just the 4-mer {ACAG}. The *extended k-spectrum* is the union of the spectrum and the k-prefix set of the k-source set, plus the k-mer $\$^k$ that is always added to avoid some corner cases.

Definition 4. *(Extended k-spectrum). The extended k-spectrum $S_k'(T_1, \ldots, T_m)$ of a set of strings $T_1, \ldots, T_m$ is the set $S_k(T_1, \ldots, T_m) \cup P_k(R_k(T_1, \ldots, T_m)) \cup \{\$^k\}$*

We are now ready to define the SBWT. The definition below corresponds to the multi-SBWT definition of Alanko et al. [2].

Definition 5. *(Spectral Burrows-Wheeler transform, SBWT) Let $\{T_1, \ldots, T_m\}$ be a set of strings from an alphabet Σ. Let x_i be the colexicographically i-th element of the extended k-spectrum $S_k'(T_1, \ldots, T_m)$ of size n. The SBWT of order k is a sequence $X_1, X_2, \ldots X_n$ of subsets of Σ. The set X_i is the empty set if $i > 1$ and $x_{i-1}[2..k] = x_i[2..k]$, otherwise $X_i = \{c \in \Sigma \mid x_j[2..k]c \in S_k'(T_1, \ldots, T_m)\}$*

The sets in the SBWT represent the labels of outgoing edges in the node-centric de Bruijn graph of the input strings, such that we only include outgoing edges from k-mers that have a different suffix of length $k-1$ than the preceding k-mer in the colexicographically sorted list. Figure 1 illustrates the SBWT and the associated de Bruijn graph. The addition of the k-prefix set of the k-source set is a technical detail necessary to make the transformation invertible and searchable.

There are many ways to represent the subset sequence of the SBWT [1,2]. In this paper, we focus on the *matrix representation*. This representation is currently the most practical version known for small alphabets, and it is used e.g. in the k-mer pseudoalignment tool Themisto [3].

Definition 6. *(Plain Matrix SBWT) The plain matrix representation of the SBWT sequence is a binary matrix M with σ rows and n columns. The value of $M[i][j]$ is set to 1 iff subset X_j includes the ith character in the alphabet.*

Figure 3 illustrates the matrix SBWT of our running example. Lastly, we define the central object of interest in this paper: the LCS array of an SBWT:

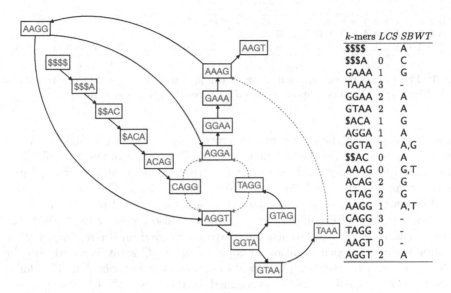

k-mers	LCS	SBWT
$$$$	-	A
$$$A	0	C
GAAA	1	G
TAAA	3	-
GGAA	2	A
GTAA	2	A
$ACA	1	G
AGGA	1	A
GGTA	1	A,G
$$AC	0	A
AAAG	0	G,T
ACAG	2	G
GTAG	2	G
AAGG	1	A,T
CAGG	3	-
TAGG	3	-
AAGT	0	-
AGGT	2	A

Fig. 1. Left: The de Bruijn graph (with $k = 4$) of the set of two strings {AGGTAAA, ACAGGTAGGAAAGGAAAGT}. Red dashed edges are pruned from the graph because the node they point to can be reached from another (black) edge. Right: The extended k-spectrum of the input strings in colexicographical order, together with the longest common suffix (LCS) array and the spectral Burrows-Wheeler transform (SBWT).

	$$$$	$$$A	GAAA	TAAA	GGAA	GTAA	$ACA	AGGA	GGTA	$$AC	AAAG	ACAG	GTAG	AAGG	CAGG	TAGG	AAGT	AGGT
A	1	0	0	0	1	1	0	1	1	1	0	0	0	1	0	0	0	1
C	0	1	0	0	0	0	0	0	0	0	0	0	0	0	0	0	0	0
G	0	0	1	0	0	0	1	0	1	0	1	1	1	0	0	0	0	0
T	0	0	0	0	0	0	0	0	0	0	1	0	0	1	0	0	0	0

Fig. 2. The binary matrix representation of the spectral Burrows-Wheeler transform (SBWT).

Definition 7. *(Longest common suffix array, LCS array) Let $\{T_1, \ldots, T_m\}$ be a set of strings and let x_i denote the colexicographically i-th k-mer of $S'_k(T_1, \ldots, T_m)$. The LCS array is an array of length $|S'_k(T_1, \ldots, T_m)|$ such that $LCS[1] = 0$ and for $i > 1$, the value of $LCS[i]$ is the length of the longest common suffix of k-mers x_i and x_{i-1}.*

In the definition above, the empty string is considered a common suffix of any two k-mers, so the longest common suffix is well-defined for any pair of k-mers. Figure 1 illustrates the LCS array of our running example.

3   Basic $O(nk)$-Time *LCS* Array Construction

Before describing how to compute the *LCS* array we are going to explain how the whole k-spectrum can be recovered from the SBWT. We can reconstruct the

0	1	2	3	4	5	6	7	8	9	10	11	12	13	14	15	16	17	18	19	20	21	22	23	24	25	26	27	28	29	30	31	32	33	34
V	A	C	G	A	A	G	A	A	G	A	A	G	T	G	G	A	T	A																
B	1	0	1	0	1	0	1	1	0	1	0	1	0	1	0	1	0	0	1	0	1	0	0	1	0	1	0	1	0	1	0	0	1	1	1	1	0

Fig. 3. The concatenated representation of the spectral Burrows-Wheeler transform (SBWT) used by the super-alphabet-based LCS construction algorithm.

full k-spectrum from the binary matrix representation M of the SBWT with σ rows and n columns and the cumulative array C included in the SBWT. Since k-mers are colexicographically sorted, they are assembled back to front. First, the last character of each k-mer is retrieved based on the C array. These last added characters will be accessed later and are then stored in an array L. In accordance with the LF mapping property, which also holds for the SBWT, the previous character of each k-mer is recursively retrieved until reaching length k as follows: First, at each iteration, a copy C' of the C array is saved and the vector P for storing the last propagated characters is initialized with a dollar symbol. Then, each column i of M is scanned. If $M[c, i]$ is 1, the first free position of the c block marked by the C' array in P is set to the character in $L[i]$. Since we are scanning every column in M, we do not need to issue rank queries, but it is instead sufficient to increase the counter $C'[c]$ by one. At the end of each iteration, the newly propagated characters are copied to L. Considering the de Brujin graph of the SBWT, with this procedure edge labels are propagated one step forward in the graph.

Calculating the LCS array from the SBWT is similar to the procedure described above. The LCS array is initialized as an array of zeros and it is updated at each round of M scanning by checking the mismatches between two adjacent newly propagated characters. Once an entry of the LCS array is updated, it is never modified again. Since for each character of the k-mers we need to traverse all columns of M once, the whole k-spectra can be retrieved in $O(n\sigma k)$-time, where n is the number of k-mers in the SBWT. Instead of scanning M k times, we could traverse the Subset Wavelet Tree of the string (see [1]) and issue a binary rank operation for every character in each subset. Repeating this for each k-mer character will result in the LCS construction in time $O(nk \log \sigma)$. This reduces to $O(nk)$ assuming a constant σ. Computing the LCS array does not alter this time complexity.

4   Faster Construction via Super-Alphabet Techniques

The super-alphabet techniques described here are based on first decoding a c-symbol suffix of each k-mer using the previous algorithm in $O(cn)$ time and subsequently computing the remaining information in $O(1 + (k - c)/c)$ rounds and $O(cn + (k - c)n/c)$ time overall with $O(n)$ extra space. Given $c = 2$, the algorithm first replicates the basic one up to the computation of the last 2 characters of each k-mer as well as their LCS values. At this point, the 2 last symbols of the ith k-mer, $P[i]$ and $L[i]$, are combined to create a super-character

Algorithm 1. Basic LCS array construction in $O(nk)$ time.
Input: SBWT matrix M with n columns and σ rows, $\Sigma = \{1, \ldots, \sigma\}$ and C array.
Output: k-bounded LCS array.

$LCS \leftarrow$ Array of length n initialized to 0
$mismatches \leftarrow$ Array of length n initialized to 0   ▷ positions set in LCS
$L \leftarrow$ Array of length n, with $\sigma + 1$ characters, initialized according to C
for $round = 0 \ldots k - 1$ **do**
    for $i = 1 \ldots n - 1$ **do**   ▷ LCS[1]=0 by definition
        if $mismatches[i + 1] = 0$ **and** $L[i + 1] \neq L[i]$ **then**
            $mismatches[i + 1] \leftarrow 1$
            $LCS[i + 1] \leftarrow round$ ▷ store the longest match length
    $P \leftarrow$ Array of length n initialized to $\$$
    $C' \leftarrow$ copy of the C array
    for $i = 1 \ldots n$ **do**
        for $c \in \Sigma$ **do**
            if $M[c, i] = 1$ **then**
                $C'[c] \leftarrow C'[c] + 1$
                $P[C'[c]] \leftarrow L[i]$
    $L \leftarrow P$
return LCS

(or meta-character) $P[i] \cdot L[i]$ which is stored in $L[i]$. A new C array is then generated from the alphabet of super-characters. The following super-characters for each k-mer are then retrieved as in the basic algorithm. The only difference is that in the present case, the algorithm uses the concatenated representation of the SBWT of super-characters instead of the plain matrix representation. The concatenated representation of the SBWT sequence[3] consists of a concatenation of the subsets characters, stored in a vector V, and an encoding of the subsets sizes stored in a bitvector B. In further detail, let $S(X_i)$ be the concatenation of characters in the subset X_i, then $V = S(X_i) \cdot S(X_2) \cdot S(X_n)$. No symbol will be stored in V if X_i is the empty set. The empty sets are represented in $B = 1 \cdot 0^{|S(X_1)|} \cdot 1 \cdot 0^{|S(X_2)|} \ldots 1 \cdot 0^{|S(X_n)|}$. The concatenated representation of a c-super-alphabet, V' and B', can be obtained from V and B, the concatenated representation of the $c/2$-(super-)alphabet. V' is filled in, scanning V, with $V[j]$ where $0 \leq j \leq |V|$ concatenated with the characters in the subset X marked by the C array entry of $V[j]$ in V. For each character in V, $1 \cdot 0^{|X|}$ is appended to B'. No rank nor select queries are necessary as it is sufficient to update a copy of the C array. Considering the de Brujin graph of the SBWT, to create a super-concatenated representation edge labels are propagated one step backward in the graph.

Similarly to the basic algorithm, the preceding super-character of each k-mer is recursively retrieved until reaching length k as follows: First, at each iteration, a copy of the super C array is stored and P is initialized with the

---

[3] A similar but different structure is described in [2].

smallest super-character $\$^c$. Then V' is scanned keeping track of the number of subsets encountered with a counter v which is increased by 1 if $B[i + v] = 1$. If $B[i + v] = 0$, $L[i]$ is assigned to P at the index corresponding to the position of the $V'[i]$ super-character block marked by the C array. As for the basic alphabet, since every subset is inspected in order, there is no need to issue rank queries, but it is instead sufficient to increase the copied C' counter for $V'[i]$ by one. At the end of each iteration, the newly propagated super-characters are stored in L. Since we are skipping nodes in the graph, the iteration number r goes from c to at most $k + c - 1$ with steps of size c.

The LCS array using super-characters is computed by checking first the presence of mismatches in the rightmost single characters with an appropriate mask and only if no mismatch is found, subsequent characters are checked. The LCS is then updated accordingly. Given a super-character with $c = 2$ at index i as $c2 \cdot c1$, the algorithm compares first $c1[i]$ and $c1[i - 1]$. In the presence of a mismatch LCS would be updated to the iteration number r. If $c2[i] \neq c2[i - 1]$, $LCS[i] = r + 1$ since 1 is, in this case, the number of matches found in the characters of the super-character. If on the contrary, $c2[i] = c2[i - 1]$, the LCS could not be updated yet. The algorithm never checks more characters than necessary as it stops at the first encountered mismatch.

5   Construction in Linear Time

Our linear-time algorithm can be seen as a generalization of the linear-time LCP algorithm of Beller et al. [4] from the regular BWT to the SBWT. When the input is the spectrum of a single string and k approaches n, the SBWT coincides with the BWT of the reverse of the input[4], and both algorithms perform the same iteration steps.

The algorithm fills in the LCS in increasing order of the values. The main loop has k iterations, such that iteration i fills in LCS values that are equal to $i - 1$. Values that are not yet computed are denoted with $\perp$.

We denote the *colexicographic interval* of string α with $[\ell, r]_\alpha$, where ℓ and r respectively are the colexicographic ranks of the smallest and largest k-mer in the SBWT that have α as a suffix. The right extensions of interval $[\ell, r]_\alpha$, denoted with EnumerateRight(ℓ, r), are those characters c such that αc is a suffix of at least one k-mer in the SBWT. The interval of right extension c from $[\ell, r]_\alpha$, denoted with ExtendRight(ℓ, r, c), can be computed using the formula $[2 + C[c] + rank_c(\ell - 1), 1 + C[c] + rank_c(r)]_{\alpha c}$, where the rank is over the subset sequence of the SBWT [2], and $C[c]$ is the number of characters in the SBWT that are smaller than c.

The input to iteration i is a list of colexicographic intervals of substrings of length $i - 1$. For each interval $[\ell, r]_\alpha$ in the list, the algorithm computes all right-extensions $[\ell', r']_{\alpha c}$. If $LCS[r' + 1]$ is not yet filled yet, the algorithm sets $LCS[r' + 1] = i - 1$ and adds $[\ell', r']_{\alpha c}$ to the list of intervals for the next round.

---

[4] Assuming the input to the BWT is terminated with a $\$-symbol, and there is an added $\$-edge from the last k-mer of the input to the root of the SBWT graph.

Otherwise, $LCS[r' + 1]$ is not modified and interval $[\ell', r']_{ac}$ is not added to the next round. Algorithm 2 lists the pseudocode. The algorithm is designed so that at the end, every value of the LCS array has been computed.

Algorithm 2. Construction in $O(n \log \sigma)$ time.
Input: SBWT with support for EnumerateRight and ExtendRight.
Output: k-bounded LCS array.

1: $LCS \leftarrow$ Array of length n initialized to $\bot$
2: $LCS[1] \leftarrow 0$ ▷ By definition.
3: $I \leftarrow ([1, n])$ ▷ List of intervals for current round.
4: $I' \leftarrow ([1, 1])$ ▷ List of intervals for the next round. Here interval of $
5: **for** $i = 1..k$ **do**
6:     **while** $|I| > 0$ **do**
7:         $[\ell, r] \leftarrow$ Pop I
8:         **for** $c \in$ EnumerateRight(ℓ, r) **do**
9:             $[\ell', r'] \leftarrow$ ExtendRight(ℓ, r, c)
10:            **if** $r' < n$ and $LCS[r' + 1] = \bot$ **then**
11:                $LCS[r' + 1] \leftarrow i - 1$
12:                Push $[\ell', r']$ to I'
13:     $I \leftarrow I'$
14:     $I' \leftarrow$ Empty list
15: **return** LCS

5.1 Correctness

To prove the correctness of the algorithm, we introduce the concept of an L-*interval*. A colexicographic interval $[\ell, r]_\alpha$ is called an L-interval iff it is the longest colexicographic interval of a string with interval endpoint r. In case there are multiple strings with the same interval $[\ell, r]$, then the α in the subscript of the notation is the shortest string with this interval. The number of L-intervals is clearly $O(n)$ because each L-interval has a distinct endpoint. LCS array can be derived from the L-intervals as follows:

Lemma 1. *If $[\ell, r]_{c\alpha}$ is an L-interval, with $\alpha \in \Sigma^*$ and $c \in \Sigma$, then $LCS[r+1] = |\alpha|$*

Proof. It must be that $LCS[r + 1] < |c\alpha|$ because otherwise the k-mer with colexicographic rank $r + 1$ should have been included in the interval $[\ell, r]_{c\alpha}$. It must be that $LCS[r + 1] \geq |\alpha|$ because otherwise the interval of α also has endpoint r, which means that $c\alpha$ is not the shortest string with interval ending at r, contradicting the initial assumption.

The L-intervals form a tree, where the children of $[\ell, r]_\alpha$ are the single-character right-extensions $[\ell', r']_{ac}$ that are L-intervals. The lemma below implies that every L-interval is reachable by right extensions by traversing only L-intervals from the interval of the empty string:

Lemma 2. *Let αc be a substring of the input such that $\alpha \in \Sigma^*$ and $c \in \Sigma$. If $[\ell, r]_{\alpha c}$ denotes an L-interval, then $[\ell', r']_\alpha$ is an L-interval.*

Proof. Suppose for a contradiction that the Lemma does not hold. Then there exists an L-interval interval $[x, r']_\beta$ with $x \leq \ell'$ such that β is a proper suffix of α. Then by the SBWT right extension formula, the interval $[\ell'', r'']_{\beta c}$ is such that $r'' = r$ and $\ell'' \leq \ell$. It can't be that $\ell'' = \ell$, or otherwise αc was not the shortest string with interval $[\ell, r]$, and it can't be that $\ell'' < \ell$ because then the starting point ℓ was not minimal for end point r. In both cases we have a contradiction, which proves the claim.

We can now prove the correctness and the time complexity of the algorithm:

Theorem 1. *Given an SBWT having n subsets of alphabet Σ with $|\Sigma| = O(1)$, Algorithm 2 correctly computes every value of the LCS array in time $O(n)$.*

Proof. The algorithm traverses the L-interval tree in breadth-first order by right-extending from the empty string and visiting the shortest string representing each L-interval. Whenever the algorithm comes across an interval $[\ell', r']$ such that $LCS[r' + 1]$ is already set, we know that endpoint r' has already been visited before with a string shorter than the current string, so either $[\ell', r']$ is not an L-interval or the current string is not the shortest representative of it, so we can ignore it. By Lemma 2, the shortest representative string of every L-interval is reachable this way. There is guaranteed to be an L-interval for every endpoint r because there is at least a singleton colexicographic interval to every endpoint. Therefore, every value of the LCS array is eventually computed, and by Lemma 1, every computed value is correct. Since the number of L-intervals is $O(n)$, and EnumerateRight and ExtendRight can be implemented in constant time for a constant-sized alphabet, the total time is $O(n)$.

For small alphabets, the call to EnumerateRight can be replaced by a process that tries all σ possible right extensions. In this case, it is enough to track only interval endpoints, halving the space and number of rank queries required.

6   Experimental Evaluation

Experimental Setup. All our experiments were conducted on a machine with four 2.10 GHz Intel Xeon E7-4830 v3 CPUs with 12 cores each for a total of 48 cores, 30 MiB L3 cache, 1.5 TiB of main memory, and a 12 TiB serial ATA hard disk. The OS was Linux (Ubuntu 18.04.5 LTS) running kernel 5.4.0–58-generic. The compiler was g++ version 10.3.0 and the relevant compiler flags were -O3 and -DNDEBUG (-march=native was not used). All runtimes were recorded by instrumenting the code with calls to std::chrono. The peak memory (RSS) was measured using the getrusage Linux system call. C++ source code of the implementations tested is available upon request from the authors.

Datasets. We experiment on three data sets representing different types of sequencing data found in genomics applications:

1. A pangenome of 3682 E. coli genomes. The data was downloaded during the year 2020 by selecting a subset of 3682 assemblies listed in ftp://ftp.ncbi.nlm. nih.gov/genomes/genbank/bacteria/assembly_summary.txt with the organism name "Escherichia coli" with date before March 22, 2016. The resulting collection is available at zenodo.org/record/6577997. It contains 745,409 sequences of a total length 18,957,578,183.
2. The human reference genome version GRCh38.p14, available at https://www. ncbi.nlm.nih.gov/assembly/GCF_000001405.40. It contains 705 sequences of total length 3,298,430,636.
3. A set of 34,673,774 paired-end Illumina HiSeq 2500 reads each of length 251 sampled from the human gut (SRA identifier ERR5035349) in a study on irritable bowel syndrome and bile acid malabsorption [9]. The total length of this data set is 8,703,117,274 bases.

We focus solely on genomic data as that is currently the main application of the SBWT. The constructed index structures include both forward and reverse DNA strands. We experiment with values $k = 16, 32, 48, 64, 80, 96, 112, 128$ and 255. For the metagenomic reads, the maximum value used was 251 since this is the length of the reads. Figure 4 shows a plot of the number of distinct k-mers for varying k.

Algorithms. The basic and linear algorithms are implemented on top of the matrix representation of the SBWT. In the linear algorithm, we apply the observation mentioned at the end of Sect. 5.1 and only track interval end points.

The super-alphabet algorithm (labelled SA-2 in the plots) first constructs the concatenated representation from the matrix representation and operates on it alone after the initial round of alphabet expansion. We experimented only with a super-alphabet of size 2, and leave a more detailed exploration, including larger super-alphabets, for future work.

Results. Figure 5 shows on the top the runtime of each algorithm as a function of the k-mer size for each of the three data sets. We observe that the super-alphabet algorithm is consistently faster than the basic and linear algorithms until k reaches 128, after which the linear algorithm is clearly fastest—roughly three times faster than the basic algorithm on the E.coli dataset.

Memory usage for the algorithms is displayed at the bottom of Fig. 5. The super-alphabet algorithm uses significantly more memory than the other two, which is partly attributable to its use of the concatenated representation of the SBWT, which it must first build from the matrix representation, increasing peak memory. Moreover, it uses a larger data type to hold the current column of the SBWT matrix (a 16-bit word per element instead of an 8-bit one used in the basic algorithm). In comparison, the basic and linear implementations use startlingly little memory, which may make them preferable on systems where memory is scarce.

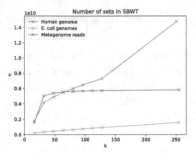

Fig. 4. The number of sets in the SBWT (approximately equal to the number of k-mers) in each dataset for various k used in our experiments.

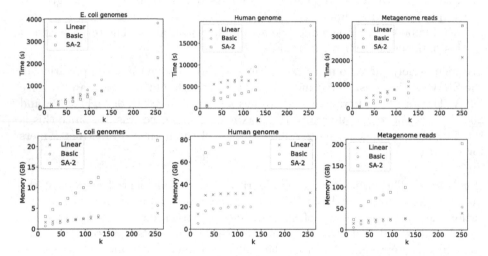

Fig. 5. Runtime and memory usage of LCS array construction algorithms versus k.

7   Concluding Remarks

We have explored the design space of longest common suffix array construction algorithms for k-spectra. In particular, we have described two algorithms that, on real genomic datasets, significantly outperform our baseline $O(nk)$-time, $O(n)$ space approach. The first exploits the smaller nucleotide alphabet to form metacharacters and reduce the number of rounds needed by the basic algorithm. The second takes linear time (assuming a constant-size alphabet) by computing the LCS values in a special order and also performs well in practice, especially when k is large.

All our algorithms have some dependency on σ and we leave removing this as an open problem. From a practical point of view, it would be interesting to develop parallel algorithms that may further accelerate LCS array construction on large data sets.

References

1. Alanko, J.N., Biagi, E., Puglisi, S.J., Vuohtoniemi, J.: Subset wavelet trees. In: Proceedings of the 21st International Symposium on Experimental Algorithms (SEA), LIPIcs. Schloss Dagstuhl - Leibniz-Zentrum für Informatik (2023)
2. Alanko, J.N., Puglisi, S.J., Vuohtoniemi, J.: Small searchable k-spectra via subset rank queries on the spectral burrows-wheeler transform. In Proceedings of SIAM Conference on Applied and Computational Discrete Algorithms (ACDA), pp. 225–236. Society for Industrial and Applied Mathematics (2023)
3. Alanko, J.N., Vuohtoniemi, J., Mäklin, T., Puglisi, S.J.: Themisto: a scalable colored k-mer index for sensitive pseudoalignment against hundreds of thousands of bacterial genomes. Bioinformatics (2023)
4. Beller, T., Gog, S., Ohlebusch, E., Schnattinger, T.: Computing the longest common prefix array based on the Burrows-Wheeler transform. J. Discrete Algorithms **18**, 22–31 (2013)
5. Boucher, C., Bowe, A., Gagie, T., Puglisi, S.J., Sadakane, K.: Variable-order de Bruijn graphs. In: Proceedings of the 25th Data Compression Conference (DCC), pp. 383–392. IEEE (2015)
6. Compeau, P.E., Pevzner, P.A., Tesler, G.: Why are de Bruijn graphs useful for genome assembly? Nat. Biotechnol. **29**(11), 987 (2011)
7. Conte, A., Cotumaccio, N., Gagie, T., Manzini, G., Prezza, N., Sciortino, M.: Computing matching statistics on Wheeler DFAs. arXiv preprint arXiv:2301.05338 (2023)
8. Holley, G., Melsted, P.: Bifrost: highly parallel construction and indexing of colored and compacted de Bruijn graphs. Genome Biol. **21**(1), 1–20 (2020)
9. Jeffery, I.B., et al.: Differences in fecal microbiomes and metabolomes of people with vs without irritable bowel syndrome and bile acid malabsorption. Gastroenterology **158**(4), 1016–1028 (2020)
10. Maillet, N., Lemaitre, C., Chikhi, R., Lavenier, D., Peterlongo, P.: Compareads: comparing huge metagenomic experiments. BMC Bioinf. **13**(19), 1–10 (2012)
11. Marchet, C., Boucher, C., Puglisi, S.J., Medvedev, P., Salson, M., Chikhi, R.: Data structures based on k-mers for querying large collections of sequencing data sets. Genome Res. **31**(1), 1–12 (2021)
12. Ondov, B.D., et al.: Mash: fast genome and metagenome distance estimation using MinHash. Genome Biol. **17**(1), 1–14 (2016)
13. Salikhov, K.: Efficient algorithms and data structures for indexing DNA sequence data. PhD thesis, Université Paris-Est; Université Lomonossov (Moscou) (2017)

On Suffix Tree Detection

Amihood Amir[1], Eitan Kondratovsky[2], and Avivit Levy[3]($\boxtimes$)

[1] Department of Computer Science, Bar-Ilan University, Ramat-Gan, Israel
amir@esc.biu.ac.il
[2] Cheriton School of Computer Science, Waterloo University, Waterloo, Canada
e2kondra@uwaterloo.ca
[3] Software Engineering Department, Shenkar College of Engineering and Design, Ramat-Gan, Israel
avivitlevy@shenkar.ac.il
https://u.cs.biu.ac.il/~amir/, https://u.cs.biu.ac.il/~kondrae

Abstract. A suffix tree is a fundamental data structure for string processing and information retrieval, however, its structure is still not well understood. The *suffix trees reverse engineering problem*, which its research aims at reducing this gap, is the following. Given an ordered rooted tree T with unlabeled edges, determine whether there exists a string w such that the unlabeled-edges suffix tree of w is isomorphic to T. Previous studies on this problem consider the relaxation of having the suffix links as well as assume a binary alphabet. This paper is the first to consider the *suffix tree detection problem*, in which the relaxation of having suffix links as input is removed. We study suffix tree detection on two scenarios that are interesting per se. We provide a suffix tree detection algorithm for *general alphabet periodic strings*. Given an ordered tree T with n leaves, our detection algorithm takes $O(n + |\Sigma|^p)$-time, where *p is the unknown in advance length* of a period that repeats at least 3 times in a string S having a suffix tree structure identical to T, if such S exists. Therefore, it is a polynomial time algorithm if p is a constant and a linear time algorithm if, in addition, the alphabet has a sub-linear size. We also show some necessary (but insufficient) conditions for *binary alphabet general strings* suffix tree detection. By this we take another step towards understanding suffix trees structure.

Keywords: Suffix tree · Reverse engineering · Suffix tree detection · Periodic string

1 Introduction

A suffix tree is a fundamental data structure for string processing and information retrieval being one of the most well-known and widely used text indexing structures. It provides a linear space full-text index of a given string and has played a central role in combinatorial pattern matching and its applications. A multitude of important problems can efficiently be solved using suffix trees [3,15].

Partly supported by ISF grant 1475/18 and BSF grant 2018141.

Despite their essential role, the structure of suffix trees is still not well understood. For instance, only almost forty years after this data structure was first introduced by Peter Weiner [26], it was proved that each internal edge in a suffix tree can contain at most one implicit suffix node [5]. For suffix arrays, which are a known alternative data structure to suffix trees, several characterization theorems are known (e.g. [4,16,20,23]). For example, Kucherov et al. [20] present a bijective characterization of suffix array permutations obtained from a characterization of Burrows-Wheeler arrays. Considering the *suffix trees reverse engineering problem*, which was first presented by [18] and studied also by [6,24], is a step towards closing this gap in the understanding of suffix trees structure.

The idea of *reverse engineering* has been studied for several data structures. For example, Franek et al. [11] verify a border array in linear time; I et al. [17] verify and enumerate a parameterized border array; Duval et al. [9] validate string automata; Bannai et al. [4] infer strings from graphs; Clément et al. [7] reverse engineer prefix tables; Gawrychowski et al. [12] validate the KMP failure function; Crochemore et al. [8] study cover array string recognition; Kärkkäinen et al. [19] infer a string from the LCP array; Nakashima et al. [22] infer a string from the Lyndon factorization and Gawrychowski et al. [13] prove some general conditions that allow string reconstruction (e.g., maximal palindromes). There are some known negative reverse engineering results, such as Gelle and Iván [14] who showed that recognizing Union-Find trees is NP-complete.

The *suffix trees reverse engineering problem* is the following. Given an ordered rooted tree T with unlabeled edges, determine whether there exists a string w such that the unlabeled-edges suffix tree of w is isomorphic to T. I et al. [18] emphasize that this problem is very challenging due to the following reasons:

1. The length of each edge string is not given;
2. The mapping from strings to edge-unlabeled suffix trees is not injective. Therefore, only relaxed versions of the problem had been solved until now.

As a first step towards solving the problem, I et al. [18] restrict the problem to a binary alphabet. In addition, they assume that suffix links of inner nodes are given as input. Under these conditions, they solve the suffix trees reverse engineering problem in linear time in the size of the input tree T. All other attempts also assume that the suffix links are given. Starikovskaya and Vildhøj [24] prove some new properties of suffix trees and show how to reverse engineer implicit suffix trees (where there is no special end-of-string symbol). Casaux and Rivals [6] assume the existence of leaf suffix links (where each leaf points to the next longer leaf) besides the existence of internal nodes suffix links. Under these assumptions they reverse engineer suffix trees over general alphabets in linear time.

Removing the relaxation of having the suffix links is quite a challenge posing additional difficulties on the suffix tree structure analysis. For example:

1. It changes the degrees of freedom for the inner nodes suffix links references from 1 to $O(n)$, potentially, where n is the number of the suffix tree leaves, as the number of inner nodes is $O(n)$.

2. Even for a binary alphabet, the ordered suffix tree structure does not necessarily obey character-symmetry laws, i.e., by flipping a binary string characters, one is not guaranteed to have a suffix tree with an isomorphic structure.

Therefore, the task of studying ordered suffix trees structure (even on binary strings) where the suffix links *are not given as input*, is a natural next step in the study of the structure of suffix trees, but also requires to cope with such difficulties.

In this paper, we consider the reverse engineering problem, in which the relaxation of having the suffix links as input is removed. Formally,

Definition 1. The Suffix Tree Detection problem *is the following:*

Input: *An ordered rooted tree T*
Output: *YES, if there exists a string S such that its unlabeled-edges ordered*
            suffix tree is isomorphic to T,
            NO, otherwise.

We study suffix tree detection on two scenarios that are interesting per se: (1) *general alphabet periodic strings* and (2) *binary alphabet general strings*. The conditions on these two classes of strings considered in this paper are indeed restrictive. Nevertheless, these classes are rich enough to retain an inherent difficulty of removing the relaxation of suffix links availability as input.

Paper Contribution. The main contributions of this paper are:

- Being the first, to our knowledge, to consider the suffix tree reverse engineering problem *without the relaxation of suffix links availability.*
- Providing a detection procedure for general alphabet periodic strings suffix tree structure. Given an ordered tree T with n leaves, our detection algorithm takes $O(n + |\Sigma|^p)$-time, where p *is the unknown in advance length* of a period that repeats at least 3 times in a string S having a suffix tree structure identical to T, if such S exists. Therefore, it is a polynomial time algorithm if p is a constant and a linear time algorithm if, in addition, the alphabet has a sub-linear size.[1]
- Proving several necessary (but insufficient) conditions for binary alphabet general strings suffix tree detection.

By this we take another step towards understanding suffix trees structure.

Paper Organization. The paper is organized as follows. In Sect. 2 we give some basic needed formal terms. In Sect. 3 we study the general alphabet periodic strings suffix tree detection problem. In Sect. 4 we prove some necessary (but insufficient) conditions for a binary alphabet general string suffix tree detection, which can be validated using a linear time algorithm. We conclude the paper in Sect. 5 with some open problems.

---

[1] Though Sect. 3 studies periodic string having period length at least 2, which we call *non-trivial periodic strings*, this complexity is also valid for detection of trivial periodic strings (i.e., unary strings), as follows from condition 2 of Theorem 2 in Sect. 4.

2   Preliminaries

In this section we describe some basic terms used in the paper. We begin with a formal definition[2] of the term *suffix tree*, which is the subject of this study.

Definition 2. Suffix Tree [21,25,26]
The suffix tree for the string S of length n is defined as a rooted tree such that:

1. *The tree has exactly n leaves numbered from 1 to n.*
2. *Except for the root, every internal node has at least two children.*
3. *Each edge is labelled with a non-empty sub-string of S.*
4. *No two edges starting out of a node can have string-labels beginning with the same character. The string obtained by concatenating all the string-labels found on the path from the root to leaf i is the suffix $S[i..n]$, $1 \le i \le n$.*

In order to ensure the existence of such a tree for any string, S is assumed to end with a terminal symbol denoted by $\$$, where $\$ \notin \Sigma$. This ensures that no suffix is a prefix of another, and that there are n leaf nodes, one for each of the $n - 1$ suffixes of S and an additional leaf for the $\$$.

Throughout the paper, we assume that the suffix tree (thus, any given tree to be checked) is ordered, i.e., the labels on edges are sorted by lexicographic order, and that a $\$$ is lexicographically smaller than every other character in Σ.

Notation. Wee denote by $|S|$ the length of a string S. Let T be a suffix tree of a string S.

- We denote by $T(\sigma_i)$ the sub-tree of T's root that is reached by following the edge-label starting with $\sigma_i \in \Sigma$, if σ_i appears in S.
- Similarly, let X be any internal node of T and T_X be the sub-tree rooted at X in T, we denote by $T_X(\sigma_i)$ the sub-tree of X that is reached by following the edge-label starting with $\sigma_i \in \Sigma$, if it exists.

Common Terms. We use some common definitions and terms, as follows.

- Trees terms, such as nodes **level** (starting from 0 for the root), **depth** or **degree**, refer to the unlabelled-edges suffix tree structure.
- **Implicit** nodes in a suffix tree T refer to any non-branching nodes that do not exist in T but can be added between the nodes of T when edge-labels longer than a single character are split to form shorter edge labels.
- **The path label** of a non-root internal node X in T refers to the concatenation of edge-labels on the path from the root to X.

A well-known term related to the suffix tree data structure is that of a *suffix link*, given in Definition 3. Suffix links are a key feature for some classical linear-time construction algorithms [21,25,26], although Farach's algorithm for suffix tree construction [10] does not make use of them. Suffix links are also used in some algorithms running on the tree.

---

[2] Assuming the basic terms: rooted trees, internal tree nodes and tree leaves are known.

Definition 3. Suffix Links
*Given a suffix tree T, let s be a path label of a non-root internal node X, and let
s' be the string s truncated by its first character. Let X' be a non-root internal
node with the path label s'. Then, the suffix tree edge from the node X to the
node X' is called a* suffix link.

In Sect. 3 we deal with periodic strings, formally defined next.

Definition 4. Periodic and Primitive Strings
Let S be a string of length n. S is called periodic *if $S = P^i \mathrm{pref}(P)$, where
$i \in \mathbb{N}$, $i \geq 2$, P is a substring of S such that $|P| \leq n/2$, P^i is the concatenation
of P to itself i times, and $\mathrm{pref}(P)$ is a prefix of P. The shortest such substring
P is called the* period *of S.*

If S is not periodic it is called aperiodic *or* primitive.

A periodic string having a period length at least 2 is called a non-trivial
periodic string.

Note that by Definition 4, the period P of a periodic string S is uniquely defined.
In addition, such a period P is a primitive string.

Notation. Let S be a periodic string having a period $P = P[1 \ldots p]$ of length
$p \geq 2$. Denote by $P^{(i)} = P[i..p]P[1..i-1]$, the i-th rotation of P, where $1 \leq i \leq p$.
Definition 5 describes a binary tree structure characterization that we use.

Definition 5. Tree/Sub-Tree Alignment
*Let T be a full binary tree, i.e. each node is either a leaf or has exactly two
children. T is called* left aligned *(respectively,* right aligned*) if every right child
(respectively, left child) of an internal node in T is a leaf. We call this the* left-
alignment condition *(respectively,* right-alignment condition*).*

A tree T with only a root and two leaves is called trivially right-aligned
(alternatively, trivially left-aligned*).*

A tree T is not aligned *if it is neither left- nor right-aligned.*

A sub-tree of T with a root t is called left aligned sub-tree *(respectively,
right aligned*) if the left-alignment condition (respectively, the right-alignment
condition) holds for the sub-tree of T rooted at t. The sub-tree of T rooted at t
is called* non-aligned sub-tree *if it is neither a left nor a right aligned sub-tree.*

Note that the left and right child of an inner node in a full binary tree representing a suffix tree, correspond to following the edge with a smaller or larger
character by lexicographic order, respectively..

For the statement of condition 2 of Theorem 2 (stating necessary conditions
for binary string suffix tree detection) in Sect. 4, we need the term *alignment
inversion* and the classification of its types given next in Definition 6.

Definition 6. Alignment Inversion
*Let T be a binary tree with root node t. Denote by t_ℓ the left child of t and by
t_r the right child of t. We say that t has an* alignment inversion *if one of the
following conditions holds:*

1. *If t_ℓ has 2 children, t_r is a leaf, whereas, the left child of t_ℓ, denoted by t_{ℓ_ℓ} is a leaf, and its right child, denoted by t_{ℓ_r}, has 2 children. We call this a left-right inversion.*
2. *If t_ℓ is a leaf, t_r has 2 children, whereas, the left child of t_r, denoted by t_{r_ℓ}, has 2 children, and its right child, denoted by t_{r_r}, is a leaf. We call this a right-left inversion.*
3. *If both t_ℓ and t_r have 2 children. We call this a balanced inversion.*

Remark. In Sect. 4 we assume that $|\Sigma| = 2$, without loss of generality, $\Sigma = \{a, b\}$.

3   Periodic Strings Suffix Tree Detection

In this section we study the structure of any ordered tree representing a suffix tree of a non-trivial periodic string, i.e., having period length at least 2.[3] We begin by discussing limitations of periodic strings suffix tree detection. Note that in this section, we refer to a string S ending with a \$ as periodic (resp. a-periodic), if it is periodic (resp. a-periodic) when the \$ is omitted.

Detection Limitations. A major limitation for periodic strings suffix tree structure characterization is non-uniqueness, i.e., there exist trees structures that *match both a periodic and a non-periodic string.* Consider the periodic string *babbbabbbabbba*\$ and the non-periodic string *bbbabbbabbbaba*\$, which have identical suffix trees structure, but the "period" is violated in the incomplete "occurrence" at the end of the non-periodic string. Another different example is the periodic string *babbbabbbabb*\$ and the non-periodic string *babbbabbbabbb*\$, in which the last "occurrence" of the period has the correct quantity of each character but their order is different, however, both strings have identical suffix tree structure. We conclude with the following periodic string: *bbabbbbabbbbabbbbabb*\$ and the non-periodic string *bbbbbbbabbbbabbbbabb*\$, which has one mismatch error with the periodic string. They don't have identical suffix tree structure, however, the suffix tree structure of the second non-periodic string is identical to the suffix tree structure of the following periodic string: *bbbbbabbbbabbbbabb*\$. Thus, violating the periodicity in a complete period occurrence, does not guarantee the in-existence of a periodic string suffix tree that is isomorphic to the resulting aperiodic string suffix tree structure. Therefore, our study is directed to differentiate between ordered trees such that *there exists* a periodic string with identical suffix tree structure and ordered trees where *there is no periodic string suffix tree* with identical structure.

Another important limitation for periodic strings suffix tree structure characterization is having a sufficient number of period repetitions. By periodicity definition (see Definition 4) there should be at least two period repetitions in a periodic string. However, we show that given an ordered tree, determining

---

[3] Note that the structure of a unary string, which is a trivial periodic string with period length 1, is characterized by Theorem 2.2 in Sect. 4.

if it is isomorphic to a periodic string suffix tree can be done for strings with at least *three* period repetitions. Having a sufficient number of repetitions is also a limitation in the *period recovery problem* [1] as well as the *cover recovery problem* [2].

Lemmas 1, 2 and 3 below specify necessary conditions on non-trivial periodic strings suffix trees structure. Omitted proofs will appear in the full version of this paper.

Lemma 1. *Let T be an ordered suffix tree of a periodic string S of length n having a period P of length $p \geq 2$. Then, every internal node at depth at least $p - 1$ has out-degree 2. Moreover, every internal node at depth at least $p - 1$ is a root of a (maybe trivially) right-aligned sub-tree.*

Proof. By the definition of a period, P is primitive, thus every $P^{(i)}$ and $P^{(j)}$, where $i \neq j \mod p$, have at least one mismatch. Therefore, every two suffixes S_i and S_j, where $i \neq j \mod p$, have at least one mismatch in their first p characters. Thus, at depth higher than $p-1$ an internal node can have out-degree more than 2 if at least 3 suffixes share a prefix, where the mismatches pairs are different.

Since there are at most p mismatches in the first p characters of every S_i and S_j, where $i \neq j \mod p$, then from depth $p - 1$ all suffixes that are still in the same sub-tree have equivalent length modulo p. The only mismatch that creates a new internal node in this set of suffixes is when a shorter suffix ends with a \$, while all the other suffixes are longer. Since all these suffixes have equivalent length modulo p, in all the longer suffixes this character is equal to $S[n - p]$.

We prove that the resulting sub-tree at depth $p-1$ is a right-aligned sub-tree, from which, in particular, we get that every internal node at such a sub-tree has out-degree 2. The order between pairs of suffixes S_i and S_j, where $i = j \mod p$, is determined by the comparison of \$ to the character $S[n - p]$. Recall that \$ is smaller and thus forms a left leaf. By periodicity, $S[n - p]$ repeats every p positions, therefore, the repeated comparison forms a right-aligned sub-tree.

Lemma 2. *Let T be an ordered suffix tree of a periodic string S of length n having a period P of length $p \geq 2$. Then, if $\lfloor (n - 1)/p \rfloor > 2$ (i.e., the period appears at least 3 times), there are exactly p sub-trees with depth at least $\lfloor (n - 1)/p \rfloor - 2$ that are all (maybe trivially) right-aligned. Moreover, for every $\sigma \in \Sigma$, the number of such sub-trees in $T(\sigma)$ is the number of occurrences of σ in P.*

Proof. From Lemma 1, it follows that after a prefix of length at most p until the closest appearance of the last p characters of S (excluding the \$), each set of suffixes that have equivalent length modulo p are in the same right-aligned sub-tree.

Each sub-tree corresponds to a different set of suffixes S_i having equivalent length modulo p. Since $\lfloor \frac{n-1}{p} \rfloor > 2$, each such set of suffixes S_i having equivalent length modulo p is not empty and has at least 2 suffixes. Thus, there must be exactly p such right-aligned sub-trees.

Moreover, the number of suffixes in each of these sets is at least: $\lfloor \frac{n-1}{p} \rfloor - 1$. In each level one such suffix ends as a \$-labelled left leaf. Therefore, the number of internal nodes in the each formed right-aligned sub-tree is at least $\lfloor \frac{n-1}{p} \rfloor - 1$. Thus, the depth of the formed right-aligned sub-tree is at least $\lfloor (n-1)/p \rfloor - 2$.

Since the period length is p, every S_i which belongs to the same sub-tree starts with the same character. That is, $S_i[1] = S_{i+p}[1] = S_{i+2p}[1] = \ldots$, for any $1 \leq i \leq p$. Since different length-equivalence classes modulo p are formed by mismatches in the first p characters of S, the number of such right-aligned sub-trees in $T(\sigma)$, for every character σ, is the number of occurrences of σ in P.

Lemma 3. *Let T be an ordered suffix tree of a periodic string S of length n having a period P of length $p \geq 2$. Then, the label of every edge:*

1. *to an inner node of T: has length at most p.*
2. *to a left leaf of T: has length 1 and its label is \$.*
3. *to a right leaf of T: has length $p+1$ and its label is the suffix $S[n-p\ldots n]$.*

Lemmas 1, 2 (and 3) are not sufficient to reject any tree structure where no periodic string has an identical suffix tree structure. However, they can be used to reject the suffix tree of the following **almost** periodic string (with one deletion error): $bbbbabbbbbbabbbbbabb\$$. Lemma 4 below rejects its structure as matching any periodic string suffix tree.

Lemma 4. *Let T be an ordered suffix tree of a string of length n, then if the following hold:*

1. *There exists p, such that $2 \leq p < n$, $\lfloor (n-1)/p \rfloor > 2$ and T has p right-aligned sub-trees.*
2. *There exists a sub-tree at depth $p-1$ that is not right-aligned.*

Then, there is no periodic string with a suffix tree structure identical to T.

Lemmas 1, 2, 3 and 4 are still not sufficient to reject any tree structure where no periodic string has an identical suffix tree structure. For example, they hold for the suffix tree of the following non-periodic string $babbabbbbbabbbbbabbb\$$, yet there is no periodic string having an identical suffix tree structure. Nevertheless, Lemma 5 below rejects its structure as matching any periodic string suffix tree.

Lemma 5. *Let T be an ordered suffix tree of a string of length n, then if the following hold:*

1. *There exists p, such that $2 \leq p < n$, $\lfloor (n-1)/p \rfloor > 2$ and T has p right-aligned sub-trees.*
2. *The difference in depths of two of these p right-aligned sub-trees is strictly greater than 1.*

Then, there is no periodic string with a suffix tree structure identical to T.

Proof. If there exists p, such that $2 \leq p < n$, $\lfloor (n-1)/p \rfloor > 2$ and T has p right-aligned sub-trees, then by Lemma 2, the period of a periodic string with a suffix tree T (if such a string exists) is of length p. We next show that the difference in the depth of any two of these right-aligned sub-trees is at most 1.

Each right-aligned sub-tree corresponds to a different set of suffixes having equivalent length modulo p (see Lemma 2's proof). Moreover, the number of suffixes in each of these sets is at least: $\lfloor \frac{n-1}{p} \rfloor - 1$. Thus, their depth is at least $\lfloor (n-1)/p \rfloor - 2$. Also, note that the number of suffixes with equivalent length modulo p in each such set cannot exceed $\lceil (n-1)/p \rceil$. It follows that the difference in the depth of any two such sub-trees is at most 2.

Assume that there exists a non-empty set of $\lfloor (n-1)/p \rfloor - 1$ suffixes having equivalent length modulo p and let S_i be the longest suffix in this set. The length of S_i is, therefore, $r + p(\lfloor (n-1)/p \rfloor - 2) + 1$ (including the \$), where r is the length of the path label of S_i's right-aligned sub-tree root. We claim that in such a case there is no other set with $\lceil (n-1)/p \rceil$ suffixes having equivalent length modulo p. Note that proving this claim concludes the proof of the lemma.

Every suffix S_j, where $i \neq j \mod p$, has a mismatch with S_i which must be encountered after at most $p - 1$ characters. Note that, $n - 1 - (p-1) \geq p\lfloor (n-1)/p \rfloor - p + 1 = p(\lfloor (n-1)/p \rfloor - 1) + 1 = p(\lfloor (n-1)/p \rfloor - 2) + p + 1$. This means that the shortest suffix with equivalent length to S_i modulo p has the same prefix of length p as S_i and, therefore, its length is at least $p + 1$ (including the \$). Thus, $r \geq p$, since this shortest suffix ends with a \$ as a left leaf of the root of S_i's right-aligned sub-tree. Moreover, the label of the edge leading to the root of S_i's right-aligned sub-tree is of length p. Therefore, $r \leq p - 1 + p < 2p$.

Now, every other set of suffixes with equivalent length modulo p that has more suffixes than $\lfloor \frac{n-1}{p} \rfloor - 1$, must have a longest suffix with length strictly greater than that of S_i. Since $r < 2p$, such a suffix must begin within the first $2p - 1$ characters of S. Therefore, it can have at most one more suffix than the number of suffixes in set of S_i. This concludes the proof.

Theorem 1. *Let T be an ordered tree with n leaves. Then, there exists a periodic string S of length n having a period P of length $p \geq 2$ that repeats at least 3 times in S s.t. T is identical to the structure of S's suffix tree, if and only if:*

1. *Every internal node at depth at least $p - 1 \geq 1$ has out-degree 2.*
2. *There are exactly p sub-trees with depth at least $\lfloor (n-1)/p \rfloor - 2 \geq 1$ that are all (maybe trivially) right-aligned. Moreover, for every $\sigma \in \Sigma$, the number of such sub-trees in $T(\sigma)$ is the number of occurrences of σ in P.*
3. *Every sub-tree at depth $p - 1$ is right-aligned.*
4. *The depths difference between every two of the p right-aligned sub-trees of T is at most 1.*
5. *Let T' be the tree T in which the p right-aligned sub-trees that T has by condition 2 are trimmed at their sub-trees roots. Then, there exists an a-periodic string $\hat{P}$ of length x, such that $p + 2 \leq x \leq 2p$, $\hat{P}[i] = \hat{P}[p + i]$, for $1 \leq i \leq x - p - 1$, $\hat{P}[x] = \$$ and the suffix tree of $\hat{P}$ is identical to T'.*

Proof. Assume that there exists a periodic string S of length n having a period P of length $p \geq 2$ that repeats at least 3 times in S such that T is identical to the structure of S's suffix tree. Then, Lemmas 1–5 ensure that conditions 1–4 hold. In addition, the set of at most $2p$ shortest suffixes of S determine the structure of T above the p right-aligned sub-trees that T has by condition 2. This is because at least one and at most p mismatches occur between the suffix of length $p+2$ to every other shorter suffix. The mismatches of the $p+2$ shortest suffixes result in inner nodes above the p right-aligned sub-trees, since these are suffixes that do not have equivalent length modulo p. Suffixes with length $p+2 < x \leq 2p$ may still create an edge to a root of the p right-aligned sub-trees. Any longer suffix has equivalent length modulo p to one of the suffixes of length at most $p+1$. Therefore, there exists x, $p+2 \leq x \leq 2p$, such that by taking $\hat{P}$ to be the x-length suffix of S, we get that the suffix tree of $\hat{P}$ is identical to T'. Note that $\hat{P}$ is a-periodic (having length at most $2p$ including the $).

Assume that conditions 1–5 hold for an ordered tree T with n leaves, we show that there exists a periodic string S of length n having a period P of length $p \geq 2$ that repeats at least 3 times in S such that T is identical to the structure of S's suffix tree. By condition 5, we know that there exists an a-periodic string $\hat{P}$ of length x, such that $p+2 \leq x \leq 2p$, $\hat{P}[i] = \hat{P}[p+i]$, for $1 \leq i \leq x-p-1$, $\hat{P}[x] = \$$ and the suffix tree of $\hat{P}$ is identical to T', where T' is the tree T in which the p right-aligned sub-trees that T has by condition 2 are trimmed at their sub-trees roots. We inductively construct the periodic string S from end to beginning as follows. We begin by assigning $S[n-x+1\ldots n] \leftarrow \hat{P}$. Then, for every i, starting from $n-x$ down to 1, we assign $S[i] \leftarrow S[i+p]$. Note that S is periodic with a period of length p. By condition 1, we have that $p \geq 2$ and by condition 2, we have that $\lfloor (n-1)/p \rfloor > 2$ (thus, the period repeats at least 3 times in S). Let T'' be the suffix tree of S. Then, Lemmas 1–5 ensure that conditions 1–4 hold for T''. Since the construction of S only added suffixes longer than x which belong to the right-aligned sub-trees, trimming the p right-aligned sub-trees of T'' at their roots gives the tree T'. Therefore, T'' is identical to T. The theorem follows.

Corollary 1. General Alphabet Periodic String Suffix Tree Detection
Given an ordered tree T with n leaves, then there exists an $O(n + |\Sigma|^p)$-time algorithm to detect if there exists a non-trivial periodic string S of length n over Σ having at least 3 period repetitions, s.t. T is identical to S's suffix tree structure, where p is the unknown in advance period length of S, if such S exists.

Proof. Given an ordered tree T with n leaves, the algorithm's steps are:

1. Scan T to find the *unique*[4] number p of right-aligned sub-trees with depth at least $\lfloor (n-1)/p \rfloor - 2 \geq 1$ and the numbers p_σ of the right-aligned sub-trees in every $T(\sigma)$ sub-tree of T.[5]
2. If $\lfloor (n-1)/p \rfloor \leq 2$ or $p < 2$, return NO.
3. If two of these p right-aligned sub-trees have depth difference strictly greater than 1, return NO.
4. Scan T to check if every internal node at depth at least $p-1$ has out-degree 2. If not, return NO.

---

[4] p is unique by condition 2 of Theorem 1.
[5] The characters $\sigma \in \Sigma$ are chosen to the edges of T's root from left to right according to the order of Σ, where the first edge is $.

5. If the scan detects a sub-tree at depth $p-1$ that is not right-aligned, return NO.
6. Let T' be the tree T in which the p right-aligned sub-trees of T are trimmed at their sub-trees roots and let x be the number of leaves in T'.
7. For every a-periodic string $\hat{P}$ of length x, such that $\hat{P}[i] = \hat{P}[p+i]$, for $1 \le i \le x-p-1$, $\hat{P}[x] = \$$, having p_σ occurrences of each $\sigma \in \Sigma$ in the last $p+1$ characters:
   (a) Construct the suffix tree $T^{\hat{P}}$ of the string $\hat{P}$.
   (b) If $T^{\hat{P}}$ is identical to T', return YES.
8. Return NO.

The algorithm correctness: Step 1 identifies the only possible candidate period length p according to condition 2 of Theorem 1. Then, step 2 rejects T if this candidate is not long enough or for this candidate there are not enough repetitions, as we only detect suffix trees of non-trivial periodic strings with at least three period repetitions. Step 3 rejects T according to condition 4 of Theorem 1. Step 4 rejects T according to condition 1 of Theorem 1. This step is performed after step 3 to prevent another unnecessary scan of T if the information of the first three steps is enough to reject T. Step 5 rejects T according to condition 3 of Theorem 1. Steps 6–8 reject or accept T according to condition 5 of Theorem 1.

The algorithm time complexity is $O(n + |\Sigma|^p)$, as steps 1– take time linear in the size of T, which is $O(n)$, step 7 takes $O(|\Sigma|^p + p)$ time (recall that $p+2 \le x \le 2p$ by condition 5 of Theorem 1) and step 8 takes $O(1)$ time.

4   Necessary Conditions on a Binary String Suffix Tree

In this section we describe some necessary conditions on any ordered tree representing a binary string suffix tree. For completeness, we state Observation 1, which follows from Definition 2 and the binary string ordered tree assumption.

Observation 1 *Let T be an ordered suffix tree of a binary string of length n. Then,*

1. *Every internal node of T has out-degree at least 2 and at most 3.*
2. *The leftmost child of the root of T has out-degree 0 (no children). This child corresponds to the $\$$ character in the string represented by T.*
3. *The edge to the leftmost child among three children of any node in T is labelled with $\$$.*

Theorem 2 below lists several necessary conditions for a tree T to be a suffix tree of a binary string.

Theorem 2. Necessary Conditions
Let T be an ordered suffix tree of a binary string of length n. Then,

1. *The maximum out-degree (number of children) of nodes in the same depth is monotonically non increasing with the node depth. I.e., if there exists an internal node with out-degree 3, then all the above levels must have at least one node with out-degree 3.*
2. *Let $d \geq 0$ be the highest level at which the maximum out-degree of the nodes is 2. Then the following hold:*
   - *If $d = 0$, the tree is right-aligned.*
   - *If $d = 1$, the root of $T(a)$ cannot have left-right inversion, however, it can have right-left or balanced inversion. On the other hand, the root of $T(b)$ can either be aligned or have left-right, right-left or balanced inversion.*
3. *The maximum number of leaves in $T_X(a)$, for an internal node X, is monotonically non-increasing with the depth of X.*
4. *The maximum number of leaves in $T_X(b)$, for an internal node X, is monotonically non increasing with the depth of X.*
5. *If all the possible strings of length 2: aa, ab, ba and bb, appear in the string represented by T, then the root of T has only one child with out-degree 3.*

The proof of Theorem 2 is deferred to the full version.

We conclude this section by referring to the algorithmic task of verifying if the conditions specified in (Observation 1 and) Theorem 2 hold for a given ordered tree T, assuming a binary string. Theorem 3 below can be achieved simply by visiting the tree.

Theorem 3. Verification Algorithm
Given an ordered tree T, there exists an $O(n)$ time algorithm to verify if each of the conditions of (Observation 1 and) Theorem 2 holds, assuming a binary string.

Remark. Theorem 2's conditions are insufficient.

5   Conclusion and Open Problems

In this paper we made another step towards understanding the structure of suffix trees by studying the suffix tree detection problem in which the relaxation of having the suffix links as input is removed. Some interesting open problems are:

- Providing a full characterization theorem describing both necessary and sufficient conditions for an ordered tree to be a suffix tree of some string.
- Improving the time complexity of our general alphabet periodic string suffix tree detection algorithm.

We believe that these problems should be further addressed due to the importance of suffix trees and their fundamental role.

References

1. Amir, A., Eisenberg, E., Levy, A., Porat, E., Shapira, N.: Cycle detection and correction. ACM Trans. Algorithms **9**(1), 1–20 (2012). https://doi.org/10.1145/2390176.2390189
2. Amir, A., Levy, A., Lewenstein, M., Lubin, R., Porat, B.: Can we recover the cover? Algorithmica **81**(7), 2857–2875 (2019). https://doi.org/10.1007/s00453-019-00559-8
3. Apostolico, A.: The myriad virtues of subword trees. In: Apostolico, A., Galil, Z. (eds.) Combinatorial Algorithms on Words. NATO ASI Series, vol. 12, pp. 85–96. Springer, Heidelberg (1985). https://doi.org/10.1007/978-3-642-82456-2_6
4. Bannai, H., Inenaga, S., Shinohara, A., Takeda, M.: Inferring strings from graphs and arrays. In: Rovan, B., Vojtáš, P. (eds.) MFCS 2003. LNCS, vol. 2747, pp. 208–217. Springer, Heidelberg (2003). https://doi.org/10.1007/978-3-540-45138-9_15
5. Breslauer, D., Italiano, G.F.: On suffix extensions in suffix trees. Theoret. Comput. Sci. **457**, 27–34 (2012)
6. Cazaux, B., Rivals, E.: Reverse engineering of compact suffix trees and links: a novel algorithm. J. Discrete Algorithms **28**, 9–22 (2014)
7. Clément, J., Crochemore, M., Rindone, G.: Reverse engineering prefix tables. In: Proceedings of the 26th International Symposium on Theoretical Aspects of Computer Science STACS. LIPIcs, vol. 3, pp. 289–300. Schloss Dagstuhl - Leibniz-Zentrum fuer Informatik, Germany (2009)
8. Crochemore, M., Iliopoulos, C.S., Pissis, S.P., Tischler, G.: Cover array string reconstruction. In: Amir, A., Parida, L. (eds.) CPM 2010. LNCS, vol. 6129, pp. 251–259. Springer, Heidelberg (2010). https://doi.org/10.1007/978-3-642-13509-5_23
9. Duval, J., Lecroq, T., Lefebvre, A.: Efficient validation and construction of border arrays and validation of string matching automata. RAIRO Theor. Inform. Appl. **43**(2), 281–297 (2009)
10. Farach, M.: Optimal suffix tree construction with large alphabets. In: Proceedings of the 38th IEEE Symposium on Foundations of Computer Science, pp. 137–143 (1997)
11. Franek, F., et al.: Verifying a border array in linear time. J. Comb. Math. Comb. Comput. **42**, 223–236 (2000)
12. Gawrychowski, P., Jez, A., Jez, L.: Validating the Knuth-Morris-Pratt failure function, fast and online. Theory Comput. Syst. **54**(2), 337–372 (2014)
13. Gawrychowski, P., Kociumaka, T., Radoszewski, J., Rytter, W., Walen, T.: Universal reconstruction of a string. Theor. Comput. Sci. **812**, 174–186 (2020)
14. Gelle, K., Iván, S.: Recognizing union-find trees is NP-complete, even without rank info. Int. J. Found. Comput. Sci. **30**(6–7), 1029–1045 (2019)
15. Gusfield, D.: Algorithms on Strings, Trees, and Sequences: Computer Science and Computational Biology. Cambridge University Press, Cambridge (1997)
16. He, M., Munro, J.I., Rao, S.S.: A categorization theorem on suffix arrays with applications to space efficient text indexes. In: SODA, vol. 5, pp. 23–32. Citeseer (2005)
17. Tomohiro, I., Inenaga, S., Bannai, H., Takeda, M.: Verifying and enumerating parameterized border arrays. Theor. Comput. Sci. **412**(50), 6959–6981 (2011)
18. Tomohiro, I., Inenaga, S., Bannai, H., Takeda, M.: Inferring strings from suffix trees and links on a binary alphabet. Discrete Appl. Math. **163**, 316–325 (2014)

19. Kärkkäinen, J., Piatkowski, M., Puglisi, S.J.: String inference from longest-common-prefix array. In: Proceedings of the 44th International Colloquium on Automata, Languages, and Programming, ICALP. LIPIcs, vol. 80, pp. 62:1–62:14. Schloss Dagstuhl - Leibniz-Zentrum für Informatik (2017)
20. Kucherov, G., Tóthmérész, L., Vialette, S.: On the combinatorics of suffix arrays. Inf. Process. Lett. **113**(22), 915–920 (2013)
21. McCreight, E.M.: A space-economical suffix tree construction algorithm. J. ACM **23**, 262–272 (1976)
22. Nakashima, Y., Okabe, T., Tomohiro, I., Inenaga, S., Bannai, H., Takeda, M.: Inferring strings from Lyndon factorization. Theor. Comput. Sci. **689**, 147–156 (2017)
23. Schürmann, K.B., Stoye, J.: Counting suffix arrays and strings. Theoret. Comput. Sci. **395**(2–3), 220–234 (2008)
24. Starikovskaya, T., Vildhøj, H.W.: A suffix tree or not a suffix tree? J. Discrete Algorithms **32**, 14–23 (2015)
25. Ukkonen, E.: On-line construction of suffix trees. Algorithmica **14**, 249–260 (1995)
26. Weiner, P.: Linear pattern matching algorithm. In: Proceedings of the 14 IEEE Symposium on Switching and Automata Theory, pp. 1–11 (1973)

Optimally Computing Compressed Indexing Arrays Based on the Compact Directed Acyclic Word Graph

Hiroki Arimura[1]([✉]) [ID], Shunsuke Inenaga[2] [ID], Yasuaki Kobayashi[1] [ID],
Yuto Nakashima[2] [ID], and Mizuki Sue[1]

[1] Graduate School of IST, Hokkaido University, Sapporo, Japan
{arim,koba,sue}@ist.hokudai.ac.jp
[2] Department of Informatics, Kyushu University, Fukuoka, Japan
inenaga@inf.kyushu-u.ac.jp, nakashima.yuto.003@m.kyushu-u.ac.jp

Abstract. In this paper, we present the first study of the computational complexity of converting an automata-based text index structure, called the Compact Directed Acyclic Word Graph (CDAWG), of size e for a text T of length n into other text indexing structures for the same text, suitable for highly repetitive texts: the *run-length BWT* of size r, the *irreducible PLCP array* of size r, and the *quasi-irreducible LPF array* of size e, as well as the *lex-parse* of size $O(r)$ and the *LZ77-parse* of size z, where $r, z \leqslant e$. As main results, we showed that the above structures can be optimally computed from either the CDAWG for T stored in read-only memory or its self-index version of size e without a text in $O(e)$ worst-case time and words of working space. To obtain the above results, we devised techniques for enumerating a particular subset of suffixes in the lexicographic and text orders using the forward and backward search on the CDAWG by extending the result by Belazzougui *et al.* in 2015.

Keywords: Highly-repetitive text · suffix tree · longest common prefix

1 Introduction

Backgrounds. Compressed indexes for repetitive texts, which can compress a text beyond its entropy bound, have received a lot of attention in the last decade in information retrieval [12]. Among them, the most popular and powerful compressed text indexing structures [12] are the *run-length Burrows-Wheeler transformation* (RLBWT) [12] of size r, the *Lempel-Ziv-parse* (LZ-parse) [13] of size z, and finally the *Compact Directed Acyclic Word Graph* (CDAWG) [5] of size e. It is known [12] that the size parameters r, z, and e can be much smaller than the information theoretic upperbound of a text for highly-repetitive texts such as collections of genome sequences and markup texts [12]. Among these repetition-aware text indexes, we focus on the CDAWG for a text T, which is a minimized compacted finite automaton with e transitions for the set of all suffixes of T [5]; It is the edge-labeled DAG obtained from the suffix tree for T by merging all isomorphic subtrees [8], and can be computed from T in linear

F. M. Nardini et al. (Eds.): SPIRE 2023, LNCS 14240, pp. 28–34, 2023.
https://doi.org/10.1007/978-3-031-43980-3_3

time and space [12]. The relationships between the size parameters r, z, and e of the RLBWT, LZ-parse, and CDAWG has been studied by, e.g. [2,4,6,10,11,14]; However, it seems that the actual complexity of converting the CDAWG into the other structures in sublinear time and space has not yet been explored [12].

Research Goal and Main Results. In this paper, we study for the first time the conversion problem from the CDAWG for T into the following compressed indexing structures for T:

(i)   the *run-length BWT* (RLBWT) [12] of size $r \leqslant e$;
(ii) the *irreducible permuted longest common prefix* (PLCP) array [9] of size r;
(iii) the *quasi-irreducible longest previous factor* (LPF) array [7] of size e (Sect. 2);
(iv) the *lex-parse* [13] with size at most $2r = O(r)$; and
(v)   LZ-parse [13] with size $z \leqslant e$.

We present in Sect. 4 and 5 algorithms for solving the aforementioned conversion problems from the CDAWG. Then, we obtain the following results.

Main Results (Theorem 4.1, 5.1, and 5.2). For any text T of length n over an integer alphabet Σ, we can solve the conversion problems from the CDAWG G of size e for T into the above compressed index array structures (i)–(v) for the same text in $O(e)$ worst-case time using $O(e)$ words of working space, where an input G is given in the form of either the CDAWG of size e for T stored in read-only memory, or its self-index version [3,15] of size $O(e)$ without a text.

Techniques. We devise in Sect. 3 techniques for enumerating a *canonical subset of suffixes* in the lexicographic and text orders using the *forward and backward DFS* at the CDAWG G extending [4]. For definitions and proofs omitted here, see the manuscript [1]. **Related work.** Belazzougui *et al.* [4] showed that $r \leqslant e$ and $z \leqslant e$ hold. As closely related work, the CDAWG G can be converted into the LZ78-parse of size $z_{78} \geq z$ in $O(e + z_{78} \log z_{78})$ time and space [2] via an $O(e)$-sized grammar [3] on G. In general, $e = \Theta(\log n)$ and $r = O(1)$ for Fibonacci words [11]. For Thue-Morse words,[1] our $O(e)$-time conversion method can run as fast as $O(r \operatorname{polylog}(n))$-time one since $e = O(\log n)$ [14] and $r = \Theta(\log n)$ [6].

2   Preliminaries

We prepare the necessary notation and definitions in the following sections. For precise definitions, see the literature [8,12] or the manuscript [1].

Basic Definitions and Notation. For any integers $i \leqslant j$, the notation $[i..j]$ or $i..j$ denotes the interval $\{i, i+1, \ldots, j\}$ of integers, and $[n]$ denotes $\{1, \ldots, n\}$. For a string $S[1..n] = S[1] \cdots S[n]$ of length n and any $i \leqslant j$, we denote $S[i..j] = S[i]S[i+1] \cdots S[j]$. Then, $S[1..j]$, $S[i..j]$, and $S[i..|S|]$ are a *prefix*, a *factor*, and a *suffix* of S, resp. The *reversal* of S is $S^{-1} = S[n] \cdots S[1]$. Throughout, we assume a string $T[1..n] \in \Sigma^n$, called a *text*, over an alphabet Σ with symbol

---

[1] The n-th Thue-Morse word is $\tau_n = \varphi^n(0)$ for the morphism $\varphi(0) = 01$ and $\varphi(1) = 10$.

order $\leqslant_\Sigma$, which is terminated by the *end-marker* $T[n] = \$$ such that $\$ \leqslant_\Sigma a$ for $\forall a \in \Sigma$. $\mathrm{Suf}(T) = \{T_1, \ldots, T_n\} \subseteq \Sigma^+$ denotes the set of all of n non-empty suffixes of T, where $T_p := T[p..n]$ is the p-th suffix with position p. For any suffix $S \in \Sigma^*$ in $\mathrm{Suf}(T)$, we define: (i) $\mathrm{pos}(S) := n+1-|S|$ gives the starting position of S. (ii) $\mathrm{rnk}(S)$ gives the lexicographic rank of S in $\mathrm{Suf}(T)$. $lcp(X, Y)$ denotes the *length of the longest common prefix* of strings X and Y. In what follows, we refer to any suffix as S, any factors of T as $X, Y, U, L, P, \ldots$, nodes of a graph as $v, w, \ldots$, and edges as $f, g, \ldots$, which are possibly subscripted. We denote by $\preccurlyeq_{\mathrm{lex}}$ the *lexicographic (lex-) order* over Σ^*, and by $\preccurlyeq_{\mathrm{pos}}$ the *text order* defined as $X \preccurlyeq_{\mathrm{pos}} Y \Leftrightarrow |X| \geq |Y|$.

The *CDAWG* [5] for a text T is an edge-labeled DAG $G = CDAWG(T) = (\mathcal{V}, \mathcal{E}, suf, root, sink)$, where $\mathcal{V}$, $\mathcal{E}$, and suf are the sets of *nodes*, *labeled edges*, and *suffix links*; *root* and *sink* $\in \mathcal{V}$ are the root and sink, resp. Each labeled edge $f = (v, X, w) \in \mathcal{E}$ goes from $\mathrm{src}(f) = v$ to $\mathrm{dst}(f) = w$ by spelling $lab(v) = X \in \Sigma^+$. The *size* of G is $e := |\mathcal{E}(G)| + |suf_G|$. For the definitions of $SA, BWT, PLCP, LPF$, their compressed arrays, and the lex-parse and LZ-parse for T, see, e.g., [1,7–9,12,13]. Throughout, all time and space complexities are measured in the worst-case and in words, resp., over an integer alphabet Σ.

3 Techniques

Our Approach. The *first idea* is to use the one-to-one correspondence between the set $\mathrm{Path}(G)$ of all n root-to-sink paths and the set $\mathrm{Suf}(T)$ of all n non-empty suffixes of T, whose elements π and S are connected by $\mathrm{str}(\pi) = S$, resp. We represent a sparse indexing array $\widetilde{A} : Dom \rightarrow Range$ with domain $Dom \subseteq [n]$ by the graph $\widetilde{A} = \{ (\mathrm{idx}(S), \mathrm{val}(S)) \mid S \in \mathcal{CS} \}$ of mapping $\widetilde{A}$ and its index set by $Dom = \{ \mathrm{idx}(S) \mid S \in \mathcal{CS} \}$ using some subset $\mathcal{CS} \subseteq \mathrm{Path}(G)$ of *canonical suffixes* and some mappings $\mathrm{idx} : \mathcal{CS} \rightarrow Dom$ and $\mathrm{val} : \mathcal{CS} \rightarrow Range$. The *second idea* is to enumerate elements $(i, \widetilde{A}[i])$ of $\widetilde{A}$ on $G = CDAWG_{\overline{\Pi}}(T)$ by doing the DFS over a spanning tree for a set of *certificates* in $\mathcal{E}$ defined under the pair $\Pi_{\mathrm{lex}}^{\mathrm{pos}} = (\preccurlyeq_{\mathrm{pos}}, \preccurlyeq_{\mathrm{lex}})$ of path orderings. Below, we explain how to enumerate elements of $\widetilde{A}$ under $\Pi_{\mathrm{lex}}^{\mathrm{pos}}$, where $\preccurlyeq_{\mathrm{pos}}$ and $\preccurlyeq_{\mathrm{lex}}$ play different roles.

(1) For the *run-length BWT* for T (Sect. 4), the set of primary edges (in the sense of [5]) w.r.t. the first ordering $\preccurlyeq_{\mathrm{pos}}$ defines a spanning tree $\mathcal{T}$ over G from the root, while the second ordering $\preccurlyeq_{\mathrm{lex}}$ specifies the order of traversal. Finally, the set of secondary edges w.r.t. $\preccurlyeq_{\mathrm{pos}}$ provides a collection $\mathcal{C}$ of target values to search in the DFS. Actually, we can extract an equal-letter run from each secondary edge in constant time.

(2) For the *quasi-irreducible LPF* (Sect. 5), we do the backward DFS of G from the sink based on the pair $\Pi = (\preccurlyeq_{\mathrm{pos}}, \preccurlyeq_{\mathrm{pos}})$. Then, the set of primary edges (defined below) w.r.t. the second ordering $\preccurlyeq_{\mathrm{pos}}$ defines a spanning tree, the first ordering $\preccurlyeq_{\mathrm{pos}}$ specifies the text order, and the set of secondary edges w.r.t. the first ordering $\preccurlyeq_{\mathrm{pos}}$ provides a collection of target values, which are the LCP values of neighboring suffixes. The PLCP array can be computed in a similar way, but with the pair $\Pi = (\preccurlyeq_{\mathrm{pos}}, \preccurlyeq_{\mathrm{lex}})$.

Ordered CDAWG. Below, we introduce necessary terminology for Sect. 4 and Sect. 5. Consider the CDAWG G with an underlying pair $\Pi = (\preccurlyeq_-, \preccurlyeq_+)$ of path orderings. Let v be any node, and let us read the elements of $\{-,+\}$ *"upper"* and *"lower"*, resp. Then, $\mathcal{U}_-(v)$ denotes the set of all paths from the root to v [4,5], called *upper paths*, in G, while $\mathcal{U}_+(v)$ denotes the set of all paths from v to the sink, called *lower paths*. For any $\delta \in \{-,+\}$, the δ-*representative* of the set $\mathcal{U}_\delta$ is the smallest element $\mathtt{repr}_\delta(v)$ of $\mathcal{U}_\delta(v)$ under $\preccurlyeq_\delta$, i.e., $\mathtt{repr}_\delta(v) := \min_{\preccurlyeq_\delta} \mathcal{U}_\delta(v)$. For example, under $\Pi_{\mathrm{lex}}^{\mathrm{pos}} = (\preccurlyeq_{\mathrm{pos}}, \preccurlyeq_{\mathrm{lex}})$, $\mathtt{repr}_-(v)$ is the longest strings in $\mathcal{U}_-(v)$, while $\mathtt{repr}_+(v)$ is the lex-first string in $\mathcal{U}_+(v)$. $\Pi_{\mathrm{pos}}^{\mathrm{pos}}$ has representatives, too.

We classify edges as follows: any δ-edge $f \in N_\delta(v)$, $\delta \in \{-,+\}$, is said to be δ-*primary* if $\mathtt{repr}_\delta(v)$ goes through f. We denote by $\mathcal{E}_\delta^\star$ and $\overline{\mathcal{E}_\delta^\star} := \mathcal{E} - \mathcal{E}_\delta^\star$ the sets of *all δ-primary* and all δ-*secondary* edges, resp. We assume that in G under $\Pi_{\mathrm{lex}}^{\mathrm{pos}}$, all incoming and outgoing edges of N_- and N_+, resp., at any node v are sorted by a pair of edge orderings $\Gamma_{\mathrm{lex}}^{\mathrm{pos}} = (\leqslant_{-,\mathrm{pos}}^E, \leqslant_{+,\mathrm{lex}}^E)$ compatible to $\Pi_{\mathrm{lex}}^{\mathrm{pos}}$. For example, $f <_{-,\mathrm{pos}}^E f'$ is defined by comparing the lengths, $U \prec_{\mathrm{pos}} U'$, of longest paths $U, U' \in \mathcal{U}_-(v)$ going through $f, f' \in N_-(v)$, while $f_1 \leqslant_{+,\mathrm{lex}}^E f_2$ is defined by the lex-order $\mathtt{lab}(f_1)[1] <_\Sigma \mathtt{lab}(f_2)[1]$ of the edge labels (see [1]).

Definition 3.1 (canonical suffix and search path). For $\delta \in \{-,+\}$, we define a δ-*canonical suffix* S, its δ-*certificate* f_δ, and its δ-*search path* P_δ for f_δ as follows, where $\pi = (f_1, \ldots, f_\ell)$ is any root-to-sink path such that $\mathtt{str}(\pi) = S$:

(a) If S is *trivial*, i.e., $S = S_\delta := \mathtt{repr}_\delta(end_\delta)$ with $end_- = sink$ and $end_+ = root$, it is δ-*canonical*, f_δ is a virtual one, $\boldsymbol{f_\delta}$, $P_\delta := \boldsymbol{S_\delta}$, and $\mathtt{cano}_\delta(f_\delta) := \boldsymbol{S_\delta}$.

(b) Otherwise, S is *non-trivial*. Then, S is δ-*canonical* if it has a δ-*canonical factoring* such that $S = \mathtt{str}(U_\delta) \cdot X_\delta \cdot \mathtt{str}(D_\delta)$ for some edge $f_\delta = f_k = (v, X_\delta, w) \in \overline{\mathcal{E}_\delta^\star}$, $k \in [\ell]$, in π that satisfies: (i) f_δ is the highest $\overline{\mathcal{E}_-^\star}$-edge in S if $\delta = (-)$ and the lowest $\overline{\mathcal{E}_+^\star}$-edge in S if $\delta = (-)$; (ii) $U_\delta = (f_1, \ldots, f_{k-1}) = \mathtt{repr}_-(v)$; (iii) $L_\delta = (f_{k+1}, \ldots, f_\ell) = \mathtt{repr}_+(w)$. Then, the δ-*certificate* is $f_\delta := f_k$, and the δ-*search path* for f is the path $P_- := U \cdot X$ for $\delta = (-)$ and the path $P_+ := X \cdot U$ for $\delta = (+)$, where $X = \mathtt{lab}(f)$. Let $\mathtt{cano}_\delta(f_\delta) = S$.

$\diamond$

We denote by $\mathcal{CS}_\delta(G) \subseteq \mathtt{Suf}(T)$ and $\mathcal{SP}_\delta(G) \subseteq (\mathcal{E})^*$ the *set of all δ-canonical suffixes* of T and the *set of all δ-search paths* of G, resp. By definition, we can reconstruct any canonical suffix S_δ from its certificate f_δ by the bijection $\mathtt{cano}_\delta$ between $\mathcal{CE}_\delta := \overline{\mathcal{E}_\delta^\star} \cup \{f_\delta\}$ and $\mathcal{CS}_\delta(G)$. Moreover, we have $\mathcal{CS}_-(G) = \mathcal{CS}_+(G)$ (see [1]), and thus, its elements are simply called *canonical suffixes* in G.

The *forward* (resp. *backward*) *search tree* $\mathcal{T}_-$ (resp. $\mathcal{T}_+$) for certificates in $\mathcal{CE}_\delta$ are constructed as the directed graph obtained by merging common prefixes (resp. suffixes) of $\mathcal{SP}_-(G)$ (resp. $\mathcal{SP}_+(G)$). We observe that (i) $\mathcal{T}_-$ is connected at the root (resp. so is $\mathcal{T}_+$ at the sink), and for every $\delta \in \{-,+\}$, (ii) $\mathcal{T}_\delta$ is spanning over $\mathcal{CE}_\delta$, and (iii) $\mathcal{T}_\delta$ contains at most e edges. Since $\mathcal{SP}_-(G)$ is prefix-free while $\mathcal{SP}_+(G)$ is suffix-free (see [1]), $\mathcal{T}_-$ and $\mathcal{T}_+$ are well-defined, resp. Finally, we note that all operations above, used in Sect. 4 and 5, can be answered in $O(1)$ time after preprocessing G in $O(e)$ time and space [3,4] (see [1] for details).

Algorithm 1: The algorithm for computing the quasi-irreducible BWT for $T[1..n]$ from the CDAWG G for T stored in read-only memory.

---

1 **Procedure** RecRBWT(v);
2 **if** $N_+(v) = \emptyset$ **then return** ('\$', 1) ; $\qquad\qquad$ ▷ *Case: trivial suffix. $T[n] = $ '\$'*
3 **else** $\qquad\qquad\qquad\qquad\qquad\qquad\qquad\qquad\qquad\qquad$ ▷ *Case: non-trivial suffix*
4 $\quad$ **for each** $f = (v, X, w) \in N_+(v)$ in order $\leqslant^E_{+,\text{lex}}$ compatible to $\preccurlyeq_{\text{lex}}$ **do**
5 $\quad\quad$ **if** is-primary$_-(f)$ **then** $\qquad\qquad\qquad\qquad$ ▷ *Case: $(-)$-primary*
6 $\quad\quad\quad$ $RBWT' \leftarrow$ RecRBWT(w);
7 $\quad\quad$ **else** $\qquad\qquad\qquad\qquad\qquad\qquad\qquad$ ▷ *Case: $(-)$-secondary*
8 $\quad\quad\quad$ $c \leftarrow$ precsym(f); $\ell \leftarrow$ nleaves$(\text{dst}(f))$; $RBWT' \leftarrow (c, \ell)$;
9 $\quad\quad$ $RBWT \leftarrow RBWT \circ RBWT'$; $\qquad\qquad$ ▷ *Concatenation of encodings*
10 $\quad$ **return** $RBWT$;

---

4 Computing Run-Length BWT

Characterizations. Under $\Pi^{\text{pos}}_{\text{lex}} = (\preccurlyeq_{\text{pos}}, \preccurlyeq_{\text{lex}})$, the set QI_{BWT} of all *quasi-irreducible ranks* is defined by the set $QI_{BWT} := \{ \text{rnk}(S) \mid S \in \mathcal{CS}(G) \} \subseteq [n]$, We observe that $|QI_{BWT}| \leqslant e$ since $|\mathcal{CS}(G)| \leqslant e$. We then have the *interpolation property* below.

Lemma 4.1 (interpolation property). *Under* $\Pi^{\text{pos}}_{\text{lex}}$, *if* $i_* \notin QI_{BWT}$, $BWT[i] = BWT[i - 1] \in \Sigma$ *holds for* $\forall i \in [n]$.

Algorithm. In Algorithm 1, we present the recursive procedure that computes the quasi-irreducible BWT for text T from either an input CDAWG G for T stored in read-only memory or its self-index $G = CDAWG^-_\Pi(T)$ when it is invoked with the root v. Let $\mathcal{I} = (I_1, \ldots, I_h)$, $h = |\overline{\mathcal{E}^*_-} \cup \{f_-\}| \leqslant e$, be the ordered partition of the rank space $[1..n]$ consisting of the SA-intervals for all $(-)$-search paths in $\mathcal{SP}_-(G)$ sorted in the lex-order $\preccurlyeq_{\text{lex}}$, where for $\forall i \in [h]$, $I_i = [sp(P_i)..ep(P_i)] \subseteq [n]$. Firstly, the next lemma characterizes the BWT.

Lemma 4.2. *Let* $T[1..n]$ *be any text. (1) $BWT[1..n] = BWT[I_1] \circ \cdots \circ BWT[I_h]$. (2)For $\forall i \in [h]$, (i) and (ii) below hold: (i)If $i = i_*$ with $SA[i_*] = 1$, then P_i is a trivial $(-)$-search path, $I_{i_*} = \{i_*\}$, and $BWT[I_{i_*}] = T[n] = $ '\$'. (ii) If $i \neq i_*$, P_i is a non-trivial $(-)$-search path with certificate $f \in \overline{\mathcal{E}^*_-}$ .Then, $BWT[I_i]$ is the equal-symbol run $c^\ell \in \Sigma^+$, where $c := T[p-1]$, $\ell := |I_i|$, and $p = \text{pos}(\text{cano}_\delta(f))$.*

In Algorithm 1, the concatenation of two run-length encodings at line 9 can be easily done in $O(1)$ time by keeping the symbols at both ends. If a read-only text T is available, the preceding symbol, precsym$(f) := T[p - 1]$, of $S = \text{cano}_-(f)$ can be computed in $O(1)$ time at line 8. In the case of the self-index, we have the next lemma, and then, the main result of this section.

Lemma 4.3. *Given the self-index of G, the preceding symbol* precsym(f) *of $S = \text{cano}_-(f)$ for $\forall f \in \overline{\mathcal{E}^*_-}$ can be computed in amortized $O(1)$ time and space.*

Algorithm 2: The algorithm for computing the quasi-irreducible $GLPF_{\preccurlyeq_+}$ array for a text T from the CDAWG for T or its self-index.

---

1 **Procedure** QIrrGLPF(v, QGL) ;   ▷ *Assume path orderings* $\Pi = (\preccurlyeq_{pos}, \preccurlyeq_+)$
2 **if** $N_-(v) = \emptyset$ **then**   ▷ *Case: trivial suffix at the root*
3     $QGL \leftarrow QGL \circ (1,0)$
4 **else**   ▷ *Case: non-trivial suffix at branching node*
5   **for each** $f = (w, X, v)$ in order $\leqslant^E_{-,pos}$ compatible to $\preccurlyeq_{pos}$ **do**
6   **if** is-primary$_+(f)$ **then**   ▷ *Case: (+)-primary*
7   QIrrGLPF(w, QGL)
8   **else**   ▷ *Case: (+)-secondary*
9           $\ell \leftarrow |\mathbf{repr}_-(w)|$; $p \leftarrow n + 1 - |\mathbf{repr}_-(w)| - |X| - |\mathbf{repr}_+(v)|$;
10           $QGL \leftarrow QGL \circ (p, \ell)$;   ▷*output : $GLPF[p] = \ell$*

Theorem 4.1. *Given a self-index version of CDAWG(T) without a text, Algorithm 1 constructs the RLBWT of size $r \leqslant e$ in $O(e)$ time and $O(e)$ space.*

5 Computing Irreducible *GLPF* Arrays

Let $\Pi = (\preccurlyeq_{pos}, \preccurlyeq_+)$ with $\preccurlyeq_+ \in \{\preccurlyeq_{lex}, \preccurlyeq_{pos}\}$. First, we define a generalization of PLCP and LPF, the *generalized longest previous factor (GLPF) array* for a text $T[1..n]$, under $\preccurlyeq_+$ by the array $GLPF_{\preccurlyeq_+}[1..n] \in \mathbb{N}^n$ such that for any $p \in [n]$, $GLPF_{\preccurlyeq_+}[p] := \max(\{ lcp(T_p, T_q) \mid T_q \prec_+ T_p, q \in [n] \} \cup \{0\})$. Then, the *quasi-irreducible GLPF array* under path orderings Π is defined by the subset $\widetilde{GLPF} := \{ (p, GLPF_{\preccurlyeq_+}[p]) \mid S \in \mathcal{CS}(G), p = pos(S) \} \subseteq [n] \times \mathbb{N}$. We observe that $|\widetilde{GLPF}| \leqslant e$, $PLCP = GLPF_{\preccurlyeq_{lex}}$, and $LPF = GLPF_{\preccurlyeq_{pos}}$ for any text T.

Lemma 5.1 (characterization of $GLPF_{\preccurlyeq_+}$ value). *For $\forall (p, \ell) \in [n] \times \mathbb{N}$, the conditions (a)–(c) below are equivalent: (a) $(p, \ell) \in \widetilde{GLPF}$. (b) $GLPF_{\preccurlyeq_+}[p] = \ell$ and its length-ℓ prefix is left-maximal in T. (c) For some $S \in \mathcal{CS}(G)$, $p = pos(S)$, and either (i) S is (+)-trivial and $\ell = 0$, or (ii) S is (+)-nontrivial, $S = cano_+(f)$ and $\ell = |\mathbf{repr}_-(w)|$ hold for some (+)-certificate $f = (w, X, v) \in \overline{\mathcal{E}^\star_+}$.*

Proposition 5.1 (interpolation property). *For any position $p \in [n]$, if $p \notin QI_{GLPF_{\preccurlyeq_+}}$ then $GLPF_{\preccurlyeq_+}[p] = GLPF_{\preccurlyeq_+}[p-1] - 1$ holds. Consequently, PLCP and LPF satisfy the interpolation property as above w.r.t. QI_{GLPF}.*

Algorithm 2 computes the quasi-irreducible GLPF array for T from the self-index G, when it is invoked with $v = sink(G)$ and $QGL = \varepsilon$. Hence, we have:

Theorem 5.1. *Given a self-index version of CDAWG(T) without a text, Algorithm 2 constructs the quasi-irreducible $GLPF_{\preccurlyeq_+}$ for T in $O(e)$ time and space.*

Theorem 5.2. *The lex-parse of size $2r = O(e)$ and the LZ-parse of size $z \leqslant r$ of a text $T[1..n]$ can be computed from a self-index version of $CDAWG(T)$ without a text for the same text in $O(e)$ time and space.*

Acknowledgments. The authors thank the anonymous reviewers for their comments which greatly improved this paper. The first author is also grateful to Hideo Bannai for information on conversion between text indexes, and to Mitsuru Funakoshi for discussion on the sensitivity of text indexes.

References

1. Arimura, H., Inenaga, S., Kobayashi, Y., Nakashima, Y., Sue, M.: Optimally computing compressed indexing arrays based on the compact directed acyclic word graph. CoRR (2023). http://arxiv.org/abs/
2. Bannai, H., Gawrychowski, P., Inenaga, S., Takeda, M.: Converting SLP to LZ78 in almost linear time. In: Fischer, J., Sanders, P. (eds.) CPM 2013. LNCS, vol. 7922, pp. 38–49. Springer, Heidelberg (2013). https://doi.org/10.1007/978-3-642-38905-4_6
3. Belazzougui, D., Cunial, F.: Representing the suffix tree with the CDAWG. In: CPM 2017. LIPIcs, vol. 78, pp. 7:1–7:13 (2017)
4. Belazzougui, D., Cunial, F., Gagie, T., Prezza, N., Raffinot, M.: Composite repetition-aware data structures. In: Cicalese, F., Porat, E., Vaccaro, U. (eds.) CPM 2015. LNCS, vol. 9133, pp. 26–39. Springer, Cham (2015). https://doi.org/10.1007/978-3-319-19929-0_3
5. Blumer, A., Blumer, J., Haussler, D., McConnell, R., Ehrenfeucht, A.: Complete inverted files for efficient text retrieval and analysis. JACM **34**(3), 578–595 (1987)
6. Brlek, S., Frosini, A., Mancini, I., Pergola, E., Rinaldi, S.: Burrows-wheeler transform of words defined by morphisms. In: Colbourn, C.J., Grossi, R., Pisanti, N. (eds.) IWOCA 2019. LNCS, vol. 11638, pp. 393–404. Springer, Cham (2019). https://doi.org/10.1007/978-3-030-25005-8_32
7. Crochemore, M., Ilie, L.: Computing longest previous factor in linear time and applications. IPL **106**(2), 75–80 (2008)
8. Gusfield, D.: Algorithms on Strings, Trees, and Sequences: Computer Science and Computational Biology. Cambridge University Press, Cambridge (1997)
9. Kärkkäinen, J., Manzini, G., Puglisi, S.J.: Permuted longest-common-prefix array. In: Kucherov, G., Ukkonen, E. (eds.) CPM 2009. LNCS, vol. 5577, pp. 181–192. Springer, Heidelberg (2009). https://doi.org/10.1007/978-3-642-02441-2_17
10. Kempa, D., Kociumaka, T.: Resolution of the burrows-wheeler transform conjecture. Commun. ACM **65**(6), 91–98 (2022)
11. Mantaci, S., Restivo, A., Rosone, G., Sciortino, M., Versari, L.: Measuring the clustering effect of BWT via RLE. Theoret. Comput. Sci. **698**, 79–87 (2017)
12. Navarro, G.: Indexing highly repetitive string collections, part ii: Compressed indexes. ACM Comput. Surv. (CSUR) **54**(2), 1–32 (2021)
13. Navarro, G., Ochoa, C., Prezza, N.: On the approximation ratio of ordered parsings. IEEE Trans. Inf. Theory **67**(2), 1008–1026 (2020)
14. Radoszewski, J., Rytter, W.: On the structure of compacted subword graphs of Thue-Morse words and their applications. JDA **11**, 15–24 (2012)
15. Takagi, T., Goto, K., Fujishige, Y., Inenaga, S., Arimura, H.: Linear-size CDAWG: new repetition-aware indexing and grammar compression. In: Fici, G., Sciortino, M., Venturini, R. (eds.) SPIRE 2017. LNCS, vol. 10508, pp. 304–316. Springer, Cham (2017). https://doi.org/10.1007/978-3-319-67428-5_26

Evaluating Regular Path Queries
on Compressed Adjacency Matrices

Diego Arroyuelo[1,2], Adrián Gómez-Brandón[1,3(✉)],
and Gonzalo Navarro[1,4]

[1] Millennium Institute for Foundational Research on Data (IMFD), Santiago, Chile
[2] Department of Informatics, Universidad Técnica Federico Santa María,
Viña del Mar, Chile
[3] CITIC Research Center, Universidade da Coruña, A Coruña, Spain
adrian.gbrandon@udc.es
[4] Department of Computer Science, University of Chile, Santiago, Chile

Abstract. Regular Path Queries (RPQs), which are essentially regular expressions to be matched against the labels of paths in labeled graphs, are at the core of graph database query languages like SPARQL. A way to solve RPQs is to translate them into a sequence of operations on the adjacency matrices of each label. We design and implement a Boolean algebra on sparse matrix representations and, as an application, use them to handle RPQs. Our baseline representation uses the same space as the previously most compact index for RPQs and excels in handling the hardest types of queries. Our more succinct structure, based on k^2-trees, is 4 times smaller and still solves complex RPQs in reasonable time.

1 Introduction and Related Work

Graph databases have emerged as a crucial tool in several applications such as web and social networks analysis, the semantic web, and modeling knowledge, among others. We are interested in labeled graph databases, where the graph edges have labels. An important kind of queries in such databases are the *regular path queries* (RPQs, for short), which search for paths of arbitrary length matching a regular expression on their edge labels [3]. For example, in the simple RDF model [23], one can represent points of interest in New York City as nodes in a graph, and have edges such as $x \xrightarrow{\text{walk}} y$ indicating that x is within a short walking distance of y, as well as edges of the form $x \xrightarrow{\text{L}} y$ if subway stations x and y are connected directly by subway line L. Then the RPQ 'Central Park walk/(N|Q|R)$^+$/walk ?y', asks for all sites ?y of interest that are

---

Supported by ANID - Millennium Science Initiative Program – Code ICN17_002, and Fondecyt Grant 1-230755, Fondecyt Grant 1221926; CITIC is funded by Xunta de Galicia and CIGUS; GAIN/Xunta de Galicia Grant ED431C 2021/53 (GRC); Xunta de Galicia/FEDER-UE Grant IN852D 2021/3; MCIN/AEI and NextGenerationEU/PRTR Grants [PID2020-114635RB-I00, TED2021-129245B-C21].

F. M. Nardini et al. (Eds.): SPIRE 2023, LNCS 14240, pp. 35–48, 2023.
https://doi.org/10.1007/978-3-031-43980-3_4

reachable from Central Park by using subway lines N, Q, or R, through one or more stations and allowing a short walk before and after using the subway.

RPQs are at the core of current graph database query languages, extending their expressiveness. In particular, the SPARQL 1.1 standard includes the support for *property paths*, that is, RPQs extended with inverse paths (known as two-way RPQs, or 2RPQs for short) and negated label sets. As SPARQL has been adopted by several systems, RPQs have become a popular feature [3]: out of 208 million SPARQL queries in the public logs from the Wikidata Query Service [22], about 24% use at least one RPQ feature [9]. Further developments like PGQL [28], Cypher [18], G-CORE [2], TigerGraph [15], and GQL [14], to name some of the most popular ones, also support RPQ-like features.

Handling (2)RPQs can be computationally expensive to evaluate as they usually involve a large number of paths [24], mostly for regular expressions using Kleene stars. There are two main algorithmic approaches to support them [33]: (1) to represent the regular expression of the 2RPQ using a finite automaton, which is then used to search over the so-called product graph [25]; and (2) to extend the relational algebra to support computing the transitive closure of binary relations to evaluate regular expressions having Kleene stars [21]. Although most theoretical results on 2RPQs have followed the first approach, property path evaluation in SPARQL has followed the second one [33].

Recent research introduced not only time- but also space-efficient solutions for evaluating graph joins [5,6,10]. With the big graphs available today, this is an important step towards in-memory processing of graph queries. In particular, the Ring data structure [6] is able to represent a labeled graph in space close to its plain representation, while supporting worst-case optimal joins (used, as we said, for BGP queries). Moreover, by using little extra space the Ring can be used to support 2RPQs efficiently [4], using the product-graph approach [25].

In this paper, we introduce a space-efficient approach for evaluating 2RPQs that, essentially, represents the subgraph corresponding to each graph label p using a sparse representation of its Boolean adjacency matrix M_p. We evaluate 2RPQs by translating them into classic operations on Boolean matrices [21]. This approach is typically disregarded because matrix sizes are quadratic on the number of graph nodes, but we exploit the sparsity of those matrices to represent them efficiently, using k^2-trees [11]. The use of k^2-trees to represent each RDF predicate is not new, for example it has been used to handle triple matching and binary joins [1] and full BGPs [5], but not 2RPQs. We show how to translate 2RPQs into matrix operations and how to handle the particularities of 2RPQs.

The result is the most space-efficient graph database representation (nearly 4 bytes per graph edge on a Wikidata graph, 4 times less than the previously most compact representation—the Ring [4]—and 14–21 times smaller than classical systems). In exchange, our structure is on average 5 times slower than the Ring, though it still solves most complex 2RPQs in a few seconds. We also implement an uncompressed baseline for sparse matrices based on the CSR and CSC formats [29, Sect. 3.4]. Its space matches that of the Ring and it excels on the most expensive 2RPQs, namely those where no graph node is specified. It is only outperformed by Blazegraph, which uses 5.5 times more space. Our new matrix-algebra-based approach stands out in the space-time tradeoff map.

2   Basic Concepts

2.1   Labeled Graphs and Regular Path Queries (RPQs)

Let $\mathcal{U}$ be a totally ordered, countably infinite set of *symbols* or *constants*, which we call the *universe*. A *directed edge-labeled graph* $G \subseteq \mathcal{U}^3$ is a finite set of triples $(s, p, o) \in \mathcal{U}^3$ encoding the graph edges $s \xrightarrow{p} o$ from vertex s to vertex o with edge label p. In the RDF model [23] (which has gained popularity in representing directed edge-labeled graphs), s is called a *subject*, p a *predicate*, and o an *object*.

For a graph G, we define its set of edge labels as $P = \{p \mid \exists\, s, o, (s, p, o) \in G\}$. Similarly, let $V = \{x \mid \exists\, y, z,\ (x, y, z) \in G \vee (z, y, x) \in G\}$ be the set of graph nodes. We assume that the graph nodes have been mapped to integers in the range $[1 .. |V|]$. A path ρ from a node x_0 to node x_n in a graph G is a string $x_0 p_1 x_1 \cdots x_{n-1} p_n x_n$ such that $(x_{i-1}, p_i, x_i) \in G$ for $1 \leq i \leq n$. Given a path ρ, we denote $\mathsf{word}(\rho) = p_1 \cdots p_n$ the string labeling path ρ. Two-way RPQs (2RPQs) also allow traversing reversed edges. Hence, we define the set of inverse labels as $\hat{}P = \{\hat{}p \mid p \in P\}$, and $P^{\leftrightarrow} = P \cup \hat{}P$ the set of predicates and their inverses. We define the *inverse graph* as $\hat{}G = \{(y, \hat{}p, x) \mid (x, p, y) \in G\}$, and its *completion* as $G^{\leftrightarrow} = G \cup \hat{}G$. A *two-way regular expression* (2RE) is then formed from the following rules: ε is a 2RE; if $c \in P^{\leftrightarrow}$, then c is a 2RE; if E, E_1 and E_2 are 2REs, then so are E^* (Kleene star), E_1/E_2 (concatenation), and $E_1 \mid E_2$ (disjunction). If E is a 2RE, we also abbreviate E^*/E as E^+ and $\varepsilon|E$ as $E^?$.

The *language* $L(E)$ of E is defined exactly as that of the regular expressions over the alphabet $P^{\leftrightarrow}$ of terminals, and we say that a path ρ *matches* a 2RE E iff $\mathsf{word}(\rho) \in L(E)$. A *two-way regular path query*, or 2RPQ for short, is a query of the form (x, E, y), which looks for all the pairs of nodes (s, o) such that there exists a path $\rho = s p_1 \cdots p_n o$ in $G^{\leftrightarrow}$ where $\mathsf{word}(\rho) \in L(E)$; x and/or y can be constants (thus fixing the value of s and/or o, respectively), or variables.

2.2   An Algebra on Boolean Matrices

Let $A = (a_{i,j})_{1 \leq i,j \leq n}$ and $B = (b_{i,j})_{1 \leq i,j \leq n}$ be square $n \times n$ Boolean matrices. We define the following operations of interest for our work:

- **Transpose:** A^T, where $a_{i,j}^T = a_{j,i}$, for $1 \leq i, j \leq n$.
- **Sum:** $A + B = C = (c_{i,j})$, where $c_{i,j} = a_{i,j} \vee b_{i,j}$, for $1 \leq i, j \leq n$.
- **Product:** $A \times B = C$, where for $1 \leq i, j \leq n$ we have $c_{i,j} = \bigvee_{1 \leq k \leq n} a_{i,k} \wedge b_{k,j}$.
- **Exponentiation:** $A^k = \prod_{i=1}^{k} A$, that is, $A \times \cdots \times A$, writing A k times.
- **Transitive closure:** $A^+ = A + A^2 + \cdots + A^n$.
- **Reflexive-transitive closure:** $A^* = I + A^+$, where I is the identity matrix.
- **Row/column restrictions:** $\langle r \rangle A$, a matrix whose row r equals row r of A; $A\langle c \rangle$, a matrix whose column c equals column c of A; and $\langle r \rangle A \langle c \rangle$, a matrix whose cell (r, c) equals entry $A[r][c]$. The remaining cells are 0.

The implementation of these operations on sparse matrix representations is relatively straightforward, except for the multiplication and transitive closures.

2.3 K^2-Trees

A k^2-tree [11] is a data structure able to space-efficiently represent binary relations, point grids, and graphs. We will use it in this paper to represent Boolean matrices, as follows. Let A be a $v \times v$ Boolean matrix, assuming v is a power of 2.[1] The root node of the k^2-tree represents the whole matrix A. Then, A is divided into 4 equally-sized quadrants, $A = \begin{pmatrix} A_0 & A_1 \\ A_2 & A_3 \end{pmatrix}$, such that submatrix A_0 is represented recursively by the first child of the root, A_1 by the second child, and so on. The process stops as soon as one gets into an empty submatrix, which is represented by a leaf node. Each node in this tree has 4 children (in general, k^2 children, yet we use $k = 2$). This order in which quadrants are represented (i.e., top-left, top-right, bottom-left, and bottom-right) is known as z-order. The resulting tree height is $\log_4 v^2 = \log_2 v$.

To represent this tree space-efficiently, we traverse the tree in level order. At each node, we write its 4-bit signature (which represents the node) indicating whether each of the 4 children represents an empty submatrix or not. For instance, the signature 0110 indicates that quadrants 0 and 3 of the submatrix represented by the current node are empty, whereas A_1 and A_2 (second and third children) are non-empty. The result is a bit vector $L[1\mathinner{.\,.}4n]$, where n is the number of internal nodes in the tree. Each tree node is represented by the first bit of its signature. Given a node i, its j-th child ($0 \le j \le 3$) is represented at position $4 \cdot \mathsf{rank}_1(L, i) + 1$, where $\mathsf{rank}(L, i)$ counts the number of 1 s in $L[1\mathinner{.\,.}i]$ in $O(1)$ time using $o(n)$ additional bits of space [12,26].

The k^2-tree representation is especially useful for representing sparse matrices. Let matrix A have a 1 s. Then, in the worst case every 1 induces a 4-bit signature in every level of the k^2-tree, for a total of $4a \log_2 v$ bits. The actual upper bound is lower because not all those signatures can be different: in the worst case all the k^2-tree nodes up to level $\lfloor \log_4 a \rfloor$ exist, and from there on each 1 of A has its own path; this adds up to $4a \log_4(v^2/a) + 4a/3 + O(1)$ bits. The figures further improve when the 1 s are clustered in A [7].

3 Evaluating RPQs Through the Boolean Matrix Algebra

For a given directed edge-labeled graph G of n edges, let P be the corresponding set of graph labels as defined in Sect. 2.1. In our approach, for every $p \in P$ we define a $|V| \times |V|$ Boolean matrix M_p, such that $M_p[x][y] = 1$ iff $(x, p, y) \in G$. We translate an RPQ into operations on those matrices, so that the resulting Boolean matrix contains all pairs (x, y) that match the regular expression. We define next the recursive formulas $\mathcal{M}$ to translate 2RPQs into matrix operations, following Losemann and Martens' work [21]. We start with the base cases: $\mathcal{M}(\varepsilon) = I$, the identity matrix; $\mathcal{M}(p) = M_p$, for $p \in P$; $\mathcal{M}(\hat{\ }p) = M_p^T$, for $p \in P$.

Next, let E_1 and E_2 be 2RPQs. We define the following recursive rules: $\mathcal{M}(E_1 \mid E_2) = \mathcal{M}(E_1) + \mathcal{M}(E_2)$; $\mathcal{M}(E_1/E_2) = \mathcal{M}(E_1) \times \mathcal{M}(E_2)$; $\mathcal{M}(E_1^+) = \mathcal{M}(E_1)^+$; $\mathcal{M}(E_1^*) = I + \mathcal{M}(E_1)^+$, where I is the corresponding identity matrix.

---

[1] If v is not a power of 2 we round it up to the next power, leaving the extended cells empty. This imposes almost no extra overhead on the k^2-tree representation.

Then, given a 2RPQ $R = (x, E, y)$, we evaluate it as follows: (1) if x and y are both variables, $\mathcal{R}(R) = \mathcal{M}(E)$; (2) If x is a variable and y is a constant, $\mathcal{R}(R) = \mathcal{M}(E)\langle y \rangle$; (3) If x is a constant and y is a variable, $\mathcal{R}(R) = \langle x \rangle \mathcal{M}(E)$; (4) If x and y are both constant, $\mathcal{R}(R) = \langle x \rangle \mathcal{M}(E)\langle y \rangle$.

4   Implementation of the Boolean Matrix Algebra

We now describe how the Boolean-matrix operations are carried out. To analyze the corresponding algorithms, we use $|M_p|$ as the number of 1s in the matrix, which is the number of edges with label p in graph G. We represent each matrix M_p using a k^2-tree of $O(\log |V|)$ levels, and each 1 in M_p induces $O(\log |V|)$ 1s in its k^2-tree representation. We will use $v = |V|$, as well as $a = |A|$ and $b = |B|$ for the number of 1s in matrices A and B. We assume $|V| = 2^i$, for $i \geq 0$.

We implement k^2-trees, and thus bitvectors with rank support, from scratch. We store the bitvector as consecutive bits packed in a 64-bit-words array. To support rank we store the cumulative sum of 1s up to every sth cell of the array. To save space, full 64-bit integers store the full sum only every 2^{16} bits, and the others are stored in relative form using 16-bit integers. To compute rank we start from the last recorded sum and use *popcount* on the full words until reaching the desired one, and a partial *popcount* on the desired word. Here s is a space-time tradeoff parameter: we use $n/1024 + n/(4s)$ additional bits of space for storing a bitvector $B[1 .. n]$, and compute rank in time $O(s)$. We use $s = 4$.

4.1   Transposition

Transposition is used to implement reversed edges, as seen in Sect. 3. Instead of materializing the transposed matrix as a k^2-tree, we note that $A^T = \begin{pmatrix} A_0^T & A_2^T \\ A_1^T & A_3^T \end{pmatrix}$. So, the k^2-tree for A^T can be obtained by interchanging the roles of the second and third children of every node. We do not materialize this interchange, but associate a *transposed* flag to every matrix, so we simply have to toggle it in order to transpose the matrix in $O(1)$ time.

4.2   Boolean Sum

The easier case to implement $A + B$ arises when no matrix is transposed. In this case we can perform a sequential pass over both k^2-tree bitvectors, so as to merge their corresponding nodes levelwise, without need of any rank operation.

We implement this traversal with a queue of tasks, which are of two types. (1) A *copy* task indicates just to copy the next node from A or B; and (2) a *merge* task indicates merging the next nodes of A and B. The queue is initialized with a merge task, the read-pointers (which indicate the next k^2-tree node to be read) at the beginning of the bitvectors of A and B, and the write-pointer at the beginning of the output k^2-tree bitvector.

To process a copy task, we append the next signature (of A or B) to the output, and enqueue its (up to) 4 children, as copy tasks for A or B, respectively.

To process a merge task, we append to the output the bitwise-or of the next 4-bit signatures of A and B, and enqueue up to 4 new elements, as follows. For i from 1 to 4, if the ith bit of the signatures of both A and B are 1, we append a merge task. If only one of them is 1, we append a copy task for the corresponding matrix. If none is 1, we do not append a task. We do not append new tasks when the corresponding nodes are k^2-tree leaves. The process finishes when the queue becomes empty. The total time is proportional to the sum of the bitvector length of both matrices, $O(a\log(v^2/a) + b\log(v^2/b)) \subseteq O((a + b)\log v)$.

Handling Transpositions. If both A and B are transposed, we just merge them as described and mark the result as transposed. When one is transposed and the other is not, we cannot anymore resort to a sequential traversal of both bitvectors. The transposed one must already have rank support built to enable k^2-tree traversals. We traverse sequentially the non-transposed k^2-tree, and include in the queue the corresponding node of the transposed one (as those nodes are not read in left-to-right order). To generate the new tasks, we must use the k^2-tree traversal operations to locate the corresponding nodes in the transposed k^2-tree.

While the time complexity is the same, summing a transposed with a non-transposed matrix is slower in practice. We always choose that the transposed matrix is the one with a shorter bitvector (we can because $A^T + B = (A + B^T)^T$), in order to minimize the non-local traversals.

4.3   Boolean Multiplication

For the multiplication $A \times B$ we use the following classic divide-and-conquer recursive procedure. Let $A = \left(\begin{smallmatrix} A_0 & A_1 \\ A_2 & A_3 \end{smallmatrix}\right)$ and $B = \left(\begin{smallmatrix} B_0 & B_1 \\ B_2 & B_3 \end{smallmatrix}\right)$ be the four submatrices into which the k^2-tree representation splits A and B. Then, we recursively compute 8 products of those submatrices in order to produce

$$A \times B = \left(\begin{array}{c:c} A_0 \times B_0 + A_1 \times B_2 & A_0 \times B_1 + A_1 \times B_3 \\ \hdashline A_2 \times B_0 + A_3 \times B_2 & A_2 \times B_1 + A_3 \times B_3 \end{array} \right). \tag{1}$$

A fortunate consequence of the k^2-tree representation is that, if any of those submatrices is empty (i.e., there is a 0 in the signature of the root of A or B), then we know that its product with any other submatrix is also zero. Further, summing a product $A_i \times B_j$ with a zero matrix does not even need to copy the product; we just reference it as the final result.

Once the k^2-tree bitvectors of the four submatrices are recursively obtained, we concatenate them levelwise. There is no need to build the rank data structures until we obtain the final matrix because the concatenation proceeds left-to-right in each level. We only take care of maintaining, for each bitvector, $O(\log v)$ pointers to the positions where the levels start.

Transpositions are handled easily, by exchanging the meaning of M_1 and M_2 in every node of the k^2-tree bitvector if $M = \left(\begin{smallmatrix} M_0 & M_1 \\ M_2 & M_3 \end{smallmatrix}\right)$ is transposed.

A Rough Analysis. One term of the multiplication cost is given by the number of recursive calls, which follows the recurrence $T(v^2) = 8 \cdot T(v^2/4)$. Since our matrices are sparse, the worst case arises when every submatrix has points up to the level ℓ where we have $4^\ell \geq \min(a,b)$ submatrices, that is, $\ell = \log_4 \min(a,b)$. From this level, the worst case is that the $\max(a,b)/\min(a,b)$ points in the submatrices of the fuller matrix distribute uniformly for $\ell' = \log_4 \frac{\max(a,b)}{\min(a,b)}$ further levels. Between those levels, the recurrence becomes $T'(v^2) = 2 \cdot T'(v^2/4)$ because the single point in the emptier submatrix can make us enter into at most two submatrices of the other. This continues until, in level $\ell + \ell'$, both submatrices contain one point each, and from there on the cost is just $\log_2 v - \ell - \ell'$ to track a single point along both submatrices. The cost up to level ℓ is then $8^\ell = \min(a,b)^{3/2}$. From each of those 8^ℓ submatrices we have a cost of $2^{\ell'} = (\max(a,b)/\min(a,b))^{1/2}$, and from each of those $8^\ell 2^{\ell'} = \min(a,b)\sqrt{\max(a,b)}$ submatrices we have $O(\log(v^2/\max(a,b)))$ additional time. The total cost of recursive calls is then $O(\min(a,b)\sqrt{\max(a,b)}\log(v^2/\max(a,b)))$.

The second part of the cost is that of summing pairs of partial submatrices. In the worst case, those matrices may add up to $a \cdot b$ points at across every level ℓ of the recursion. Since summing submatrices in level ℓ costs $O(\ell)$ per element, the total cost of summing partial results is in $O(ab \log^2 v)$. Since this is an utterly pessimistic upper bound, we offer an average-case time analysis for matrices with uniformly distributed 1 s. We multiply 8^ℓ pairs of $v/2^\ell \times v/2^\ell$ submatrices in level ℓ. On average, each has $a/4^\ell$ 1 s in A and $b/4^\ell$ cells in B. Every such a_{ik} will pair with every such $b_{k'j}$ iff $k = k'$, which occurs with probability $1/(v/2^\ell)$, so on average there will be $8^\ell(a/4^\ell)(b/4^\ell)(2^\ell/v) = ab/v$ 1 s to sum per level ℓ, with a maximum of v^2. This leads to a total average cost upper bounded by $O(\min(a,b)\sqrt{\max(a,b)}\log v + \min(v^2, (ab/v))\log^2 v)$.

4.4 Closure

We opted for a simple transitive closure algorithm for now. The closure A^+ is obtained by iteratively computing $A \leftarrow A + A \times A$ until no change occurs in A [19]. This occurs at most after $\log_2 v$ iterations, so the time complexity is $O(\log v)$ times that of multiplying A by itself (note that a grows in every iteration, so the time complexity becomes bounded by $O(|A^+|^{3/2}\log^3 v)$). The transitive closure is computed as $A^* = I + A^+$, where I is the identity matrix.

Needless to say, unrestricted closure operations are the most expensive, both in time complexity and in practice, so we aim to avoid them as much as possible.

4.5 Restrictions

Restrictions indicate that we only want to retrieve a column or a row of the matrix after the operations, or even just a cell. A naive way to implement them is to first obtain the full matrix M and then traverse the desired row or column. Yet, restrictions give an important opportunity of optimizing all the other operations.

Sums. For $\langle r\rangle(A+B)\langle c\rangle$ (where only $\langle r\rangle$ or only $\langle c\rangle$ could be present as well), we restrict the traversal of both matrices, acting as if the submatrices not intersecting the desired row and/or column were empty. That is, we implement the restricted sum as $\langle r\rangle A\langle c\rangle + \langle r\rangle B\langle c\rangle$. We cannot, however, simply traverse both k^2-tree bitvectors and write the output left-to-right, as in Sect. 4.2, because now we do not know beforehand whether a submatrix (or the merge of two submatrices) will be nonempty after restricting it to some row/column, even if it intersects the row/column. Our solution is then recursive, similar to the multiplication algorithm (yet still considerably simpler).

Products. A restricted product $\langle r\rangle(A \times B)\langle c\rangle$ is handled as $((\langle r\rangle A) \times (B\langle c\rangle))$, where again only one of the restrictions may be present. We consider the column or row restrictions along the whole recursion, pretending that the submatrices that do not intersect the desired row or column are empty.

Closures. Operation $A^+\langle c\rangle$ is implemented as $S \leftarrow (E+A)\langle c\rangle$, where E is the empty matrix, and then repeatedly doing $P \leftarrow A\times S$ and $S \leftarrow S+P$ until S does not change. Note that the only nonzero column of P and S is c. To implement $A^*\langle c\rangle$ we start with $S = (I + A)\langle c\rangle$ instead. A row restriction $\langle r\rangle A^+$ is handled analogously, starting with $S = \langle r\rangle(A + E)$ and then iterating over $P \leftarrow S \times A$ and $S \leftarrow S + P$, or using the initial step $S \leftarrow \langle r\rangle(I + A)$ for $\langle r\rangle A^*$.

Note that this iteration does not make the path lengths grow exponentially for the transitive closure, but linearly. Therefore, we could need up to v iterations to compute the closure. In practice, the closure is reached much sooner and the operations are significantly faster, leading to a much better solution.

When both row and column are restricted, we only want a cell of the transitive closure. We then choose the row/column with fewer elements in A and run a row-restricted or column-restricted closure, whichever is emptier. At each step, we check if the desired cell is full, stopping immediately if so.

4.6   Query Plan

We first build the syntax tree of the 2RE E of the 2RPQ (x, E, y). In principle, we can simply traverse the syntax tree and solve it in postorder in the standard way, interpreting the leaves p as the matrix M_p, $\hat{}p$ as M_p^T, and ε as I, and interpreting the internal nodes as the corresponding operations on the matrices resulting from their children, according to the translations of Sect. 3. Our particular application, however, enables some relevant optimizations.

Let us first assume that both x and y are variables. A first simple optimization is that the closures are idempotent, so a sequence of closures is reduced to one. More precisely, $(A^*)^* = (A^*)^+ = (A^+)^* = A^*$ and $(A^+)^+ = A^+$. Sums and products yield more important optimizations, though.

Sums. We exploit the fact that the Boolean sum is commutative and associative to carry out a sequence of consecutive sums, $E_1 \mid \ldots \mid E_m$, in the best possible order. Since the cost of computing $A + B$ is proportional to $|A| + |B|$, if it were the case that $|A + B| = |A| + |B|$, the best possible order would be given by building the Huffman tree [20] of the matrices $A_i = \mathcal{M}(E_i)$ using $|A_i|$ as their weight. Since, instead, it holds that $\max(|A|, |B|) \leq |A + B| \leq |A| + |B|$, we opt for a heuristic that simulates Huffman's algorithm on the actual size of the matrices as they are produced. Concretely, we start with $\{A_1, \ldots, A_m\}$ and iteratively remove from the set the two matrices A_i and A_j with the smallest sizes, sum them, and return $A_i + A_j$ to the set, until it has a single matrix.

Products. Matrix multiplication is not commutative but still associative, so we can decide the order in which the sequence of multiplications to compute $E_1 / \cdots / E_m$ is carried out. We cannot apply the well-known optimal algorithm to choose the order for dense matrices [13, Sect. 15.2] because the time complexity of our sparse matrix multiplications depends on the number of 1s in the matrices. Further, this number of 1s can increase or decrease after a multiplication. We then opt for a heuristic analogous to the one we use for sums: we start from the sequence $A_1, \ldots, A_m = \mathcal{M}(E_1), \ldots, \mathcal{M}(E_m)$ and iteratively choose the consecutive pair A_i, A_{i+1} that minimizes $|A_i| + |A_{i+1}|$, multiply them, and replace the pair by $A_i \times A_{i+1}$, until the sequence has a single element.

Handling Restrictions. When x (resp., y) is a constant we are restricting a row (resp., column) of the matrix after the operations. For efficiency, then, we apply the restricted operations of Sect. 4.5. Regarding the sums, because $\langle r \rangle (A + B) \langle c \rangle = \langle r \rangle A \langle c \rangle + \langle r \rangle B \langle c \rangle$, we can restrict all the involved matrices at the same time. Consequently, the sum can be computed in any order, and the plan still focuses on looking for the best order based on Huffman's algorithm. In the restriction on products, we obtain a sequence $\langle r \rangle A_1 \times \cdots \times A_m \langle c \rangle$ (where only $\langle r \rangle$ or only $\langle c \rangle$ could be present). Consider the case $\langle r \rangle A_1 \times \cdots \times A_m$. The number of 1s reduces faster when multiplying the pair that contains the restricted matrix, so we compute $A' = \langle r \rangle A_1 \times A_2$. The matrix A' already has all zeros except in row r, so we can continue left-to-right in the sequence with normal matrix multiplications, $A' \times A_3$, and so on. The case $A_1 \times \cdots \times A_m \langle c \rangle$ is analogous, starting with $A' = A_{m-1} \times A_m \langle c \rangle$ and then completing the multiplications right to left. When both restrictions are present, we choose an end and proceed as explained until the final multiplication, $\langle r \rangle A' \times A'' \langle c \rangle$, which is carried out with the restricted multiplication algorithm to enforce the other restriction.

Some restrictions can be inherited by the operands of a node, which speeds up processing. Since $\langle r \rangle (A + B) \langle c \rangle = \langle r \rangle A \langle c \rangle + \langle r \rangle B \langle c \rangle$, both children of a sum inherit the same restrictions. Instead, the product $\langle r \rangle (A \times B) \langle c \rangle = (\langle r \rangle A) \times (B \langle c \rangle)$, thus only the left child inherits a row restriction and only the right child inherits a column restriction. Closures do not inherit their restrictions to their operand, because $\langle r \rangle A^* \langle c \rangle \neq (\langle r \rangle A \langle c \rangle)^*$ and $\langle r \rangle A^+ \langle c \rangle \neq (\langle r \rangle A \langle c \rangle)^+$. Restrictions are not inherited to leaves of the syntax tree, however, because internal operands handle them more efficiently than leaves. On the other hand, they are removed from parents when inherited to children because the nonrestricted operands run faster than those of Sect. 4.5 when their operands have already been restricted.

Finally, we create a special implementation for the case $A^+ \times B\langle c \rangle$ that avoids computing the full closure A^+, as a kind of restricted positive closure that starts instead with $S \leftarrow A \times B\langle c \rangle$. To handle $A^* \times B\langle c \rangle$ we start with $S \leftarrow (E + B)\langle c \rangle$. The cases $\langle r \rangle A \times B^{*/+}$ are handled analogously, as well as the cases with both restrictions. The parser is enhanced to detect those cases.

5   Experimental Results

We implemented our scheme in C++11 and ran our experiments on an Intel(R) Xeon(R) CPU E5-2630 at 2.30GHz, with 6 cores, 15 MB of cache, and 384 GB of RAM. We compiled using g++ with flags -std=c++11, -O3, and -msse4.2.

5.1   A Baseline

We implemented a baseline representation of sparse matrices, which combines (and adapts to the Boolean case) the well-known CSR and CSC formats [29, Sect. 3.4] in order to speed up multiplications. It stores a vector of nonempty row numbers and a similar vector of their starting positions in a third, larger, vector. This third vector stores, for each nonempty row, the increasing sequence of the columns of its nonempty cells. Similar (redundant) vectors are stored for the column-wise view of the matrix.

Transpositions are carried out in $O(1)$ time by just exchanging the row-view and the column-view vectors. The Boolean sum $A + B$ merges the nonempty rows, and when the same row appears in both matrices it merges their nonempty columns. The column-view is computed analogously, thus the sum takes time $O(a + b)$. For the Boolean multiplication $A \times B$, we use Schoor's algorithm [30], whose average time is $O(ab/v)$ if the 1s are uniformly distributed. Our implementation, which is more space-efficient, takes $O(ab \log(v)/v)$ time.

Row and/or column restrictions are handled by restricting the above algorithms to the given row/column; note that finding the desired rows/columns takes just $O(\log v)$ time with the baseline format. Closure operations and their restrictions are performed as for the k^2-tree based representation. The parser and its optimizations are also exactly the same.

5.2   Benchmark

We used a Wikidata graph [32] of $n = 958{,}844{,}164$ edges, $v = 348{,}945{,}080$ nodes, and 5,419 predicates. Separating the edges by predicate and representing the two nodes of each edge as two 32-bit integers, the data set requires 8.5 GB.

We compared our implementations with the following systems:

- *Ring*: A compact data structure that supports RPQs in labeled graphs [4].
- *Jena*: A reference implementation of the SPARQL standard.
- *Virtuoso*: A popular graph database that hosts the public DBpedia endpoint, among others [17].

Table 1. Index space (in bytes per triple), indexing time (in hours), and some statistics on the query times (in seconds). Row "Timeouts" counts queries that take over 60 s or are rejected by the planner as too costly. 2RPQs with some constant node are indicated by c, and without by ¬c.

	k^2-tree	Baseline	Ring	Jena	Virtuoso	Blazegraph
Index space	4.33	16.45	16.41	95.83	60.07	90.79
Index time	0.3	5.5	7.5	37.4	3.0	39.4
Average	8.40	5.67	1.68	5.26	3.87	3.58
Median	1.38	2.46	0.08	0.20	0.14	0.13
Timeout	83	48	22	105	55	46
Average c	7.47	5.37	0.65	3.83	2.98	3.30
Median c	1.32	2.48	0.08	0.17	0.11	0.13
Timeout c	57	37	2	63	37	39
Average ¬c	24.19	10.75	19.22	29.59	18.95	8.35
Median ¬c	13.52	0.63	5.53	4.50	7.98	0.19
Timeout ¬c	26	11	20	42	18	7

– *Blazegraph*: The graph database system [31] hosting the official Wikidata Query Service [22].

To evaluate complex real-world 2RPQs, we extracted all 2RPQs that were not simple labels, from the code-500 (timeout) sections of the seven intervals of the Wikidata Query Logs [22]. We then normalized variable names and removed disrupting queries: duplicated queries and queries producing more than 10^6 results for compatibility with Virtuoso. The result was 1,589 unique queries.

We ran the queries in each system with a timeout limit of 60 s. Table 1 summarizes the space usage and time performance of all the systems. Notably, our approach, using k^2-trees, yields the most compact structure, requiring only 4.33 bytes per triple (bpt). This is less than half the space of the described plain representation of the raw data, and nearly a fourth of the space used by the next smallest representations that support 2RPQs (Ring and our baseline). Classical systems use 14–21 times more space than our k^2-trees. Note also that the k^2-tree representation is orders of magnitude faster to build than the others.

Our reduced space is paid in terms of time performance. Our structure is around 5 times slower than the Ring, on average, and 1.5–2.5 times slower than the classical systems. Still, we solve these complex 2RPQs in less than 10 s on average. Among our matrix-based methods, our structure is 50% slower than the baseline, which uses 4 times more space.

On 2RPQs with some constant, our structure is 11.5 times slower than the Ring and 2.0–2.5 times slower than the classical systems. The gap is considerably narrowed, however, on 2RPQs with both variables, where our structure is just 25% slower than the Ring. Blazegraph is the fastest system in this case, being around 3 times faster than our structure, yet this comes at the expense of using

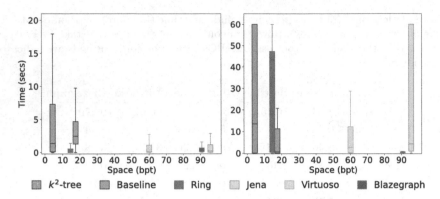

Fig. 1. Space and query time distribution of the systems in general (left) and for the 2RPQs with no constants (right). The baseline and the Ring use almost the same space.

21 times more space. Our baseline yields the best tradeoff for these queries, as it uses 5.5 times less space than Blazegraph and is only 30% slower.

Figure 1 displays the space and query time distribution of all the systems. It can be seen that the k^2-trees and the Ring are the dominant representations in general. When it comes to handling the hardest types of 2RPQs (i.e., without constants), the dominant representations are the two matrix-algebra-based solutions we have introduced and Blazegraph.

6   Conclusions

We have explored the use of a matrix algebra to implement Regular Path Queries (RPQs) on graph databases. This path is usually disregarded because the matrix sizes are quadratic on the number of graph nodes, but we exploit their sparsity to sidestep this issue. Our experiments show that even our baseline (i.e., uncompressed) sparse matrix representation uses the same space of the most compact among previous representations, and outperforms them on the most difficult RPQs (i.e., those with no constant ends). We also develop a more compressed sparse matrix representation based on k^2-trees, which is four times smaller than the baseline and, although slower, it still handles the RPQs within a few seconds.

Immediate extensions to our work are the implementation of negated labels, which require a nonexpensive way to represent and handle submatrices full of 1 s. Such extensions of k^2-trees have been proposed [8], but they have not been adapted to handle Boolean matrix operations. We also plan to implement more efficient transitive closure algorithms [27]. Finally, we plan to strenghten our query optimizer in order to detect common subexpressions and exploit a number of identities of the Boolean algebra we have disregarded for now.

This work can be combined with Qdags [6] to support multijoins as well. We also plan to extend it to a complete algebra for sparse matrices, Boolean and possibly numeric [29]. Such matrices arise, for example, in ML applications [16].

An extended version can be found in https://arxiv.org/abs/2307.14930.

References

1. Álvarez-García, S., Brisaboa, N.R., Fernández, J., Martínez-Prieto, M., Navarro, G.: Compressed vertical partitioning for efficient RDF management. Knowl. Inf. Syst. **44**(2), 439–474 (2015)
2. Angles, R., et al.: G-CORE: a core for future graph query languages. In: SIGMOD International Conference on Management of Data, pp. 1421–1432. ACM (2018). https://doi.org/10.1145/3183713.3190654
3. Angles, R., Arenas, M., Barceló, P., Hogan, A., Reutter, J.L., Vrgoc, D.: Foundations of modern query languages for graph databases. ACM Comput. Surv. **50**(5), 68:1–68:40 (2017). https://doi.org/10.1145/3104031
4. Arroyuelo, D., Hogan, A., Navarro, G., Rojas-Ledesma, J.: Time- and space-efficient regular path queries. In: Proceedings of the 38th IEEE International Conference on Data Engineering (ICDE), pp. 3091–3105 (2022)
5. Arroyuelo, D., Navarro, G., Reutter, J.L., Rojas-Ledesma, J.: Optimal joins using compressed quadtrees. ACM Trans. Database Syst. **47**(2), article 8 (2022)
6. Arroyuelo, D., Hogan, A., Navarro, G., Reutter, J., Rojas-Ledesma, J., Soto, A.: Worst-case optimal graph joins in almost no space. In: ACM International Conference on Management of Data (SIGMOD), pp. 102–114 (2021)
7. de Bernardo, G., Gagie, T., Ladra, S., Navarro, G., Seco, D.: Faster compressed quadtrees. J. Comput. Syst. Sci. **131**, 86–104 (2023)
8. de Bernardo, G., Álvarez-García, S., Brisaboa, N.R., Navarro, G., Pedreira, O.: Compact querieable representations of raster data. In: Kurland, O., Lewenstein, M., Porat, E. (eds.) SPIRE 2013. LNCS, vol. 8214, pp. 96–108. Springer, Cham (2013). https://doi.org/10.1007/978-3-319-02432-5_14
9. Bonifati, A., Martens, W., Timm, T.: Navigating the maze of Wikidata query logs. In: The World Wide Web Conference (WWW), pp. 127–138. ACM (2019)
10. Brisaboa, N., Cerdeira-Pena, A., de Bernardo, G., Fariña, A., Navarro, G.: Space/time-efficient RDF stores based on circular suffix sorting. J. Supercomput. **79**, 5643–5683 (2023)
11. Brisaboa, N.R., Ladra, S., Navarro, G.: Compact representation of web graphs with extended functionality. Inf. Syst. **39**(1), 152–174 (2014)
12. Clark, D.R.: Compact PAT trees. Ph.D. thesis, University of Waterloo, Canada (1996)
13. Cormen, T.H., Leiserson, C.E., Rivest, R.L., Stein, C.: Introduction to Algorithms, 3rd edn. MIT Press, Cambridge (2009)
14. Deutsch, A., et al.: Graph pattern matching in GQL and SQL/PGQ. In: Proceedings of the International Conference on Management of Data (SIGMOD), pp. 2246–2258 (2022)
15. Deutsch, A., Xu, Y., Wu, M., Lee, V.E.: Aggregation support for modern graph analytics in TigerGraph. In: SIGMOD International Conference on Management of Data, pp. 377–392. ACM (2020). https://doi.org/10.1145/3318464.3386144
16. Elgohary, A., Boehm, M., Haas, P.J., Reiss, F.R., Reinwald, B.: Compressed linear algebra for declarative large-scale machine learning. Commun. ACM **62**(524), 83–91 (2019)
17. Erling, O., Mikhailov, I.: RDF support in the Virtuoso DBMS. In: Pellegrini, T., Auer, S., Tochtermann, K., Schaffert, S. (eds.) Networked Knowledge - Networked Media. Studies in Computational Intelligence, vol. 221, pp. 7–24. Springer, Heidelberg (2009). https://doi.org/10.1007/978-3-642-02184-8_2

18. Francis, N., et al.: Cypher: an evolving query language for property graphs. In: SIGMOD International Conference on Management of Data, pp. 1433–1445. ACM (2018)
19. Furman, M.E.: Application of a method of fast multiplication of matrices in the problem of Finding the transitive closure of a graph. Sov. Math. Dokl. **11**(5), 1252 (1970)
20. Huffman, D.A.: A method for the construction of minimum-redundancy codes. Proc. Inst. Electr. Radio Eng. **40**(9), 1098–1101 (1952)
21. Losemann, K., Martens, W.: The complexity of evaluating path expressions in SPARQL. In: Proceedings of the 31st Symposium on Principles of Database Systems (PODS), pp. 101–112. ACM (2012)
22. Malyshev, S., Krötzsch, M., González, L., Gonsior, J., Bielefeldt, A.: Getting the most out of Wikidata: semantic technology usage in Wikipedia's knowledge graph. In: Vrandečić, D., et al. (eds.) ISWC 2018. LNCS, vol. 11137, pp. 376–394. Springer, Cham (2018). https://doi.org/10.1007/978-3-030-00668-6_23
23. Manola, F., Miller, E.: RDF primer. W3C Recommendation (2004). http://www.w3.org/TR/rdf-primer/
24. Martens, W., Niewerth, M., Popp, T., Rojas, C., Vansummeren, S., Vrgoc, D.: Representing paths in graph database pattern matching. Proc. VLDB Endow. **16**(7), 1790–1803 (2023). https://www.vldb.org/pvldb/vol16/p1790-martens.pdf
25. Mendelzon, A.O., Wood, P.T.: Finding regular simple paths in graph databases. SIAM J. Comput. **24**(6), 1235–1258 (1995)
26. Munro, J.I.: Tables. In: Chandru, V., Vinay, V. (eds.) FSTTCS 1996. LNCS, vol. 1180, pp. 37–42. Springer, Heidelberg (1996). https://doi.org/10.1007/3-540-62034-6_35
27. Penn, G.: Efficient transitive closure of sparse matrices over closed semirings. Theoret. Comput. Sci. **354**(1), 72–81 (2006)
28. van Rest, O., Hong, S., Kim, J., Meng, X., Chafi, H.: PGQL: a property graph query language. In: International Workshop on Graph Data Management: Experiences and Systems (GRADES), p. 7. ACM (2016)
29. Saad, Y.: Iterative Methods for Sparse Linear Systems. SIAM (2003)
30. Schoor, A.: Fast algorithm for sparse matrix multiplication. Inf. Process. Lett. **15**(2), 87–89 (1982)
31. Thompson, B.B., Personick, M., Cutcher, M.: The bigdata®RDF graph database. In: Linked Data Management, pp. 193–237. Chapman and Hall/CRC (2014)
32. Vrandecic, D., Krötzsch, M.: Wikidata: a free collaborative knowledge base. Commun. ACM **57**(10), 78–85 (2014)
33. Yakovets, N., Godfrey, P., Gryz, J.: Query planning for evaluating SPARQL property paths. In: SIGMOD International Conference on Management of Data, pp. 1875–1889. ACM (2016)

Approximate Cartesian Tree Matching: An Approach Using Swaps

Bastien Auvray[1]([✉]), Julien David[1,2], Richard Groult[1,3], and Thierry Lecroq[1,3]

[1] CNRS NormaSTIC FR 3638, Caen, Le Havre, Rouen, France
bastien.auvray@etu.univ-rouen.fr
[2] Normandie University, UNICAEN, ENSICAEN, CNRS, GREYC, Caen, France
[3] Univ Rouen Normandie, LITIS UR 4108, 76000 Rouen, France

Abstract. Cartesian tree pattern matching consists of finding all the factors of a text that have the same Cartesian tree than a given pattern. There already exist theoretical and practical solutions for the exact case. In this paper, we propose the first algorithm for solving approximate Cartesian tree pattern matching. We consider Cartesian tree pattern matching with one swap: given a pattern of length m and a text of length n we present two algorithms that find all the factors of the text that have the same Cartesian tree of the pattern after one transposition of two adjacent symbols. The first algorithm uses a characterization of a linear representation of the Cartesian trees called parent-distance after one swap and runs in time $\Theta(mn)$ using $\Theta(m)$ space. The second algorithm generates all the parent-distance tables of sequences that have the same Cartesian tree than the pattern after one swap. It runs in time $\mathcal{O}((m^2 + n)\log m)$ and has $\mathcal{O}(m^2)$ space complexity.

Keywords: Cartesian tree · Approximate pattern matching · Swap · Transposition

1 Introduction

In general terms, the pattern matching problem consists of finding one or all the occurrences of a pattern in a text. When both the pattern and the text are strings the problem has been extensively studied and has received a huge number of solutions [5]. Searching time series or list of values for patterns representing specific fluctuations of the values requires a redefinition of the notion of pattern. The question is to deal with the recognition of peaks, breakdowns, or more features. For those specific needs one can use the notion of Cartesian tree.

Cartesian trees have been introduced by Vuillemin in 1980 [16]. They are mainly associated to strings of numbers and are structured as heaps from which original strings can be recovered by symmetrical traversals of the trees. It has been shown that they are connected to Lyndon trees [3], to Range Minimum

---

Supported by the NormaSTIC Federation https://www.normastic.fr/.

F. M. Nardini et al. (Eds.): SPIRE 2023, LNCS 14240, pp. 49–61, 2023.
https://doi.org/10.1007/978-3-031-43980-3_5

Queries [4] or to parallel suffix tree construction [14]. Recently, Park *et al.* [12] introduced a new metric of generalized matching, called Cartesian tree matching. It is the problem of finding every factor of a text t which has the same Cartesian tree as that of a given pattern p. Cartesian tree matching can be applied, for instance, to finding patterns in time series such as share prices in stock markets or gene sample time data.

Park *et al.* introduced the parent-distance representation which is a linear form of the Cartesian tree and that has a one-to-one mapping with Cartesian trees. They gave linear-time solutions for single and multiple pattern Cartesian tree matching, utilizing this parent-distance representation and existing classical string algorithms, i.e., Knuth-Morris-Pratt [9] and Aho-Corasick [1] algorithms. More efficient solutions for practical cases were given in [15]. Recently, new results on Cartesian pattern matching appeared [6,8,10,13].

All these previous works on Cartesian tree matching are concerned with finding exact occurrences of patterns consisting of contiguous symbols. The only results known on non-contiguous symbols presents an algorithm for episode matching [11] (finding all minimal length factors of t that contain p as a subsequence) in Cartesian tree framework.

To the best of our knowledge, no result is known about approximate pattern matching in this Cartesian tree framework. However in real life applications data are often noisy and it is thus important to find factors of the text that are similar, to some extent, to the pattern. In this paper, we present the first results in this setting by considering approximate Cartesian tree pattern matching with one transposition (aka swap) of one symbol with the adjacent symbol. Swap pattern matching has received a lot of attention in classical sequences since the first paper in 1997 [2] (see [7] and references therein). Swaps are common in real life data and it seems natural to consider them in the Cartesian pattern matching framework. We are able to design two algorithms for solving the Cartesian tree pattern matching with at most one swap. The first one runs in time $\Theta(mn)$ and uses a characterization of a linear representation of Cartesian trees while the second one runs in $\mathcal{O}((m^2 + n)\log m)$ and uses a graph to generate all the linear representations of Cartesian trees of sequences that match the pattern after one swap.

The remaining of the article is organized as follows: Sect. 2 presents the basic notions and notations used in the paper. Given a sequence x, in Sect. 3 we give a characterization of the parent-distance representations of Cartesian trees that correspond to sequences x after one swap. Section 4 presents the swap graph where vertices are Cartesian trees and there is an edge between two vertices if both Cartesian trees can be obtained from the other using one swap. In Sect. 5 we give our two algorithms for Cartesian tree pattern matching with swaps. Section 6 contains our perspectives.

2 Preliminaries

2.1 Basic Notations

We consider sequences of integers with a total order denoted by $<$. For a given sequence x, $|x|$ denotes the length of x, $x[i]$ is the i-th element of x and $x[i \ldots j]$ represents the factor of x starting at the i-th element and ending at the j-th element. For simplicity, we assume all elements in a given sequence to be distinct and numbered from 1 to $|x|$.

2.2 Cartesian Tree Matching

Given a sequence x of length n, its Cartesian tree $C(x)$ is recursively defined as follows (see example Fig. 1):

- if x is empty, then $C(x)$ is the empty tree;
- if $x[1 \ldots n]$ is not empty and $x[i]$ is the smallest value of x, $C(x)$ is the Cartesian tree with i as its root, the Cartesian tree of $x[1 \ldots i-1]$ as the left subtree and the Cartesian tree of $x[i+1 \ldots n]$ as the right subtree.

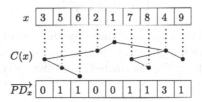

Fig. 1. A sequence $x = (3, 5, 6, 2, 1, 7, 8, 4, 9)$, its Cartesian tree $C(x)$ and its corresponding parent-distance table $\overrightarrow{PD}_x$.

We will denote by $x \approx_{CT} y$ if sequences x and y share the same Cartesian tree. For example, $x = (3, 5, 6, 2, 1, 7, 8, 4, 9) \approx_{CT} (3, 4, 8, 2, 1, 7, 9, 5, 6)$.

The Cartesian tree matching (CTM) problem consists in finding all factors of a text which share the same Cartesian tree as a pattern. Formally, Park *et al.* [12] define it as follows:

Definition 1 (Cartesian tree matching). *Given two sequences $p[1 \ldots m]$ and $t[1 \ldots n]$, find every $1 \leq i \leq n-m+1$ such that $t[i \ldots i+m-1] \approx_{CT} p[1 \ldots m]$.*

In order to solve CTM without building every possible Cartesian tree, an efficient representation of these trees was introduced by Park *et al.* [12], the parent-distance representation (see example Fig. 1):

Definition 2 (Parent-distance representation).   *Given a sequence* $x[1 \ldots n]$, *the parent-distance representation of* x *is an integer sequence* $\overrightarrow{PD}_x[1 \ldots n]$, *which is defined as follows:*

$$\overrightarrow{PD}_x[i] = \begin{cases} i - max_{1 \leq j < i}\{j \mid x[j] < x[i]\} & \text{if such } j \text{ exists} \\ 0 & \text{otherwise} \end{cases}$$

Since the parent-distance representation has a one-to-one mapping with Cartesian trees, it can replace them without loss of information.

2.3   Approximate Cartesian Tree Matching

In order to define an approximate version of Cartesian tree matching, we use the following notion of transposition on sequences:

Definition 3 (Swap).   *Let* x *and* y *be two sequences of length* n, *and* $i \in \{1, \ldots, n-1\}$, *we denote* $y = \tau(x, i)$ *to describe a swap, that is:*

$$y = \tau(x, i) \text{ if } \begin{cases} x[j] = y[j], \forall j \notin \{i, i+1\} \\ x[i] = y[i+1] \\ x[i+1] = y[i] \end{cases}$$

This kind of transposition is the one made by the Bubble Sort algorithm. It is therefore a natural operation on permutations and sequences. For the Cartesian tree point of view, see Fig. 2. We use the notion of swap to define the approximate Cartesian tree matching.

Definition 4. (CT_τ Matching).   *Let* x *and* y *be two sequences of length* n, *we have* $x \overset{\tau}{\approx}_{CT} y$ *if:*

$$\begin{cases} x \approx_{CT} y, \text{ or} \\ \exists\, x',\; y', \exists\, i \in \{1, \ldots, n-1\}, x' \approx_{CT} x, y' \approx_{CT} y, x' = \tau(y', i) \text{ and } y' = \tau(x', i) \end{cases}$$

Figure 2 shows an example of sequences that CT_τ match.

Lastly, in order to fully characterize the approximate Cartesian tree matching, we introduce the notion of reverse parent-distance of a sequence that we compute as if read from right to left.

Definition 5 (Reverse parent-distance).   *Given a sequence* $x[1 \ldots n]$, *the reverse parent-distance representation of* x *is an integer sequence* $\overleftarrow{PD}_x[1 \ldots n]$, *which is defined as follows:*

$$\overleftarrow{PD}_x[i] = \begin{cases} min_{i < j \leq n}\{j \mid x[i] > x[j]\} - i & \text{if such } j \text{ exists} \\ 0 & \text{otherwise} \end{cases}$$

$$x' = (4, 5, 6, 1, 2, 7, 8, 3, 9) \qquad y' = (4, 5, 6, 2, 1, 7, 8, 3, 9)$$

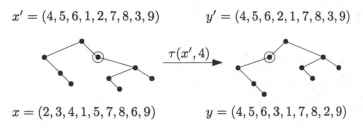

$$x = (2, 3, 4, 1, 5, 7, 8, 6, 9) \qquad y = (4, 5, 6, 3, 1, 7, 8, 2, 9)$$

Fig. 2. The sequence x CT_τ matches y. A swap at position 4 moves the circled node from the right subtree of the root to the left one. In general, a swap at position i consists either in moving the leftmost descendant of the right subtree to a rightmost position in the left subtree (that is if $x[i] < x[i + 1]$), or the opposite, in moving the rightmost descendant of the left subtree to a leftmost position of the right subtree of its parent. Note that we also have $x \overset{\tau}{\approx}_{CT} y'$, $x' \overset{\tau}{\approx}_{CT} y$ and of course $x' \overset{\tau}{\approx}_{CT} y'$.

3   Characterization of the Parent-Distance Tables When a Swap Occurs

In this section, we describe how the parent-distances $\overrightarrow{PD_x}$ and $\overleftarrow{PD_x}$ of a sequence x of length n are modified into tables $\overrightarrow{PD_y}$ and $\overleftarrow{PD_y}$ when $y \overset{\tau}{\approx}_{CT} x$, with a swap occurring at position i. Figure 3 sums up the different parts of the parent-distance tables we are going to characterize. Those results will be used in Sect. 5 to obtain an algorithm that solves the CT_τ matching problem.

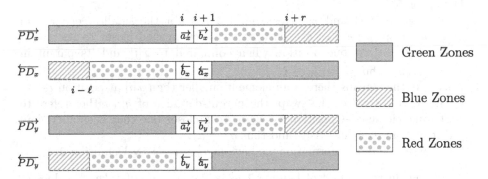

Fig. 3. This figure sums up the different Lemmas of this section. For instance, the green zones correspond to Definition 6 and Lemma 3. The values $\overrightarrow{a_x}$, $\overrightarrow{b_x}$, ..., are the 8 values found in the parent-distance tables of x and y at position i and $i + 1$, that is $\overrightarrow{PD_x}[i] = \overrightarrow{a_x}$, $\overrightarrow{PD_x}[i + 1] = \overrightarrow{b_x}$, ... Values $i - \ell$ and $i + r$ respectively denote the last and first position of each blue zone. (Color figure online)

First, we describe how the parent-distances are modified at positions i and $i + 1$.

Lemma 1. *Suppose that $x[i] < x[i+1]$, then the following properties hold:*

1. $\overleftarrow{b_y} = 1$

2. $\overrightarrow{b_y} = \begin{cases} 0 & \text{if } \overrightarrow{a_x} = 0 \\ \overrightarrow{a_x} + 1 & \text{otherwise} \end{cases}$

3. $\overleftarrow{a_y} = \begin{cases} 0 & \text{if } \overleftarrow{b_x} = 0 \\ \overleftarrow{b_x} - 1 & \text{otherwise} \end{cases}$

4. $\overrightarrow{a_y} \leq \begin{cases} i - 1 & \text{if } \overrightarrow{a_x} = 0 \\ \overrightarrow{a_x} & \text{otherwise} \end{cases}$

Proof. Suppose $x[i] < x[i+1]$, we have $\overrightarrow{b_x} = 1$ by definition of the parent-distance (Definition 2) and $\overleftarrow{b_x} \neq 1$ by definition of the reverse parent-distance (Definition 5). Then, if a swap occurs at position i, $y[i] > y[i+1]$ and we have:

1. $\overleftarrow{b_y} = 1$ by Definition 5.
2. If $\overrightarrow{a_x} = 0$, $x[i]$ is the smallest element in $x[1\ldots i]$ by Definition 2. Which implies $y[i+1]$ is the smallest element in $y[1\ldots i+1]$ and thus $\overrightarrow{b_y} = 0$ by Definition 2.
   Otherwise, $x[i]$ (resp. $y[i+1]$) is not the smallest element in $x[1\ldots i]$ (resp. $y[1\ldots i+1]$). $y[i+1]$ has been pushed away from its parent in $y[1\ldots i-1]$ by one position compared to $x[i]$ and its parent in $x[1\ldots i-1]$. Thus, $\overrightarrow{b_y} = \overrightarrow{a_x} + 1$.
3. If $\overleftarrow{b_x} = 0$, $x[i]$ is the smallest element in $x[i\ldots n]$ by Definition 5. Which implies $y[i+1]$ is the smallest element in $y[i+1\ldots n]$, and thus $\overleftarrow{a_y} = 0$ by Definition 5.
   Otherwise, $\overleftarrow{b_x} > 1$ and $x[i]$ (resp. $y[i+1]$) is not the smallest element in $x[i\ldots n]$ (resp. $y[i+1\ldots n]$). $y[i+1]$ has been pushed closer to its parent in $y[i+2\ldots n]$ by one position when compared to $x[i]$ and its parent in $x[i+2\ldots n]$. Thus, $\overleftarrow{a_y} = \overleftarrow{b_x} - 1$.
4. If $\overrightarrow{a_x} > 0$, that means there is an element smaller than $x[i]$ at position $i - \overrightarrow{a_x}$ by Definition 2. After the swap, the parent-distance of $y[i]$ either refers to that same element at position $i - \overrightarrow{a_x}$ or to a closer one that is smaller than $y[i]$ if such an element exists, and thus $\overrightarrow{a_y} \leq \overrightarrow{a_x}$.
   Otherwise, the only information we have is $\overrightarrow{a_y} \leq i - 1$ by Definition 2.

Note that in the item 4 of Lemma 1 $\overrightarrow{a_y} \leq \overrightarrow{a_x}$, $\overrightarrow{a_y}$ cannot take all values in $\{1, \ldots, \overrightarrow{a_x}\}$, since some won't produce a valid parent-distance table.

Lemma 2. *Suppose that $x[i] > x[i+1]$, then the following properties hold:*

1. $\overrightarrow{b_y} = 1$

2. $\overleftarrow{b_y} = \begin{cases} 0 & \text{if } \overleftarrow{a_x} = 0 \\ \overleftarrow{a_x} + 1 & \text{otherwise} \end{cases}$

3. $\overrightarrow{a_y} = \begin{cases} 0 & \text{if } \overrightarrow{b_x} = 0 \\ \overrightarrow{b_x} - 1 & \text{otherwise} \end{cases}$

4. $\overleftarrow{a_y} \leq \begin{cases} i-1 & \text{if } \overleftarrow{a_x} = 0 \\ \overleftarrow{a_x} & \text{otherwise} \end{cases}$

The proof is similar to the one of Lemma 1. In the following, we define the green and blue zones of the parent-distances tables, which are equal, meaning that they are unaffected by the swap. Also, we define the red zones whose values differ by at most 1. Due to a lack of space, proofs are omitted. We strongly invite the reader to use Fig. 3 to get a better grasp of the definitions.

Definition 6 (The green zones). *Given a sequence x and a position i, the green zone of $\overrightarrow{PD_x}$ is $\overrightarrow{PD_x}[1 \dots i-1]$ and the green zone of $\overleftarrow{PD_x}$ is $\overleftarrow{PD_x}[i + 2 \dots n]$.*

Lemma 3 (The green zones). *The green zones of $\overrightarrow{PD_x}$ and $\overrightarrow{PD_y}$ (resp. $\overleftarrow{PD_x}$ and $\overleftarrow{PD_y}$) are equal.*

Definition 7 (The blue zones). *Given a sequence x and a position i, the blue zone of $\overrightarrow{PD_x}$ is $\overrightarrow{PD_x}[i + r \dots n]$ where:*

$$r = \begin{cases} \overleftarrow{b_x} & \text{if } x[i] < x[i+1] \text{ and } \overleftarrow{b_x} > 1 \\ \overleftarrow{a_x} + 1 & \text{if } x[i] > x[i+1] \text{ and } \overleftarrow{a_x} > 0 \\ n - i + 1 & \text{otherwise} \end{cases}$$

The blue zone of $\overleftarrow{PD_x}$ is $\overleftarrow{PD_x}[1 \dots i - \ell]$ where:

$$\ell = \begin{cases} \overrightarrow{a_x} & \text{if } x[i] < x[i+1] \text{ and } \overrightarrow{a_x} > 0 \\ \overrightarrow{b_x} - 1 & \text{if } x[i] > x[i+1] \text{ and } \overrightarrow{b_x} > 1 \\ i & \text{otherwise} \end{cases}$$

Note that in the last cases, the blue zones are empty.

Lemma 4 (The blue zones). *The blue zones of $\overrightarrow{PD_x}$ and $\overrightarrow{PD_y}$ (resp. $\overleftarrow{PD_x}$ and $\overleftarrow{PD_y}$) are equal.*

Definition 8 (The red zones). *Given a sequence x and a position i, if the blue zone of $\overrightarrow{PD_x}$ is $\overrightarrow{PD_x}[i+r \dots n]$, then the right red zone is $\overrightarrow{PD_x}[i+2 \dots i+r-1]$. Conversely, if the blue zone of $\overleftarrow{PD_x}$ is $\overleftarrow{PD_x}[1 \dots i - \ell]$, then the left red zone is $\overleftarrow{PD_x}[i - \ell + 1 \dots i - 1]$.*

Lemma 5 (The red zones). *We distinguish two symmetrical cases:*

1. $x[i] < x[i+1]$:
   (a) *For each position j in the red zone of $\overrightarrow{PD_x}$, we have either $\overrightarrow{PD_y}[j] = \overrightarrow{PD_x}[j]$ or $\overrightarrow{PD_y}[j] = \overrightarrow{PD_x}[j] - 1$.*
   (b) *For each position j in the red zone of $\overleftarrow{PD_x}$, we have either $\overleftarrow{PD_y}[j] = \overleftarrow{PD_x}[j]$ or $\overleftarrow{PD_y}[j] = \overleftarrow{PD_x}[j] + 1$.*

2. $x[i] > x[i+1]$:

   (a) *For each position j in the red zone of $\overrightarrow{PD}_x$, we have either $\overrightarrow{PD}_y[j] = \overrightarrow{PD}_x[j]$ or $\overrightarrow{PD}_y[j] = \overrightarrow{PD}_x[j] + 1$.*

   (b) *For each position j in the red zone of $\overleftarrow{PD}_x$, we have either $\overleftarrow{PD}_y[j] = \overleftarrow{PD}_x[j]$ or $\overleftarrow{PD}_y[j] = \overleftarrow{PD}_x[j] - 1$.*

We now show that swaps at different positions produce different Cartesian trees.

Lemma 6. *Let x be a sequence of length n and $i, j \in \{1, \ldots, n-1\}$, with $i \neq j$. Then $\tau(x, i) \not\approx_{CT} \tau(x, j)$.*

Proof. Suppose without loss of generality that $j > i$. If $j > i+1$, then according to Lemma 3, we have:

$$\forall k < j, \overrightarrow{PD}_x[k] = \overrightarrow{PD}_{\tau(x,j)}[k] = \overrightarrow{PD}_{\tau(x,i)}[k]$$

And according to Lemma 1, we have that $\overrightarrow{PD}_x[i+1] \neq \overrightarrow{PD}_{\tau(x,i)}[i+1]$, which leads to a contradiction.

Then suppose that $j = i+1$, then it is sufficient to consider what happens on a sequence of length 3: having local differences on the parent-distance tables implies having different parent-distances and therefore do not CT match. One can easily check that the lemma is true for each sequence of length 3.

4   Swap Graph of Cartesian Trees

Let $\mathcal{C}_n$ be the set of Cartesian trees with n nodes, which is equal to the set of binary trees with n nodes. Also $\mathcal{C} = \bigcup_{n \geq 0} \mathcal{C}_n$. Let $\mathcal{G}_n = (\mathcal{C}_n, E_n)$ be the Swap Graph of Cartesian trees, where $\mathcal{C}_n$ is its set of vertices and E_n the set of edges. Let x and y be two sequences, we have $\{C(x), C(y)\} \in E_n$ if $x \overset{\tau}{\approx}_{CT} y$. Figure 4 shows the Swap Graph with n smaller than 4.

In the following, we study the set of neighbors a vertex can have in the Swap Graph. Let $T \in \mathcal{C}_n$ be a Cartesian tree of size n and $ng(T)$ be its set of neighbors in the Swap Graph. Also, for $i < n$, we note $ng(T, i)$ the set of trees obtained by doing a swap at position i on the associated sequences, that is

$$ng(T, i) = \{C(y) \in \mathcal{C}_n \mid \exists\, x \text{ such that } T = C(x) \text{ and } y = \tau(x, i)\}$$

Also, we have

$$ng(T) = \bigcup_{i=1}^{n-1} ng(T, i)$$

where all unions are disjoint according to Lemma 6.

Lemma 7. *Let $T \in \mathcal{C}_n$ be a Cartesian tree of size n, with a left-subtree A of size $k-1$ and B a right subtree of size $n-k$. We have*

$$|ng(T)| = |ng(A)| + |ng(B)| + |ng(T, k-1)| + |ng(T, k)|$$

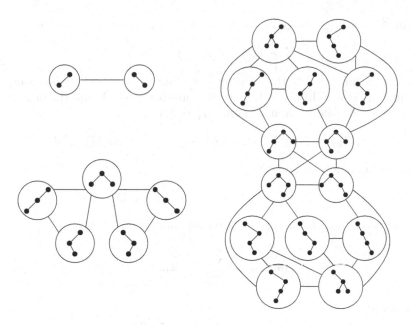

Fig. 4. Swap Graph of Cartesian trees of size $2, 3$ and 4.

Proof. The result follows from Lemma 6 and the definition of $ng(T)$. Indeed, we have $|ng(A)| = |\bigcup_{i=1}^{k-2} ng(T, i)|$ and $|ng(B)| = |\bigcup_{i=k+1}^{n-1} ng(T, i)|$.

Let $lbl : C \mapsto \mathbb{N}$ be the function that computes the length of the left-branch of a tree such that

$$\forall\, T \in C, lbl(T) = \begin{cases} 0, & \text{if } T \text{ is empty} \\ 1 + lbl(A), & \text{where } A \text{ is the left-subtree of } T. \end{cases}$$

The equivalent lbr function can be defined to compute the length of the right-branch.

Lemma 8. *Let $T \in C_n$ be a Cartesian tree with a root i, a left-subtree A and a right-subtree B.*

$$|ng(T, i-1)| = lbl(B) + 1 \text{ and } |ng(T, i)| = lbr(A) + 1$$

Proof. We only prove that $|ng(T, i-1)| = lbl(B) + 1$ since the rest of the proof uses the same arguments. Let x be a sequence such that $C(x) = T$. As stated in the definition section (see Fig. 2), the swap $\tau(x, i-1)$ moves the rightmost node of A into a leftmost position in B. Let $j_1, \ldots, j_{lbl(B)}$ be the positions in the sequence x that corresponds to the nodes of the left branch of B. For each $\ell < lbl(B)$, there always exists a sequence $y = \tau(x, i)$ such that $y[i] < y[j_1] < \cdots < y[j_\ell] < y[i-1] < y[j_{\ell+1}] < \cdots < y[j_{lbl(B)}]$. Therefore, there exist exactly $lbl(B) + 1$ possible output trees when applying such a swap.

Lemma 9. *Let T_h be the complete binary tree of height $h > 0$, we have*

$$|ng(T_h)| = 6(2^h - 1) - 2h$$

Proof. Let us first remark that a complete binary tree of height 0 is simply a leaf, thus $ng(T_0) = \emptyset$. In the following, we consider $h > 0$. Using the fact we are dealing with complete trees and Lemma 7, we have:

$$|ng(T_h)| = 2|ng(T_{h-1})| + |ng(T_h, 2^h - 1)| + |ng(T_h, 2^h)|$$

Also, using Lemma 8 we have

$$|ng(T_h)| = 2|ng(T_{h-1})| + 2(h + 1)$$

Using the telescoping technique we obtained that

$$|ng(T_h)| = \sum_{i=0}^{h-1}(i + 2)2^{h-i}$$

Which simplifies into

$$|ng(T_h)| = 6(2^h - 1) - 2h$$

This lemma can be reformulated: let $n = 2^{h+1} - 1$ be the number of nodes in the complete tree, we have

$$|ng(T_h)| = \lceil 3(n - 1) - 2(\log_2(n + 1) - 1)\rceil$$

Lemma 10. *For every Cartesian tree T of size n, we have*

$$n - 1 \le |ng(T)| \le \lceil 3(n - 1) - 2(\log_2(n + 1) - 1)\rceil$$

Proof. The first part of the inequality is given by Lemma 6: each tree has at least $n - 1$ neighbors, since each transposition at a position in $\{1, \ldots, n - 1\}$ produces different Cartesian trees. The upper bound is obtained by the following reasoning: each Cartesian tree can be obtained starting from a complete binary tree by iteratively removing some leaves. Let T be a Cartesian tree for which the upper bound holds and T' the tree obtained after removing a leaf from T.

- If the leaf i had a right sibling (resp. left sibling), then $|ng(T, i)| = 2$ (resp. $|ng(T, i - 1)| = 2$) and $|ng(T, i + 1)| \ge 2$ (resp. $|ng(T, i)| \ge 2$). In T', the neighbors from $ng(T, i)$ (resp. $ng(T, i-1)$) are lost and $|ng(T', i)| = |ng(T, i+1)|-1$ (resp. $|ng(T', i-1)| = |ng(T, i)|-1$). Therefore at $|ng(T')| \le |ng(T)|-3$.
- If the leaf i had no sibling, then there are two possibilities:
  • either the leaf were at an extremity of the tree (the end of the left/right branch). In this case, removing the leaf will only remove 2 neighbors. This can only be made twice for each level of the complete tree (only one time for the first level), that is $log_2(n + 1) - 1$ times.
  • or the leaf is in "the middle" of the tree, in this case the tree is at least of height 2 and we can remove at least 3 neighbors.

We use the previous lemma to obtain a lower bound on the diameter of the Swap Graph.

Lemma 11. *The diameter of the Swap Graph $\mathcal{G}_n$ is $\Omega(\frac{n}{\ln n})$.*

Proof. The number of vertices in the graph is equal to the number of binary trees enumerated by the Catalan numbers, that is $\frac{\binom{2n}{n}}{n+1}$. Since the maximal degree of a vertex is less than $3n$ according to Lemma 10, the diameter is lower bounded by the value k such that:

$$(3n)^k = \frac{\binom{2n}{n}}{n+1}$$

$$\implies k = \frac{\ln\left(\frac{\binom{2n}{n}}{n+1}\right)}{\ln(3) + \ln(n)}$$

$$\implies k = \frac{2n\ln(2n) - 2n\ln(n) - \ln(n+1)}{\ln(3) + \ln(n)}$$

By decomposing $2n\ln(2n)$ into $2n\ln 2 + 2n\ln n$ we obtain

$$\implies k = \frac{2n\ln(2)}{\ln(3) + \ln(n)} - \frac{\ln(n+1)}{\ln(3) + \ln(n)}$$

$$\implies k = \Omega\left(\frac{n}{\ln n}\right)$$

5 Algorithms

In this section, we present two algorithms to compute the number of occurrences of $\overset{\tau}{\approx}_{CT}$ matching of a given pattern in a given text. The first algorithm uses two parent-distance tables and the set of Lemmas presented in Sect. 3. The second algorithm uses Lemma 10 and the fact that each sequence has a bounded number of Cartesian trees that can be obtained from a swap to build an automaton.

5.1 An Algorithm Using the Parent-Distance Tables

Algorithm 1, below, is based on Lemmas 1 to 6. The function candidateSwapPosition at line 8 computes the exact location of the swap, using the green zones. There can be up to two candidates for the position i of the swap but only one can be valid according to Lemma 6.

Theorem 1. *Given two sequences p and t of length m and n, Algorithm 1 has a $\Theta(mn)$ worst-case time complexity and a $\Theta(m)$ space complexity.*

Proof. The space complexity is obtained from the length of the parent-distance tables. Regarding the time complexity, in the worst-case scenario, one has to check the zones of each parent-distance table, meaning that the number of comparisons is bounded by $2m$ for each position in the sequence t. This worst case can be reached by taking a sequence t whose associated Cartesian tree is a comb and a sequence p obtained by doing a swap on a sequence associated to a comb.

Algorithm 1: *DoubleParentDistanceMethod(p, t)*

Input : Two sequences p and t

Output: The number of positions j such that $t[j \ldots j + p - 1] \overset{\tau}{\approx}_{CT} p$

1 $occ \leftarrow 0$;

2 $(\overrightarrow{PD_p}, \overleftarrow{PD_p}) \leftarrow$ Compute the parent-distance tables of p;

3 **for** $j \in \{1, \ldots, |t| - |p| + 1\}$ **do**

4 | $(\overrightarrow{PD_x}, \overleftarrow{PD_x}) \leftarrow$ Compute the parent-distance tables of $x = t[j \ldots j + p - 1]$;

5 | **if** $\overrightarrow{PD_p} = \overrightarrow{PD_x}$ **then**

6 | | $occ \leftarrow occ + 1$;

7 | **else**

8 | | **foreach** $i \in candidateSwapPosition(\overrightarrow{PD_p}, \overleftarrow{PD_p}, \overrightarrow{PD_x}, \overleftarrow{PD_x})$ **do**

9 | | | **if** *Lemmas 1, 2, 4 and 5 hold for p, x and i* **then**

10 | | | | $occ \leftarrow occ + 1$;

11 **return** occ;

5.2 An Aho-Corasick Based Algorithm

The idea of the second method is to take advantage of the upper bound on the size of the neighborhood of a given Cartesian tree in the Swap Graph. Given a sequence p, we compute the set of its neighbors $ng(C(p))$, then we compute the set of all parent-distance tables and build the automaton that recognizes this set of tables using the Aho-Corasick method for multiple Cartesian tree matching [12]. Then, for each position in a sequence t, it is sufficient to read the parent-distance table of each factor $t[j \ldots j + |p| - 1]$ into the automaton and check if it ends on a final state. The following theorem can be obtained from Section 4.2 in [12].

Theorem 2. *Given two sequences p and t of length m and n, the Aho-Corasick based algorithm has an $\mathcal{O}((m^2 + n) \log m)$ worst-case time complexity and an $\mathcal{O}(m^2)$ space complexity.*

6   Perspectives

From the pattern matching point of view, the first step would be to generalize our result to sequences with a partial order instead of a total one. Then, it could be interesting to obtain a general method, where the number of swaps is given as a parameter. Though, we fear that if too many swaps are applied, the result loses its interest, even though the complexity might grow rapidly.

The analysis of both algorithms should be improved. An amortized analysis could be done on the first algorithm and on the computation of the set of parent-distance tables for the second algorithm. In the second algorithm the upper bounds on the time and space complexity could be improved by studying the size of the minimal automaton.

References

1. Aho, A.V., Corasick, M.J.: Efficient string matching: an aid to bibliographic search. Commun. ACM **18**(6), 333–340 (1975)
2. Amir, A., Aumann, Y., Landau, G.M., Lewenstein, M., Lewenstein, N.: Pattern matching with swaps. In: 38th Annual Symposium on Foundations of Computer Science, FOCS 1997, Miami Beach, Florida, USA, 19–22 October 1997, pp. 144–153. IEEE Computer Society (1997)
3. Crochemore, M., Russo, L.M.: Cartesian and Lyndon trees. Theor. Comput. Sci. **806**, 1–9 (2020)
4. Demaine, E.D., Landau, G.M., Weimann, O.: On Cartesian trees and range minimum queries. Algorithmica **68**(3), 610–625 (2014)
5. Faro, S., Lecroq, T.: The exact online string matching problem: a review of the most recent results. ACM Comput. Surv. **45**(2), 13 (2013)
6. Faro, S., Lecroq, T., Park, K., Scafiti, S.: On the longest common Cartesian substring problem. Comput. J. **66**(4), 907–923 (2023)
7. Faro, S., Pavone, A.: An efficient skip-search approach to swap matching. Comput. J. **61**(9), 1351–1360 (2018)
8. Kim, S.H., Cho, H.G.: A compact index for Cartesian tree matching. In: CPM, Wrocław, Poland, pp. 18:1–18:19 (2021)
9. Knuth, D.E., Morris, J.H., Jr., Pratt, V.R.: Fast pattern matching in strings. SIAM J. Comput. **6**(1), 323–350 (1977)
10. Nishimoto, A., Fujisato, N., Nakashima, Y., Inenaga, S.: Position heaps for Cartesian-tree matching on strings and tries. In: SPIRE, Lille, France, pp. 241–254 (2021)
11. Oizumi, T., Kai, T., Mieno, T., Inenaga, S., Arimura, H.: Cartesian tree subsequence matching. In: Bannai, H., Holub, J. (eds.) CPM. LIPIcs, Prague, Czech Republic, vol. 223, pp. 14:1–14:18. Schloss Dagstuhl - Leibniz-Zentrum für Informatik (2022)
12. Park, S., Amir, A., Landau, G., Park, K.: Cartesian tree matching and indexing. In: CPM, Pisa, Italy, vol. 16, pp. 1–14 (2019)
13. Park, S.G., Bataa, M., Amir, A., Landau, G.M., Park, K.: Finding patterns and periods in Cartesian tree matching. Theor. Comput. Sci. **845**, 181–197 (2020)
14. Shun, J., Blelloch, G.E.: A simple parallel Cartesian tree algorithm and its application to parallel suffix tree construction. ACM Trans. Parallel Comput. **1**(1), 20 (2014)
15. Song, S., Gu, G., Ryu, C., Faro, S., Lecroq, T., Park, K.: Fast algorithms for single and multiple pattern Cartesian tree matching. Theor. Comput. Sci. **849**, 47–63 (2021)
16. Vuillemin, J.: A unifying look at data structures. Commun. ACM **23**(4), 229–239 (1980)

Optimal Wheeler Language Recognition

Ruben Becker[1], Davide Cenzato[1], Sung-Hwan Kim[1], Bojana Kodric[1],
Alberto Policriti[2], and Nicola Prezza[1(✉)]

[1] Ca' Foscari University of Venice, Venice, Italy
{rubensimon.becker,davide.cenzato,sunghwan.kim,
bojana.kodric,nicola.prezza}@unive.it
[2] University of Udine, Udine, Italy
alberto.policriti@uniud.it

Abstract. A Wheeler automaton is a finite state automaton whose states admit a total *Wheeler order*, reflecting the co-lexicographic order of the strings labeling source-to-node paths). A *Wheeler language* is a regular language admitting an accepting Wheeler automaton. Wheeler languages admit efficient and elegant solutions to hard problems such as automata compression and regular expression matching, therefore deciding whether a regular language is Wheeler is relevant in applications requiring efficient solutions to those problems. In this paper, we show that it is possible to decide whether a DFA with n states and m transitions recognizes a Wheeler language in $O(mn)$ time. This is a significant improvement over the running time $O(n^{13} + m \log n)$ of the previous polynomial-time algorithm (Alanko et al. Information and Computation 2021). A proof-of-concept implementation of this algorithm is available in a public repository. We complement this upper bound with a conditional matching lower bound stating that, unless the strong exponential time hypothesis (SETH) fails, the problem cannot be solved in strongly subquadratic time. The same problem is known to be PSPACE-complete when the input is an NFA (D'Agostino et al. Theoretical Computer Science 2023). Together with that result, our paper essentially closes the algorithmic problem of Wheeler language recognition.

Keywords: Wheeler Languages · Regular Languages · Finite Automata

1 Introduction

Wheeler automata were introduced by Gagie et al. in [10] as a natural generalization of prefix-sorting techniques (standing at the core of the most successful

---

The full version of this paper can be found in [3]. Ruben Becker, Davide Cenzato, Sung-Hwan Kim, Bojana Kodric, and Nicola Prezza are funded by the European Union (ERC, REGINDEX, 101039208. Views and opinions expressed are however those of the author(s) only and do not necessarily reflect those of the European Union or the European Research Council. Neither the European Union nor the granting authority can be held responsible for them. Alberto Policriti is supported by project National Biodiversity Future Center-NBFC (CN_00000033, CUP G23C22001110007) under the National Recovery and Resilience Plan of Italian Ministry of University and Research funded by European Union–NextGenerationEU.

F. M. Nardini et al. (Eds.): SPIRE 2023, LNCS 14240, pp. 62–74, 2023.
https://doi.org/10.1007/978-3-031-43980-3_6

string processing algorithms) to labeled graphs. Informally speaking, an automaton on alphabet Σ is Wheeler if the co-lexicographic order of the strings labeling source-to-states paths can be "lifted" to a *total* order of the states (a formal definition is given in Definition 5). As shown by the authors of [10], Wheeler automata can be encoded in just $O(\log |\Sigma|)$ bits per edge and they support near-optimal time pattern matching queries (i.e. finding all nodes reached by a path labeled with a given query string). These properties make them a powerful tool in fields such as bioinformatics, where one popular way to cope with the rapidly-increasing number of available fully-sequenced genomes, is to encode them in a pangenome graph: aligning short DNA sequences allows one to discover whether the sequences at hand contain variants recorded (as sub-paths) in the graph [8].

Wheeler languages—that is, regular languages recognized by Wheeler automata—were later studied by Alanko et al. in [1]. In that paper, the authors showed that Wheeler DFAs (WDFAs) and Wheeler NFAs (WNFAs) have the same expressive power. As a matter of fact, the class of Wheeler languages proved to possess several other remarkable properties, in addition to represent the class of regular languages for which efficient indexing data structures exist. Such properties motivated them to study the following decisional problem (as well as the corresponding variant on NFAs/regular expressions):

Definition 1 (WHEELERLANGUAGEDFA). *Given a DFA $\mathcal{A}$, decide if the regular language $\mathcal{L}(\mathcal{A})$ recognized by $\mathcal{A}$ is Wheeler.*

Alanko et al. [1] provided the following characterization: a language $\mathcal{L}$ is Wheeler if and only if, for any co-lexicographically monotone sequence of strings $\alpha_1 \prec \alpha_2 \prec \ldots$ (or with reversed signs $\succ$) belonging to the prefix-closure of $\mathcal{L}$, on the minimum DFA for $\mathcal{L}$ there exists some $N \in \mathbb{N}$ and state u such that by reading α_i from the source state we end up in state u for all $i \geq N$. This characterization allowed them to devise a polynomial-time algorithm solving WHEELERLANGUAGEDFA. This result is not trivial for two main reasons: (1) the smallest WDFA for a Wheeler language $\mathcal{L}$ could be *exponentially larger* than the smallest DFA for $\mathcal{L}$ [1], and (2) the corresponding WHEELERLANGUAGENFA problem (i.e., the input $\mathcal{A}$ is an NFA) is PSPACE-complete [7].

Our Contributions. Despite being polynomial, the algorithm of Alanko et al. has a prohibitive time complexity: $O(n^{13} + m \log n)$, where m and n are the number of transitions and states of the input DFA[1]. In this paper, we present a much simpler parameterized (worst-case quadratic) algorithm solving WHEELERLANGUAGEDFA. The complexity of our algorithm depends on a parameter p—the *co-lex width* of the minimum DFA $\mathcal{A}_{min}$ for the language [6] (Definition 11), which is never larger than n and which measures the "distance" of $\mathcal{A}_{min}$ from being Wheeler; e.g., if $\mathcal{A}_{min}$ is itself Wheeler, then $p = 1$. We prove:

Theorem 1. WHEELERLANGUAGEDFA *can be solved in* $O(mp + m \log n) \subseteq O(mn)$ *time on any DFA $\mathcal{A}$ with n states and m edges, where $p \leq n$ is the co-lex width of the minimum automaton $\mathcal{A}_{min}$ equivalent to $\mathcal{A}$.*

---

[1] While the authors only claim $m^{O(1)}$ time, a finer analysis yields this bound.

The intuition behind Theorem 1 is the following. Starting from the characterization of Wheeler languages of Alanko et al. [1] based on monotone sequences, we show that $\mathcal{L}(\mathcal{A})$ is not Wheeler if and only if the *square automaton* $\mathcal{A}_{min}^2 = \mathcal{A}_{min} \times \mathcal{A}_{min}$ contains a cycle $(u_1, v_1) \rightarrow (u_2, v_2) \rightarrow \cdots \rightarrow (u_k, v_k) \rightarrow (u_1, v_1)$ such that, for all $i = 1, \ldots, k$, the following two properties hold: (i) $u_i \neq v_i$ and (ii) the co-lexicographic ranges of strings reaching u_i and v_i intersect. As a result, after computing $\mathcal{A}_{min}$ ($O(m \log n)$ time by Hopcroft's algorithm) and directly building this "pruned" version of $\mathcal{A}_{min}^2$ in $O(mp + m \log n)$ time using recent techniques described in [2,13], testing its acyclicity yields the answer. A proof-of-concept implementation of the algorithm behind Theorem 1 is available at http://github.com/regindex/Wheeler-language-recognizer.

We complement the above upper bound with a matching conditional lower bound. Our lower bound is obtained via a reduction from the following problem:

Definition 2 (Orthogonal Vectors problem (OV)). *Given two sets A and B, each containing N vectors from $\{0,1\}^d$, decide whether there exist $a \in A$ and $b \in B$ such that $a^T b = 0$.*

By a classic reduction [15], for $d \in \omega(\log N)$ OV cannot be solved in time $O(N^{2-\eta} \operatorname{poly}(d))$ for any constant $\eta > 0$ unless the strong exponential time hypothesis [12] (SETH) fails. We prove[2]:

Theorem 2. *If WHEELERLANGUAGEDFA can be solved in time $O(m^{2-\eta})$ for some $\eta > 0$ on a DFA with m transitions on a binary alphabet, then the Orthogonal Vectors problem with $d \in \Omega(\log N)$ can be solved in time $O(N^{2-\eta} \operatorname{poly}(d))$.*

To prove Theorem 2, we adapt the reduction used by Equi et al. [9] to study the complexity of the *pattern matching on labeled graphs* problem. The intuition is the following. Our new characterization of Wheeler languages states that we essentially need to find two distinct equally-labeled cycles in the minimum DFA for the language (in addition to checking some other properties on those cycles) in order to solve WHEELERLANGUAGEDFA. Given an instance of OV, we build a DFA (minimum for its language) having one (non-simple) cycle for each vector in the instance, such that the strings labeling two such cycles match if and only if the two corresponding vectors in the OV instance are orthogonal. As a result, a subquadratic-time solution of WHEELERLANGUAGEDFA on this DFA yields a subquadratic-time solution for the OV instance.

2   Preliminaries

Strings. Let Σ be a finite alphabet. A *finite string* $\alpha \in \Sigma^*$ is a finite concatenation of characters from Σ. The notation $|\alpha|$ indicates the length of the string α. The symbol ε denotes the empty string. The notation $\alpha[i]$ denotes the i-th character from the beginning of α, with indices starting from 1. Letting $\alpha, \beta \in \Sigma^*$,

---

[2] Our lower bound states that there is no algorithm solving *all* instances in $O(m^{2-\eta})$ time. On sparse DFAs ($m \in \Theta(n)$) our algorithm runs in $O(mn) = O(m^2)$ time.

$\alpha \cdot \beta$ (or simply $\alpha\beta$) denotes the concatenation of strings. The notation $\alpha[i..j]$ denotes $\alpha[i] \cdot \alpha[i+1] \cdot \ldots \cdot \alpha[j]$. An ω-*string* $\beta \in \Sigma^\omega$ (or *infinite string/string of infinite length*) is an infinite numerable concatenation of characters from Σ. In this paper, we work with *left-infinite* ω-strings, meaning that $\beta \in \Sigma^\omega$ is constructed from the empty string ε by prepending an infinite number of characters to it. In particular, the operation of appending a character $a \in \Sigma$ at the end of a ω-string $\alpha \in \Sigma^\omega$ is well-defined and yields the ω-string αa. The notation α^ω, where $\alpha \in \Sigma^*$, denotes the concatenation of an infinite (numerable) number of copies of string α. The co-lexicographic (or co-lex) order $\prec$ of two strings $\alpha, \beta \in \Sigma^* \cup \Sigma^\omega$ is defined as follows. (i) $\varepsilon \prec \alpha$ for every $\alpha \in \Sigma^+ \cup \Sigma^\omega$, and (ii) if $\alpha = \alpha'a$ and $\beta = \beta'b$ (with $a, b \in \Sigma$ and $\alpha', \beta' \in \Sigma^* \cup \Sigma^\omega$), $\alpha \prec \beta$ holds if and only if $(a \prec b) \vee (a = b \wedge \alpha' \prec \beta')$. In this paper, the symbols $\prec$ and $\preceq$ will be used to denote the total order on the alphabet and the co-lexicographic order between strings/ω-strings. Notation $[N]$ indicates the set of integers $\{1, 2, \ldots, N\}$.

DFAs, WDFAs, and Wheeler Languages. In this paper, we work with deterministic finite state automata (DFAs):

Definition 3 (DFA). *A DFA $\mathcal{A}$ is a quintuple $(Q, \Sigma, \delta, s, F)$ where Q is a finite set of states, Σ is an alphabet set, $\delta : Q \times \Sigma \to Q$ is a transition function, $s(\in Q)$ is a source state, and $F(\subseteq Q)$ is a set of final states.*

For $u, v \in Q$ and $a \in \Sigma$ such that $\delta(u, a) = v$, we define $\lambda(u, v) = a$. We extend the domain of the transition function to words $\alpha \in \Sigma^*$ as usual, i.e., for $a \in \Sigma$, $\alpha \in \Sigma^*$, and $u \in Q$: $\delta(u, a \cdot \alpha) = \delta(\delta(u, a), \alpha)$ and $\delta(u, \varepsilon) = u$.

In this work, $n = |Q|$ denotes the number of states and $m = |\delta| = |\{(u, v, a) \in Q \times Q \times \Sigma : \delta(u, a) = v\}|$ the number of transitions of the input DFA.

The notation I_u indicates the set of words *reaching* u from the initial state:

Definition 4. *Let $\mathcal{A} = (Q, \Sigma, \delta, s, F)$ be a DFA. If $u \in Q$, let I_u be defined as:*

$$I_u = \{\alpha \in \Sigma^* : u = \delta(s, \alpha)\};$$

The *language* $\mathcal{L}(\mathcal{A})$ recognized by $\mathcal{A}$ is defined as $\mathcal{L}(\mathcal{A}) = \cup_{u \in F} I_u$.

A classic result in language theory [14] states that the minimum DFA (denoted with $\mathcal{A}_{min}$) recognizing the language $\mathcal{L}(\mathcal{A})$ of any DFA $\mathcal{A}$ is unique. The DFA $\mathcal{A}_{min}$ can be computed from $\mathcal{A}$ in $O(m \log n)$ time [11].

Wheeler automata were introduced in [10] as a generalization of prefix sorting from strings to labeled graphs. We consider the following particular case:

Definition 5 (Wheeler DFA). *A Wheeler DFA (WDFA) [10] $\mathcal{A}$ is a DFA for which there exists a total order $< \subseteq Q \times Q$ (called Wheeler order) satisfying:*

(i) $s < u$ for every $u \in Q - \{s\}$.
(ii) If $u' = \delta(u, a)$, $v' = \delta(v, b)$, and $a \prec b$, then $u' < v'$.
(iii) If $u' = \delta(u, a) \neq \delta(v, a) = v'$ and $u < v$, then $u' < v'$.

The symbol $<$ will indicate both the total order of integers and the Wheeler order among the states of a Wheeler DFA. The meaning of symbol $<$ will always be clear from the context. Definition 5 defines the Wheeler order in terms of *local* axioms. On DFAs, an equivalent *global* definition is the following [1]:

Definition 6. *Let u, v be two states of a DFA $\mathcal{A}$. Let $u <_{\mathcal{A}} v$ if and only if $(\forall \alpha \in I_u)(\forall \beta \in I_v) (\alpha \prec \beta)$.*

Lemma 1 ([1]). *$\mathcal{A}$ is Wheeler if and only if $<_{\mathcal{A}}$ is total, if and only if $<_{\mathcal{A}}$ is the (unique) Wheeler order of $\mathcal{A}$.*

In fact, when a Wheeler order exists for a DFA, this order is unique [1] (as opposed to the NFA case). The class of languages recognized by Wheeler automata is of particular interest:

Definition 7 (Wheeler language). *A regular language $\mathcal{L}$ is said to be Wheeler if and only if there exists a Wheeler NFA $\mathcal{A}$ such that $\mathcal{L} = \mathcal{L}(\mathcal{A})$, if and only if there exists a Wheeler DFA $\mathcal{A}'$ such that $\mathcal{L} = \mathcal{L}(\mathcal{A}')$.*

The equivalence between WNFAs and WDFAs was established in [1]. In the same paper [1], the authors provided a *Myhill-Nerode theorem* for Wheeler languages that is crucial for our results. Their result can be stated in terms of the minimum accepting DFA for $\mathcal{L}$. We first need the following definition:

Definition 8 (Entanglement [5]). *Given a DFA $\mathcal{A}$, two distinct states $u \neq v$ of $\mathcal{A}$ are said to be entangled if there exists a monotone infinite sequence $\alpha_1 \prec \beta_1 \prec \cdots \prec \alpha_i \prec \beta_i \prec \alpha_{i+1} \prec \beta_{i+1} \prec \cdots$ (or with reversed sign $\succ$) such that $\alpha_i \in I_u$ and $\beta_i \in I_v$ for every $i \geq 1$.*

The characterization of Wheeler languages of Alanko et al. [1] states that:

Lemma 2 ([1]). *For a DFA $\mathcal{A}$, $\mathcal{L}(\mathcal{A})$ is not Wheeler if and only if there exist entangled states u and v in its minimum DFA $\mathcal{A}_{min}$.*

Lemma 2 is at the core of our algorithm for recognizing Wheeler languages.

2.1 Infima and Suprema Strings

Lemma 1 suggests that the Wheeler order can be defined by looking just at the lower and upper bounds of I_u for each state $u \in Q$. Let us define:

Definition 9 (Infimum and supremum [13]). *For a DFA $\mathcal{A} = (Q, \Sigma, \delta, s, F)$, let $u \in Q$ be a state of $\mathcal{A}$. The infimum string $\inf I_u$ and supremum string $\sup I_u$ are the greatest lower bound and the least upper bound, respectively, of I_u:*

$$\inf I_u = \gamma \in \Sigma^* \cup \Sigma^\omega \ s.t. \ (\forall \beta \in \Sigma^* \cup \Sigma^\omega \ s.t. \ (\forall \alpha \in I_u \ \beta \preceq \alpha) \ \beta \preceq \gamma)$$
$$\sup I_u = \gamma \in \Sigma^* \cup \Sigma^\omega \ s.t. \ (\forall \beta \in \Sigma^* \cup \Sigma^\omega \ s.t. \ (\forall \alpha \in I_u \ \alpha \preceq \beta) \ \gamma \preceq \beta)$$

Kim et al. [13] and Conte et al. [4] use the above definition to give yet another equivalent definition of Wheeler order:

Lemma 3 ([4,13]). *Let u, v be two states of a WDFA $\mathcal{A}$. Let $u < v$ if and only if $\sup I_u \preceq \inf I_v$. Then $<$ is the Wheeler order of $\mathcal{A}$.*

Following Lemma 3, it is convenient to represent each state $u \in Q$ as an *open interval* $\mathcal{I}(u) = (\inf I_u, \sup I_u)$, i.e., the subset of $\Sigma^* \cup \Sigma^\omega$ containing all strings co-lexicographically strictly larger than $\inf I_u$ and strictly smaller than $\sup I_u$. Note that, for two states $u, v \in Q$, if $|I_u|, |I_v| > 1$ then $\mathcal{I}(u) \cap \mathcal{I}(v) = \emptyset$ if and only if $\sup I_u \preceq \inf I_v$ or $\sup I_v \preceq \inf I_u$. If $|I_u| = 1$ (analogously for $|I_v| = 1$), then $\mathcal{I}(u) = \emptyset$ so $\mathcal{I}(u) \cap \mathcal{I}(v)$ is always empty.

Following [13], in the rest of the paper the intervals $\mathcal{I}(u) = (\inf I_u, \sup I_u)$ are encoded as pairs of integers: the co-lexicographic ranks of $\inf I_u$ and $\sup I_u$ in $\{\inf I_u : u \in Q\} \cup \{\sup I_u : u \in Q\}$. Using this representation, the check $\mathcal{I}(u) \cap \mathcal{I}(v) \neq \emptyset$ can be trivially performed in constant time. The authors of [2] show that the relative co-lexicographic ranks of all infima and suprema strings of a DFA can be computed efficiently:

Lemma 4 ([2, Sec. 4]). *Given a DFA $\mathcal{A} = (Q, \Sigma, \delta, s, F)$, we can sort the set $\{\inf I_u : u \in Q\} \cup \{\sup I_u : u \in Q\}$ co-lexicographically in $O(|\delta| \log |Q|)$ time.*

We conclude this section by mentioning two useful properties of infima and suprema strings which will turn out useful later on in this work.

Lemma 5. *Let u be a state of a DFA $\mathcal{A}$, and $\gamma \in \Sigma^*$ be a finite string. Then the following holds:*

1. *If $\inf I_u$ ($\sup I_u$) is finite, then $\inf I_u \in I_u$ ($\sup I_u \in I_u$).*
2. *For any finite suffix α' of $\inf I_u$ or $\sup I_u$, there exists $\alpha \in I_u$ suffixed by α'.*
3. *I_u is a singleton if and only if $\inf I_u = \sup I_u$.*
4. *If $\inf I_u \prec \gamma^\omega$, then there exists $\alpha \in I_u$ such that $\alpha \prec \gamma^\omega$; similarly, if $\gamma^\omega \prec \sup I_u$, then there exists $\alpha \in I_u$ such that $\gamma^\omega \prec \alpha$.*

Proof. (1)–(3) See [13, Observation 8]. (4) Assume $\inf I_u \prec \gamma^\omega$. If $\inf I_u$ is finite, then $\inf I_u \in I_u$ by (1) and the claim follows by setting $\alpha = \inf I_u$. Let us assume $\inf I_u$ has infinite length. Let α' be the shortest suffix of $\inf I_u$ such that $\alpha' \prec \gamma^\omega$ and α' is not a suffix of γ^ω; note that α' is finite, otherwise $\alpha' = \inf I_u = \gamma^\omega$ by definition of $\prec$, which contradicts the assumption $\inf I_u \prec \gamma^\omega$. Then by (2), there exists $\alpha \in I_u$ suffixed by α'. By definition of α', any string suffixed by α' is smaller than γ^ω, hence $\alpha \prec \gamma^\omega$. The case with $\gamma^\omega \prec \sup I_u$ is analogous.

3   Recognizing Wheeler Languages

In this section, we present our algorithm to decide if the language accepted by a DFA $\mathcal{A} = (Q, \Sigma, \delta, s, F)$ is Wheeler. Let $\mathcal{A}_{min} = (Q_{min}, \Sigma, \delta_{min}, s_{min}, F_{min})$ be the minimum-size DFA accepting $\mathcal{L}(\mathcal{A})$.

68   R. Becker et al.

Definition 10 (Square automaton). *The* square automaton $\mathcal{A}^2_{min} = \mathcal{A}_{min} \times \mathcal{A}_{min} = (Q^2_{min} = Q_{min} \times Q_{min}, \Sigma, \delta', (s_{min}, s_{min}), F^2_{min} = F_{min} \times F_{min})$ *is the automaton whose states are pairs of states of* $\mathcal{A}_{min}$ *and whose transition function is defined as* $\delta'((u, v), a) = (\delta_{min}(u, a), \delta_{min}(v, a))$ *for* $u, v \in Q_{min}$ *and* $a \in \Sigma$.

We are ready to prove our new characterization of Wheeler languages. The characterization states that $\mathcal{A}^2_{min}$ can be used to detect repeated cycles in $\mathcal{A}_{min}$, and that we can use this fact to check if $\mathcal{L}(\mathcal{A}_{min}) = \mathcal{L}(\mathcal{A})$ is Wheeler:

Theorem 3. *For a DFA* $\mathcal{A}$, $\mathcal{L}(\mathcal{A})$ *is not Wheeler if and only if* $\mathcal{A}^2_{min}$ *contains a cycle* $(u_1, v_1) \to (u_2, v_2) \to \cdots \to (u_k, v_k) \to (u_1, v_1)$ *such that, for* $1 \leq \forall i \leq k$, *the following hold: (i)* $u_i \neq v_i$ *and (ii)* $\mathcal{I}(u_i) \cap \mathcal{I}(v_i) \neq \emptyset$.

Proof. ($\Leftarrow$) Assume that $\mathcal{A}^2_{min}$ contains such a cycle $(u_1, v_1) \to (u_2, v_2) \to \cdots \to (u_k, v_k) \to (u_1, v_1)$ where k is the cycle length. Then, by definition of $\mathcal{A}^2_{min}$ there exist cycles $u_1 \to u_2 \to \cdots \to u_k \to u_1$ and $v_1 \to v_2 \to \cdots \to v_k \to v_1$ in $\mathcal{A}_{min}$, both of which are labeled by the same string $\gamma = \lambda(u_1, u_2) \cdots \lambda(u_{k-1}, u_k)\lambda(u_k, u_1) = \lambda(v_1, v_2) \cdots \lambda(v_{k-1}, v_k)\lambda(v_k, v_1)$ of length k.

Let $\gamma_1 = \max\{\inf I_{u_1}, \inf I_{v_1}\}$ and $\gamma_2 = \min\{\sup I_{u_1}, \sup I_{v_1}\}$. First, we claim $\gamma_1 \prec \gamma_2$. To see this, without loss of generality, assume $\inf I_{u_1} \preceq \inf I_{v_1}$. Observe $|I_{u_1}|, |I_{v_1}| > 1$ because u and v are on two cycles (hence I_{u_1} and I_{v_1} contain an infinite number of strings). By Lemma 5.3 both $\inf I_{u_1} \prec \sup I_{u_1}$ and $\inf I_{v_1} \prec \sup I_{v_1}$ hold. Since $\mathcal{I}(u_1) \cap \mathcal{I}(v_1) \neq \emptyset$, $\inf I_{v_1} \prec \sup I_{u_1}$ also holds. Then, $\inf I_{u_1} \preceq \inf I_{v_1} \prec \sup I_{u_1}, \sup I_{v_1}$. Therefore $\gamma_1 = \inf I_{v_1} \prec \min\{\sup I_{u_1}, \sup I_{v_1}\} = \gamma_2$.

As a consequence, we can see that at least one of the following must hold: (i) $\gamma_1 \prec \gamma^\omega$ and (ii) $\gamma^\omega \prec \gamma_2$; note that the complement of the case (i) is $\gamma^\omega \preceq \gamma_1$, which implies $\gamma^\omega \prec \gamma_2$ because $(\gamma^\omega \preceq)\gamma_1 \prec \gamma_2$. Therefore, by Lemma 5.4, there must exist $\alpha \in I_{u_1}$ and $\beta \in I_{v_1}$ such that either $\alpha, \beta \prec \gamma^\omega$ or $\gamma^\omega \prec \alpha, \beta$ hold.

Note that it holds $\alpha \neq \beta$ since $\mathcal{A}_{min}$ is deterministic. Without loss of generality, assume $\alpha \prec \beta$. We consider the case $\alpha \prec \beta \prec \gamma^\omega$; the other case $(\gamma^\omega \prec \alpha \prec \beta)$ is symmetric. Let l be any integer such that $\max\{|\alpha|, |\beta|\} < l \cdot |\gamma|$. Then we can see that, for every $d \geq 0$, the following three properties hold: (i) $\alpha(\gamma^l)^d \prec \beta(\gamma^l)^d \prec \alpha(\gamma^l)^{d+1} \prec \beta(\gamma^l)^{d+1}$, (ii) $\alpha(\gamma^l)^d \in I_{u_1}$ (because $\alpha \in I_{u_1}$ and γ labels a cycle from u_1, so $\delta_{min}(u_1, \gamma^k) = u_1$ for any integer $k \geq 0$) and, similarly, (iii) $\beta(\gamma^l)^d \in I_{v_1}$. Properties (i-iii) imply that there is an infinite monotone nondecreasing sequence of strings alternating between I_{u_1} and I_{v_1}, i.e., u_1 and v_1 are entangled (Definition 8) and, by Lemma 2, $\mathcal{L}(\mathcal{A}_{min}) = \mathcal{L}(\mathcal{A})$ is not Wheeler.

($\Rightarrow$) Assume that $\mathcal{L}(\mathcal{A}_{min}) = \mathcal{L}(\mathcal{A})$ is not Wheeler. By Lemma 2, there exist entangled states $u_0 \neq v_0$ in $\mathcal{A}_{min}$ (in particular, $\mathcal{I}(u_0) \cap \mathcal{I}(v_0) \neq \emptyset$). Without loss of generality, we can assume that there is an infinite nondecreasing sequence $S_0 = \alpha_1 \prec \beta_1 \prec \alpha_2 \prec \beta_2 \prec \cdots$ such that, for every $i \geq 1$, $\alpha_i \in I_{u_0}$ and $\beta_i \in I_{v_0}$ (the other case with the reversed sign is analogous).

Observe that, since the alphabet is finite, S_0 must ultimately (i.e., from a sufficiently large index i) contain strings α_i, β_i sharing the last character. We can therefore assume without loss of generality that all strings in S_0 end with the same character a. Then, there exist u_1, v_1 such that $\delta_{min}(u_1, a) = u_0$ and $\delta_{min}(v_1, a) = v_0$. Note that, by the determinism of $\mathcal{A}_{min}$, it must be $u_1 \neq v_1$.

Moreover, we can choose two *entangled* such u_1, v_1. To see this, let $u_1^1, \ldots, u_1^s$ and $v_1^1, \ldots, v_1^r$ be the s and r predecessors of u_0 and v_0, respectively, such that $\delta_{min}(u_1^i, a) = u_0$ and $\delta_{min}(v_1^j, a) = v_0$ for all $1 \leq i \leq s$ and $1 \leq j \leq r$. Assume for the purpose of contradiction that u_1^i and v_1^j are not entangled for all pairs u_1^i, v_1^j. Then, by definition of entanglement any monotone sequence $\mu_1 \prec \mu_2 \prec \cdots \in I_{u_1^i} \cup I_{v_1^j}$ ultimately ends up in *just one* of the two sets: there exists $N \in \mathbb{N}$ such that either $\mu_N, \mu_{N+1}, \cdots \in I_{u_1^i}$ or $\mu_N, \mu_{N+1}, \cdots \in I_{v_1^j}$. Since this is true for *any* pair u_1^i, v_1^j, any monotone sequence $\mu_1 \prec \mu_2 \prec \cdots \in \bigcup_{i=1}^{s} I_{u_1^i} \cup \bigcup_{j=1}^{r} I_{v_1^j}$ ultimately ends up in *either* (i) $\bigcup_{i=1}^{s} I_{u_1^i}$ or (ii) $\bigcup_{j=1}^{r} I_{v_1^j}$. But then, this implies that sequence S_0 cannot exist: any monotone sequence $\mu_1 a \prec \mu_2 a \prec \cdots \in I_{u_0} \cup I_{v_0}$ ultimately ends up in either (i) I_{u_0} or (ii) I_{v_0}.

Summing up, we found $u_1 \neq v_1$ such that $\delta_{min}(u_1, a) = u_0$, $\delta_{min}(v_1, a) = v_0$, and u_1, v_1 are entangled (in particular, $\mathcal{I}(u_1) \cap \mathcal{I}(v_1) \neq \emptyset$). We iterate this process for $k = |Q_{min}|^2$ times; this yields two paths $u_k \to u_{k-1} \to \cdots \to u_0$ and $v_k \to v_{k-1} \to \cdots \to v_0$ labeled with the same string of length k, with $u_i \neq v_i$ and $\mathcal{I}(u_i) \cap \mathcal{I}(v_i) \neq \emptyset$ for all $0 \leq i \leq k$. But then, since we chose $k = |Q_{min}|^2$, by the pigeonhole principle there must exist two indices $j \leq i$ such that $(u_i, v_i) = (u_j, v_j)$. In particular, there exists $k' \leq k$ such that $u_i \to u_{i-1} \to \cdots \to u_{i-k'+1} \to u_i$ and $v_i \to v_{i-1} \to \cdots \to v_{i-k'+1} \to v_i$ are two cycles of the same length k', labeled with the same string, such that $u_t \neq v_t$ and $\mathcal{I}(u_t) \cap \mathcal{I}(v_t) \neq \emptyset$ for all indices $i - k' + 1 \leq t \leq i$. This yields our main claim. $\square$

4   The Algorithm

Theorem 3 immediately gives a quadratic algorithm for WHEELERLAN-GUAGEDFA:

1. Compute $\mathcal{A}_{min} = (Q_{min}, \Sigma, \delta_{min}, s_{min}, F_{min})$ by Hopcroft's algorithm [11].
2. On $\mathcal{A}_{min}$, compute intervals $\mathcal{I}(u)$ for each $u \in Q_{min}$, using[3] [2, Sec. 4].
3. Compute $\mathcal{A}_{min}^2$.
4. Remove from $\mathcal{A}_{min}^2$ all states (u, v) (and incident transitions) such that either $u = v$ or $\mathcal{I}(u) \cap \mathcal{I}(v) = \emptyset$. Let $\hat{\mathcal{A}}_{min}^2$ be the resulting pruned automaton.
5. Test acyclicity of $\hat{\mathcal{A}}_{min}^2$. If $\hat{\mathcal{A}}_{min}^2$ is acyclic, return "$\mathcal{L}(\mathcal{A})$ is Wheeler". Otherwise, return "$\mathcal{L}(\mathcal{A})$ is not Wheeler".

Since, by its definition, $\mathcal{A}_{min}$ cannot be larger than $\mathcal{A}$, in the rest of the paper we will for simplicity assume that $\mathcal{A}_{min}$ has n nodes and m transitions. Steps (1) and (2) run in $O(m \log n)$ time. Note that, for each transition $\delta_{min}(u, a) = u'$ and for each node $v \neq u$, by the determinism of $\mathcal{A}_{min}$ there exists at most one transition $\delta_{min}(v, a) = v'$ labeled with a and originating in v; such a pair of transitions define one transition of $\mathcal{A}_{min}^2$. It follows that the number of transitions (thus the size) of $\mathcal{A}_{min}^2$ is $O(mn)$, therefore steps (3–5) run in $O(mn)$ time (acyclicity can be tested in $O(|\mathcal{A}_{min}^2|)$ time using, for example, Kahn's topological sorting algorithm). Overall, the algorithm runs in $O(mn)$ time.

---

[3]   In the full version [3], we discuss more in detail how to apply [2, Sec. 4] on $\mathcal{A}_{min}$.

4.1 A Parameterized Algorithm

Our algorithm can be optimized by observing that we can directly build $\hat{\mathcal{A}}^2_{min}$, which could be much smaller than $\mathcal{A}^2_{min}$. For example, if $\mathcal{A}_{min}$ is Wheeler, then $\mathcal{I}(u) \cap \mathcal{I}(v) = \emptyset$ for all states $u \neq v$ of $\mathcal{A}_{min}$ (see Definition 6 and Lemma 1), so $\hat{\mathcal{A}}^2_{min}$ is empty. As a matter of fact, we show that the size of $\hat{\mathcal{A}}^2_{min}$ depends on the *width* of the (partial [6]) order $<_{\mathcal{A}_{min}}$, i.e., the size of the largest antichain:

Definition 11 ([6]). *The co-lex width $width(\mathcal{A})$ of a DFA $\mathcal{A}$ is the width of the order $<_{\mathcal{A}}$ defined in Definition 6.*

The co-lex width is an important measure parameterizing problems such as pattern matching on graphs and compression of labeled graphs [5,6]. Note that $width(\mathcal{A}_{min}) = 1$ if and only if $\mathcal{A}_{min}$ is Wheeler. We show:

Lemma 6. *Let $p = width(\mathcal{A}_{min})$. Then, $\hat{\mathcal{A}}^2_{min}$ has at most $2n(p-1)$ states and at most $2m(p-1)$ transitions and can be built from $\mathcal{A}_{min}$ in $O(mp)$ time.*

The intuition behind Lemma 6 is that $\hat{\mathcal{A}}^2_{min}$ contains only states (u, v) such that $\mathcal{I}(u) \cap \mathcal{I}(v) \neq \emptyset$. By Lemma 3, this holds if and only if u and v are incomparable by the order $<_{\mathcal{A}_{min}}$. Since the width of this order is (by definition) p, the bounds follow easily. To build $\hat{\mathcal{A}}^2_{min}$, we sort the states of $\hat{\mathcal{A}}_{min}$ by the strings $\inf I_u$ and observe that incomparable states are adjacent in this order. It follows that we can easily build $\hat{\mathcal{A}}^2_{min}$ in time proportional to its size, $O(mp)$. The details can be found in the full version [3]. Theorem 1 follows.

Implementation. We implemented the algorithm of Theorem 1. The code is available at http://github.com/regindex/Wheeler-language-recognizer. It takes in input either a regular expression or a DFA and checks if the recognized language is Wheeler. We tested our algorithm on two random DFA datasets:

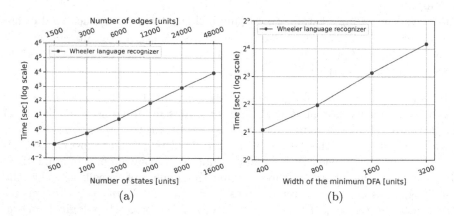

Fig. 1. Wall clock time for our algorithm on different random DFA datasets (a) different n, $m = 3n$, and p is similar to n; (b) different p with fixed m.

(i) one with different combinations of number of states and transitions where $n = \{500 \cdot 2^i : i = 0, \ldots, 5\}$ and $m = 3n$ to show the quadratic running time, and (ii) the other with a fixed number of transitions, $m = 16 \cdot 10^3$, and different widths $p = \{400, 800, 1600, 3200\}$ of the minimum DFAs to show the running time with respect to p. Our experiments were run on a server with Intel(R) Xeon(R) W-2245498 CPU @ 3.90 GHz with 8 cores and 128 gigabytes of RAM running Ubuntu 18.04 LTS 64-bit. As expected, our experimental results show that the running time grows linearly in mp. It is worth noting that, on our input instances in the first dataset, p is roughly similar to n, and we double n at each step, the running time shows a quadratic growth (Fig. 1(a)). On the other hand, when the number of transitions is fixed, the running time grows linearly to the width p of the minimum DFA (Fig. 1(b)). This can be measured from the slopes of the fitted lines on the log-log plots, which are 2.03 and 1.04, respectively.

5 A Matching Conditional Lower Bound

In this section, we show that an algorithm for WHEELERLANGUAGEDFA with running time $O(m^{2-\eta})$, yields an algorithm for the Orthogonal Vectors problem (see Definition 2) with running time $O(N^{2-\eta} \operatorname{poly}(d))$, thus contradicting SETH. This is our second main theorem (Theorem 2) formulated in the introduction. We prove this theorem using Theorem 3 and the following proposition, which reduces an instance of the OV problem with two sets of N d dimensional vectors each into an instance of our problem with a minimum DFA of size $\Theta(Nd)$.

Proposition 1. *For an instance of the OV problem, we can in $O(N(d + \log N))$ time construct a DFA $\mathcal{A}$ with $m \in O(N(d + \log N))$ edges that is minimum for its language $\mathcal{L}(\mathcal{A})$ such that the OV instance is a YES-instance if and only if $\mathcal{A}^2$ contains a cycle $(u_1, v_1) \rightarrow (u_2, v_2) \rightarrow \cdots \rightarrow (u_k, v_k) \rightarrow (u_1, v_1)$ such that, for $1 \leq \forall i \leq k$, the following hold: (i) $u_i \neq v_i$ and (ii) $\mathcal{I}(u_i) \cap \mathcal{I}(v_i) \neq \emptyset$.*

Once this proposition is established, we can take an OV instance with sets of size N containing vectors of dimension $d \in \omega(\log N)$ and construct the DFA $\mathcal{A}$ of size $\Theta(m) = \Theta(N(d + \log N)) = \Theta(Nd)$. Now assume that we can solve WHEELERLANGUAGEDFA in $O(m^{2-\eta})$ on $\mathcal{A}$. Using Theorem 3 and Proposition 1, we can thus solve the OV instance in $O((Nd)^{2-\eta}) = O(N^{2-\eta} \operatorname{poly}(d))$ time, as the OV instance is a YES instance if and only if the language recognized by $\mathcal{A}$ is not Wheeler. This shows Theorem 2. The rest of this section is dedicated to illustrate how we prove Proposition 1. The details can be found in the full version [3].

Construction of $\mathcal{A}$. For a given instance $A = \{a_1, \ldots, a_N\}$ and $B = \{b_1, \ldots, b_N\}$ of the OV problem, we build a DFA $\mathcal{A} = (Q, \Sigma, \delta, s, F)$ with the properties in Proposition 1 by adapting a technique of Equi et al. [9]. We first notice that we can, w.l.o.g., make the following assumptions on the OV instance: (1) The vectors in A are distinct, (2) N is a power of two, say $N = 2^\ell$. We describe the construction of $\mathcal{A}$ based on a small example, while the general description can be found in the full version [3]. We let $\Sigma = \{0, 1, \#\}$. Later we show how to reduce the alphabet's size to 2.

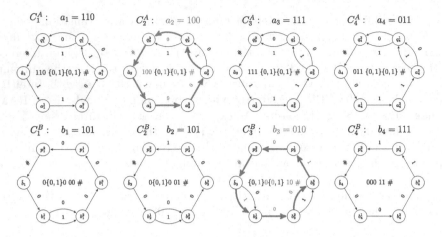

Fig. 2. Illustration of the cycles generated for the bit vectors in the example $A = \{110, 100, 111, 011\}$ and $B = \{101, 101, 010, 111\}$. The two cycles C_2^A and C_3^B generate a match, i.e., can read the same string, as the vectors a_2 and b_3 are orthogonal.

Let $A = \{110, 100, 111, 011\}$ and $B = \{101, 101, 010, 111\}$ be the given instance of the OV problem. Thus $N = 4$ and $\ell = \log_2 N = 2$. Notice that the only pair of orthogonal vectors in A and B are $a_2 = 100$ and $b_3 = 010$. In our DFA $\mathcal{A}$ we build (non-simple) cycles C_i^A and C_j^B for every vector a_i and b_j in A and B respectively. As an example, for a_2 we build the cycle C_2^A labeled with $100\{0,1\}\{0,1\}\#$, i.e. the bit string 100 of a_2, followed by a sub-graph recognizing any bit string of length $\ell = 2$, followed by $\#$. For b_3 we build the (non-simple) cycle C_3^B labeled with $\{0,1\}0\{0,1\}10\#$, i.e., the bits 0 of b_3 are converted to a sub-graph recognizing both 0 and 1, and the bits 1 of b_3 are converted to an edge recognizing 0; this subgraph is followed by a path of length ℓ spelling 10, which is the 3rd smallest among the length-ℓ binary strings (i.e. the identifier for C_3^B to prevent it from any match with C_j^B for $j \neq 3$ while allowing matches with C_i^A's), which is followed by an edge labeled with $\#$. Notice that these two cycles indeed generate a match (underlined characters indicate the match): $\underline{100}\{0,\underline{1}\}\{\underline{0},1\}\underline{\#}$ and $\{0,\underline{1}\}\underline{0}\{\underline{0},1\}\underline{10\#}$. We can see C_i^A and C_j^B built in this way will match if and only if the two corresponding vectors are orthogonal. Characters $\#$ are introduced to synchronize the match (otherwise, other rotations of the cycles could match). The subgraphs between the part corresponding to the input vectors and the character $\#$ are introduced to avoid that two distinct cycles C_i^B and C_j^B ($i \neq j$) generate a match. Note that, since we assume that A contains distinct vectors, distinct cycles C_i^A and C_j^A ($i \neq j$) will never generate a match.

The remaining details of the reduction ensure that (1) the graph is a connected DFA, (2) corresponding nodes (i.e. same distance from $\#$ in the cycles) u, v in any pair of cycles C_i^A and C_j^B that correspond to orthogonal vectors a_i and b_j, respectively, have a non-empty co-lexicographic intersection $\mathcal{I}(u) \cap \mathcal{I}(v)$,

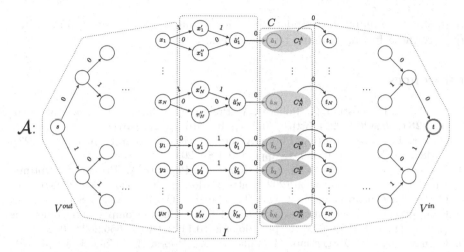

Fig. 3. Illustration of our construction of $\mathcal{A}$ for an arbitrary instance $A = \{a_1, \ldots, a_N\}$ and $B = \{b_1, \ldots, b_N\}$ of the OV problem. The cycles $C_1^A, \ldots, C_N^A, C_1^B, \ldots, C_N^B$ are expanded in Fig. 2 on a particular OV instance.

(3) the DFA is indeed minimum for its recognized language, and (4) the alphabet can be reduced to $\{0, 1\}$ by an opportune mapping.

An illustration of the overall construction for an arbitrary instance of the OV problem can be found in Fig. 3. The complete proof can be found in the full version [3]. We proceed with a sketch on how we achieve the above properties. Property (1) is achieved by connecting the above described cycles (we call the set of all cycles' nodes C), to the source node s through a binary out-tree of logarithmic depth (we call these nodes V^{out}). Property (2) is achieved by connecting V^{out} to C through the nodes in I that ensure that nodes $u \in C_i^A, v \in C_j^B$ in the same relative positions (i.e. same distance from #) in their cycles that correspond to orthogonal vectors a_i and b_j are reached by strings of alternating co-lexicographic order, i.e., for a suitable string τ, I_u contains two strings suffixed by 00τ and 11τ, while I_v contains a string suffixed by 01τ. This implies $\mathcal{I}(u) \cap \mathcal{I}(v) \neq \emptyset$. Property (3) is instead achieved by connecting one node from each cycle C_i^A and C_j^B (the one with out-edge #) with an edge labeled 0 to a complete binary in-tree (we call these nodes V^{in}) with root being the only accepting state t. Observe that the reversed automaton (i.e. the automaton obtained by reversing the direction of all the edges) is *deterministic*; this ensures that any two nodes in the graph can reach t through a distinct binary string, witnessing that $\mathcal{A}$ is indeed minimal (by the Myhill-Nerode characterization of the minimum DFA [14]). Property (4) can be easily satisfied by transforming the instance as follows. Edges labeled 0 (1) are replaced with a directed path labeled with 00 (11), while edges labeled # are replaced with a directed path labeled 101. The pattern 101 then appears only on the paths that originally corresponded to # and thus two transformed cycles match if and only if they used to match before the transformation. We note that also forward- and reverse- determinism (thus minimality) are maintained under this transformation (the nodes $\hat{a}'_i$ and $\hat{b}'_j$ in I are introduced for maintaining reverse-determinism).

References

1. Alanko, J., D'Agostino, G., Policriti, A., Prezza, N.: Wheeler languages. Inf. Comput. **281**, 104820 (2021). https://doi.org/10.1016/j.ic.2021.104820
2. Becker, R., et al.: Sorting Finite Automata via Partition Refinement (2023). 10.48550/arXiv. 2305.05129, to appear at ESA 2023
3. Becker, R., Cenzato, D., Kim, S.H., Kodric, B., Policriti, A., Prezza, N.: Optimal wheeler language recognition (2023). https://doi.org/10.48550/arXiv.2306.04737
4. Conte, A., Cotumaccio, N., Gagie, T., Manzini, G., Prezza, N., Sciortino, M.: Computing matching statistics on wheeler DFAs. In: Data Compression Conference (DCC), pp. 150–159 (2023). https://doi.org/10.1109/DCC55655.2023.00023
5. Cotumaccio, N., D'Agostino, G., Policriti, A., Prezza, N.: Co-lexicographically ordering automata and regular languages - part I. J. ACM **70**, 1–73 (2023). https://doi.org/10.1145/3607471, (online available)
6. Cotumaccio, N., Prezza, N.: On indexing and compressing finite automata. In: Proceedings of the 32nd Annual ACM-SIAM Symposium on Discrete Algorithms (SODA), pp. 2585–2599 (2021). https://doi.org/10.1137/1.9781611976465.153
7. D'Agostino, G., Martincigh, D., Policriti, A.: Ordering regular languages and automata: complexity. Theoret. Comput. Sci. **949**, 113709 (2023). https://doi.org/10.1016/j.tcs.2023.113709
8. Eizenga, J.M., et al.: Pangenome graphs. Ann. Rev. Genomics Hum. Genet. **21**(1), 139–162 (2020). https://doi.org/10.1146/annurev-genom-120219-080406, pMID: 32453966
9. Equi, M., Grossi, R., Mäkinen, V., Tomescu, A.I.: On the complexity of string matching for graphs. In: Proceedings of the 46th International Colloquium on Automata, Languages, and Programming (ICALP), pp. 55:1–55:15 (2019). https://doi.org/10.4230/LIPIcs.ICALP.2019.55
10. Gagie, T., Manzini, G., Sirén, J.: Wheeler graphs: a framework for BWT-based data structures. Theoret. Comput. Sci. **698**, 67–78 (2017). https://doi.org/10.1016/j.tcs.2017.06.016
11. Hopcroft, J.: An $n \log n$ algorithm for minimizing states in a finite automaton. In: Proceedings of an International Symposium on the Theory of Machines and Computations, pp. 189–196 (1971). https://doi.org/10.1016/B978-0-12-417750-5.50022-1
12. Impagliazzo, R., Paturi, R.: On the complexity of k-SAT. J. Comput. Syst. Sci. **62**(2), 367–375 (2001). https://doi.org/10.1006/jcss.2000.1727
13. Kim, S.H., Olivares, F., Prezza, N.: Faster prefix-sorting algorithms for deterministic finite automata. In: Proceedings of the 34th Annual Symposium on Combinatorial Pattern Matching (CPM), pp. 16:1–16:16 (2023). https://doi.org/10.4230/LIPIcs.CPM.2023.16
14. Nerode, A.: Linear automaton transformations. In: Proceedings of the American Mathematical Society, vol. 9, no. 4, pp. 541–544 (1958). https://doi.org/10.2307/2033204
15. Williams, R.: A new algorithm for optimal 2-constraint satisfaction and its implications. Theor. Comput. Sci. **348**(2–3), 357–365 (2005). https://doi.org/10.1016/j.tcs.2005.09.023

Approximation and Fixed Parameter Algorithms for the Approximate Cover Problem

Guillaume Blin[1] , Alexandru Popa[2] , Mathieu Raffinot[1] ,
and Raluca Uricaru[1(✉)]

[1] CNRS, Bordeaux INP, LaBRI, UMR 5800, Univ. Bordeaux, 33400 Talence, France
{guillaume.blin,mathieu.raffinot,raluca.uricaru}@u-bordeaux.fr
[2] Faculty of Mathematics and Computer Science, University of Bucharest,
Bucharest, Romania
alexandru.popa@fmi.unibuc.ro

Abstract. Amir et al. (CPM 2017) introduce the *approximate string cover problem* (**ACP**) motivated by applications including molecular biology, coding, automata theory, formal language theory and combinatorics. A *cover* of a string T is a string C for which every letter of T lies within some occurrence of C. The input of the **ACP** consists of a string T and the goal is to find a string C of length less than the length of T that covers a string T', which is as close to T as possible (under some predefined distance). Amir et al. study this problem for the Hamming distance and show that it is NP-hard.

In this paper we continue the work of Amir et al. and show the following results for the cover length relaxation of the **ACP**. After observing that the NP-hardness proof by Amir et al. (CPM 2017, TCS 2019) suffers from several lapses, we propose an amendment to the proof. We then introduce an approximation algorithm for a variant of the **ACP**, in which we aim to maximize the length of the input string minus the distance to the string covered by the approximate cover returned by the algorithm. This problem is naturally as hard as the **ACP**. We prove an asymptotic approximation ratio of $\mathcal{O}(\sqrt{|T|})$, where $|T|$ is the size of the input string. Finally, we present an FPT algorithm with respect to the alphabet size and the size of the cover based on a dynamic programming framework.

Keywords: Approximate Cover · NP-hardness · Approximation
algorithm · FPT algorithm

1 Introduction

Motivation and Problem Definition. Redundancy is a common aspect of all data and was intensively studied over the years [29,32]. Data, which can be

This work was supported by a grant of the Ministry of Research, Innovation and Digitization, CNCS - UEFISCDI, project number PN-III-P1-1.1-TE-2021-0253, within PNCDI III. We also thank the PHC Brincusi project between Univ. of Bordeaux and Univ. of Bucharest that facilitated the bilateral visits leading to this work.
M. Raffinot—Currently at Apple.

F. M. Nardini et al. (Eds.): SPIRE 2023, LNCS 14240, pp. 75–88, 2023.
https://doi.org/10.1007/978-3-031-43980-3_7

highly redundant or repetitive, usually comes with some key patterns or regularities [34]. One type of redundancy is the periodicity, which is a crucial phenomenon when analyzing physical data like an analog signal. Periodicity has been thoroughly studied in various fields such as Signal Processing [33], Bioinformatics [15], Dynamical Systems [23] and Control Theory [10].

It is of importance to expand the study of redundancy to other types of repeated patterns than periods, one common type of repeated pattern in strings being a *cover*. Simply put, a *cover* of a string T is a string C such that every letter of T lies within some occurrence of C in T. Note that the notion of *cover* is a generalization of the one of *period*, as indeed every periodic string admits a cover. However, the main difference between these two notions resides in the fact that in the case of periods, occurrences do not overlap. A string may admit zero, one or several covers. However, as for periods, one is interested in the shortest cover. The problem of determining the shortest cover of a string, known as the **Minimal String Cover Problem**, was shown to be solvable in linear time [9].

However, the notion of cover may still not be sufficiently general to capture repetitive signal in strings like those coming from biological sequences, which would benefit from considering approximate repeats. In this paper we elaborate on the work of Amir et al. [1–5] who introduce and study *approximate string covers*, which can be briefly defined as string covers in the presence of errors. Intuitively, given an original string, the idea is to cover a second string at a minimal *distance* from the original one. As in [1], here we will consider the Hamming distance, which implies that the covered string has the same length as the original one (*i.e.*, errors come from mismatches only). Below we will refer to the Hamming distance between two strings s_1 and s_2 of equal length as $h(s_1, s_2)$.

Related Work. Our exposition of related results follows mainly [1–3,5]. Quasi-periodicity was introduced by Ehrenfeucht in 1990 (according to [7]) in a Tech Report for Purdue University, even though in was not published until 1993 [8]. The notion of quasi-periodicity is first considered by Apostolico, Farach and Iliopoulos [9]. They define the quasi-period of a string to be the length of its shortest cover and present a linear (time and space) algorithm for computing it [9]. This notion attracted the attention of numerous researchers [12,13,28,30, 31]. The following surveys summarize the first decade of results: [7,25,26].

However, quasi-periodicity takes many forms, depending on the type of patterns one wants to recover. Further work dealt with different variants such as seeds [22,24], λ-seeds [20], the maximum quasi-periodic substring [14], k-covers [17], λ-covers [21], enhanced covers [19], partial covers [25]. Another variation point is the context, e.g. indeterminate strings [6,18] or weighted sequences [11,16]. Also Landau and Schmidt study a weaker form of quasi-periodicity and focus on approximate tandem repeats [27].

Our Results. In this paper we follow up on the work of Amir et al. [1,3] and present several algorithmic results and structural properties of the **Approximate Cover Problem (ACP)** and more precisely of one of its relaxations, namely the cover-length relaxation. The remaining of the paper is organised as follows.

After giving the definitions and necessary notations in Sect. 2, we present an amended proof of the NP-hardness of the **ACP** in Sect. 3. One should note that while Amir et al. [3,4] have the result right (the problem is indeed NP-hard, as we will prove), their proof was incorrect. Second, in Sect. 4, we present a polynomial time, $\mathcal{O}(\sqrt{|T|})$-approximation algorithm for the maximization version of the cover-length relaxation of the **ACP**, *Max Similarity **ACP***. The key idea is to split the input instances into two groups according to the size of the cover with respect to the size of T and then to design a cover for each group. Finally, we propose an FPT algorithm for the minimization version of the cover length relaxation of the **ACP**, based on a dynamic programming framework.

2 Preliminaries and Problem Definitions

In this section we give all notations and formal definitions necessary for the approximate string cover related problems that are tackled in this paper.

String Cover. Let T and C be strings over an alphabet Σ, with n the length of T, respectively m the length of C. We say that C is a *cover* of the text T if we have $m < n$ and if there is a succession of possibly overlapping occurrences of C in T, such that every character of T belongs to at least one of these occurrences.

Let us take several examples of strings and their covers: in the case of a string $T = abcd$, no cover exists; if we consider the string $T = ababaaba$ we get the cover aba; the string $T = ababab$ admits a cover $abab$, and a shorter one ab (that is also a period of T).

Tiling. Let T be a text of length n over alphabet Σ, and C a cover of this text of length m ($m < n$). We call a **tiling** of C over T, a list of at least two strictly ascending ordered indices $\mathcal{L} = [i_1, \ldots, i_l]$ with $0 \leq i_k \leq n - m$ and for all $i_k \in \mathcal{L}$ there is an occurrence of C in T at position i_k.

A tiling is said to be **valid** with respect to the text and the cover if $i_1 = 0$, $i_l = n - m$ and $\forall i_k$ with $k < l$, we have $i_{k+1} - i_k \leq m$. For example, the list $[0, 2, 5]$ is a valid tiling for the text $T = ababaaba$ and the cover aba.

Moreover, given a cover C of length m, there is an infinity of strings with lengths superior to m that can be covered by C (meaning that each one of these strings admits a valid tiling with respect to the cover C). However, for a fixed n and a given tiling (that is correct with respect to C), there exists a unique string covered by C and respecting the tiling.

Approximate String Cover. Let T be a string of length n over an alphabet Σ. We call C' an *approximate cover* of T if and only if: (i) C' covers one of the strings T' being at minimal distance from T (here the Hamming distance, hence $h(T, T')$ has to be minimal) among all strings of length n admitting a cover; (ii) C' is of minimal length among the covers of strings T'.

Note that, given the definition above, approximate covers are required to be primitive, meaning that they cannot be covered by a shorter cover.

Approximate Cover Problem. The Approximate String Cover Problem (**ACP**) takes a string T of length n as input and computes an approximate cover of T that is an exact cover for a string T', as well as the distance from T' to the original text T (here the Hamming distance).

Amir et al. deduce the NP-hardness of the **ACP** with respect to the Hamming distance [3] by proving the NP-hardness of a relaxation of this problem called the **cover-length relaxation**. In this variant of the **ACP** the size of the cover, m, is fixed and specified in the input. The idea is to find an m-length cover of a string T' ($|T| = |T'|$) with the smallest $h(T, T')$. Note that the algorithmic results on the cover-length relaxation of the **ACP** naturally extend to the **ACP**. For example, if we have a polynomial-time approximation algorithm with a c factor for the cover-length relaxation, we can design an algorithm with the same factor for the general **ACP** by simply calling the approximation algorithm with all the possible cover length values and selecting the best solution.

Though the cover-length relaxation problem is defined as a minimization problem (*i.e.*, minimize $h(T, T')$), it can be tackled from the opposite point of view, meaning as a maximization problem. This time the goal is to maximize the number of positions on which T and T' match (*i.e.*, maximize $|T| - h(T, T')$). Here, we name this variant the *Max Similarity ACP*. It is straightforward that from the complexity point of view, the two versions are equally hard.

Note that by abuse of notation, except when explicitly stated, when we write **ACP** we refer to the cover-length relaxation of the **ACP**.

3   NP-Hardness of the ACP

As stated previously, the original hardness proof provided in [3,4] is, in the current version, wrong. We will here provide some key elements to show the current problem of the construction and a way of fixing it. In order to avoid redundancy with the original proof, we will keep the presentation simple and let the reader refer to the original proof for further details.

The proof is a reduction from 3-SAT: we are given a 3-CNF formula φ on N variables, $x_1, \ldots, x_N$, with l clauses such that, without loss of generality, the literals in each clause are sorted by the index of their variables. Amir et al. define a text T over an alphabet $\Sigma = \{x_i, \overline{x}_i | i \in [1..N]\} \bigcup \{p, y_1, y_2, y_3, y_4\}$ as follows[1].

The text T is the result of the concatenation of two parts: a so-called header (left part) and a body (right part). The header part is composed of β copies of the following string $B = p^\alpha \ y_1 \ y_2 \ \overline{x}_N \ldots \overline{x}_1 \ y_3 \ y_4 \ p^\alpha \ y_1 \ y_2 \ x_N \ldots x_1 \ y_3 \ y_4 \ p^\alpha$ where p^α is a run of α characters p. While the p-runs will be used as a padding gadget, the characters $\{y_1, y_2, y_3, y_4\}$ are additional dummy variables that will be helpful in the specific case where a clause is satisfied by one of $\{x_1, \overline{x_1}, x_N, \overline{x_N}\}$ in a given truth assignment satisfying φ.

The body part of T is defined directly from the formula φ as $p^\gamma \ L_1^1 \ L_2^1 \ L_3^1 \ p^\gamma \ L_1^2 \ L_2^2 \ L_3^2 \ p^\gamma \ldots p^\gamma \ L_1^l \ L_2^l \ L_3^l \ p^\gamma$ where for all $1 \le j \le l$, L_i^j is the ith literal of the jth clause. Finally, Amir et al. define the size of the cover as $m = 2\alpha + N + 4$. In the original construction, $\alpha = N + 7$, $\beta = l(N + 3)$ and $\gamma = 2N + 14$.

---

[1] In the original construction $y_4 = x_0$, $y_3 = x_{-1}$, $y_2 = x_{N+1}$ and $y_1 = x_{N+2}$.

For the sake of clarity, we provide a full example of T with $\alpha = 11$, $\beta = 21$, $\gamma = 22$, $m = 30$ for $\varphi = (x_1 \vee x_2 \vee x_3) \wedge (x_2 \vee x_3 \vee \overline{x}_4) \wedge (x_1 \vee \overline{x}_3 \vee \overline{x}_4)$:

$$[p^\alpha y_1 y_2 \overline{x}_4 \overline{x}_3 \overline{x}_2 \overline{x}_1 y_3 y_4 p^\alpha y_1 y_2 x_4 x_3 x_2 x_1 y_3 y_4 p^\alpha]^\beta p^\gamma x_1 x_2 x_3 p^\gamma x_2 x_3 \overline{x}_4 p^\gamma x_1 \overline{x}_3 \overline{x}_4 p^\gamma$$

Now, one has to prove the correctness of the reduction.

Lemma 1. *For a given δ, φ is satisfiable if and only if T has an m-approximate cover with at most δ errors*

Proof. In the original proof, $\delta = l(N + 3)(N + 1)$. Let us first prove why the original reduction does not hold and propose a tuning allowing the correctness of this last.

Given a truth assignment $\mathcal{A}$ satisfying φ, let us build an m-length string $C = p^\alpha \, y_1 \, y_2 \, z_N \, \ldots z_1 \, y_3 \, y_4 \, p^\alpha$ where for all $1 \leq i \leq N, z_i = x_i$ if x_i belongs to $\mathcal{A}$; $z_i = \overline{x}_i$ otherwise. Considering our toy instance, let us take $\mathcal{A} = \{x_1, \overline{x}_2, \overline{x}_3, \overline{x}_4\}$ which is indeed satisfying φ. Now, one has to prove that T may indeed be covered with C with at most δ errors. This should be straightforward and guaranteed by the construction; which is not the case in the original proof. Let us have a closer look to a possible cover of T using C. The key idea is that the runs of p characters in C will allow us to cover from α to 2α characters in T depending on the overlap of two consecutive occurrences of C in the corresponding tiling. The overall shape of the corresponding tiling is illustrated afterwards as horizontal lines.

$$[p^\alpha y_1 y_2 \overline{x}_4 \overline{x}_3 \overline{x}_2 \overline{x}_1 y_3 y_4 \underline{p^\alpha y_1 y_2 x_4 x_3 x_2 x_1 y_3 y_4 p^\alpha}]^\beta p^\gamma x_1 x_2 x_3 p^\gamma x_2 x_3 \overline{x}_4 p^\gamma x_1 \overline{x}_3 \overline{x}_4 p^\gamma$$

Clearly, the head part of T can be covered by 2β occurrences of C with an overlap of α characters p between each occurrence for each B. Formally, the first 2β indices of the corresponding valid tiling are $\{i|B| + 1, i|B| + m - (\alpha - 1)|0 \leq i < \beta\}$. As illustrated below, and stated in the original construction, the corresponding cover of the head induces exactly N errors (enlighted in bold in the following illustration) for each copy of B, leading to an overall of $l(N + 3)N$ errors.

$$[p^\alpha y_1 y_2 \overline{x}_4 \overline{x}_3 \overline{x}_2 \mathbf{x}_1 y_3 y_4 p^\alpha y_1 y_2 \mathbf{x}_4 \mathbf{x}_3 \mathbf{x}_2 x_1 y_3 y_4 p^\alpha]^\beta$$

$$[p^\alpha y_1 y_2 \overline{x}_4 \overline{x}_3 \overline{x}_2 \mathbf{x}_1 y_3 y_4 p^\alpha y_1 y_2 \overline{\mathbf{x}}_4 \overline{\mathbf{x}}_3 \overline{\mathbf{x}}_2 x_1 y_3 y_4 p^\alpha]^\beta$$

As stated in the original proof, since the body part of T is constructed according to φ, given the assignment $\mathcal{A}$, at least one of the three literals of each clause should be present in the cover C. Therefore, it should be possible to cover T while not inducing an error for exactly one of the literals for each clause. Due to the reverse order of the literals in C compared to the one in the body part, only one literal can be a match per clause. All the other characters of C except the p ones, will be aligned to characters of T inducing errors. To do so, in the corresponding valid tiling, one will have to set precisely the overlapping of the occurrences of C using the runs of p present as prefix and suffix of C. Considering our toy instance, one would obtain the following approximate cover:

$$p^\gamma \qquad x_1 \mathbf{x}_2 \mathbf{x}_3 p^\gamma \; \mathbf{x}_2 \mathbf{x}_3 \overline{x}_4 \qquad\qquad p^\gamma \qquad x_1 \overline{\mathbf{x}}_3 \overline{\mathbf{x}}_4 \qquad p^\gamma$$

$$\overbrace{p^{\alpha_1} y_1 y_2 \overline{\mathbf{x}}_4 \overline{\mathbf{x}}_3 \overline{\mathbf{x}}_2} x_1 y_3 y_4 p^{\alpha_2} y_1 y_2 \overline{x}_4 \overbrace{\overline{\mathbf{x}}_3 \overline{\mathbf{x}}_2 x_1 y_3 y_4 p^{\alpha_3}} y_1 y_2 \overline{x}_4 \overbrace{\overline{\mathbf{x}}_3 \overline{\mathbf{x}}_2 x_1 y_3 y_4 p^\alpha}$$

Let us show that, in the current construction, this cover cannot be built. Indeed, in order for the tiling to be valid, one need that $\alpha_2 = \gamma$, $\alpha_3 + N + 1 = \gamma$, $\gamma \le \alpha_1 + N + 1 \le \gamma + \alpha$ (by construction, there are α characters p on the left of the body part coming from the head part in T) and $\alpha + N + 1 = \gamma$. While the first two conditions can be satisfied by setting $\alpha_2 = 2\alpha$ and $\alpha_3 = \gamma - N - 1$; the two last conditions cannot be satisfied in the current setting. Indeed, considering the head part, α_1 should be defined between 0 (fully overlapping to the left) and α (not overlapping at all). Then by definition, $\alpha_1 + N + 1$ is at most γ. One way to fix this first problem is to decrease the size of the first run of p in the body part to α. Indeed, since there is already another run of α characters p at the end of the header part, the third condition will hold. The last condition is harder to fulfil since the size of the last run of characters p in the body part that will be covered by C will depend on the literal satisfying the last clause. In order to fix this second problem, one has to increase the size of the last run of characters p in order to ensure that in every case, all the characters of T will be covered. In order for the construction to keep the original property regarding the number of allowed errors, one can move one occurrence of B from the head part to the end of the body part.

Let us now consider that the body part starts with a run of α characters p and ends by a run of α characters p followed by an occurrence of B that has been removed from the head part. Then, considering our toy instance, one would indeed by able to obtain the following approximate cover:

$$B\ p^\alpha \qquad x_1\mathsf{x}_2\mathsf{x}_3 p^\gamma\ \mathsf{x}_2\mathsf{x}_3\overline{x}_4 \qquad p^\gamma \qquad x_1\overline{\mathsf{x}}_3\overline{x}_4 \qquad p^\alpha\ B$$

$$\overbrace{p^{\alpha_1}\mathsf{y}_1\mathsf{y}_2\overline{\mathsf{x}}_4\overline{\mathsf{x}}_3\overline{\mathsf{x}}_2}\ x_1\mathsf{y}_3\mathsf{y}_4 p^{\alpha_2}\mathsf{y}_1\mathsf{y}_2\overline{x}_4\ \overbrace{\overline{\mathsf{x}}_3\overline{\mathsf{x}}_2 x_1\mathsf{y}_3\mathsf{y}_4 p^{\alpha_3}}\ \mathsf{y}_1\mathsf{y}_2\overline{x}_4\ \overbrace{\overline{\mathsf{x}}_3\overline{\mathsf{x}}_2 x_1\mathsf{y}_3\mathsf{y}_4 p^{\alpha_4}}$$

Depending on the literal satisfying the last clause, α_4 will be between α and $\alpha + N + 1$ which can be satisfied since B starts with a run of α characters p. On the whole, T has now the following shape:

$$B^{\beta - 1}\ p^\alpha\ L_1^1 L_2^1 L_3^1\ p^{2\alpha}\ L_1^2 L_2^2 L_3^2\ p^{2\alpha}\ \ldots p^{2\alpha}\ L_1^l L_2^l L_3^l\ p^\alpha\ B$$

The key benefit of this modification is that the original proof is not changed but is indeed valid now. $\square$

4   A Polynomial-Time Approximation Algorithm for the ACP

In this section we present a polynomial-time approximation algorithm for the *Max Similarity* variant of the **ACP**, where the goal is to maximize $|T| - h(T, T')$. Nevertheless, in order to define the algorithm and to prove its approximation ratio we first give some definitions and show some lemmas. First we introduce a new notion that measures the efficiency of a cover solution with respect to the optimal approximate cover.

Definition 1 (Cover efficiency). *Let $T \in \Sigma^*$ be a string that is the input of the* Max Similarity **ACP**. *Let C^* be the optimal approximate cover of length m and let C' be a cover produced by an approximation algorithm. Respectively, let*

T' be a string covered by C' and T^* by C^*, where T' and T^* have the same length as T. Note that C^* is optimal in the sense that the text T^* is at the smallest Hamming distance from T among all n-length texts covered by m-length covers. We define the cover efficiency of C' with respect to the optimal C^* as follows:

$$\eta = \frac{|T| - h(T, T')}{|T| - h(T, T^*)}$$

Observe that $\eta \in \,]0, 1]$, with 1 corresponding to C' being optimal and $\eta > 0$ as a cover should match at least one character of T. An algorithm for the *Max Similarity* **ACP** has an approximation ratio of $f(|T|)$ if and only if the solution produced by the algorithm is within a factor of $f(|T|)$ of an optimal solution, here $\frac{1}{\eta} \leq f(|T|)$.

Before giving an approximation algorithm, we need to compute an upper bound for the optimal solution. For the following, we denote by α the most frequent character in T and by $freq_{max}(T)$ its number of occurrences in T. Intuitively, the next observation is based on the fact that given that a cover of length m has at most m different characters, in the best case the cover matches all the occurrences of these m characters (each of them occurring at most $freq_{max}(T)$ times) and none more so.

Observation 1. *An m-length optimal approximate cover of a text T cannot match more than the m most frequent characters in T. Therefore,*

$$|T| - h(T, T^*) \leq m \cdot freq_{max}(T).$$

Next, we propose an approximation algorithm (Algorithm 1) and then, in Theorem 1, we prove it to compute in polynomial time, an approximate cover giving a $\mathcal{O}(\sqrt{|T|})$-approximation, *i.e.*, referred to as $\mathcal{O}(\sqrt{|T|})$-approximate cover.

Let us analyze Algorithm 1. First we tackle the case when $m > \lfloor |T|/3 \rfloor$.

Lemma 2. *For $m > \lfloor |T|/3 \rfloor$ each approximate cover of T, of length m, admits an optimal valid tiling of length ≤ 4. All longer tilings are necessarily redundant of a tiling of length ≤ 4, i.e. all additional occurrences of the cover are redundant as they cover positions already covered by other occurrences.*

Proof. Note that as $m > \lfloor |T|/3 \rfloor$, it results that $\lfloor |T|/m \rfloor < 3$, and so an m-length cover has at most 2 non-overlapping occurrences (that we will denote as complete), and eventually 1 overlapping occurrence (denoted as incomplete). Therefore, 3 occurrences are enough to cover the text with a m-length cover. Moreover, for the case with 2 complete occurrences, *i.e.* $\lfloor |T|/m \rfloor = 2$, one may build non-redundant 4-length tilings.

Here we show that in this case one cannot build non-redundant tilings longer than 4, whatever the value of $\lfloor |T|/m \rfloor$. For this, in the proof below we will consider a valid 5-length tiling for a cover of length m and show that it is necessarily redundant of a 4-length tiling.

Algorithm 1: ACP Approximation algorithm

Input: T a text of length n and m the length of the cover
Output: C' an approximate cover

1 **if** $m > \lfloor |T|/3 \rfloor$ **then**
2   **for** *each tiling $\mathcal{L}_k$ of length ≤ 4* **do**
3   build a cover C'_k (with $\mathcal{L}_k$ valid) and its corresp. text T'_k `// with the`
         `Histogram Greedy and the Primitivity Coercion Algos. [4]`
4   return C' covering a text T', such that $h(T, T')$ is minimal

5 **else if** $m > \sqrt{|T|}$ **then**
6   return $C' = \underbrace{\beta \ldots \beta}_{\lceil m/3 \rceil \text{ times}} T_{\lceil m/3 \rceil} \ldots T_{m - \lceil m/3 \rceil - 1} \underbrace{\beta \ldots \beta}_{\lceil m/3 \rceil \text{ times}}$ `// ` $\beta \in \Sigma$
7   and the corresp. tiling from Lemma 3

8 **else**
9   compute α `// the most frequent character in ` T
10   return $C' = \alpha^{\lfloor m/2 \rfloor} \beta \alpha^{m - \lfloor m/2 \rfloor - 1}$ or $\alpha^{\lfloor m/2 \rfloor + 1} \beta \alpha^{m - \lfloor m/2 \rfloor - 2}$ with $h(T, T')$
     minimal, and the corresp. tiling from Lemma 5

Let us take the following valid 5-length tiling $[i_0, i_1, i_2, i_3, i_4]$. Note that this tiling is valid for a cover of length m if $i_0 = 0$, $i_4 = |T| - m$, and the following inequalities stand: $i_{k+1} - i_k \leq m$, for k from 0 to 3.

Now, for the tiling to be non-redundant, we need the i_0 and the i_2 occurrences to be non-adjacent and non-overlapping. Intuitively this leaves at least one position to be exclusively covered by the i_1 occurrence. Otherwise the i_1 occurrence would necessarily be redundant.

Let us consider i_1 to be the closest possible to i_0, thus $i_1 = 1$, and i_2 to be adjacent to i_1, thus $i_2 = m + 1$. Therefore the i_2 occurrence ends on position $2m$ on T. As $m > \lfloor |T|/3 \rfloor$, the occurrence starting on i_2 will inevitably overlap the m last characters of T, thus the occurrence starting on $i_4 = |T| - m$. Therefore the occurrence starting on position i_3 is necessarily redundant of the ones on positions i_2 and i_4 (*i.e.*, the positions covered by i_3 are already covered by i_2 and i_4). It is straightforward that however we place i_1, i_2 and i_3 we still get one of the 3 occurrences as redundant. We therefore proved that a valid 5-length tiling is necessarily redundant of a 4-length tiling. $\qquad \square$

We will now consider the cases where $m \leq \lfloor |T|/3 \rfloor$, first with $m > \sqrt{|T|}$ and finally $m \leq \sqrt{|T|}$.

For the following we will denote $\lfloor \frac{|T|}{m} \rfloor$ as p and $|T| - m * p$ as r, with $r < m$. Intuitively, r corresponds to the number of characters of the part of T' that cannot be covered by p non-overlapping occurrences of the m-length cover C'. To be able to cover these remaining r characters with an additional occurrence, we need tilings allowing a total of $m - r$ overlaps. For this we will use the padding composed of characters β that borders the cover C', thus making the number of overlaps flexible.

Lemma 3. *If $m > \sqrt{|T|}$ and $m \le \lfloor |T|/3 \rfloor$ then there exists a valid tiling such that the cover $C' = \underbrace{\beta \ldots \beta}_{\lceil m/3 \rceil \ times} T_{\lceil m/3 \rceil} T_{\lceil m/3 \rceil+1} \cdots T_{m-\lceil m/3 \rceil-1} \underbrace{\beta \ldots \beta}_{\lceil m/3 \rceil \ times}$, with $|C'| = m$ and $\beta \in \Sigma$, exactly covers a text T' with $|T| = |T'|$.*

Proof. From $m \le \lfloor |T|/3 \rfloor$, it directly follows that $p \ge 3$. We will now proceed with the analysis of three different cases with respect to r and show that a valid tiling can be built for each of these cases:

1. For $2\lceil m/3 \rceil \le r < m$ there should be $0 < m - r \le \lceil m/3 \rceil$ characters overlapping and therefore the following tiling

$$[0, m, 2m, \ldots, (p-1)m, (p-1)m + r = |T| - m]$$

is valid. Indeed, the occurrence of C' starting at position $n - m$ overlaps the previous occurrence (starting at position $(p-1)m$) on exactly $m - r$ characters, meaning $\underbrace{\beta \ldots \beta}_{m-r \ times}$. This is feasible for the given cover, given that $m - r \le \lceil m/3 \rceil$.

2. In case of $\lceil m/3 \rceil \le r < 2\lceil m/3 \rceil$, the following valid tiling can be built:

$$[0, \ldots, (p-2)m, (p-2)m + 2\lceil m/3 \rceil, |T| - m].$$

Indeed, the occurrence of C' starting at position $(p-2)m + 2\lceil m/3 \rceil$ overlaps the previous one on exactly $\lceil m/3 \rceil$ positions, and together with the overlap of length $m - r - \lceil m/3 \rceil$ between the last 2 occurrences, we get the necessary overlap. As above, given that $p \ge 3$ and that $\lceil m/3 \rceil \le r < 2\lceil m/3 \rceil$, this is feasible for the given cover.

3. Given $0 < r < \lceil m/3 \rceil$, a tiling with an overlap of $m - \lceil m/3 \rceil < m - r < m$ characters is needed. In this case we build the following tiling

$$[0, \ldots, (p-3)m, (p-3)m + 2\lceil m/3 \rceil, (p-2)m + \lceil m/3 \rceil, |T| - m].$$

Similarly to the previous cases we obtain the necessary number of overlapping characters from the occurrence of C' starting at position $(p-3)m$ that overlaps the one at $(p-3)m + 2\lceil m/3 \rceil$ on $\lceil m/3 \rceil$ characters, the same for $(p-3)m + 2\lceil m/3 \rceil$ and $(p-2)m + \lceil m/3 \rceil$, and finally the last two occurrences that overlap on $m - r - 2\lceil m/3 \rceil$ characters. □

Lemma 4. *In the case where $m > \sqrt{|T|}$ and $m \le \lfloor |T|/3 \rfloor$ the cover*

$$C' = \underbrace{\beta \ldots \beta}_{\lceil m/3 \rceil \ times} T_{\lceil m/3 \rceil} T_{\lceil m/3 \rceil+1} \cdots T_{m-\lceil m/3 \rceil-1} \underbrace{\beta \ldots \beta}_{\lceil m/3 \rceil \ times}$$

with $\beta \ne T_{\lceil m/3 \rceil}, \ldots, T_{m-\lceil m/3 \rceil-1}$ gives a $3\sqrt{|T|}$-approximate cover C' of T.

Proof. Given $\beta \ne T_{\lceil m/3 \rceil}, \ldots, T_{m-\lceil m/3 \rceil-1}$, C' is clearly primitive. Moreover, C' covers at least $\lceil m/3 \rceil$ characters of T (from position $\lceil m/3 \rceil$ to $m - \lceil m/3 \rceil - 1$ in T) and since $m \ge \sqrt{|T|}$, thus $\lceil m/3 \rceil \ge \lceil \sqrt{|T|}/3 \rceil$, we have:

$$\frac{1}{\eta} = \frac{|T| - h(T, T^*)}{|T| - h(T, T')} \leq \frac{|T|}{\lceil \sqrt{|T|}/3 \rceil} = 3\sqrt{|T|}$$

□

Finally, we examine the case where $m \leq \sqrt{|T|}$.

Lemma 5. *If* $m \leq \sqrt{|T|}$ *then one of the strings* $\alpha^{\lfloor m/2 \rfloor} \beta \alpha^{m - \lfloor m/2 \rfloor - 1}$ *or* $\alpha^{\lfloor m/2 \rfloor + 1} \beta \alpha^{m - \lfloor m/2 \rfloor - 2}$, *with* α *the most frequent character in* T *and* $\beta \neq \alpha$, *gives* C' *a* $\mathcal{O}(\sqrt{|T|})$-*approximate cover of* T.

Proof. Due to space limitations and given that the proof is similar to the one in the previous case, below we give the intuition and not the complete proof.

First, observe that C' is primitive thanks to the character $\beta \neq \alpha$. Then, intuitively and similarly to the previous case, as m is small ($m \leq \sqrt{|T|}$) and given the paddings, it can be proved that we will always be able to build a valid tiling for C' (covering the remaining r characters as in Lemma 3). Now, observe that C' covers a text T' such that $|T| - h(T, T') \geq freq_{max}(T)/2$. This holds given how the cover is being built: we take the best among two possible covers composed of $m - 1$ characters α and a single β shifted by one position. Thus, all α in T are exactly matched, except the ones covered by βs. In the worst case, half of the αs in T are covered by β for the first possible cover, and half by the second. Finally, with Observation 1 we get:

$$\frac{1}{\eta} = \frac{|T| - h(T, T^*)}{|T| - h(T, T')} \leq \frac{2m \cdot freq_{max}(T)}{freq_{max}(T)} = 2m \leq 2\sqrt{|T|}$$

□

From the previous results we obtain Theorem 1 for the **ACP**.

Theorem 1. *Algorithm 1 computes a* $\mathcal{O}(\sqrt{|T|})$-*approximation for the* Max Similarity **ACP** *problem in polynomial time.*

Proof. The following proof will consider the 3 cases in Algorithm 1.

1. From Lemma 2, we have that the optimal non-redundant tiling for a cover of length $m > \lfloor |T|/3 \rfloor$ is necessarily of length less than 5. By brute force we build all the tilings of length ≤ 4, which can be done in $\mathcal{O}(|T|^2)$. Indeed as the first and the last position of the tiling are fixed (0, respectively $|T| - m$), there are at most 2 positions to fix in the tiling. For each such tiling, we look for an approximate cover for which this tiling is valid. This can be done in polynomial time with the algorithms described in [4] (the Histogram Greedy followed by the Primitivity Coercion Algorithm). Thus, in this case, we can build an optimal approximate cover in $\mathcal{O}(|T|^2(|T| + m|\Sigma|))$ time.
2. From Lemma 3 and Lemma 4 we obtain that in the case where $m \leq \lfloor |T|/3 \rfloor$, the cover

$$C' = \underbrace{\beta \ldots \beta}_{\lceil m/3 \rceil \text{ times}} T_{\lceil m/3 \rceil} T_{\lceil m/3 \rceil + 1} \cdots T_{m - \lceil m/3 \rceil - 1} \underbrace{\beta \ldots \beta}_{\lceil m/3 \rceil \text{ times}}$$

is a $\mathcal{O}(\sqrt{|T|})$-approximate cover of T. Given a tiling from Lemma 3 we can compute in linear time the text T' approximately covered by C' as well as the Hamming distance $h(T, T')$.
3. Based on Lemma 5, if $m \leq \sqrt{|T|}$, the string giving the best cover among $\alpha^{\lfloor m/2 \rfloor} \beta \alpha^{m-\lfloor m/2 \rfloor - 1}$ and $\alpha^{\lfloor m/2 \rfloor + 1} \beta \alpha^{m-\lfloor m/2 \rfloor - 2}$ is a $\mathcal{O}(\sqrt{|T|})$-approximate cover of T. Moreover, as α can be computed in linear time, then so can the **ACP** solution.   □

5   An FPT Algorithm for the ACP

In this section we present an FPT algorithm for the cover-length relaxation of the Approximate String Cover Problem. Note that, unlike the previous section, here we consider the original minimization version of the **ACP**. Recall that m is the length of the cover we are looking for. Our algorithm works as follows. We build every candidate approximate cover of length m. For each candidate cover $C \in \Sigma^*$, we determine the tiling $\mathcal{L}$ such that the text produced by the cover for this tiling, denoted by $\mathcal{L}(C)$, minimizes $h(\mathcal{L}(C), T)$. This optimal tiling given the cover C can be computed by dynamic programming as detailed below.

Given C a candidate approximate cover of T, let us define the function $D(k)$, with $0 \leq k < |T|$, as

$$D(k) = \min_{\mathcal{L} \text{ a valid tiling for } C} (h(\mathcal{L}(C)_{0,\ldots,k}, T_{0,\ldots,k})).$$

In other words, $D(k)$ is the minimum Hamming distance that can be obtained with a cover C up to the position k in T (*i.e.*, for the string $T_{0,\ldots,k}$), among all possible valid tilings covering the text up to this position. Note that such tiling may not exist in the case where the cover C (*i.e.*, its length and the way C overlaps with itself) is not compatible with the length $k+1$ of the string $T_{0,\ldots,k}$.
We have the following recurrence equation for $m - 1 \leq k < |T|$:

$$D(k) = \min_{k-m \leq k' \leq k-1} \begin{cases} D(k') + h(T_{k'+1,\ldots,k}, C) & \text{if } k' = k - m \\ D(k') + h(T_{k'+1,\ldots,k}, C_{m-k+k',\ldots,m-1}) & \text{if } C_{0,\ldots,m-k+k'-1} \\ & = C_{k-k',\ldots,m-1} \\ \infty & \text{otherwise,} \end{cases}$$

where $D(k) = \infty$ for $0 \leq k < m - 1$ and $D(-1) = 0$. Note that by convention the Hamming distance with respect to the empty string, here corresponding to $D(-1)$, is 0.

In the previous equation $D(k')$, from the window of size m preceding k, are reused in order to compute $D(k)$, provided that a tiling having $i' = k'-m+1$ and $i = k-m+1$ as consecutive indices is valid, *i.e.*, the suffix and the prefix of length $m - (k - k')$ of the cover C are equal. Therefore, computing the optimal tiling given the cover builds upon overlapping subproblems, which suits the dynamic programming framework. However, for the previous recurrence equation to hold and thus for the dynamic programming to be applicable, the problem should exhibit optimal substructure. In Lemma 6, we show that this is true.

Lemma 6. *Let us take k a position on the text with $m - 1 \le k < |T|$. Given C a candidate cover, for all positions k' with $k - m \le k' \le k - 1$, let us denote with $\mathcal{L}_{k'} = [0, \dots, i' = k' - m + 1]$ an optimal valid tiling up to position k' in T. Then the tiling $\mathcal{L}_k = [0, \dots, i'_{min} = k'_{min} - m + 1, i = k - m + 1]$ with*

$$
k'_{min} = \underset{k-m \le k' \le k-1}{\arg\min}
\begin{cases}
D(k') + h(T_{k'+1,\dots,k}, C) & \text{if } k' = k - m \\
D(k') + h(T_{k'+1,\dots,k}, C_{m-k+k',\dots,m-1})) & \text{if } C_{0,\dots,m-k+k'-1} \\
& \quad = C_{k-k',\dots,m-1} \\
\infty & \text{otherwise}
\end{cases}
$$

is optimal among tilings up to position k in T.

Proof. Here we consider the case where there exists a valid tiling up to position k in the text, otherwise the proof is straightforward. Suppose, for the purpose of contradiction, that $\mathcal{L}_k = [0, \dots, i'_{min}, i]$ is not optimal for position k. This means that there exists another valid tiling up to k, and let us consider i^* and k^* such that $\mathcal{L}_k^{opt} = [0, \dots, i^* = k^* - m + 1, i]$ is optimal among tilings ending on position k in T with $i - m \le i^* \le i - 1$ and $i^* \ne i'_{min}$.

Given that $\mathcal{L}_k^{opt}$ is a valid tiling, we therefore have $h(\mathcal{L}_k^{opt}(C), T_{0,\dots,k}) = h(\mathcal{L}_{k^*}(C), T_{0,\dots,k^*}) + h(T_{k-m+1,\dots,k}, C)$. Now $\mathcal{L}_k^{opt}$ being optimal implies that $h(\mathcal{L}_{k^*}(C), T_{0,\dots,k^*})$ is minimal up to position $k - 1$ on the text, thus contradicting the supposition that $i^* \ne i'_{min}$.   □

Theorem 2. *The **ACP** can be solved in $\mathcal{O}\left(|\Sigma|^m m^2 |T|\right)$ time with the dynamic programming strategy presented above.*

Proof. First, building every candidate approximate cover of length m can be done in $|\Sigma|^m$ time. Second, in order to compute the function D we have to iterate over all $m - 1 \le i < |T|$, and for each i we have to iterate over all i' with $i - m \le i' \le i - 1$, and for each i' a Hamming distance between two substrings of length at most m is computed. This takes $\mathcal{O}(m^2 |T|)$ time. Thus, the running time of our algorithm follows.   □

6   Conclusions and Future Work

In this paper we continue the work of Amir et al. and present several results regarding the cover length relaxation of the Approximate Cover Problem (**ACP**).

Nevertheless, several intriguing open questions remain. Perhaps the most challenging question is to provide a better approximation algorithm for the problem, or to show the hardness of the approximation result. Also, we conjecture that it is possible to find a fixed parameter algorithm only with respect to the cover length, that is, independently of the size of the alphabet.

References

1. Amir, A., Levy, A., Lewenstein, M., Lubin, R., Porat, B.: Can we recover the cover? In: 28th Annual Symposium on Combinatorial Pattern Matching, CPM 2017, Warsaw, Poland, 4–6 July 2017, pp. 25:1–25:15 (2017)
2. Amir, A., Levy, A., Lewenstein, M., Lubin, R., Porat, B.: Can we recover the cover? Algorithmica **81** (2019)
3. Amir, A., Levy, A., Lubin, R., Porat, E.: Approximate cover of strings. In: 28th Annual Symposium on Combinatorial Pattern Matching, CPM 2017, Warsaw, Poland, 4–6 July 2017, pp. 26:1–26:14 (2017)
4. Amir, A., Levy, A., Lubin, R., Porat, E.: Approximate cover of strings. Theor. Comput. Sci. **793**, 59–69 (2019)
5. Amir, A., Levy, A., Porat, E.: Quasi-periodicity under mismatch errors. In: Annual Symposium on Combinatorial Pattern Matching, CPM 2018, Qingdao, China, 2–4 July 2018, pp. 4:1–4:15 (2018)
6. Antoniou, P., Crochemore, M., Iliopoulos, C.S., Jayasekera, I., Landau, G.M.: Conservative string covering of indeterminate strings. In: Stringology, pp. 108–115 (2008)
7. Apostolico, A., Breslauer, D.: Of periods, quasiperiods, repetitions and covers. In: Mycielski, J., Rozenberg, G., Salomaa, A. (eds.) Structures in Logic and Computer Science. LNCS, vol. 1261, pp. 236–248. Springer, Heidelberg (1997). https://doi.org/10.1007/3-540-63246-8_14
8. Apostolico, A., Ehrenfeucht, A.: Efficient detection of quasiperiodicities in strings. Theor. Comput. Sci. **119**(2), 247–265 (1993)
9. Apostolico, A., Farach, M., Iliopoulos, C.S.: Optimal superprimitivity testing for strings. Inf. Process. Lett. **39**(1), 17–20 (1991)
10. Bacciotti, A., Rosier, L.: Liapunov Functions and Stability in Control Theory. Springer, Heidelberg (2006)
11. Barton, C., Kociumaka, T., Pissis, S.P., Radoszewski, J.: Efficient index for weighted sequences. In: 27th Annual Symposium on Combinatorial Pattern Matching, CPM 2016, Tel Aviv, Israel, 27–29 June 2016, pp. 4:1–4:13 (2016)
12. Breslauer, D.: An on-line string superprimitivity test. Inf. Process. Lett. **44**(6), 345–347 (1992)
13. Breslauer, D.: Testing string superprimitivity in parallel. Inf. Process. Lett. **49**(5), 235–241 (1994)
14. Brodal, G.S., Pedersen, C.N.S.: Finding maximal quasiperiodicities in strings. In: Giancarlo, R., Sankoff, D. (eds.) CPM 2000. LNCS, vol. 1848, pp. 397–411. Springer, Heidelberg (2000). https://doi.org/10.1007/3-540-45123-4_33
15. Brodzik, A.K.: Quaternionic periodicity transform: an algebraic solution to the tandem repeat detection problem. Bioinformatics **23**(6), 694–700 (2007)
16. Christodoulakis, M., Iliopoulos, C., Mouchard, L., Perdikuri, K., Tsakalidis, A., Tsichlas, K.: Computation of repetitions and regularities of biologically weighted sequences. J. Comput. Biol. **13**(6), 1214–1231 (2006)
17. Cole, R., Ilopoulos, C.S., Mohamed, M., Smyth, W.F., Yang, L.: The complexity of the minimum k-cover problem. J. Autom. Lang. Comb. **10**(5–6), 641–653 (2005)
18. Crochemore, M., Iliopoulos, C.S., Kociumaka, T., Radoszewski, J., Rytter, W., Walen, T.: Covering problems for partial words and for indeterminate strings. Theor. Comput. Sci. **698**, 25–39 (2017)
19. Flouri, T., et al.: Enhanced string covering. Theor. Comput. Sci. **506**, 102–114 (2013)

20. Guo, Q., Zhang, H., Iliopoulos, C.S.: Computing the λ-seeds of a string. In: Cheng, S.-W., Poon, C.K. (eds.) AAIM 2006. LNCS, vol. 4041, pp. 303–313. Springer, Heidelberg (2006). https://doi.org/10.1007/11775096_28
21. Guo, Q., Zhang, H., Iliopoulos, C.S.: Computing the λ-covers of a string. Inf. Sci. **177**(19), 3957–3967 (2007)
22. Iliopoulos, C.S., Moore, D.W., Park, K.: Covering a string. Algorithmica **16**(3), 288–297 (1996)
23. Katok, A., Hasselblatt, B.: Introduction to the Modern Theory of Dynamical Systems, vol. 54. Cambridge University Press, Cambridge (1997)
24. Kociumaka, T., Kubica, M., Radoszewski, J., Rytter, W., Waleń, T.: A linear-time algorithm for seeds computation. ACM Trans. Algorithms **16**(2) (2020)
25. Kociumaka, T., Pissis, S.P., Radoszewski, J., Rytter, W., Waleń, T.: Fast algorithm for partial covers in words. Algorithmica **73**(1), 217–233 (2015)
26. Kolpakov, R., Kucherov, G.: Finding approximate repetitions under Hamming distance. Theor. Comput. Sci. **303**(1), 135–156 (2003)
27. Landau, G.M., Schmidt, J.P., Sokol, D.: An algorithm for approximate tandem repeats. J. Comput. Biol. **8**(1), 1–18 (2001)
28. Li, Y., Smyth, W.F.: Computing the cover array in linear time. Algorithmica **32**(1), 95–106 (2002)
29. Ming, L., Vitányi, P.M.: Kolmogorov complexity and its applications. In: Algorithms and Complexity, pp. 187–254. Elsevier (1990)
30. Moore, D., Smyth, W.F.: An optimal algorithm to compute all the covers of a string. Inf. Process. Lett. **50**(5), 239–246 (1994)
31. Moore, D., Smyth, W.F.: A correction to "An optimal algorithm to compute all the covers of a string". Inf. Process. Lett. **54**(2), 101–103 (1995)
32. Muchnik, A., Semenov, A., Ushakov, M.: Almost periodic sequences. Theor. Comput. Sci. **304**(1–3), 1–33 (2003)
33. Sethares, W.A., Staley, T.W.: Periodicity transforms. IEEE Trans. Signal Process. **47**(11), 2953–2964 (1999)
34. Timmermans, M., Heijmans, R., Daniels, H.: Cyclical patterns in risk indicators based on financial market infrastructure transaction data. De Nederlandsche Bank Working Paper (558) (2017)

Data Structures for SMEM-Finding in the PBWT

Paola Bonizzoni[1]([✉])[ID], Christina Boucher[2][ID], Davide Cozzi[1][ID],
Travis Gagie[3][ID], Dominik Köppl[4][ID], and Massimiliano Rossi[2][ID]

[1] University of Milano-Bicocca, Milano, Italy
paola.bonizzoni@unimib.it, d.cozzi@campus.unimib.it
[2] University of Florida, Gainesville, FL, USA
{christinaboucher,rossi.m}@ufl.edu
[3] Dalhousie University, Halifax, NS, Canada
travis.gagie@dal.ca
[4] University of Muenster, Muenster, Germany
dominik.koeppl@uni-muenster.de

Abstract. The positional Burrows–Wheeler Transform (PBWT) was presented as a means to find set-maximal exact matches (SMEMs) in haplotype data via the computation of the divergence array. Although run-length encoding the PBWT has been previously considered, storing the divergence array along with the PBWT in a compressed manner has not been as rigorously studied. We define two queries that can be used in combination to compute SMEMs, allowing us to define smaller data structures that support one or both of these queries. We combine these data structures, enabling the PBWT and the divergence array to be stored in a manner that allows for finding SMEMs. We estimate and compare the memory usage of these data structures, leading to one data structure that is most memory efficient. Lastly, we implement this data structure and compare its performance to prior methods using various datasets taken from the 1000 Genomes Project data.

1 Introduction

The positional Burrows–Wheeler Transform (PBWT) was defined by Durbin [5] as a means for analyzing haplotype datasets. Hence, the input consists of h sequences $S = \{S_1, \ldots, S_h\}$ containing w biallelic sites corresponding to the same loci. The main idea is that specific loci are sequenced and it is determined if the position contains the major allele (denoted as 1) or has an alternate allele (denoted as 0). We will represent S as a $h \times w$ binary matrix that is denoted as M. The PBWT of M is another $h \times w$ binary matrix, denoted as $\mathsf{PBWT}[1..h][1..w]$, where the first column is the same as the first column of M, and the j-th column of PBWT is obtained by stably sorting the rows of $\mathsf{M}[1..h][1..j-1]$ in co-lexicographic order (starting at column $j-1$) and then taking the final column of the result. We can define this more formally by first defining the *j-th prefix array* (PA_j), which is the co-lexicographic ordering of the prefixes $S_1[1..j], \ldots, S_h[1..j]$.

F. M. Nardini et al. (Eds.): SPIRE 2023, LNCS 14240, pp. 89–101, 2023.
https://doi.org/10.1007/978-3-031-43980-3_8

It follows that an equivalent definition of the PBWT is $\mathsf{col}(\mathsf{PBWT})_1 = \mathsf{col}(\mathsf{M})_1$ and $\mathsf{col}(\mathsf{PBWT})_j[i] = \mathsf{col}(\mathsf{M})_j[\mathsf{PA}_{j-1}[i]]$ for all $i = 1..h$ and $j = 2..w$, where $\mathsf{col}(\mathsf{PBWT})_j$ denotes the j-th column of the PBWT. As noted by Durbin, the PBWT is highly run-length compressible [5] when read in column-major order.

One of the main motivations for the invention of the PBWT is that it enables set-maximal exact matches (SMEMs) to be found efficiently in haplotype data. Given the input sequences $S = \{S_1, \ldots, S_h\}$ and a pattern $P[1..w]$, we define $P[i..j]$, where $1 \leq i \leq j \leq w$, to be a SMEM if it occurs in one of the sequences in S and one of the following holds: i) $i = 1$ and $j = w$; ii) $i = 1$ and $P[1..j+1]$ does not occur in S; iii) $j = w$ and $P[i-1..w]$ does not occur in S; iv) $P[i-1..j]$ and $P[i..j+1]$ does not occur in S. Given the PBWT, a pattern $P[1..m]$, and a starting column ℓ, the PBWT allows us to efficiently find all the sequences in S that contain P between columns ℓ and $\ell + m - 1$. If there are no such sequences then all the sequences that contain $P[1..m']$ between columns ℓ and $\ell + m' - 1$, where $P[1..m']$ is the longest prefix of P that occurs in S, are returned. Durbin demonstrated that SMEMs can be identified in $\mathcal{O}(hw)$-time via the computation of the *divergence arrays*. Here, the j-th *divergence array* (DA) stores, for each $i > 0$, the length of the longest common suffix between the i-th and $(i-1)$-st rows of M when the rows of M are sorted according to the co-lexicographic order of their prefixes up to the j-th column.

Although Durbin (and others, i.e., Li [10]) showed that run-length compressing PBWT leads to significant space improvement, there are only a few methods for storing the divergence array in a compressed manner. Cozzi et al. [3] provided one such approach that is based on casting the problem of finding SMEMs to computing matching statistics, and showing that computing matching statistics can be accomplished via building a data structure that mirrors that of Rossi et al. [13]. However, it is largely open how to store the PBWT alongside the auxiliary data structures needed to efficiently find SMEMs. In this paper, we generalize the approach of Cozzi et al. [3] by describing two queries (start and extend) that can be used in combination to find SMEMs in the PBWT, and address the prevailing gap in the literature by providing a comprehensive list of smaller data structures that can be used to efficiently support start and/or extend. We show that these data structures can be combined in various ways to create data structures that store the PBWT in a manner that supports SMEM-finding.

We study the theoretical bounds of each data structure, and benchmark their memory consumption under a practical setting. This benchmarking leads to a solution that is deemed most performant. We fully implement this approach and compare it to the methods of Cozzi et al. [3] and Durbin [5] by building the data structure on increasingly larger haplotype datasets from the 1,000 genomes project data. We compare both the construction time and space, and the time and space to find SMEMs, allowing us to conclude about the practicality of the methods.

2   Preliminaries

String Definitions. We assume that all input strings are binary strings since our application is biallelic haplotype data. Given a binary character b, we denote the negation of b as $\neg b = 1 - b$. We denote the empty string as ε. We denote the length of S by $|S|$. We denote the i-th prefix of S as $S[1..i]$, the i-th suffix as $S[i..n]$, and the substring spanning position i through j as $S[i..j]$, with $S[i..j] = \varepsilon$ if $i > j$. Given two strings S and T, we say that S is lexicographically smaller than T if either S is a proper prefix of T or there exists an index $i \geq 1$ such that $S[1..i] = T[1..i]$ and $S[i+1]$ occurs before $T[i+1]$. Lastly, for a string S, a binary character c, and an integer j, we define rank query $S.\mathsf{rank}_c(j)$ as the number of occurrences of c in $S[1..j]$, and the select query $S.\mathsf{select}_c(j)$ as the position of the j-th c in S.

RMQ, PSV, and NSV. Given an array $\mathsf{A}[1..n]$ of integers, a range minimum query (RMQ) for two positions $i \leq j$ asks for the position k of the minimum in $\mathsf{A}[i..j]$, i.e., $k = \operatorname{argmin}_{k' \in [i..j]} A[k']$. We denote this query by $\mathrm{RMQ}_\mathsf{A}(i,j)$. Given a position i in A, we define the previous-smaller-value (PSV) as $\mathrm{PSV}_\mathsf{A}(i) = \max(\{j : j < i, \mathsf{A}[j] < \mathsf{A}[i]\} \cup \{0\})$. We define the next-smaller-value (NSV) as $NSV_\mathsf{A}(i) = \min(\{j : j > i, \mathsf{A}[j] < \mathsf{A}[i]\} \cup \{n + 1\})$.

LCP and LCE. Given two strings S and T, we denote the length of the longest common prefix between S and T as $\mathsf{lcp}(S,T)$. Using this notation, we define the longest common prefix array of an input string S of length n (given its Suffix Array SA_S) as $\mathrm{LCP}[1..n]$ where $\mathrm{LCP}[i] = \mathsf{lcp}(S[\mathsf{SA}_S[i]..n], S[\mathsf{SA}_S[i-1]..n])$ for all $i > 1$, and $\mathrm{LCP}[1] = 0$. Given an input string S of length n and two integers $1 \leq i \leq n$ and $1 \leq j \leq n$, the Longest Common Extension (LCE) is the longest substring of S that starts at both i and j. Moreover, as we will discuss in this work, there are multiple data structures that efficiently support LCE queries for a string S.

SLPs. The concept of straight-line programs (SLPs) will be used in our work. An SLP is a representation of the input as a context-free grammar whose language is precisely the input string [11].

Matching Statistics in the PBWT. Cozzi et al. showed that the problem of finding SMEMs in the PBWT can be cast into computing matching statistics for P, which is a generalization of Bannai et al. [1]. Given a pattern $P[1..w]$, the matching statistics of P with respect to S is an array $A[1..w]$ of (row, len) pairs such that for each position $1 \leq j \leq w$: (1) $S_{A[j].\mathsf{row}}[j - A[j].\mathsf{len} + 1..j] = P[j - A[j].\mathsf{len}+1..j]$, and (2) $P[j - A[i].\mathsf{len}..j]$ is not a suffix of $S_1[1..j], \ldots, S_h[1..j]$. Informally, for each position j of the pattern P, $A[j].\mathsf{row}$ is one row of the input matrix M where a longest shared common suffix (of length $A[j].\mathsf{len}$) ending in position j in the pattern P and in $S_{A[i].\mathsf{row}}$ occurs.

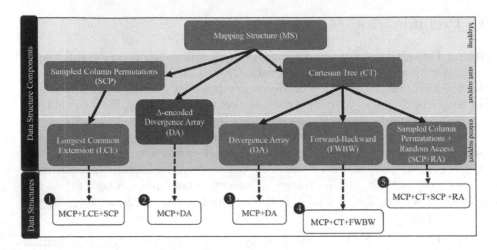

Fig. 1. An illustration of our components and data structures. The components are shown in colored boxes, and the data structures are shown in white boxes at the bottom.

3 Building Blocks for a PBWT Data Structure

In this section, we begin by defining two queries, called (start and extend), that are used to compute matching statistics in the PBWT. Then we define the smaller data structures, which we call components, that support start or extend—and in one case (i.e., the Δ-encoded divergence array), can support both start and extend. These components are used to build data structures for SMEM finding in the PBWT. We show both the components and data structures in Fig. 1. We note that we will frequently use $n = h \cdot w$ throughout this section to bound the time and space.

3.1 Queries Needed to Support SMEM-Finding

We define two queries, referred to as start and extend, which can be used in combination to compute matching statistics, and hence, find SMEMs. If we let $i, j \in [1..w]$ be two integers such that $P[i..j]$ is a suffix of one of $S_1[1..j], \ldots, S_h[1..j]$ then the extend query returns that there exists the match of $P[i..j]$ to $P[i..j + 1]$ if and only if $j < w$ and $P[i..j + 1]$ is a suffix of one of $S_1[1..j + 1], \ldots, S_h[1..j + 1]$. The start query finds the smallest integer $i' \in [i..j]$ such that $P[i'..j]$ is a suffix of one of $S_1[1..j], \ldots, S_h[1..j]$. Hence, we compute the matching statistics as follows. We assume that we computed the matching statistics up to position $i \in [1..w]$, and use the start query to find the smallest $i' \in [i..w]$ such that $P[i'..i' + A[i].\mathsf{len}]$ is a suffix of one of $S_1[1..i' + A[i].\mathsf{len}], \ldots, S_h[1..i' + A[i].\mathsf{len}]$. By minimality of i', we can set $A[j].\mathsf{len} = A[j - 1].\mathsf{len} - 1$ for all $j \in [i + 1..i' - 1]$. Then we find the longest prefix $P[i'..k]$ that is also a suffix of one of $S_1[1..k], \ldots, S_h[1..k]$ using the extend query. We set $A[i'].\mathsf{len} = k - i' + 1$. Since $i' > i$, we can proceed by induction to compute the whole array of matching statistics.

3.2 Top Level: Mapping Structure

All the data structures that we present require a component data structure, which we call a mapping structure, that when given the position in PBWT of a bit PBWT$[i][j]$ can return:

- the position in col(PBWT)$_{j+1}$ of the bit immediately to the right of PBWT$[i][j]$ in M;
- the position in col(PBWT)$_j$ of the last occurrence of ¬PBWT$[i][j]$ above PBWT$[i][j]$ (if it exists);
- the position in col(PBWT)$_j$ of the first occurrence of ¬PBWT$[i][j]$ below PBWT$[i][j]$ (if it exists).

The first query corresponds to LF-mapping in the BWT and allows us to step from one column to the next one (to the right) in the PBWT, staying in the same row in M. The second and third queries correspond to how Rossi et al. [13] jump up or down, respectively, in the BWT when they find a mismatch. We implemented the mapping structure as a run-length compressed bitvector, occupying roughly $\mathcal{O}(r \log(n/r))$ bits and answering queries in $\mathcal{O}(\log\log n)$-time, where r is the total number of runs in the columns of the PBWT.

3.3 Second Level: Start Support

In this subsection, we provide a comprehensive discussion of all the data structures that support **start** queries.

Sampled Column Permutations. If we use a Cartesian tree but neither the divergence array itself (encoded or unencoded) nor two instances of each component data structure, then it seems we need a way to find at least one occurrence of each SMEM in order to determine its length. We can use the sampled column permutations together with an analogue of Policriti and Prezza's [12] Toehold Lemma: for the bits at either end of each run in a column in the PBWT, we store which rows in the input matrix they came from, using a total of roughly $2r \lg h$ bits; whenever we reach the right end of a SMEM and expand our search interval, the expanded interval must contain the first or last bit in some run of the bits we seek, and we learn from which row of the input matrix it came.

Cartesian Trees. If we store a representation of the shape of a Cartesian tree built upon the divergence array, then we can support RMQ, PSV and NSV queries on the divergence array. Yet, we note that these queries return only a position, and cannot easily support random access. We consider three representations of the tree shape: (1) an augmented balanced-parentheses (BP) representation occupying roughly $2n+o(n)$ bits [6] and answering queries in constant time; (2) a simple DAG-compressed representation (with each non-terminal storing the size of its expansion) answering queries in time bounded by its height; and (3) an interval-tree storing selected intervals corresponding to nodes in the Cartesian tree and answering queries in constant time. We do not include Gawrychowski et al.'s compressed RMQ data structure [8] because we are not aware of an implementation and we see no easy way to estimate its space usage for the divergence array.

The constructions of the BP representation and DAG-compressed representations are standard, but our interval-tree structure needs some explanation. We query the Cartesian tree only while forward stepping through the PBWT and when our search interval contains only 0's and we want a 1, or when it contains only 1's and we want a 0. To proceed, we must ascend the tree and widen our interval (like ascending a prefix tree, discarding the early bits of our pattern) until it contains a copy of the bit we want. Notice our query interval corresponds to a node v in the Cartesian tree, and the PBWT interval we seek corresponds to the lowest ancestor u of v whose interval is not unary. It follows that we need to store in our interval-tree only the PBWT intervals for nodes u in the Cartesian tree such that the interval for at least one of u's children is unary but u's interval is not unary.

Since our intervals can nest but not otherwise overlap, we can store our interval-tree in a more space-efficient manner than usual: we write out a string with open-parens, close-parens and 0s, with each open-close pair indicating an interval and the number of 0s before, between and after them indicating its starting point, length and ending point; we encode that string as one bitvector with 0s indicating 0s and 1s indicating parens, and another bitvector with 0s indicating open-parens and 1s indicating close-parens (so the combination of the bitvectors is a wavelet tree for the string); and we store a BP representation of the tree structure of the stored intervals. If we store k intervals, then our first bitvector has $n + 2k$ bits and $2k$ copies of 1, our second bitvector has $2k$ bits with k copies of 0 and k copies of 1, and the tree structure has k nodes and so its BP representation takes $2k + o(k)$ bits. This means we use roughly $2k \lg \frac{n+2k}{2k} + 2k + 2k + o(k) = 2k \lg(n/k) + o(k \lg(n/k))$ bits, and can answer queries in constant time. We note that, since even our query intervals can nest but not contain or otherwise overlap any of our stored intervals, we can query with a single endpoint instead of a whole interval.

3.4   Third Level: Extend Support

In this subsection, we discuss all components that can support extend queries.

Divergence Array. The simplest possible data structure to support finding the length of each SMEM is to store the uncompressed divergence array, which was proposed by Durbin. The shortcoming of this is the large space requirements—as it would occupy space in bits roughly equal to the sum of the base-2 logarithms of all entries (with 2 added to each entry).

Longest Common Extension. We consider the addition of an LCE data structure. Suppose we have arrived at column $j+1$ and we know that the longest suffix of pattern $P[1..j]$ that occurs in M ending at column j has an occurrence immediately followed by $\mathsf{PBWT}[i][j + 1]$, and we know which row of M that bit $\mathsf{PBWT}[i][j + 1]$ comes from. If $P[j + 1] = \mathsf{PBWT}[i][j + 1]$ then the longest suffix of $P[1..j + 1]$ that occurs in M ending at column $j + 1$ has an occurrence ending with $\mathsf{PBWT}[i][j + 1]$. Therefore, we assume $P[j + 1] \neq \mathsf{PBWT}[i][j + 1]$. By the definition of the PBWT, there is an occurrence of the longest suffix of $P[1..j + 1]$ that occurs in M ending at column $j + 1$, ending either at the last occurrence of $P[j + 1] = \neg\mathsf{PBWT}[i][j + 1]$ above $\mathsf{PBWT}[i][j + 1]$ (if it exists), or at the first occurrence of that bit below $\mathsf{PBWT}[i][j + 1]$ (if it exists). We recall that the mapping structure allows us to quickly find these occurrences of that bit.

Forward-Backward. Suppose we use a Cartesian tree to maintain the invariant that our search interval in the PBWT contains all the bits immediately following occurrences of the longest suffix of the prefix of the pattern that we have processed so far, that occur in the desired columns of the PBWT. If that search interval contains a copy of the next bit of the pattern, then we proceed by forward stepping, without consulting the Cartesisan trees. The only time we query the Cartesian trees is when the search interval does not contain a copy of the next bit of the pattern, meaning we have reached the right end of a SMEM. It follows that, using the mapping structure and the Cartesian trees, we can find the right endpoints of all the SMEMs. If we keep instances of all our components for the reversed input matrix, we can also find all the left endpoints of the SMEMs. Since SMEMs are maximal, they cannot nest, so we can easily pair up the endpoints and obtain the SMEMs. This doubles the time and space.

Random Access. Lastly, the simplest possible component is a compressed data structure of the original input that provides efficient random access to the input, which can obviously be used to find the length of a given SMEM. Although the total length of the SMEMs can be quadratic in the length of the pattern, the fact they cannot nest implies we need only a linear number of random accesses. In fact, if we combine a random access data structure with Cartesian trees then the number of random accesses is equal to the number of SMEMs, and the total length of the sequence that we extract from M is linear in the length of the SMEMs. There are many data structures that support random access to the input matrix M, two notable ones are (a) an SLP of M (read row-wise) answering queries in $\mathcal{O}(\log n)$ time, and (b) a plain representation of M (with 8 bits packed into each byte) occupying roughly n bits and allowing access to each bit in constant time.

3.5 Δ-Encoded Divergence Array

The last component we discuss is Δ-encoded divergence array. As illustrated in Fig. 1, we leave this component last since it can support both queries. To differentially encode (Δ-encode) the divergence array, we store each entry of $\mathsf{DA}[i][j]$ with $i > 1$ as the difference $\mathsf{DA}[i][j] - \mathsf{DA}[i - 1][j]$; for $i = 0$ we always have

$\mathsf{DA}[i][j] = 0$. If the PBWT is highly run-length compressible, read in column-major order, then the Δ-encoded DA is small. To see why, consider that if $\mathsf{PBWT}[i..i + \ell - 1][j]$ is a run of equal bits in the j-th column of the PBWT and $\mathsf{col}(\mathsf{PBWT})_{j+1}[i'..i' + \ell - 1]$ are the bits immediately to their right in the input matrix, then $\mathsf{DA}[i' + k][j + 1] = \mathsf{DA}[i + k][j] + 1$ for $1 \leq k \leq \ell - 1$. Therefore,

$$\mathsf{DA}[i' + k][j + 1] - \mathsf{DA}[i' + k - 1][j + 1] = \mathsf{DA}[i + k][j] - \mathsf{DA}[i + k - 1][j]$$

for $1 \leq k \leq \ell - 1$, so the Δ-encoded of $\mathsf{DA}[i'+1..i'+\ell-1][j+1]$ is the same as that of $\mathsf{DA}[i + 1..i + \ell - 1][j]$. It follows that, if there are r runs in the columns of the PBWT, then the (linearized) Δ-encoded DA has a string attractor of size $\mathcal{O}(r)$ and, thus, it can be represented as a straight-line program occupying $\mathcal{O}(r \log^2 n)$ bits [9, Lemma 3.14].

Increasing the size of this SLP by a small constant factor, we can store at each non-terminal the length, sum, and minimum prefix sum of its expansion, and thus support random access, RMQ, PSV, and NSV queries on the divergence array in $\mathcal{O}(\log n)$-time. This is similar to how Gagie et al. [7, Lemma 6.2] used an SLP for their Δ-encoded LCP array.

4   Composite Data Structures for the PBWT

We already described two data structures that efficiently support finding SMEMs in the previous section, namely, the "Mapping Structure + Cartesian Tree + Forward-backward" and "Mapping Structure + Cartesian Tree + Sample Column Permutations + Random Access". There are three other data structures that will be evaluated (Table 1), namely: (1) Mapping Structure + Δ-Encoded Divergence Array; (2) Mapping Structure + Cartesian Tree + Divergence Array; and (3) Mapping Structure + LCE + Sampled Column permutations.

5   Experiments and Discussion

In this section, we provide experimental evaluations of our presented data structures. We begin by benchmarking the memory usage of our data structures. Based on these experiments, we fully implement one of these data structures and show the scalability of them on real data.

5.1   Comparison of Data Structures

Experimental Set-Up. We replicate the simulated dataset used by Durbin [5]. In particular, we run the Markovian coalescent simulator MaCS [2] with command line parameters `100000 2e7 -t 0.001 -r 0.001` to generate a haplotype matrix with 100,000 individuals and 360,000 sites. Next, we subsample the dataset with a parameter ξ such that, given a column of length h having o ones, we skip this column if $o/h < \xi$. We set ξ to be equal to 0.01, 0.03, 0.05,

Table 1. The estimated size in bits of the combinations of component data structures that support SMEM-finding in the PBWT. M denotes megabytes and G denotes gigabytes. In boldface the best performance. We do not list Forward-Backward here as it is not a new data structure, only two instances of the Mapping Structure and the Cartesian Tree.

Component	Sample Parameter ξ				
	0.01	0.03	0.05	0.08	0.10
Mapping structure	57M	53M	52M	51M	51M
Δ-encoded divergence array	479M	452M	435M	426M	418M
Cartesian tree	472M	472M	458M	420M	402M
Longest common extension	96M	88M	88M	80M	80M
Sampled column permutations	80M	76M	76M	76M	76M
Divergence array	125G	92G	77G	64G	58G
Random access	96M	88M	88M	80M	80M
Data Structure					
MAP + LCE + PERM	**233M**	**217M**	**216M**	**207M**	**207M**
MAP + DEDA	536M	505M	487M	477M	469M
MAP + CT + FWBW	1.1G	1.1G	1.0G	942M	906M
MAP + CT + DA	126G	93G	78G	64G	58G
MAP + CT + PERM + RA	705M	689M	674M	627M	609M

0.08, and 0.10, which results in the datasets having varying degrees of repetitiveness. See Table 1 for the size of the datasets. The haplotype matrix is publicly available at http://dolomit.cs.tu-dortmund.de/tudocomp/pbwt_matrix.xz. We ran all benchmarks on an Intel Core i3-9100 CPU (3.60 GHz) with 128 GB RAM, running Debian 11.

Implementation. We implemented all methods in C/C++. The mapping structure was implemented using sparse bitvectors with a number of set bits equal to the number of runs. The differentially-encoded divergence array and the Cartesian tree were implemented with grammar compression. Forward and backward was implemented by building the data structures in both the forward and backward directions of the mapping structure. The sampled column permutations were obtained by sampling at run boundaries. The LCE data structure and the random access were implemented with as an SLP that answers LCE queries. All data structures are publicly available at https://github.com/koeppl/pbwt.

Results. We give the estimated sizes of the data structures that are compositions of these components in Table 1. We witness that MAP+LCE+PERM was the most performant, which was followed by MAP+DEDA. The performance of MAP+DEDA was somewhat surprising since it is similar to a structure suggested by Gagie et al. [7] that was but not implemented because it was thought to be impractical. We note that as ξ increases the size of all the components and data structures decreases—this is intuitive since the datasets become less repetitive, resulting in fewer columns being selected. Hence, we see that the compression suffers for all the methods as ξ increases but MAP+LCE+PERM maintains a lead.

5.2   Comparison of Methods on 1000 Genomes Project Data

Experimental Set-Up. We implement and evaluate MAP+LCE+PERM on the 1000 Genomes Project data by downloading the VCF files for the 1000 genomes project data and then converting these files to biallelic using bcftools view -m2 -M2 -v snps [4]. We consider increasingly larger datasets by selecting the panels for Chromosomes 22, 20, 18, 16 and 1 which have 5008 samples and a number of biallelic sites that range from ∼1 million to ∼6 millions. All datasets correspond to 4,908 individuals. All data are available at https://ftp.1000genomes. ebi.ac.uk/vol1/ftp/release/20130502/. We ran all experiments in this subsection on a machine with an Intel Xeon CPU E5-2640 v4 (2.40GHz) with 756 GB RAM and 768 GB of swap, running Ubuntu 20.04.4 LTS.

Competing Methods. We compared against the PBWT implementation of Durbin [5], which are available at https://github.com/richarddurbin/pbwt. In detail, we ran both the matchIndexed and matchDynamic algorithms. We refer to these methods as PBWT-index and PBWT-dynamic, respectively. In addition, we compared against the methods of Cozzi et al. [3], which implements a mapping structure with sampled column permutations with the Thresholds data structure of Rossi et al. [13]. The method is referred to as μ-PBWT. The computation of the matching statistics is analogous to Rossi et al. with one slight modification: the inverse of the mapping function is used to compute the lengths of the matching statistics. We refer to Cozzi et al. [3] for these details.

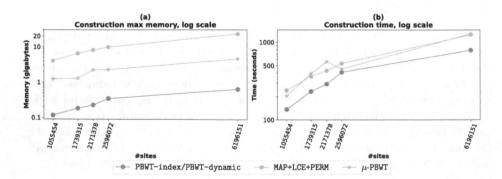

Fig. 2. Memory (a) and time (b) to construct the data structures underlying all methods for increasingly larger number of biallelic sites. Memory is reported in GB and time is reported in seconds.

Results. We give the maximum memory usage and time for constructing all the data structures in Fig. 2. We note that the construction for PBWT-index and PBWT-dynamic is the same so it is reported once (as PBWT) in Fig. 2, and that the constructed data structure is incomplete, meaning that additional indexes are needed for SMEM finding. This explains why the memory required for construction of the PBWT is small. We see that the method of μ-PBWT requires more memory than the PBWT but less memory than MAP+LCE+PERM. In terms of construction time, there was negligible difference between the performance of MAP+LCE+PERM and μ-PBWT. PBWT had the most efficient construction time.

Next, we evaluated the performance of SMEMs-finding by first extracting 100 sequences from the input panels to be used query sequences. We illustrate the memory usage and the time required for SMEM-finding when all the query strings were given as input at once, which is shown in Fig. 3(a) and (b). Figure 3(c) shows mean of the time required when each is given as an individual query, i.e., executing 100 queries one at a time. The peak memory usage to query all the sequences at once surpasses that of querying them individually. We obtained the following average number of SMEMs per 100 queries: 1,184 SMEMs for chromosome 22 (1,055,454 sites), 1,416 SMEMs for chromosome 20 (1,739,315 sites), 1,708 SMEMs for chromosome 18 (2,171,378 sites), 2,281 SMEMs chromosome 16 (2,596,072 sites) and 4,953 SMEMs for chromosome 1 (6,196,151 sites). PBWT-dynamic used the least memory when querying the whole set of queries but had opposing behavior doing one query at a time. It was fastest when the queries were given at once but slowest when the queries were given individually. Opposingly, PBWT-indexed required more memory than all other competing methods, requiring up to 20 times more memory. PBWT-indexed was the second slowest method when the queries were given at once but fastest when the methods were given individually. We see that MAP+LCE+PERM used less memory than PBWT-MatchIndexed and was at most 10 times slower than PBWT-MatchIndexed when the queries were given individually but was slightly slower than PBWT-MatchIndexed when the queries were given at all once. In addition, MAP+LCE+PERM was faster than PBWT-MatchDynamic when the queries were at once but slower than PBWT-MatchDynamic when queries were given individually. With respect to μ-PBWT, MAP+LCE+PERM used slightly more memory and query time than μ-PBWT. We note that MAP+LCE+PERM also has the advantage not requiring two passes on the query string, which makes it appropriate for online settings when the SMEMs can be found as the input is read in.

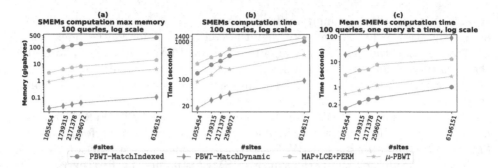

Fig. 3. Memory (a), time (b) and mean time for one query at a time (c) to compute SMEMs with 100 queries. In (c) the standard deviation values are very small so the corresponding error bars are omitted.

6   Conclusions

We presented and benchmarked a number of data structures that support SMEM-finding in the PBWT. Our experiments revealed that MAP+LCE+PERM was the most memory-efficient out of all data structures we presented. After fully implementing it, we showed that it is slightly slower and uses more memory than the method of Cozzi et al. [3]; however, we note that it has the advantage that it only requires one-pass over the query string, making it appropriate for the calculation of SMEMs in an online format.

Acknowledgements. TR, MR, and CB were supported by NIH/NHGRI No. R01HG011392, and NSF IIBR No. 2029552. TR was supported NSERC No. RGPIN-07185-2020. CB was supported by NSF SCH No. 2013998. DK was supported by JSPS KAKENHI with No. JP21K17701 and JP23H04378. PB and DC were supported by Horizon 2020 with No. 872539.

References

1. Bannai, H., Gagie, T., Tomohiro, I.: Refining the R-index. Theor. Comput. Sci. **812**, 96–108 (2020)
2. Chen, G.K., Marjoram, P., Wall, J.D.: Fast and flexible simulation of DNA sequence data. Genome Res. **19**(1), 136–142 (2009)
3. Cozzi, D., Rossi, M., Rubinacci, S., Köppl, D., Boucher, C., Bonizzoni, P.: μ-PBWT: enabling the storage and use of UK biobank data on a commodity laptop. bioRxiv, pp. 2023–02 (2023)
4. Danecek, P., et al.: Twelve years of SAMtools and BCFtools. GigaScience **10**(2) (2021)
5. Durbin, R.: Efficient haplotype matching and storage using the positional Burrows-Wheeler transform (PBWT). Bioinformatics **30**(9), 1266–1272 (2014)
6. Fischer, J., Heun, V.: Space-efficient preprocessing schemes for range minimum queries on static arrays. SIAM J. Comput. **40**(2), 465–492 (2011)

7. Gagie, T., Navarro, G., Prezza, N.: Fully functional suffix trees and optimal text searching in BWT-runs bounded space. J. ACM **67**(1), 2:1–2:54 (2020)
8. Gawrychowski, P., Jo, S., Mozes, S., Weimann, O.: Compressed range minimum queries. Theor. Comput. Sci. **812**, 39–48 (2020)
9. Kempa, D., Prezza, N.: At the roots of dictionary compression: string attractors. In: Proceedings of ACM SIGACT Symposium on Theory of Computing (STOC), pp. 827–840 (2018)
10. Li, H.: BGT: efficient and flexible genotype query across many samples. Bioinformatics **32**(4), 590–592 (2016)
11. Lohrey, M.: Algorithmics on SLP-compressed strings: a survey. Groups - Complex. - Cryptol. **4**(2), 241–299 (2012)
12. Policriti, A., Prezza, N.: LZ77 computation based on the run-length encoded BWT. Algorithmica **80**(7), 1986–2011 (2018)
13. Rossi, M., Oliva, M., Langmead, B., Gagie, T., Boucher, C.: MONI: a pangenomic index for finding maximal exact matches. J. Comput. Biol. **29**(2), 169–187 (2022)

Compressibility Measures
for Two-Dimensional Data

Lorenzo Carfagna$^{(\boxtimes)}$ and Giovanni Manzini

University of Pisa, Pisa, Italy
lorenzo.carfagna@gmail.com, giovanni.manzini@unipi.it

Abstract. In this paper we extend to two-dimensional data two recently introduced one-dimensional compressibility measures: the γ measure defined in terms of the smallest string attractor, and the δ measure defined in terms of the number of distinct substrings of the input string. Concretely, we introduce the two-dimensional measures γ_{2D} and δ_{2D} as natural generalizations of γ and δ and study some of their properties. Among other things, we prove that δ_{2D} is monotone and can be computed in linear time, and we show that although it is still true that $\delta_{2D} \leq \gamma_{2D}$ the gap between the two measures can be $\Omega(\sqrt{n})$ for families of $n \times n$ matrices and therefore asymptotically larger than the gap in one-dimension. Finally, we use the measures γ_{2D} and δ_{2D} to provide the first analysis of the space usage of the two-dimensional block tree introduced in [Brisaboa *et al.*, Two-dimensional block trees, *The computer Journal*, 2023].

Keywords: Data compression · Repetitivenes Measures · Block Tree

1 Introduction

Since the recent introduction of the notion of string attractor [6] different measures of string repetitiveness have been proposed or revisited [8,10]. It has been shown that such measures are more appropriate than the classical statistical entropy for measuring the compressibility of highly repetitive strings. In addition, these measures have been used to devise efficient compressed indices for highly repetitive string collections [11] an important setting which is hard for traditional entropy-based compressed indices.

In this paper we generalize the notion of attractor to two dimensional data, i.e. (square) matrices of symbols, and we initiate the study of the properties of the measure $\gamma_{2D}(M)$ defined as the size of the smallest attractor for the matrix M (Definition 1). As in the one-dimensional case, we introduce also the measure $\delta_{2D}(M)$ defined in terms of the number of distinct square submatrices (Definition 2) and we study the relationship between γ_{2D} and δ_{2D}. We prove that some properties that hold for strings are still valid in the two-dimensional case: for example computing γ_{2D} is NP-complete while δ_{2D} can be computed in linear time, and for every matrix M it is $\delta_{2D}(M) \leq \gamma_{2D}(M)$. However, the

F. M. Nardini et al. (Eds.): SPIRE 2023, LNCS 14240, pp. 102–113, 2023.
https://doi.org/10.1007/978-3-031-43980-3_9

gap between the two measures is larger than in one-dimension case since there are families of $n \times n$ matrices with $\delta_{2D} = O(1)$ and $\gamma_{2D} = \Omega(\sqrt{n})$, whereas for strings it is always $\gamma = O(\delta \log \frac{n}{\delta})$.

The study of the measures γ_{2D} and δ_{2D} is motivated by the fact that for two-dimensional data there is no clear definition of "context" of a symbol and therefore there is no universally accepted notion of statistical entropy. Therefore, alternative compressibility measures based on combinatorial properties such as γ_{2D} and δ_{2D} are worthwhile investigating. Indeed, in Sect. 3 we use the measures γ_{2D} and δ_{2D} to provide the first analysis of the size of the two-dimensional block tree introduced in [2]. In particular we show that the space used by a two-dimensional block tree for an $n \times n$ matrix M with delta measure δ_{2D} is bounded by $O((\delta_{2D} + \sqrt{n\delta_{2D}}) \log \frac{n}{\delta_{2D}})$, and that this space is optimal within a multiplicative factor $O(\log n)$.

For the rest of the paper, the RAM model of computation is assumed, with word size $w = \Theta(\log n)$ bits. Space is measured in words so when $O(x)$ space is indicated, the actual space occupancy in bits is $O(x \log n)$.

2   Two-Dimensional Compressibility Measures

We consider a square matrix $M \in \Sigma^{n \times n}$ of size $n \times n$ where each of the n^2 symbols $M[i][j]$ of M are drawn from the alphabet Σ with $|\Sigma| = \sigma$. Every symbol in Σ is assumed to appear in M otherwise Σ is properly restricted. A submatrix of M with topmost left cell $M[i, j]$ is said to start at position (i, j) of M. An $a \times b$ submatrix of M starting at position (i, j) is written as $M[i : i+a-1][j : j+b-1]$, meaning that it includes any cell with row index in the range $[i, i + a - 1]$ and column index in $[j, j + b - 1]$. In this section two new repetitiveness measures for square matrices called γ_{2D} and δ_{2D} are proposed, as the generalisations of the γ and δ measures for strings respectively introduced in [6] and [3,13].

Definition 1. *An attractor Γ_{2D} for a square matrix $M \in \Sigma^{n \times n}$ is a set of positions of M: $\Gamma_{2D} \subseteq \{1, ..., n\} \times \{1, ..., n\}$ such that any square submatrix has an occurrence crossing (including) a position $p = (i, j) \in \Gamma_{2D}$. The measure $\gamma_{2D}(M)$ is defined as the cardinality of a smallest attractor for M.*

We say that a position $p = (i, j) \in \Gamma_{2D}(M)$ *covers* a submatrix I of M if there exists an occurrence of I which crosses p, and that a set of positions *covers* I if it includes a position p which *covers* I; when clear from the context, the parameter M is omitted from $\Gamma_{2D}(M)$ expression.

As a first result we show that, not surprisingly, the problem of finding the size of a smallest attractor is *NP-complete* also in two dimensions. The NP-completeness proof is done considering the decision problem "is there an attractor of size k for the given input?".

Lemma 1. *Given a string $S \in \Sigma^n$, let $R^S \in \Sigma^{n \times n}$ be the square matrix where each row is equal to the string S. Then there exists an (1-dim) attractor for S of size k if and only if there exists a (2-dim) attractor of size k for R^S.*

Proof. Given S and the corresponding R^S, the following observations hold: 1) any submatrix of R^S has an occurrence starting at the same column but on the first row of R^S; 2) any two $\ell \times \ell$ submatrices of R^S are equal if and only if the two respective substrings of S composing their rows are equal, formally: $R^S[i : i + \ell - 1][j : j + \ell - 1] = R^S[i' : i' + \ell - 1][j' : j' + \ell - 1]$ if and only if $S[j, j + \ell - 1] = S[j', j' + \ell - 1]$. From 1) and 2) the lemma follows: given a string attractor $\Gamma(S)$ for S of size k, the set $\Gamma_{2D} = \{(1, j) : j \in \Gamma(S)\}$ of size k is a two dimensional attractor for R^S and, vice versa, a string attractor Γ for S could be obtained from a matrix attractor $\Gamma_{2D}(R^S)$ for R^S projecting each couple by column index, formally, $\Gamma = \{j : (i, j) \in \Gamma_{2D}(R^S)\}$ is a one dimensional attractor for S. Note that if $\Gamma_{2D}(R^S)$ is a smallest attractor it does not include two positions on the same column, because, any distinct submatrix crossing one position has an occurrence (starting in the same column but at a different row) which crosses the other, hence in this case the projection does not generate any column index collision and $|\Gamma| = |\Gamma_{2D}(R^S)| = k$, otherwise, in case of collision, Γ could be completed with any $k - |\Gamma|$ positions not in Γ to reach size $k = |\Gamma_{2D}(R^S)|$. □

As an immediate consequence of the above lemma we have the following result.

Theorem 1. *Computing γ_{2D} is NP complete.* □

It is easy to see that $\gamma_{2D} \geq \sigma$ and γ_{2D} is insensitive to transpositions but, as for strings [9], γ_{2D} is not monotone. We show this by providing a family of matrices, built using the counterexample in [9] to disprove the monotonicity of γ, containing a submatrix with smaller γ_{2D}.

Lemma 2. *γ_{2D} is not monotone.*

Proof. Let w be the string $ab\underline{b}\underline{b}a^n\underline{a}ab$ with $n > 0$, having $\gamma(w) = 3$ minimal for the subset of positions $\Gamma(w) = \{2, 4, n + 5\}$ underlined in w. The string $w \cdot b = ab\underline{b}\underline{b}a^n\underline{a}bb$ obtained concatenating the letter b to w has a smaller compressibility measure $\gamma(w \cdot b) = 2$ corresponding to $\Gamma(w \cdot b) = \{4, n + 5\}$ [9], as the prefix $w[1, 3] = abb$ occurring as a suffix of $w \cdot b$ is already covered by position $n + 5$ in $\Gamma(w \cdot b)$. Consider $R^{w \cdot b}$ of size $(n + 7) \times (n + 7)$, from Lemma 1 follows that $\gamma_{2D}(R^{w \cdot b}) = \gamma(w \cdot b) = 2$, but the submatrix $R^{w \cdot b}[1 : n + 6][1 : n + 6]$ equal to R^w has a greater $\gamma_{2D}(R^w) = \gamma(w) = 3$. □

2.1   The Measure δ_{2D}

The measure $\delta(S)$ for a string S, formally defined in [3] and previously introduced in [13] to approximate the output size of the Lempel-Ziv parsing, is the maximum over $k \in [1, |S|]$ of the expression $d_k(S)/k$ where $d_k(S)$ is the number of distinct substrings of length k in S. We now show how to generalize this measure to two dimensions, by introducing the measure δ_{2D} which is defined in a similar way, considering $k \times k$ submatrices instead of length-k substrings.

Definition 2. *Given* $M \in \Sigma^{n \times n}$, *let* $d_{k \times k}(M)$ *be the number of distinct* $k \times k$ *submatrices of* M, *then*

$$\delta_{2D}(M) = \max\{d_{k \times k}(M)/k^2 : k \in [1, \, n]\}. \tag{1}$$

The measure δ_{2D} preserves some good properties of δ: δ_{2D} is invariant through transpositions and decreases or grows by at most 1 after a single cell edit since any $d_{k \times k}$ of the updated matrix could differ at most by k^2 from the initial one. δ_{2D} is monotone: given a submatrix M' of M having size $\ell \times \ell$ with $\ell \le n$ any submatrix of M' appears somewhere in M then $d_{k \times k}(M') \le d_{k \times k}(M)$ for any $k \in [1, \ell] \subseteq [1, n]$.

The next lemma shows that, as in the one-dimensional setting, δ_{2D} is upper bounded by γ_{2D}.

Lemma 3. $\delta_{2D}(M) \le \gamma_{2D}(M)$ *for any matrix* $M \in \Sigma^{n \times n}$.

Proof. Let Γ_{2D} be a least size attractor for M i.e. $|\Gamma_{2D}| = \gamma_{2D}$. For any $k \in [1, n]$ an attractor position $p \in \Gamma_{2D}$ is included in at most k^2 distinct $k \times k$ submatrices, then we need at least $d_{k \times k}(M)/k^2$ distinct positions in Γ_{2D} to cover all $k \times k$ submatrices of M, formally, $|\Gamma_{2D}| \ge d_{k \times k}(M)/k^2$ holds for any $k \in [1, n]$ in particular for $k^* \in [1, n]$ such that $\delta_{2D} = d_{k^* \times k^*}(M)/(k^*)^2$. $\qquad \square$

One of the main reasons for introducing δ was that it can be computed efficiently: [3] describes a linear algorithm to compute $\delta(S)$ with a single visit of the Suffix tree of S. We now show that an efficient algorithm for computing δ_{2D} can be derived in a similar way using the *Isuffix tree* introduced in [7] which can be built in $O(n^2)$ time, which is linear in the size of the input. A somewhat simpler algorithm can be obtained using the *Lsuffix tree* [4,5] but its construction takes $O(n^2 \log n)$ time.

The *Isuffix tree* $IST(A)$ of a matrix $A \in \Sigma^{n \times m}$ generalises the Suffix Tree to matrices: $IST(A)$ is a compacted trie representing all square submatrices of A. The Isuffix trees adopts a linear representation of a square matrix $C \in \Sigma^{q \times q}$: let $I\Sigma = \bigcup_{i=1}^{\infty} \Sigma^i$, each string in $I\Sigma$ is considered as an atomic *Icharacter*, the unique *Istring* associated to matrix C is $I_C \in I\Sigma^{2q-1}$ where $I_C[2i+1]$ with $i \in [0, q)$ is the $(i+1)^{th}$ *column-type Icharacter* $C[1:i+1][i+1]$ and $I_C[2i]$ with $i \in [1, q)$ is the ith *row-type Icharacter* $C[i+1][1:i]$. See Fig. 1 for an example. The kth *Iprefix* of C is defined as the concatenation of the first k Icharacters $I_C[1] \cdot I_C[2] \cdot \ldots \cdot I_C[k] = I_C[1, k]$ of I_C. Note that an Iprefix ending in an odd position k is the Istring of the $\ell \times \ell$ square submatrix with $\ell = \lceil k/2 \rceil$ starting at C's top-left corner, that is, $C[1:\ell][1:\ell]$. For the example in Fig. 1, the 3rd Iprefix of C is the Istring "a a ba" which corresponds to the submatrix $C[1:2][1:2]$.

Given $A \in \Sigma^{n \times n}$, for $1 \le i, j \le n$, the *Isuffix* $I_{A_{ij}}$ of A is defined as the Istring of the largest square submatrix A_{ij} of A with upper left corner at position (i, j). From the above definitions it is clear that the Istring of any square submatrix of A, is an Iprefix (ending in a odd position) of some Isuffix $I_{A_{ij}}$. To ensure that no Isuffix $I_{A_{ij}}$ is Iprefixed by another Isuffix, A is completed with

		[1]	[3]	[5]	[7]	[9]
		a	b	a	a	$5
	[2]	a	a	a	b	$6
	[4]	a	c	a	a	$7
	[6]	b	a	a	c	$8
	[8]	$1	$2	$3	$4	$9

a	b	a	a	$5
a	a	a	b	$6
a	c	a	a	$7
b	a	a	c	$8
$1	$2	$3	$4	$9

Matrix C $I_C = $ a a ba ac aaa baa abac ...

Fig. 1. A square matrix C on the left, and its Istring I_C on the right (last two Icharacters are omitted)

an additional bottom row and right column containing $2n + 1$ distinct new symbols $\$_1, \ldots \$_{2n+1}$. For simplicity in the following we refer as A the input matrix already enlarged with $\$_i$ symbols. See Fig. 2 for an example.

The Isuffix tree $IST(A)$ is a compacted trie over the alphabet $I\Sigma$ representing all the n^2 distinct Isuffixes $I_{A_{ij}}$ of A with, among others, the following properties [7]: 1) each edge is labeled with a non empty Isubstring $I_{A_{ij}}[\ell_1, \ell_2]$ of an Isuffix $I_{A_{ij}}$, that label is represented in constant space as the quadruple $\langle i, j, \ell_1, \ell_2 \rangle$, the Isubstrings on any two sibling edges start with different Icharacters; 2) each internal node has at least two children and there are exactly n^2 leaves representing all the Isuffixes of A: let $L(u)$ be the Istring obtained concatenating the Isubstrings on the path from the root to a node u, for any leaf l_{ij}, the Istring $L(l_{ij})$ is equal to the linear representation $I_{A_{ij}}$ of the unique suffix A_{ij}; 3) The Isuffix tree satisfies the *common prefix constraint*: square submatrices of A with a common Iprefix share the same initial path in the tree; 4) The Isuffix tree satisfies the *completeness constraint* since all square submatrices of A are represented in $IST(A)$ as an Iprefix of some Isuffix of A.

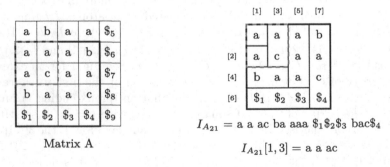

a	b	a	a	$5
a	a	a	b	$6
a	c	a	a	$7
b	a	a	c	$8
$1	$2	$3	$4	$9

	[1]	[3]	[5]	[7]
	a	a	a	b
[2]	a	c	a	a
[4]	b	a	a	c
[6]	$1	$2	$3	$4

Matrix A

$I_{A_{21}} = $ a a ac ba aaa $\$_1\$_2\$_3$ bac$\$_4$

$I_{A_{21}}[1, 3] = $ a a ac

Fig. 2. The submatrix $A[2:5][1:4] = A_{21}$ with solid black border on the left and its Istring $I_{A_{21}}$ on the right. The Istring of the submatrix $A[2:3][1:2]$ (in red) is the third Iprefix of $I_{A_{21}}$. (Color figure online)

Theorem 2. δ_{2D} *can be calculated in optimal time and space* $O(n^2)$.

Proof. Our algorithm is a generalization of the ideas used in [3] to compute the measure δ in linear time using a suffix tree. Given $A \in \Sigma^{n \times n}$, we build the array $d[1 : n]$ which stores at position k the number of distinct $k \times k$ submatrices of A then we obtain δ_{2D} as $\max_k d[k]/k^2$. Initially the Isuffix Tree $IST(A)$ of A is built in time $O(n^2)$ [7], then $IST(A)$ is visited in depth first order. Let u be a node such that the path from the root to u contains $|L(u)|$ Icharacters. Let e be an edge outgoing from u labeled with $q_e = \langle i, j, \ell_1, \ell_2 \rangle$ where $\ell_1 = |L(u)| + 1$. The Istring of a distinct square submatrix is obtained whenever appending a prefix of the Isubstring $I_{A_{ij}}[\ell_1, \ell_2]$ labelling the edge e to $L(u)$ yields an Istring of odd length. Because the traversing of e may yield new square submatrices, $d[\cdot]$ must be updated accordingly. Let $s = \lceil \frac{\ell_1 - 1}{2} \rceil + 1$ and $t = \lceil \frac{\ell_2}{2} \rceil$. Every $d[k]$ with $k \in [s, t]$ should be increased by one: to do this in constant time we set $d[s] = d[s] + 1$ and $d[t + 1] = d[t + 1] - 1$ and we assume that each value stored in an entry $d[i]$ is *implicitly* propagated to positions $i + 1, i + 2, \ldots n$: so the $+1$ is propagated from s up to t and the propagation is canceled by the -1 added at the position $t + 1$. At the end of the Isuffix tree visit, for each $k \in [1, n - 1]$ we set $d[k + 1] = d[k + 1] + d[k]$ so that $d[k]$ contains the number of distinct $k \times k$ matrices encountered during the visit and we can compute δ_{2D} as $\max_k d[k]/k^2$.

Note that when leaf l_{ij} is reached via the edge e with label $q_e = \langle i, j, \ell_1, \ell_2 \rangle$, all the Iprefixes of $I_{A_{ij}}[\ell_1, \ell_2]$ that have an Icharacter which includes some $\$_x$ symbol should not be counted. The range of well formed Iprefixes can be determined in constant time since it suffices to access one symbol in each of the last two trailing Icharacters of $I_{A_{ij}}[\ell_1, \ell_2]$ to check whether these two contains any $\$_x$. Since the Isuffix Tree can be constructed and visited in $O(n^2)$ time the overall time and space complexity of the above algorithm is $O(n^2)$. $\qquad\square$

We now study how large can be the gap between the two measures γ_{2D} and δ_{2D}, recalling that by Lemma 3 it is $\delta_{2D} \leq \gamma_{2D}$. In [8] Kociumaka et al. establish a separation result between measures δ and γ by showing a family of strings with $\delta = O(1)$ and $\gamma = \Omega(\log n)$. This bound is tight since they also prove that $\gamma = O(\delta \log \frac{n}{\delta})$. The next theorem proves that the gap between the two measures in two dimensions is much bigger: δ_{2D} can be (asymptotically) smaller than γ_{2D} up to a $\sqrt{n}$ factor.

Lemma 4. *There exists a family of* $n \times n$ *matrices with* $\delta_{2D} = O(1)$ *and* $\gamma_{2D} = \Omega(\sqrt{n})$.

Proof. Consider the matrix M of size $n \times n$ where the first row is the string S composed by $\sqrt{n}/2$ consecutive blocks of size $2\sqrt{n}$ each. The ith block S_i with $i = 1, \ldots, \sqrt{n}/2$ is the string $1^i 0^{(2\sqrt{n} - i)}$, so S_i contains (from left to right) i initial ones and the remaining positions are zeros. The remaining rows of the matrix are all equals to $\#^n$. Note that for any size k all distinct submatrices start in the first row or are equal to $\#^{(k \times k)}$. Let δ_k be $d_{k \times k}/k^2$, so that δ_{2D} can be rewritten as $\max\{\delta_k \mid k \in [1, n]\}$. We compute δ_k for each possible k. For $k = 1$, we have $\delta_1 = |\Sigma| = 3$. For $k \geq \sqrt{n}$ it is $\delta_k = O(1)$ since $k^2 \geq n$ and there at most

$(n-k+1)+1 \leq n$ distinct $k \times k$ matrices. Now consider δ_k with $k \in [2, \ldots, \sqrt{n})$. All distinct $k \times k$ submatrices (excluded the $\#^{(k \times k)}$ one) are those having as first row a distinct substring of length k of S. All those substrings are included in the language $0^a 1^b 0^c$ with $a \in [0, \ldots, k], b \in [0, \ldots, k-a], c \in [0, \ldots, k-a-b]$ such that $a + b + c = k$, to see this note that no substring of length $k < \sqrt{n}$ can contain any two non adjacent (and non empty) groups of ones since there is a group of at least $\sqrt{n} > k$ consecutive zeros between each of them in S. Fixed k, to count the strings in $0^a 1^b 0^c$ is enough to count the possible choices for the starting/ending positions of the middle 1^b block: which are $O(k^2)$, then for $k \in [2, \ldots, \sqrt{n})$, $\delta_k = \frac{O(k^2)}{k^2} = O(1)$. This proves that $\delta_{2D} = O(1)$.

To estimate γ_{2D} consider the ith block on the first row: $S_i = 1^i 0^{(2\sqrt{n}-i)}$. Each S_i with $i = 1, \ldots, \sqrt{n}/2$ is a unique occurrence since the sequence 1^i occurs only inside blocks S_j with $j \geq i$ which begins with at least i ones, but inside S_j the sequence 1^i is followed by $2\sqrt{n}-j < 2\sqrt{n}-i$ zeros, so the copy of S_i will intersect the $(j+1)^{th}$ block where no leading zeros are present. As a consequence each submatrix M_i of size $2\sqrt{n} \times 2\sqrt{n}$ having S_i as first row is a unique occurrence too. As each M_i does not overlap any other M_j with $j \neq i$ at least $\sqrt{n}/2$ positions are needed in Γ_{2D} to cover them. This proves that $\gamma_{2D} = \Omega(\sqrt{n})$. □

Given a set S, the worst-case entropy [8] of S defined as $\lceil \log_2 |S| \rceil$ is the minimum number of bits needed to encode all the elements in S. In the following Lemma, we extend the construction of Lemma 4 to define a family $\mathcal{F}$ of matrices with constant δ_{2D} and worst-case entropy $\Omega(\sqrt{n} \log n)$.

Lemma 5. *There exists a family of square matrices on a constant size alphabet Σ with common measure $\delta_{2D} = O(1)$ and worst-case entropy $\Omega(\sqrt{n} \log n)$.*

Proof. Consider again the matrix M of Lemma 4. Each of the $(\sqrt{n}/2)!$ matrices obtained permuting the $\sqrt{n}/2$ blocks S_i on the first row of M has still $\delta_{2D} = O(1)$. On the other hand, every encoding algorithm to distinguish among these matrices needs at least $\log((\sqrt{n}/2)!) = \Theta(\sqrt{n} \log n)$ bits. □

3 Space Bounds for Two-Dimensional Block Trees

Brisaboa et. al. [2] generalized the Block Tree concept [1] to two dimensional data providing a compressed representation for discrete repetitive matrices that offers direct access to any compressed submatrix in logarithmic time. Given a matrix $M \in \Sigma^{n \times n}$ and an integer parameter $k > 1$, assuming for simplicity that n is a power of k, i.e. $n = k^\alpha$, M is split into k^2 non overlapping submatrices, called blocks, each of size $(n/k) \times (n/k) = k^{\alpha-1} \times k^{\alpha-1}$. Each of these blocks corresponds to a node at level $\ell = 1$ in the *2D-BT* and the root of the tree at level $\ell = 0$ represents the whole matrix M. A tree is obtained by splitting (some of) the blocks at level ℓ, which have size $(n/k^\ell) \times (n/k^\ell)$, into k^2 non overlapping blocks of size $(n/k^{\ell+1}) \times (n/k^{\ell+1})$. At any level ℓ, nodes whose corresponding submatrix intersects the first occurrence, in row major order, of a $(n/k^\ell) \times (n/k^\ell)$ submatrix (including themselves) are internal nodes, referred in the following as *marked ones*; all others nodes are the level-ℓ leaves of the *2D-BT*, and referred in the following as *unmarked nodes*. Only marked nodes are recursively split and expanded at level $\ell + 1$; instead an unmarked node corresponding to a block X points to the marked nodes in the same level corresponding to the level-ℓ blocks overlapping the first (in row major order) occurrence O of X, and stores the relative offset $\langle O_x, O_y \rangle$ of O inside the top left of such blocks (see Fig. 3). The splitting process ends when explicitly storing blocks is more convenient than storing pointers to marked blocks. The resulting tree-shaped data structure has height $h = O(\log_k n)$. In the following the block related to node u in the tree is named B_u, and a block B_u is said to be marked (unmarked) if the corresponding node u is marked (unmarked). Note that if X is unmarked, then the (up to) four blocks intersecting the first occurrence O of X are marked by construction. If we call D the $(2n/k^\ell) \times (2n/k^\ell)$ submatrix formed by these four blocks, we observe that this is also a first occurrence (otherwise we would have another occurrence of X preceding O) and therefore the up to four blocks at level $\ell - 1$ containing D will be marked. Repeating this argument shows that an unmarked node points to marked nodes which always exist in the same level since none of their ancestors has been pruned in a previous level. Note that our marking scheme is slightly different than the one in [2] in which if a submatrix is pruned at some level its content is seen as all 0 s in the subsequent levels. This approach removes the issue of possibly pointing to pruned nodes, but makes it difficult to estimate the number of marked nodes in terms of the matrix content, which is our next objective.

It has already been proved [8] that one-dimensional Block Trees are worst case optimal in terms of δ in the following sense: a *BT* on a string $S \in \Sigma^n$ uses $O(\delta \log \frac{n \log \sigma}{\delta \log n})$ space and there exist string families requiring that amount of space to be stored. No space analysis of the *2D-BT* was given in [2]. In the following we show that a *2D-BT* built on a matrix $M \in \Sigma^{n \times n}$ occupies $O((\delta_{2D} + \sqrt{n \delta_{2D}}) \log \frac{n}{\delta_{2D}})$ space.

Lemma 6. *The number of marked nodes in any level of a 2D block tree is* $O(\delta_{2D} + \sqrt{n \delta_{2D}})$.

Proof. Consider a generic tree level, and assume the blocks in this level have size $k^\ell \times k^\ell$. In the following the term *block* denotes a $k^\ell \times k^\ell$ submatrix of M whose upper left corner is an entry of the form $M[1 + \lambda k^\ell, 1 + \mu k^\ell]$ with λ, μ integers. For any distinct $k^\ell \times k^\ell$ submatrix in M, let O be the first occurrence of that submatrix in row major order. O intersects $m \in \{1, 2, 4\}$ blocks $B_1, \ldots, B_m$ that are therefore marked. Let D be a $2k^\ell \times 2k^\ell$ submatrix built including all the m blocks $B_1, \ldots, B_m$, and therefore containing O. We call D a *superblock*. If $m = 4$, D is unique, otherwise $4 - m$ more blocks are chosen arbitrarily to reach the desired size. Let u be the number of superblocks constructed in this way; the number of marked blocks at this level is at most $4u$, hence we proceed bounding u. We partition the superblocks into 3 types: 1) those on a corner of M, i.e. including one of the entries $M[1, 1], M[1, n], M[n, 1]$ or $M[n, n]$, these are at most four; 2) those not on a corner but including an entry in the first/last

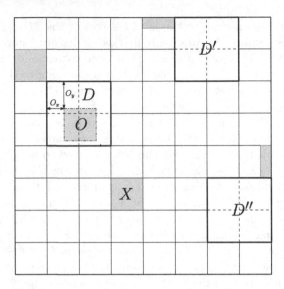

Fig. 3. The four blocks inside region D are marked since they overlap the first occurrence O, in row major order, of X. For the argument in the proof of Lemma 6 D is a type 3 superblock: by considering every entry in the red block above it as an upper left corner we obtain $k^{2\ell}$ distinct $3k^\ell \times 3k^\ell$ matrices containing D. The type 2 superblock D' borders the upper edge; by considering the k^ℓ entries in the first row marked in red we obtain k^ℓ distinct $3k^\ell \times 3k^\ell$ matrices containing D'. We also show a type 2 superblock D'' bordering the right edge; we obtain k^ℓ distinct $3k^\ell \times 3k^\ell$ matrices containing D'' by considering the $3k^\ell \times 3k^\ell$ matrices with the upper *right* corner in the portion of the last column marked in red. (Color figure online)

row/column; 3) those not including any entry in the first/last row/column. Let u_i be the number of superblocks of type i: clearly $u = u_1 + u_2 + u_3 = O(u_2 + u_3)$.

Given a superblock D of the third type, we observe that D is included into $k^{2\ell}$ distinct $3k^\ell \times 3k^\ell$ submatrices starting at any position in the $k^\ell \times k^\ell$ block touching the top left corner of D (see Fig. 3). Summing over all type 3 superblocks, we have a total of $u_3 k^{2\ell}$ submatrices of size $3k^\ell \times 3k^\ell$ starting in distinct positions inside M. These submatrices are distinct: each submatrix contains, by construction, the first occurrence of some block; if two of those matrices were equal we would have two first occurrences of the same block starting in different positions which is impossible. Since by definition the number of distinct submatrices of size $3k^\ell \times 3k^\ell$ is at most $(3k^\ell)^2 \delta_{2D}$ we have

$$u_3 k^{2\ell} \leq 9k^{2\ell} \delta_{2D} \quad \Longrightarrow \quad u_3 \leq 9\delta_{2D}$$

Consider now a superblock D' of the second type bordering the upper edge of M (the other 3 cases are treated similarly, see superblock D'' in Fig. 3). Any $3k^\ell \times 3k^\ell$ matrix which starts in the same row of D', but, in any of the k^ℓ columns preceding D' is distinct by the same argument presented before (see again Fig. 3). Reasoning as above we find $u_2 k^\ell$ distinct $3k^\ell \times 3k^\ell$ matrices which implies $u_2 \leq$

$9k^\ell \delta_{2D}$. Since it is also $u_2 \le n/k^\ell$, we have $u_2 \le \min(9k^\ell \delta_{2D}, n/k^\ell) = O(\sqrt{n\delta_{2D}})$. We conclude that the number of marked blocks at any level is $O(u) = O(u_1 + u_2 + u_3) = O(\delta_{2D} + \sqrt{n\delta_{2D}})$. □

Theorem 3. *The 2D-BT takes $O((\delta_{2D} + \sqrt{n\delta_{2D}}) \log \frac{n}{\delta_{2D}})$ space. This space is optimal within a multiplicative factor $O(\log n)$.*

Proof. The 2D-BT as described at the beginning of the section has height $\log_k n$. Such height can be reduced choosing a different size for the blocks at level $\ell = 1$. Assuming for simplicity $n = \sqrt{\delta_{2D}}k^\alpha$, M is initially divided into δ_{2D} blocks of size $k^\alpha \times k^\alpha$. In this way the height of the tree became $O(\log_k \frac{n}{\delta_{2D}})$, using the bound of the Lemma 6, we get an overall number of marked nodes $O((\delta_{2D} + \sqrt{n\delta_{2D}}) \log_k \frac{n}{\delta_{2D}})$. Each marked node produces at most k^2 unmarked nodes on the next level, hence the tree has at most $O(k^2(\delta_{2D} + \sqrt{n\delta_{2D}}) \log_k \frac{n}{\delta_{2D}})$ nodes which is $O((\delta_{2D} + \sqrt{n\delta_{2D}}) \log \frac{n}{\delta_{2D}})$ for $k = O(1)$. To prove the worst-case quasi-optimality: let $\mathcal{F}$ be the set of matrices having $\delta_{2D} = O(1)$ from Lemma 5, for any coder $C : \mathcal{F} \to \{0,1\}^*$ representing all the matrices in $\mathcal{F}$, there exist a matrix W such that $|C(W)| = \Omega(\sqrt{n} \log n)$ bits while the 2D-BT takes $O(\sqrt{n} \log^2 n)$ bits of space for any matrix in $\mathcal{F}$. □

The following result shows that the bound in Lemma 6 cannot be substantially improved at least when $\delta_{2D} = O(1)$. Since the proof of Lemma 6 shows that the number of marked blocks *at the interior* of the matrix is bounded by $O(\delta_{2D})$, we consider a family of matrices that have a hard to compress first row.

Lemma 7. *There exists an infinite family of matrices $M \in \Sigma^{n \times n}$ with $\delta_{2D} = O(1)$, such that 2D-BT for M has $\Omega(\sqrt{n})$ marked nodes on a single level.*

Proof. Let $M \in \Sigma^{n \times n}$ be the matrix of Lemma 4 with $n = k^{2\alpha}$ so that n is both a power of k and a perfect square. We have already proven that $\delta_{2D}(M) = O(1)$. Consider the 2D-BT built on M: note that for block size larger than $4\sqrt{n} \times 4\sqrt{n}$ each block on the upper edge of M includes entirely in its first row at least one of the strings S_i of the form $1^i 0^{(2\sqrt{n}-i)}$ composing S. Since each S_i is unique, any of those blocks is a first occurrence and hence marked. In particular, at level $\ell = \alpha - \lceil \log_k 4 \rceil$, counting levels from the root ($\ell = 0$) to the leaves, all $\Theta(\sqrt{n})$ blocks on M upper side are marked and the lemma follows. □

In [12] the authors introduced a variant of the one-dimensional block tree, called Γ-tree, in which, given a not necessarily minimum string attractor Γ, the marked nodes at each level are those close to an attractor position. The Γ-tree is then enriched with additional information that makes it a compressed full text index using $O(\gamma \log(n/\gamma))$ space where $\gamma = |\Gamma|$ is the size of the string attractor. Following the ideas from [12], we now show how to modify the construction of the 2D-BT assuming we have available a, not necessarily minimum, 2D-attractor Γ_{2D}.

To simplify the explanation, again we assume that $n = k^\alpha$ for some $\alpha > 0$, given a matrix $M \in \Sigma^{n \times n}$ and an attractor $\Gamma_{2D} = \{(i,j)_1, \ldots, (i,j)_\gamma\}$ for M, the splitting process is unchanged but at level ℓ we mark any node u corresponding to a $n/k^\ell \times n/k^\ell$ block B_u which includes a position $p \in \Gamma_{2D}$ and all (the at most 8) nodes of the blocks adjacent to B_u. Remaining nodes are unmarked and store a pointer ptr_B to the marked block B on the same level ℓ where an occurrence O of their corresponding submatrix that spans a position $p \in \Gamma_{2D}$ begins, as well as the relative offset of O within B. The claimed occurrence O crossing $p \in \Gamma_{2D}$ exists otherwise Γ_{2D} would not be an attractor for M, and all the (at most) 4 blocks intersecting O are ensured to be marked as they contain p or are adjacent to a block containing p. We also point out that overlapping between an unmarked block B' and the pointed occurrence O containing $p \in \Gamma_{2D}$ is impossible as if this happens B' would be adjacent to a block B with $p \in B$, hence B' would be marked as well.

If n is not a power of k, blocks on right and bottom edges of M won't be squared, but no special treatment is needed as all the previous essential properties are still valid: consider a rectangular block B of size $a \times b$ on the edge, if B is marked, B is recursively split into smaller blocks (some of those with rectangular shape), if B is unmarked, the squared matrix B' of size $c \times c$ with $c = max(a,b)$ including B is squared and will occur somewhere else crossing an attractor position while spanning at most four marked blocks, then B would occur as well. Note that B, contrary to B', may not cross any attractor position but will certainly point to marked blocks only, avoiding unmarked blocks point to other unmarked ones.

Theorem 4. *Given $M \in \Sigma^{n \times n}$ and an attractor $\Gamma_{2D}(M) = \{(i,j)_1, \ldots, (i,j)_\gamma\}$ of size γ, the 2D-BT built using Γ_{2D} takes $O(\gamma \log \frac{n}{\gamma})$ space.*

Proof. Each position $p \in \Gamma$ could mark at most 9 distinct blocks per level: the block B including p and the (up to) eight blocks adjacent to B. Hence the number of marked blocks per level is $\leq 9\gamma$. Assuming again $n = \sqrt{\gamma}k^\alpha$, dividing initially M into blocks of size $k^\alpha \times k^\alpha$ we get a shallower tree of height $O(\log_k \frac{n}{\gamma})$ with $O(\gamma)$ nodes on the second level ($\ell = 1$) and an overall number of marked nodes $O(\gamma \log_k \frac{n}{\gamma})$. Since any marked node produces at most k^2 unmarked nodes on the next level, for any $k = O(1)$ the 2D-BT built using any attractor Γ_{2D} of size γ takes $O(\gamma \log \frac{n}{\gamma})$ space. $\square$

Access to a single symbol $M[i][j]$ is quite as in the one dimensional Γ-Tree: assume the node u at level ℓ is reached with a local offset $\langle O_x, O_y \rangle$, if u is a marked node, the child c of u where the searched cell falls is determined, the coordinates $\langle O_x, O_y \rangle$ are translated to local coordinates on c where the search routine proceeds in the next level. If instead node u is unmarked, the marked node v on the same level is reached via the pointer ptr_v stored in u, the actual offset inside the block B_v is determined using the offset $\langle O'_x, O'_y \rangle$ stored in u and the access procedure continues on marked node v. The descending process halts when a marked block on the deepest level is reached and the corresponding explicit symbol is retrieved. Access procedure costs $O(\log n)$ as we visit at most 2 nodes on the same level before descend to next one.

Funding Information. This research was partially supported by MIUR-PRIN project "Multicriteria Data Structures and Algorithms: from compressed to learned indexes, and beyond" grant n. 2017WR7SIIII, and by the PNRR ECS00000017 Tuscany Health Ecosystem, Spoke 6 "Precision medicine & personalized healthcare", CUP I53C22000780001, funded by the European Commission under the NextGeneration EU programme.

References

1. Belazzougui, D., et al.: Block trees. J. Comput. Syst. Sci. **117**, 1–22 (2021)
2. Brisaboa, N., Gagie, T., Gómez-Brandón, A., Navarro, G.: Two-dimensional block trees. Comput. J. (2023, to appear)
3. Christiansen, A.R., Ettienne, M.B., Kociumaka, T., Navarro, G., Prezza, N.: Optimal-time dictionary-compressed indexes. ACM Trans. Algorithms **17**(1), 8:1–8:39 (2021)
4. Giancarlo, R.: A generalization of the suffix tree to square matrices, with applications. SIAM J. Comput. **24**(3), 520–562 (1995)
5. Giancarlo, R., Grossi, R.: On the construction of classes of suffix trees for square matrices: algorithms and applications. Inf. Comput. **130**(2), 151–182 (1996)
6. Kempa, D., Prezza, N.: At the roots of dictionary compression: string attractors. In: Diakonikolas, I., Kempe, D., Henzinger, M. (eds.) Proceedings of the 50th Annual ACM SIGACT Symposium on Theory of Computing, STOC 2018, Los Angeles, CA, USA, 25–29 June 2018, pp. 827–840. ACM (2018)
7. Kim, D.K., Na, J.C., Sim, J.S., Park, K.: Linear-time construction of two-dimensional suffix trees. Algorithmica **59**(2), 269–297 (2011)
8. Kociumaka, T., Navarro, G., Prezza, N.: Toward a definitive compressibility measure for repetitive sequences. IEEE Trans. Inf. Theory **69**(4), 2074–2092 (2023)
9. Mantaci, S., Restivo, A., Romana, G., Rosone, G., Sciortino, M.: A combinatorial view on string attractors. Theor. Comput. Sci. **850**, 236–248 (2021)
10. Navarro, G.: Indexing highly repetitive string collections, part I: repetitiveness measures. ACM Comput. Surv. **54**(2), article 29 (2021)
11. Navarro, G.: Indexing highly repetitive string collections, part II: compressed indexes. ACM Comput. Surv. **54**(2), article 26 (2021)
12. Navarro, G., Prezza, N.: Universal compressed text indexing. Theoret. Comput. Sci. **762**, 41–50 (2019)
13. Raskhodnikova, S., Ron, D., Rubinfeld, R., Smith, A.D.: Sublinear algorithms for approximating string compressibility. Algorithmica **65**(3), 685–709 (2013)

From de Bruijn Graphs to Variation Graphs – Relationships Between Pangenome Models

Adam Cicherski[ID] and Norbert Dojer[(⊠)][ID]

University of Warsaw, Warsaw, Poland
dojer@mimuw.edu.pl

Abstract. Pangenomes serve as a framework for joint analysis of genomes of related organisms. Several pangenome models were proposed, offering different functionalities, applications provided by available tools, their efficiency etc. Among them, two graph-based models are particularly widely used: variation graphs and de Bruijn graphs.

In the current paper we propose an axiomatization of the desirable properties of a graph representation of a collection of strings. We show the relationship between variation graphs satisfying these criteria and de Bruijn graphs. This relationship can be used to efficiently build a variation graph representing a given set of genomes, transfer annotations between both models or compare the results of analyzes based on each model.

Keywords: pangenome · de Bruijn graphs · variation graphs

1 Introduction

The term pangenome was initially proposed as a single data structure for joint analysis of a group of bacterial genes [25]. In the presence of a variety of whole genome sequences available it has evolved and currently it refers to a model of joint analysis of genomes of related organisms. Numerous pangenome models were proposed, ranging from collections of unaligned sequences to sophisticated models that require complex preprocessing of sequence data [1,4,8,20]. The models significantly differ in their properties, applications provided by available tools, their efficiency, and even the level of precision of the definition of the optimal graph representing given genomes.

De Bruijn graphs (dBG) have nodes uniquely labeled with k-mers and edges representing their overlaps of length $k - 1$. Since the structure of a dBG for a given set of genomes is strictly determined by the parameter k, its construction is straightforward and may be performed in linear time. Applications of dBG-based pangenome models include read mapping [12,17], variant calling [15], taxonomic classification and quantification of metagenomic sequencing data [22] and even pangenome-wide association studies [18].

© The Author(s), under exclusive license to Springer Nature Switzerland AG 2023
F. M. Nardini et al. (Eds.): SPIRE 2023, LNCS 14240, pp. 114–128, 2023.
https://doi.org/10.1007/978-3-031-43980-3_10

Variation graphs (VG) have nodes labeled with DNA sequences of arbitrary length. Genomic sequences are represented in the graph by paths, for which the concatenation of labels form the respective sequences. Such structure allows to avoid the redundancy of the dBG representation, where a single residuum occurs in labels of several nodes. Consequently, this model provides an intuitive common coordinate system, making it convenient to annotate, which is crucial in many applications. The basis for a wide spectrum of VG-based analyzes was provided by indexing methods for efficient subsequence representation and searching [5,23, 24]. Then sequencing read mapping algorithms were proposed [10,21], opening the door for variant calling and genotyping tools [6,7,13]. Moreover, variation graphs were applied to the inference of precise haplotypes [2,3] and genome graph-based peak calling [11].

The construction of VG models is more computationally resource-intensive than the construction of dBGs. Given 100 human genomes, the respective dBG can be built in a few hours [19,26], while the construction of the VG requires several days [9,10,14,16]. Moreover, there are many possible variation graphs that represent a particular collection of genomes. Two completely uninformative extremes are: an empty graph with each node labeled with one of the genome sequences, and a graph with 4 nodes labeled with single letters A, C, G, T. Existing tools tend to strike a balance between these extremes, but to the best of our knowledge, no strict definition of the desired properties of such a graph (e.g. minimum requirements or optimization criteria) have been proposed.

The aim of the current paper is to fill this gap. We propose an axiomatization of the desired properties of variation graphs. Moreover, we show the relationships between VGs satisfying proposed criteria and related dBGs. Finally, we show how these relationships may be used to transform de Bruijn graphs into variation graphs.

2   Representing String Collections with Graphs

In this section we propose the concept of string graph – a common abstraction of de Bruijn and variation graphs. Then we formalize the notion of a graph representing given set of strings and define postulated properties of such representations. Finally, we show that these properties are always satisfied in de Bruijn graphs and impose strict constraints on the structure of variation graphs.

2.1   String Graphs

A *string graph* is a tuple $G = \langle V, E, l \rangle$, where:

- V is a set of vertices,
- $E \subseteq V^2$ is a set of directed edges,
- $l : V \to \Sigma^+$ is a function labeling vertices with non-empty strings over alphabet Σ.

A *path* in a string graph is a sequence of vertices $\langle v_1, \ldots, v_m \rangle$ such that $\langle v_j, v_{j+1} \rangle \in E$ for every $j \in \{1, \ldots, m-1\}$. The set of all paths in G will be denoted by $\mathcal{P}(G)$. Given path $p = \langle v_1, \ldots, v_m \rangle$, the set of *intervals* of p is defined by formula $Int(p) = \{\langle j_1, j_2 \rangle \mid 1 \le j_1 \le j_2 \le m\}$. Given two intervals $\langle j_1, j_2 \rangle, \langle j_1', j_2' \rangle \in Int(p)$, we say that $\langle j_1, j_2 \rangle$ is a *subinterval* of $\langle j_1', j_2' \rangle$ if $j_1 \ge j_1'$ and $j_2 \le j_2'$. A *subpath* of p defined on interval $\langle j_1, j_2 \rangle$ is a path $p[j_1..j_2] = \langle v_{j_1}, \ldots, v_{j_2} \rangle$. We use similar terminology and notation for strings: given string S, the set of its *intervals* is defined by $Int(S) = \{\langle j_1, j_2 \rangle \mid 1 \le j_1 \le j_2 \le |S|\}$, where $|S|$ is the length of S, and $S[j_1..j_2]$ denotes the substring of S indicated by interval $\langle j_1, j_2 \rangle$.

In order to represent strings longer than labels of single vertices, the labeling function l must be extendable to function $\hat{l}$ defined on paths. The extension should be *subpath-compatible* in the following sense: every path p should induce an injective function $\Psi_p : Int(p) \to Int(\hat{l}(p))$ satisfying the condition

$$\Psi_p(j_1, j_2) = \langle j_1', j_2' \rangle \Rightarrow \hat{l}(p[j_1..j_2]) = \hat{l}(p)[j_1'..j_2'].$$

In the following subsection we use this concept to introduce the definition of a graph representing a set of strings and formulate appropriate properties of such representation. In subsequent subsections we describe two different implementations that are realized in de Bruijn graphs and variation graphs, respectively.

2.2 Representations of Collections of Strings

Given a set of strings $\mathcal{S} = \{S_1, \ldots, S_n\}$, a string graph G with subpath-compatible labeling extension $\hat{l}$ and $\pi : \mathcal{S} \to \mathcal{P}(G)$, we say that $\langle G, \pi \rangle$ is a *representation* of $\mathcal{S}$ iff the following conditions are satisfied:

- $\hat{l}(\pi(S_i)) = S_i$ for every $i \in \{1, \ldots, n\}$,
- every vertex in G occurs in some path $\pi(S_i)$,
- every edge in G joins two consecutive vertices in some path $\pi(S_i)$.

We define the set of *positions* in π as $Pos(\pi) = \{\langle i, j \rangle \mid 1 \le i \le n \wedge 1 \le j \le |\pi(S_i)|\}$. The set of π-*occurrences* of a vertex v is defined as $Occ_\pi(v) = \{\langle i, j \rangle \in Pos(\pi) \mid \pi(S_i)[j] = v\}$. The injections $\Psi_{\pi(S_i)} : Int(\pi(S_i)) \to Int(S_i)$ will be denoted by Ψ_i for short.

Below we define two properties of a representation: *k-completeness* and *k-faithfulness*. Intuitively, the representation is *k*-complete if every *k*-mer in $\mathcal{S}$ is depicted by the same path in the graph, and is *k*-faithful if all multiple occurrences of vertices are essential to satisfy *k*-completeness (see Fig. 1). However, the actual definitions are more elaborate because we don't assume that the functions Ψ_i are surjective, so not all $\mathcal{S}$-substrings are depicted in the graph by paths.

Let $S_i, S_{i'}$ be two (not necessarily different) strings from $\mathcal{S}$ and assume that $S_i[p..p+k-1] = S_{i'}[p'..p'+k-1]$ is a common *k*-mer of S_i and $S_{i'}$. We say that this common *k*-mer is *reflected* by a common subpath $\pi(S_i)[q..q+m] = \pi(S_{i'})[q'..q'+m]$ of $\pi(S_i)$ and $\pi(S_{i'})$ iff $\langle p, p+k-1 \rangle$ is a subinterval of $\Psi_i(q, q+m)$

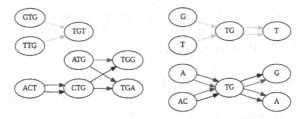

Fig. 1. The set of strings $\{GTGT, TTGT, ATGG, ATGA, ACTGG, ACTGA\}$ represented by a de Bruijn graph (left) and a variation graph (right). Different colors of edges correspond to paths representing particular strings. Both representations are 3-complete (all common 3-mers are reflected in the graphs) and 3-faithful (the paths representing strings share solely the vertices necessary to fulfill the previous condition). For example, in the VG the common 3-mer CTG of strings $ACTGG$ and $ACTGA$ is reflected by the common subpath $\langle AC, TG \rangle$ of the purple and black paths. Besides, the common 2-mer TG from the strings $ATGG$, $ATGA$, $ACTGG$, and $ACTGA$ is represented by a common vertex because the following pairs of its occurrences are directly 3-extendable: on the blue and red paths (extendable to the common 4-mer $ACTG$), on the red and black paths (common 3-mer TGG), and on the blue and purple paths (common 3-mer TGA). On the other hand, the occurrences of the same 2-mer TG in the strings $GTGT$ and $TTGT$ cannot be represented by the same vertex because they are not commonly extendable with any other occurrence.

and $\langle p', p'+k-1 \rangle$ is a subinterval of $\Psi_{i'}(q', q'+m)$. We say that $\langle G, \pi \rangle$ represents S k-*completely* iff all common k-mers in S are reflected by respective subpaths.
We say the pair of π-occurrences $\langle i, j \rangle, \langle i', j' \rangle$ of a vertex v is:

- *directly k-extendable* iff $\pi(S_i)[j - m..j + m'] = \pi(S_{i'})[j' - m..j' + m']$ for $m, m' \geq 0$ satisfying $|\hat{l}(\pi(S_i)[j - m..j + m'])| \geq k$, i.e. these occurrences extend to intervals of $\pi(S_i)$ and $\pi(S_{i'})$, respectively, indicating their common subpath labeled with a string of length $\geq k$,
- *k-extendable* if there is a sequence of occurrences of v that starts from $\langle i, j \rangle$, ends at $\langle i', j' \rangle$ and each two consecutive occurrences in that sequence are directly k-extendable.

We say that $\langle G, \pi \rangle$ represents S k-*faithfully* if every pair of occurrences of a vertex is k-extendable.

2.3 De Bruijn Graphs

A *de Bruijn graph* of length k is a string graph satisfying the following conditions:

- $|l(v)| = k$ for every $v \in V$,
- $l(v) = l(w) \Rightarrow v = w$ for all $v, w \in V$,
- $l(v)[2..k] = l(w)[1..k-1]$ for every $\langle v, w \rangle \in E$.

In other words: vertices are labeled with unique k-mers and edges may connect vertices having labels overlapping with $k-1$ characters.

We define the extension $\hat{l}$ of the de Bruijn graph labeling function in the following way: the labeling of a path $p = \langle v_1, \ldots, v_m \rangle$ is a concatenation of the labels of its vertices with deduplicated overlaps, i.e. $\hat{l}(p) = l(v_0) \cdot l(v_1)[k] \cdot \ldots \cdot l(v_m)[k]$.

Consider $\langle j_1, j_2 \rangle \in Int(p)$. The label of the subpath of p indicated by $\langle j_1, j_2 \rangle$ satisfies the equation $\hat{l}(p[j_1..j_2]) = \hat{l}(p)[j_1..j_2 + k - 1]$, so the function $\Psi_p(j_1, j_2) = \langle j_1, j_2 + k - 1 \rangle$ ensures subpath-compatibility.

Proposition 1. *Given set of strings $S = \{S_1, \ldots, S_n\}$ such that $|S_i| \geq k$ for every $i \in \{1, \ldots, n\}$, there is a unique up to isomorphism representation of S by a de Bruijn graph of length k. Moreover this representation is k-complete and k-faithfull.*

Proof. The set of vertices of a dBG of length k representing S is actually determined by the set of different k-mers in S. A mapping between the sets of vertices of two such graphs that preserves the labels of those vertices, must be an isomorphism. Uniqueness of vertex labels implies also k-completeness. Finally, k-faithfulness is obvious from the fact that the graph has no vertices with labels shorter than k.

2.4   Variation Graphs

In *variation graphs* vertices may be labeled with strings of any length. The extension of the labeling function to paths is defined as the concatenation of the labels of consecutive vertices, i.e. $\hat{l}(p) = l(v_1) \cdot \ldots \cdot l(v_m)$ for $p = \langle v_1, \ldots, v_m \rangle$.

Consider a path $p = \langle v_1, \ldots, v_m \rangle$ and let $s_j = \sum_{i=1}^{j} |l(v_i)|$ for $j \in \{0, \ldots, m\}$. The label of the subpath of p indicated by $\langle j_1, j_2 \rangle \in Int(p)$ satisfies the equation $\hat{l}(p[j_1..j_2]) = \hat{l}(p)[s_{j_1-1} + 1..s_{j_2}]$, so the function $\Psi_p(j_1, j_2) = \langle s_{j_1-1} + 1, s_{j_2} \rangle$ ensures subpath-compatibility. When $|l(v)| = 1$ for every vertex v, the graph is called *singular*. In this case $Int(p) = Int(\hat{l}(p))$, $s_j = j$ for all $j \in \{0, \ldots, m\}$ and the above formula simplifies to $\Psi_p(j_1, j_2) = \langle j_1, j_2 \rangle$.

Two representations of the set of strings S are *equivalent* if they reflect exactly the same m-mers for all $m > 0$. Note that equivalent singular graph representations must be isomorphic. Moreover, each variation graph representation is equivalent to a singular graph representation obtained by replacing each vertex with an unbranched path whose vertices are labeled with consecutive characters of the original vertex label.

Below we show that k-completeness and k-faithfulness properties determine the structure of the VG-representation up to equivalence.

Lemma 1. *Assume that variation graph representations $\langle G, \pi \rangle$ and $\langle G', \pi' \rangle$ of a set of strings $S = \{S_1, \ldots, S_n\}$ are k-complete. Let $\langle i_1, j_1 \rangle, \langle i_2, j_2 \rangle$ be a k-extendable pair of π-occurrences of a vertex v. Then every vertex v' of G' having such π'-occurrence $\langle i_1, j_1' \rangle$ that S_{i_1}-intervals $\Psi_{\pi'(S_{i_1})}(j_1', j_1')$ and $\Psi_{\pi(S_{i_1})}(j_1, j_1)$ overlap, has also such π'-occurrence $\langle i_2, j_2' \rangle$ that S_{i_2}-intervals $\Psi_{\pi'(S_{i_2})}(j_2', j_2')$ and $\Psi_{\pi(S_{i_2})}(j_2, j_2)$ overlap.*

Proof. From the definition of k-extendability there exists a sequence of π-occurrences of v, in which each two consecutive occurrences extend to intervals indicating common subpath labeled with a string of length $\geq k$. Every such extension reflects a common k-mer in $\mathcal{S}$. In the case of the first pair of π-occurrences of v, the S_{i_1}-interval indicating the occurrence of the k-mer overlaps $\Psi_{\pi'(S_{i_1})}(j_1', j_1')$. From k-completeness of $\langle G', \pi' \rangle$, by induction, v' has π'-occurrences satisfying such condition for every π-occurrence of v in the sequence, in particular for $\langle i_2, j_2 \rangle$.

Theorem 1. *Assume that variation graph representations $\langle G, \pi \rangle$ and $\langle G', \pi' \rangle$ of a set of strings $\mathcal{S} = \{S_1, \ldots, S_n\}$ are k-complete and k-faithfull. Then $\langle G, \pi \rangle$ and $\langle G', \pi' \rangle$ are equivalent.*

Proof. Assume that the common m-mer $S_{i_1}[p..p+m-1] = S_{i_2}[p'..p'+m-1]$ of S_{i_1} and S_{i_2} is reflected in one of the representations, say $\langle G, \pi \rangle$. Thus there exists a common subpath of $\pi(S_{i_1})$ and $\pi(S_{i_2})$ covering the considered occurrences of the m-mer. For each vertex on this subpath the pair of its occurrences on $\pi(S_{i_1})$ and $\pi(S_{i_2})$ is k-extendable due to k-faithfulness of $\langle G, \pi \rangle$. For every position $\langle i_1, j' \rangle$ on the minimal subpath of $\pi'(S_{i_1})$ covering $S_{i_1}[p..p+m-1]$, S_{i_1}-interval $\Psi_{\pi'(S_{i_1})}(j', j')$ overlaps $\Psi_{\pi(S_{i_1})}(j, j)$, where $\langle i_1, j \rangle$ is one of the positions on the $\pi(S_{i_1})$-subpath covering the m-mer. Thus, due to Lemma 1, all vertices on the minimal subpath of $\pi'(S_{i_1})$ covering $S_{i_1}[p..p+m-1]$ have corresponding occurrences on $\pi'(S_{i_2})$, which, combined together, form a subpath covering $S_{i_2}[p'..p'+m-1]$. Therefore the considered common m-mer is reflected in representation $\langle G', \pi' \rangle$ too.

The above theorem shows that whenever a k-complete and k-faithful variation graph representation of a given set of strings exists, it is determined up to equivalence. In the next section we complement this result by showing how such variation graph can be built.

3   Graph Transformation

In this section we show how to transform a de Bruijn graph representing a given set of strings into a corresponding variation graph. The transformation algorithm consists of 3 steps (see Fig. 2):

1. Split – conversion of vertices of the de Bruijn graph to unbranched paths with each vertex labeled with a single character.
2. Merge – a series of local modifications that merge incident edges inherited from the de Bruijn graph.
3. Collapse – a series of local modifications that removes isolated edges inherited from the de Bruijn graph.

Each modification of the graph is accompanied by a corresponding adjustment of the associated representation functions.

We begin this section by introducing *transition graph* – another type of string graph that will be used in intermediate stages of the transformation. Then we describe consecutive transformation steps and show that they preserve desirable properties. Finally, we show that the whole transformation yields a singular variation graph that is both k-complete and k-faithful.

3.1   Transition Graphs

A *transition graph* of length k over an alphabet Σ is a string graph $G = \langle V, E, l \rangle$, in which $l : V \to \Sigma$ (i.e. every vertex is labeled with a single character) and E is a disjoint union of two subsets:

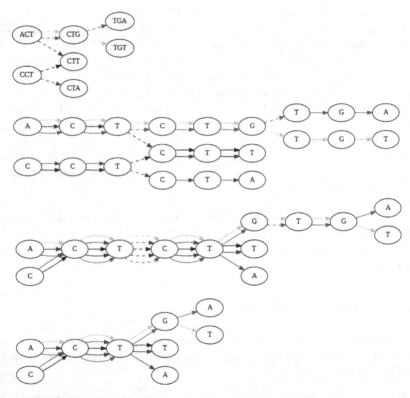

Fig. 2. Steps of the graph transformation algorithm for the graph representation of the set of strings $S = \{ACTGA, ACTGT, ACTT, CCTT, CCTA\}$. From top: input de Bruijn graph for $k = 3$ and transition graphs resulting from Split, Merge and Collapse transformations, respectively. Solid lines represent V-edges, dashed lines represent B-edges.

- E_V – the set of *variation edges* (or *V-edges* for short),
- E_B – the set of *de Bruijn edges* (or *B-edges* for short),

A path is called V-*path* if all its edges are V-edges. Labeling function l is extended to V-paths as in variation graphs, i.e. $\hat{l}(\langle v_1, \ldots, v_m \rangle) = l(v_1) \cdot \ldots \cdot l(v_m)$ for a V-paths $\langle v_1, \ldots, v_m \rangle$.

Every path p with n B-edges can be split into a sequence of $n + 1$ *maximal sub-V-paths* $\langle p_0, \ldots, p_n \rangle$ such that for every $i \in \{1, \ldots, n\}$ the last vertex of p_{i-1} and the first vertex of p_i are connected by a B-edge. When this decomposition satisfies the following conditions:

- $|p_0| \geq k - 1$, $|p_n| \geq k - 1$ and $|p_i| \geq k$ for $i \in \{1, \ldots, n-1\}$,
- $\hat{l}(p_{i-1})[|p_{i-1}| - k + 2..|p_{i-1}|] = \hat{l}(p_i)[1..k-1])$ for $i \in \{1, \ldots, n\}$,

the extension of labeling function is defined by formula $\hat{l}(p) = \hat{l}(p_0) \cdot \hat{l}(p_1)[k..|\hat{l}(p_1)|] \cdot \ldots \cdot \hat{l}(p_n)[k..|\hat{l}(p_n)|]$. If moreover $|p_0| \geq k$ and $|p_n| \geq k$, we call p a *valid* path. Interval $\langle j_1, j_2 \rangle \in Int(p)$ is called *valid* iff $p[j_1..j_2]$ is a valid path. The set of all valid intervals of a valid path p will be denoted by $VInt(p)$. Since the labeling function is not defined on all possible paths, the notion of subpath-compatibility is adapted accordingly: the domain of Ψ_p injection is restricted to $VInt(p)$.

Let $p = \langle v_1, \ldots, v_m \rangle$ be a valid path and let $S = \hat{l}(p)$. Let b_j denote the number of B-edges in p preceding its j-th vertex. The function $\psi_p : \{1, \ldots, m\} \to \{1, \ldots, |S|\}$ defined by the formula $\psi_p(j) = j - (k-1)b_j$ indicates residues in S corresponding to particular vertices in p, i.e. the residues for which the condition $S[\psi_p(j)] = l(v_j)$ must be satisfied. Thus the function $\Psi_p : VInt(p) \to Int(S)$ defined by formula $\Psi_p(j_1, j_2) = \langle \psi_p(j_1), \psi_p(j_2) \rangle$ ensures subpath-compatibility.

All the definitions from Sect. 2.2 apply to transition graphs with the only restriction that strings can be represented by valid paths only.

Let $S_i, S_{i'}$ be two (not necessarily different) strings from S and assume that their representations have a common B-edge $\pi(S_i)[j..j+1] = \pi(S_{i'})[j'..j'+1]$. We say that these B-edge occurrences are *consistent* iff they are extandable to a common subpath $\pi(S_i)[j - k + 2..j + k - 1] = \pi(S_{i'})[j' - k + 2..j' + k - 1]$. Representation $\langle G, \pi \rangle$ is *consistent* iff all common occurrences of B-edges are consistent. When this condition holds for every pair of valid paths in G (i.e. not necessarily representing S-sequences), we say that G is *consistent*.

3.2   Transformation 1: Split

The Split operation transforms a dBG representation $\langle G, \pi \rangle$ of a string collection $S = \{S_1, \ldots, S_n\}$ into a transition graph representation $\langle G', \pi' \rangle$ by splitting each vertex v of G into k vertices labeled with single characters. More formally, $G' = \langle V', E_B, E_V, l' \rangle$, where:

- $V' = \bigcup_{v \in V} \{v_1, \ldots, v_k\}$,
- $E_V = \bigcup_{v \in V} \{\langle v_1, v_2 \rangle, \ldots, \langle v_{k-1}, v_k \rangle\}$,
- $E_B = \{\langle v_k, w_1 \rangle : \langle v, w \rangle \in E\}$,
- $l'(v_j) = l(v)[j]$ for all $v \in V$ and $j \in \{1, \ldots, k\}$,

and paths $\pi'(S_i)$ are constructed from $\pi(S_i)$ by replacing each vertex v with a sequence $v_1, \ldots, v_k$.

Lemma 2. *G' is a consistent transition graph of length k and $\langle G', \pi' \rangle$ represents S k-completely and k-faithfully.*

Proof. Consistency follows from the fact that every B-edge $\langle v_k, w_1 \rangle$ is preceded and followed by unbranched V-paths $\langle v_1, \ldots, v_k \rangle$ and $\langle w_1, \ldots, w_k \rangle$, respectively. Every common k-mer W in S was reflected in $\langle G, \pi \rangle$ by the unique vertex $v \in V$ such that $l(v) = W$, and thus is also reflected in $\langle G', \pi' \rangle$ by the path $\langle v_1, \ldots, v_k \rangle$, which proves that the representation is k-complete. Finally, k-faithfullness is due to the fact that every occurrence of a vertex v_i in any valid path extends to the same subpath $\langle v_1, \ldots, v_k \rangle$.

3.3   Transformation 2: Merge

The second step of our algorithm consists of a series of local graph modifications merging B-edges sharing either origin or target vertices. Below we describe the first case, the other one is symmetric.

Let v be a vertex having $m > 1$ outgoing B-edges. From the consistency condition each one is preceded by the same V-path $\langle v_2, \ldots, v_{k-1}, v \rangle$ and followed by another V-path of length $k-1$ and the same sequence of labels. Let us denote the paths following the B-edge by $\langle w_1^1, \ldots, w_{k-1}^1 \rangle, \ldots, \langle w_1^m, \ldots, w_{k-1}^m \rangle$.

The Merge operation for each $n \in \{1, \ldots, k-1\}$ merges all vertices $w_n^1, \ldots, w_n^m$ into a single vertex w_n (note that these vertices have the same label, namely $l(v_{n+1})$). Edges incident with merged vertices are replaced with edges joining respective new vertices. Consequently, all B-edges outgoing from v are merged into a single B-edge $\langle v, w_1 \rangle$.

The transformed paths $\pi'(S_i)$ are obtained from $\pi(S_i)$ by replacing all occurrences of vertices $w_n^1, \ldots, w_n^m$ with w_n.

Lemma 3. *The Merge operation preserves the consistency of the graph, as well as the k-completeness and k-faithfulness of the representation.*

Proof. Consistency follows from the fact that merged B-edges are followed in a new graph by a unique path $\langle w_1, \ldots, w_{k-1} \rangle$.

If a common k-mer is reflected in $\langle G, \pi \rangle$ by a V-path p containing the vertex w_n^i, then after the merge, it is reflected in $\langle G', \pi' \rangle$ by a path obtained from p by replacing w_n^i with w_n. Therefore, the representation is still k-complete.

Let $\langle i, j \rangle, \langle i', j' \rangle$ be two π'-occurrences of the same vertex u' in paths $\pi'(S)$, i.e. $u' = \pi'(S_i)[j] = \pi'(S_{i'})[j']$. If corresponding paths in π also have the same vertex u on respective positions (i.e. $u = \pi(S_i)[j] = \pi(S_{i'})[j']$), the sequence supporting the k-extendability of these π-occurrences of u supports the k-extendability of the π'-occurrences of u' too. The only case where the π-paths have different vertices on these positions is when $\pi(S_i)[j] = w_n^m$ and $\pi(S_{i'})[j'] = w_n^{m'}$ for $m \neq m'$.

If both $\pi'(S_i)[j]$ and $\pi'(S_{i'})[j']$ can be extended to the common subpaths of the form $\langle v, w_1, w_2, \ldots, w_{k-1} \rangle$, we have $v = \pi'(S_i)[j - n] = \pi'(S_{i'})[j' - n]$ and this pair of π'-occurrences of v is k-extendable, because vertex v was unaffected by the transformation. Thus there exists a sequence $\langle i, j - n \rangle = \langle i_0, j_0 \rangle, \langle i_1, j_1 \rangle, \ldots, \langle i_l, j_l \rangle = \langle i', j' - n \rangle$ of π'-occurrences of v supporting the k-extendability of this pair. for some $1 < l' < l$, the occurrence $\langle i_{l'}, j_{l'} \rangle$ can be skipped in this sequence, because in this case occurrences $\langle i_{l'-1}, j_{l'-1} \rangle$ and $\langle i_{l'+1}, j_{l'+1} \rangle$ must be directly k-extendable. Therefore we can assume without loss of generality that $j_{l'} < |\pi'(S_{i_{l'}})|$ for all $1 \le l' \le l$. Thus all these occurrences of v are followed in paths $\pi(S_{i_1}), \ldots, \pi(S_{i_1})$ by occurrences of vertices $w_1, w_2, \ldots, w_{k-1}$. Consequently, the sequence $\langle i_0, j_0 + n \rangle, \ldots, \langle i_l, j_l + n \rangle$ supports the k-extendability of the pair of π'-occurrences $\langle i, j \rangle, \langle i', j' \rangle$ of vertex w_n.

The proof is concluded with the observation that each occurrence of w_n originating from w_n^m (respectively $w_n^{m'}$) is k-extendable with some occurrence of this vertex belonging to a subpaths of the form $\langle v, w_1, w_2, \ldots, w_{k-1} \rangle$, and all occurrences belonging to such subpath are k-extendable one with each other.

3.4 Transformation 3: Collapse

The third step of our transformation consists of a series of local modifications, each of which removes one B-edge.

Consider a B-edge $\langle v_{k-1}, w_1 \rangle$. Due to the consistency of the representation all the occurrences of this edge in paths representing S extend to the same subpath $\langle v_1, \ldots, v_{k-1}, w_1, \ldots, w_{k-1} \rangle$ satisfying $l(v_1) = l(w_1), \ldots, l(v_{k-1}) = l(w_{k-1})$. Collapse operation on the edge $\langle v_{k-1}, w_1 \rangle$ consists of:

- merging each pair of vertices v_n, w_n into a new vertex u_n,
- removing the B-edge $\langle v_{k-1}, w_1 \rangle$,
- replacing other edges incident with merged vertices with new edges incident with respective new vertices,
- replacing $\langle v_1, \ldots, v_{k-1}, w_1, \ldots, w_{k-1} \rangle$ with $\langle u_1, \ldots, u_{k-1} \rangle$ in paths representing S,
- replacing other occurrences of v_n and w_n with u_n in paths representing S.

It may happen that a π-path traverses B-edge $\langle v_{k-1}, w_1 \rangle$ several times such that $k - 1$ vertices preceding and $k - 1$ vertices following this edge overlap, i.e. $v_1 = w_{t+1}, v_2 = w_{t+2}, \ldots, v_{k-1-t} = w_{k-1}$ for some $1 < t < k - 1$ (see Fig. 3). In this case:

- in addition, vertices $u_n, u_{t+n}, u_{2t+n}, \ldots$ for $1 \le n < t$ are merged (as a transitive consequence of merging $v_n = w_{t+n}, v_{t+n} = w_{2t+n}, \ldots$),
- a subpath $\langle v_1, \ldots, w_{k-1} \rangle$ in $\pi(S_i)$ traversing B-edge $\langle v_{k-1}, w_1 \rangle$ m times in this way is replaced in $\pi'(S_i)$ with subpath $\langle u_1, u_2 \ldots u_{k-1+(m-1)t} \rangle$, where $u_{n+t} = u_n$.

We will denote by U_n the set of all vertices in representation $\langle G, \pi \rangle$ that will be merged into a single vertex u_n in representation $\langle G', \pi' \rangle$ resulting from the Collapse operation on a B-edge $\langle v_{k-1}, w_1 \rangle$.

It must contain v_n and w_n, but there can be more elements in this set. If $v_n = v_{n+t}$ (and thus path $\langle v_1, \ldots v_{k-1} \rangle$ has a cycle), then also w_{n+t} belongs to U_n and $U_n = U_{n+t}$. Similarly, if $w_n = w_{n+t}$ then v_{n+t} has to belong to U_n. Finally, if $\langle v1, \ldots, v_{k-1} \rangle$ and $\langle w_1, \ldots, w_{k-1} \rangle$ share common vertices $v_n = w_{m_1}, v_{m_1} = w_{m_2}, v_{m_2} = w_{m_3} \ldots, v_{m_{r-1}} = w_{m_r}$ for some positive integer r, then $v_{m_1}, v_{m_2}, \ldots v_{m_r}$ also belongs to U_n and $U_n = U_{m_1} = U_{m_2} = \ldots = U_{m_r}$.

Lemma 4. *The Collapse operation preserves consistency, k-faithfulness and k-completeness of the representation.*

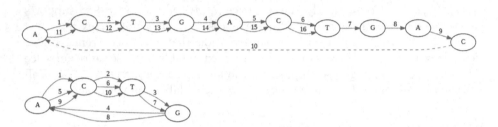

Fig. 3. Example of Collapse transformation applied to a B-edge, in which $k-1$ vertices following this edge overlap with $k-1$ vertices preceding it. $S = \{ACTGACTGACT\}$, $k = 7$, labels above edges indicate their order in the path. Top: B-edge 10 is followed by $k-1$ vertices connected by edges 11–15 and preceded by $k-1$ vertices connected by edges 5–9, the overlap forms a subpath with vertices A and C connected by edge labeled with 5 before the B-edge and 15 after the B-edge. Bottom: after the Collapse operation the overlapping subpath is merged with two other subpaths with vertices A and C: preceding the B-edge and following it.

Proof. Consistency follows from the fact that Collapse does not modify the labels of the vertices.

Similarly, k-completeness follows from the fact that if a common k-mer was reflected in $\langle G, \pi \rangle$ by a valid V-path p, then it is also reflected in $\langle G', \pi' \rangle$ by a path obtained from p by replacing occurrences of v_n (or w_n, respectively) with u_n.

In order to prove k-faithfulness, consider two occurrences $\langle i, j \rangle$ and $\langle i', j' \rangle$ of the same vertex in representation $\langle G', \pi' \rangle$ originating from transformation of $\langle i, q \rangle$ and $\langle i', q' \rangle$ in representation $\langle G, \pi \rangle$. If before the modification of the graph these paths had on respective positions the same vertex $\pi(S_i)[q] = \pi(S_{i'})[q']$, then their extensions are preserved and support the k-extendability of $\langle i, j \rangle$ and $\langle i', j' \rangle$.

If paths in $\langle G, \pi \rangle$ had different vertices on respective positions, then both these vertices belong to the set U_n.

First, consider situation when one of them was v_n and the other was w_n. We know that there is at least one path $\pi(S_i)$ that traverses the whole subpath $\langle v_1, \ldots, v_{k-2}v_{k-1}, w_1, w_2, \ldots, w_{k-1} \rangle$. Thus in $\pi'(S_i)$ we have an occurrence of u_n that corresponds to occurrences of both v_n and w_n in $\pi(S_i)$ on positions separated by $k-2$ vertices. Therefore this occurrence of u_n is k-extendable with all other occurrences originating from v_n, and also with all other occurrences originating from w_n.

For similar reasons all occurrences originating from v_{n+t} have to be k-extendable with the ones originating from w_{n+t}. Thus if $v_n = v_{n+t}$, all occurrences originating from different vertices w_n and w_{n+t} are also k-extendable (and symmetrically: all occurrences originating from different vertices v_n and v_{n+t} are k-extendable if $w_n = w_{n+t}$).

The same argument applies to the case when $v_n = w_{m_1}, v_{m_1} = w_{m_2}, v_{m_2} = w_{m_3}, \ldots, v_{m_{r-1}} = w_{m_r}$: all occurrences of u_n originating from v_n are k-extendable with those originating from v_{m_1}, since $w_{m_1} = v_n$. Because $w_{m_2} = v_{m_1}$, these occurrences are k-extendable with those originating from v_{m_2}. Applying this reasoning repeatedly, we conclude that all occurrences of u_n originating from the vertex v_n are k-extendable with those originating from any vertex v_{m_s} for $s \in \{1, \ldots r\}$.

Thus we showed that all occurrences of vertex u_n in $\pi'(S_i)$ originating from different vertices in $\pi(S_i)$ are always k-extendable one with each other.

3.5   Correctness of the Algorithm

We can now formulate the main result of this section.

Theorem 2. *Given a de Bruijn graph of length k representing a collection of strings S, the transformation algorithm always terminates resulting in a k-complete and k-faithful variation graph representing S.*

Proof. Algorithm terminates, because Split operation is applied only once for each vertex and every execution of Merge or Collapse reduce the number of B-edges in the graph.

The transformations ensure that the final transition graph has no B-edges, so it is in fact a singular variation graph. Lemmas 2-4 guarantee k-completeness and k-faithfulness.

Theorems 1 and 2 can be summarized by the following statement.

Corollary 1. *Let $S = \{S_1, \ldots, S_n\}$ be a set of strings such that $|S_i| \geq k$ for every $i \in \{1, \ldots, n\}$, Then the k-complete and k-faithful variation graph representation of S exists and is unique up to equivalence.*

The algorithm builds a singular graph representation. One can additionally join unbranched paths into single vertices with concatenated labels. Such postprocessing results in an equivalent graph of reduced size.

4 Conclusion

In this article we proposed an axiomatization of the graph representation of a collection of strings. Our axiomatization is based on two properties: k-completeness and k-faithfulness that are always satisfied in de Bruijn graphs and determine up to equivalence the structure of variation graphs. Furthermore, we showed the relationship between variation graphs satisfying the above conditions and de Bruijn graphs. This relationship can be used not only to build a variation graph representing a given set of sequences, but also to provide a direct method of transferring annotations between both pangenome models.

The proposed axiomatization may be further developed. For example, one can formulate properties of a variation graph that express desirable differences from the structure associated with the corresponding de Bruijn graph.

Acknowledgements. This research was funded in whole by National Science Centre, Poland, grant no. 2022/47/B/ST6/03154.

References

1. Baaijens, J.A.: Computational graph pangenomics: a tutorial on data structures and their applications. Nat. Comput. **21**(1), 81–108 (2022). https://doi.org/10.1007/s11047-022-09882-6
2. Baaijens, J.A., Van der Roest, B., Köster, J., Stougie, L., Schönhuth, A.: Full-length de novo viral quasispecies assembly through variation graph construction. Bioinformatics **35**(24), 5086–5094 (2019). https://doi.org/10.1093/bioinformatics/btz443
3. Baaijens, J.A., Stougie, L., Schönhuth, A.: Strain-aware assembly of genomes from mixed samples using variation graphs. BioRxiv (2019). https://doi.org/10.1101/645721. http://biorxiv.org/lookup/doi/10.1101/645721
4. C.P.G. Consortium: Computational pan-genomics: status, promises and challenges. Brief Bioinform. **19**(1), 118–135 (2018). https://doi.org/10.1093/bib/bbw089
5. Durbin, R.: Efficient haplotype matching and storage using the positional burrows-wheeler transform (PBWT). Bioinformatics **30**(9), 1266–1272 (2014). https://doi.org/10.1093/bioinformatics/btu014
6. Eggertsson, H.P., et al.: Graphtyper enables population-scale genotyping using pangenome graphs. Nat. Genet. **49**(11), 1654–1660 (2017). https://doi.org/10.1038/ng.3964
7. Eggertsson, H.P., et al.: GraphTyper2 enables population-scale genotyping of structural variation using pangenome graphs. Nat. Commun. **10**(1), 5402 (2019). https://doi.org/10.1038/s41467-019-13341-9
8. Eizenga, J.M., et al.: Pangenome graphs. Annu. Rev. Genomics Hum. Genet. **21**, 139–162 (2020). https://doi.org/10.1146/annurev-genom-120219-080406
9. Garrison, E., Guarracino, A.: Unbiased pangenome graphs. Bioinformatics **39**(1) (2023). https://doi.org/10.1093/bioinformatics/btac743
10. Garrison, E., Sirén, J., Novak, A.M., et al.: Variation graph toolkit improves read mapping by representing genetic variation in the reference. Nat. Biotechnol. **36**(9), 875–879 (2018). https://doi.org/10.1038/nbt.4227. http://www.nature.com/doifinder/10.1038/nbt.4227

11. Grytten, I., Rand, K.D., Nederbragt, A.J., Storvik, G.O., Glad, I.K., Sandve, G.K.: Graph peak caller: calling ChIP-seq peaks on graph-based reference genomes. PLoS Comput. Biol. **15**(2), e1006731 (2019). https://doi.org/10.1371/journal.pcbi. 1006731

12. Heydari, M., Miclotte, G., Van de Peer, Y., Fostier, J.: BrownieAligner: accurate alignment of Illumina sequencing data to de Bruijn graphs. BMC Bioinform. **19**(1), 311 (2018). https://doi.org/10.1186/s12859-018-2319-7

13. Hickey, G., et al.: Genotyping structural variants in pangenome graphs using the vg toolkit. Genome Biol. **21**(1), 35 (2020). https://doi.org/10.1186/s13059-020-1941-7

14. Hickey, G., et al.: Pangenome graph construction from genome alignment with minigraph-cactus. BioRxiv (2022). https://doi.org/10.1101/2022.10.06.511217. http://biorxiv.org/lookup/doi/10.1101/2022.10.06.511217

15. Iqbal, Z., Caccamo, M., Turner, I., Flicek, P., McVean, G.: De novo assembly and genotyping of variants using colored de Bruijn graphs. Nat. Genet. **44**(2), 226–232 (2012). https://doi.org/10.1038/ng.1028

16. Li, H., Feng, X., Chu, C.: The design and construction of reference pangenome graphs with minigraph. Genome Biol. **21**(1), 265 (2020). https://doi.org/10.1186/s13059-020-02168-z. https://genomebiology.biomedcentral.com/articles/10.1186/s13059-020-02168-z

17. Limasset, A., Cazaux, B., Rivals, E., Peterlongo, P.: Read mapping on de Bruijn graphs. BMC Bioinform. **17**(1), 237 (2016). https://doi.org/10.1186/s12859-016-1103-9

18. Manuweera, B., Mudge, J., Kahanda, I., Mumey, B., Ramaraj, T., Cleary, A.: Pangenome-wide association studies with frequented regions. In: Proceedings of the 10th ACM International Conference on Bioinformatics, Computational Biology and Health Informatics, pp. 627–632. ACM, New York, NY, USA (2019). https://doi.org/10.1145/3307339.3343478

19. Minkin, I., Pham, S., Medvedev, P.: TwoPaCo: an efficient algorithm to build the compacted de Bruijn graph from many complete genomes. Bioinformatics **33**(24), 4024–4032 (2017). https://doi.org/10.1093/bioinformatics/btw609

20. Paten, B., Novak, A.M., Eizenga, J.M., Garrison, E.: Genome graphs and the evolution of genome inference. Genome Res. **27**(5), 665–676 (2017). https://doi.org/10.1101/gr.214155.116

21. Rautiainen, M., Marschall, T.: GraphAligner: rapid and versatile sequence-to-graph alignment. Genome Biol. **21**(1), 253 (2020). https://doi.org/10.1186/s13059-020-02157-2. https://genomebiology.biomedcentral.com/articles/10.1186/s13059-020-02157-2

22. Schaeffer, L., Pimentel, H., Bray, N., Melsted, P., Pachter, L.: Pseudoalignment for metagenomic read assignment. Bioinformatics **33**(14), 2082–2088 (2017). https://doi.org/10.1093/bioinformatics/btx106

23. Sirén, J.: Indexing variation graphs. In: Fekete, S., Ramachandran, V. (eds.) 2017 Proceedings of the Ninteenth Workshop on Algorithm Engineering and Experiments (ALENEX), pp. 13–27. Society for Industrial and Applied Mathematics, Philadelphia, PA (2017). https://doi.org/10.1137/1.9781611974768.2. http://epubs.siam.org/doi/10.1137/1.9781611974768.2

24. Sirén, J., Välimäki, N., Mäkinen, V.: Indexing graphs for path queries with applications in genome research. IEEE/ACM Trans. Comput. Biol. Bioinform. **11**(2), 375–388 (2014). https://doi.org/10.1109/TCBB.2013.2297101

25. Tettelin, H., et al.: Genome analysis of multiple pathogenic isolates of streptococcus agalactiae: implications for the microbial "pan-genome". Proc. Natl. Acad. Sci. USA **102**(39), 13950–13955 (2005). https://doi.org/10.1073/pnas.0506758102
26. Yu, C., Mao, K., Zhao, Y., Chang, C., Wang, G.: Stliter: a novel algorithm to iteratively build the compacted de Bruijn graph from many complete genomes. IEEE/ACM Trans. Comput. Biol. Bioinform. **19**(4), 2471–2483 (2022). https://doi.org/10.1109/TCBB.2021.3062068

CAGE: Cache-Aware Graphlet Enumeration

Alessio Conte(ID), Roberto Grossi(ID), and Davide Rucci$^{(\boxtimes)}$(ID)

University of Pisa, Pisa, Italy
{alessio.conte,roberto.grossi}@unipi.it, davide.rucci@phd.unipi.it

Abstract. When information is (implicitly or explicitly) linked in its own nature, and is modeled as a network, retrieving patterns can benefit from this linked structure. In networks, "graphlets" (connected induced subgraphs of a given size k) are the counterparts of textual n-grams, as their frequency and shape can give powerful insights in the structure of a network and the role of its nodes. Differently from n-grams, the number of graphlets increases dramatically with their size k. We aim to push the exact enumeration of graphlets as far as possible, as enumeration (contrary to counting or approximation) gives the end-user the flexibility of arbitrary queries and restrictions on the graphlets found. For this, we exploit combinatorial and cache-efficient design strategies to cut the computational cost. The resulting algorithm CAGE (Cache-Aware Graphlet Enumeration) outperforms existing enumeration strategies by at least an order of magnitude, exhibiting a low number of L1-L2-L3 cache misses in the CPU. It is also competitive with the fastest known counting algorithms, without having their limitations on k.

Keywords: Graph algorithms · network analysis · graphlets · cache-aware algorithms · enumeration

1 Introduction

Information retrieval for textual documents can benefit from the discovery of patterns underlying the text, e.g. using n-grams (also q-grams, contiguous sequences of n items in a text) [25]. As most modern information is linked in nature, it can be modeled as a network. Patterns themselves can inherit this linked structure, and one of the fundamental tools are the so-called motifs, recurrent and statistically relevant patterns inside a network, i.e., frequent subgraphs: from the seminal papers of Milo et al. [17] and Pržulj [21] these are generally encoded in the form of k-graphlets (or, simply, graphlets), where k is usually kept small ($k \leq 5$). They are the natural candidates for having a role in network analysis, such as n-grams do in text analysis. The frequency and shape of graphlets can give powerful insights in the structure of a network and the role of its nodes.

Graphlets have found a remarkably wide range of applications; among many are finding the important nodes of a graph [3], comparing large graphs [26], comparing temporal networks [2] or understanding the role of genes in pathways and cancer mechanisms [32]. There is also increasing interest over graphlets in the field of machine learning: one can extract information on the number of graphlets involving a node, or where a node occurs in it (i.e., its *orbit*), to produce graph kernels [26], or embeddings [6].

Formally, for a given undirected graph $G = (V, E)$, each set S of k nodes in G induces a subgraph $G[S] = (S, \{(x, y) \in E \mid x, y \in S\})$ by taking the corresponding edges from G. A k-graphlet of G is a set S of k nodes that induces a *connected* subgraph, namely, it satisfies the conditions $S \subseteq V$, $|S| = k$, and $G[S]$ is connected. A *graphlet enumeration algorithm* has G as input, and yields all sets S of G that are k-graphlets as output. An output example is in Fig. 1a.

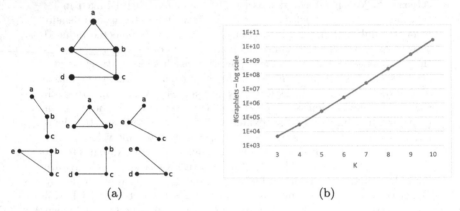

(a)   (b)

Fig. 1. (a) an example graph with its 3-graphlets below; (b) k-graphlet count in the *Brady* network.

The key question addressed by this paper is "how far can we push graphlet enumeration algorithms?". The question addresses the purest version of graphlet discovery, and its relevance is in building an efficient and general tool that can be exploited in any graphlet-related application. We will show a hard limit faced by current enumeration strategies, and how to overcome it.

The main challenge in this problem is the high computational cost involved. Differently from n-grams, the number of graphlets increases dramatically with the size k. Figure 1b shows how graphlets increase exponentially with k, even in a small graph like the Brady network [14] (with 1,117 nodes and 1,330 edges). Existing approaches deal with this problem by keeping k small [1, 4, 11, 12, 16, 19, 20] (e.g., 3 or 4) if graphs are not very small, or by giving up exact results in favor of estimation [5, 7, 26, 29, 30]. Indeed, according to a recent survey on motifs [10], papers focused on estimation are becoming popular in recent years. For more related approaches, we refer the reader to the in-depth survey in [22].

1.1   Results

In this paper we develop a practical output-sensitive graphlet enumeration algorithm that is based on the classical binary partition scheme, a technique that induces a recursive tree, whose nodes are the recursive calls, and solutions are output in the leaves. We refine this strategy to take full advantage of cache memory, while also cutting down the size of the recursion tree beyond what is possible with current enumeration algorithms. Our contributions are as follows:

1. We empirically observed that current enumeration methods cannot further reduce their enumeration tree as the number of *failure* leaves, leaves which do not report a graphlet, is negligible (9% or less) even with simple strategies.
2. Based on this, we aim to go beyond the already good performance of simple enumerators: we achieve this by designing an algorithm that *better exploits cache memory* in the CPU, as we show using the Intel VTune Profiler. Intel CPUs have three cache levels (L1-L2-L3): L3 cache is faster but smaller than RAM (and typically shared among the CPU cores); L2 and L1 are even smaller and faster (typically one private L1-L2 per core). Our algorithm exhibits zero or few L3 cache misses (compared to RAM loads and stores) and L2-L3 cache-bound times have tiny values, in the range 0–5%.
3. Finally, we break through the hard limit on the number of recursive calls, *collapsing the three lowest levels of the recursion tree*. As the tree grows exponentially, this affects the largest levels. We also generate sets of solutions at once, in a compressed – yet easy to access – format.

We call the resulting algorithm CAGE (Cache-Aware Graphlet Enumeration), which constitutes an evolution of the reference enumeration algorithm [12], hereafter called KS-Simple (see Sect. 2), and outperforms it by *over an order of magnitude*. We evaluate CAGE against the state of the art, namely, KS-Simple, Kavosh [11], and FaSE [19], showing dramatically improved performance and scalability. We also include fast counting approaches [1,20] that follow a more analytical method, showing they are more competitive; however, by design they cannot run for $k > 5$, whereas CAGE does not share this limitation.

In the rest of this paper, for an undirected graph $G = (V, E)$, the neighborhood of a node u is $N(u) = \{v \in V \mid (u, v) \in E\}$, the degree of u is $|N(u)|$, and Δ is the maximum degree among all nodes. For a set of nodes $S \subseteq V$, we define the neighborhood of S as $N(S) = \cup_{u \in S} N(u) \setminus S$, the distance-2 neighborhood as $N^2(S) = N(N(S)) \setminus (S \cup N(S))$ and the distance-3 neighborhood as $N^3(S) = N(N^2(S)) \setminus (S \cup N(S) \cup N^2(S))$.

2   Baseline Algorithm

The classical approach to list all k-graphlets in a graph $G = (V, E)$ is based on binary partition. Intuitively, for a graph $G = (V, E)$ and a node $v \in V$, we initialize $S = \{v\}$, and enumerate the k-graphlets containing v by considering a node $u \in N(S)$ and two cases: the k-graphlets containing u, and those that do not. In the first case, enumeration continues with $S := S \cup \{u\}$, and k reduced to $k - 1$; in the second, with the same S and the k-graphlets in $G \setminus \{u\}$.[1] In

---

[1] Here $G \setminus \{u\}$ is G without u and its incident edges.

particular, after setting an arbitrary scanning order for the nodes $v \in V$, each graphlet is built by recursively adding a member of $N(S)$ to S, after the initial call with $S = \{v\}$. We get to a recursive leaf when we reach one of the following:

- *(success leaf)* when $|S| = k$;
- *(failure leaf)* when $|S| < k$ and $|N(S)| = 0$.

One of the first algorithm able to enumerate all k-graphlets is ESU [31], while the current state-of-the-art algorithm is that of Komusiewicz and Sommer [12], hereafter called KS-Simple, that follows the above binary partition scheme, but with a clever optimization summarized below in Property 1.

Property 1 ([12]). *If a k-graphlet containing S does not exist in G after taking $u \in N(S)$ then no k-graphlet can exist in $G \setminus \{u\}$ with the same S.*

As an example, consider the graph in Fig. 1a while enumerating all 4-graphlets containing vertex b in $G \setminus \{a\}$. After the output of $\{b, c, d, e\}$, vertex e is removed and no more 4-graphlets (in particular containing b) exist. Thanks to Property 1 we can stop the computation as soon as the next recursive call, which will obviously fail. This property allows KS-Simple to have a running time of $O(k^2 \Delta)$ per graphlet.

3   The CAGE Algorithm

In this section we present our main contribution, the CAGE algorithm. We give the pseudocode and discuss its cache-awareness, carefully considering all the scenarios that the algorithm may encounter during its execution, and show how we can effectively compress the last three levels of its recursion tree.

3.1   Addressing a Hard Combinatorial Limit

Before describing our algorithm, we introduce the principles behind its design.

First, we opted for a data-driven design and empirically observed that – on a dataset of over 150 graphs – the total number of failure leaves in the binary partition scheme is in fact orders of magnitude smaller than the total number of leaves, never surparssing 9%.

Second, we wanted to better exploit the cache in the CPU, the main motivation for this paper. For example, if we need to check whether an edge (x, y) exists and the adjacency list $N(x)$ is already in cache, whereas we have no such a guarantee for $N(y)$, it is clearly more efficient to test membership $y \in N(x)$ rather than $x \in N(y)$. Therefore, we carefully orchestrated the loading of graph data during execution (see comments in the pseudocode).[2]

---

[2] We are not directly controlling the cache, but rather allowing the algorithm to run cache-friendly by making standard assumptions on the associative cache [8].

Third, once some data is in the cache in the CPU, we would like to exploit it at its best before it is released to load further data. As $|S| < k$ and k is small, it is reasonable to assume that both S and $N(S)$ fit into the cache most of the times, as $|N(S)| < k\Delta$ (exceptions may occur for large Δ). Based on this, we stop the recursion when $|S| = k - 3$, and deduce all the possible extensions of S into graphlets by adding three further nodes. This greatly truncates the recursion tree, as each level of calls can be Δ times more populated than the previous one (potentially reducing the total number of calls by a $O(\Delta^3)$ factor).[3] The last two principles are intimately connected to each other, and give a speedup of at least one order of magnitude in our experiments in Sect. 4.

3.2 Pseudocode

Algorithm 1 implements our principles stated above. It is called Cache-Aware Graphlet Enumeration (CAGE), and is a non-trivial and cache-aware extension of KS-Simple [12].

According to the first principle, we replaced the ENUM() function in KS-Simple with the one in Algorithm 1. We decided to keep the for loop of KS-Simple at lines 20–24, and fixed the scanning order of the vertices top-level loop for efficient memory usage: as we only look for the graphlets in which v is the smallest node of the order, we know all vertices smaller than v are in X, so we do not need to store them explicitly. The second and third principles are implemented at lines 2–19, as the base case. We observe that the base case is no more $|S| = k$ as in KS-Simple, but $|S| = k - 3$ as shown in Sect. 3.3. This is the outcome of several iterative improvements over KS-Simple.

3.3 Exploiting What is Already Cached

Consider S of size $|S| = k - 3$, and let v be the smallest node in S according to an arbitrary ordering (we use label ordering) of the vertices of G. As we have to extend S in all possible ways with 3 nodes, we pick them from $N(S)$, $N^2(S)$, and $N^3(S)$, ignoring the nodes belonging to X (or only reachable from S via X). Specifically, X is composed of the nodes $v' \in V$ such that $v' < v$, and of the nodes that have already been examined in the loop at lines 20–24 in the parent call. We only need to store explicitly the latter nodes.

When picking the 3 nodes, we have to deal with 4 different cases. They are discussed below with an accompanying drawing shown in Fig. 2.

Case 1: The 3 nodes come from $N(S) \setminus X$ (see line 4).

Case 2: A node u is from $N(S) \setminus X$, and the other 2 nodes are chosen from $N^2(S) \setminus X$ so that they are u's neighbors (see lines 8 and 16).

Case 3: Distinct nodes u, v are from $N(S) \setminus X$, and the remaining node z is chosen from $N^2(S) \setminus X$, so that

- *Case 3a:* z is a common neighbor of both u and v (see lines 13 and 17), or

---

[3] This idea can be generalized to $k - 4$, $k - 5$, and so on, but there is no payoff going further than $k - 3$ in practice as there are too many cases to handle.

Algorithm 1: CAGE: cache-efficient ENUM() algorithm, *solutions* is a global variable.

---

1 **Function** ENUM(G,S,k,X)// Hp. $k > 3$ (case $k = 3$ is simpler)
  // find all graphlets in $G \setminus X$ containing all nodes in S
2   **if** $|N(S) \setminus X| = 0$ **then return false** // failure leaf
3   **if** $|S| = k - 3$ **then** // exploit the cache to complete S
4     $localcount \leftarrow \binom{|N(S) \setminus X|}{3}$ // (1) zero if $|N(S) \setminus X| < 3$
5   **forall** $u \in N(S) \setminus X$ **do** // $N(S)$ loaded in cache
6       $udeg \leftarrow duplicated \leftarrow 0$
7   **forall** $z \in (N(u) \setminus (N(S) \cup S)) \setminus X$ **do** // $z \in N^2(S) \setminus X$
        // $N(u)$ and S loaded in cache
8         $udeg \leftarrow udeg + 1$ // needed for (2)
9   **forall** $w \in (N(z) \setminus (N(S) \cup S)) \setminus X$ **do** // (4) see Fig. 2
          // $N(S), N(u)$, S cached, and $N(z)$ loaded
10   **if** $w \notin N(u)$ **then** $localcount \leftarrow localcount + 1$
11   **forall** $v \in (N(S) \setminus \{u,z\}) \setminus X$ **do** // (3) $N(S), N(z)$ cached
12   **if** $v \in N(z)$ **then** // (3a) counted twice, from v and z
13             $duplicated \leftarrow duplicated + 1$
14   **else** // (3b)
15             $localcount \leftarrow localcount + 1$
16       $localcount \leftarrow localcount + \binom{udeg}{2}$ // (2) see Section 3.4
17       $localcount \leftarrow localcount + duplicated/2$ // (3a) duplicates counted once
18     $solutions \leftarrow solutions + localcount$
19   **return** ($localcount > 0$)
20   $found \leftarrow$ **false**
21   **forall** $u \in N(S) \setminus X$ **do** // here $|S| < k - 3$. Vertices sorted by label
22   **if** ENUM($G, S \cup \{u\}, k, X$) **then** $found \leftarrow$ **true else break**
23     $X \leftarrow X \cup \{u\}$
24   **return** $found$

- *Case 3b:* z is a neighbor of only u (symmetrically, only v) (see line 15).

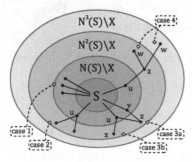

Case 4: A node u is from $N(S) \setminus X$, a node z from $(N^2(S) \setminus X) \cap N(u)$, and the remaining node w from $N(z) \setminus (N(S) \cup S \cup N(u)) \setminus X$, which comes either from $N^3(S)$ or $N^2(S) \setminus N(u)$ (see line 10). As a result, CAGE is actually pruning the leaves, their parents, and their grandparents in the recursion tree, each time reducing the total number of calls roughly by a factor of Δ.

Fig. 2. Cases 1–4

As each case involves the same or fewer steps than KS-Simple [12], CAGE shares its worst-case running time guarantee:

Theorem 1. *CAGE reports all k-graphlets including the nodes in S, but not the nodes in X, in $O(k^2\Delta)$ time per graphlet using $O(\min\{k\Delta, |V|\})$ space.*

In order to make our implementation as cache-friendly as possible, we coupled the pruning of the recursion tree with some cache-friendly data structures so that S, $N(S) \setminus X$, and $N(u)$ are stored as vectors, $N(S)$ is stored as a cuckoo hash table [18], and X is not explicitly stored due to vertex ordering. We have:

Lemma 1. *Given an associative CPU cache [9], if its size is $\Omega(k\Delta)$ words of memory, then the base case $|S| = k - 3$ in CAGE loads S, $N(S)$, $N(u)$, and $N(z)$ once for distinct nodes u and z, and then keeps them in the cache when they are accessed.*

Comments in the pseudocode at lines 5–17 give indications when S, $N(S)$, $N(u)$, and $N(z)$ are loaded into cache, or already cached. This predicted behavior has been carefully planned during our experimental study. Moreover, the cache words used in practice are likely to be far less than the worst case $O(k\Delta)$, as real-world networks are sparse, with most vertices having very small degree.

3.4 Remarks

We did not discuss the case $k = 3$ as CAGE is mainly aimed at larger values of k, and it can be easily obtained as a simplification of the pseudocode. Moreover, for simplicity, we only report the number of solutions found, and the graphlets are not output. All algorithms in the experiments also do not make explicit output, as is usual when comparing enumeration algorithms [24]. Furthermore, CAGE can produce a compressed output, efficient to both read and write and a common practice in enumeration (see, e.g., [27,28]): it suffices to output the $k - 3$ nodes in S, as well as the sets from which to choose the completions, e.g., for *case 1*, write the contents of S, followed by "plus any 3 elements from", followed by the contents of $N(S) \setminus X$. When dealing with case 3a, we can decide to output only the choice $u < v$, to avoid the duplication when z is connected to both u and v.

4 Experimental Results

This section provides the details and results of our extensive experimental phase.

4.1 Environment and Dataset

Our experiments were carried on a dual-processor Intel Xeon Gold 5318Y Icelake @ 2.10GHz machine, with 48 physical cores each and 1TB of shared RAM; private cache L1 per core: 48K, private cache L2 per core: 1.25MB, shared cache L3: 36MB, running Ubuntu Server 22.04 LTS, Intel C++ compiler `icpx`, version 2022.1.0, and cache analysis was done with Intel VTune profiler, version 2022.2.0.

The dataset consists of 155 graphs taken from public repositories LASAGNE [14], Network Repository [23], and SNAP [13]. These files range in size from very small (hundreds of nodes and edges) to very large (millions of nodes and half a billion edges). Table 1 shows a relevant subset of the dataset, and the 10 graphs mentioned in this section can be downloaded from GitHub[4], as well as the source code of our algorithm and the re-implemented KS-Simple.

Table 1. A sample from our dataset, sorted by $|E|$

Graph	Type	$|V|$	$|E|$	Δ
Brady	Biological	1,117	1,330	28
ca-GrQc	Collaboration	5,242	14,484	81
cti	Mesh	16,840	48,232	6
Wing	Mesh	62,032	121,544	4
Roadnet-TX	Road Network	1,379,917	1,921,660	12
Roadnet-CA	Road Network	1,965,206	2,766,607	12
Auto	Mesh	448,695	3,314,611	37
Hugetrace-00	DIMACS10	4,588,486	6,879,133	4
IMDB	Movies	913,201	37,588,613	11,941
Arabic-2005	Web Crawl	22,744,080	553,903,073	575,628

4.2 CAGE Implementation and Comparison Methodology

We implemented CAGE in C++, compiled with the highest optimization flag -O3. The cache-friendly nature discussed in Algorithm 1 is enhanced by the likely and unlikely macros to help the compiler with branch prediction, and we do not physically delete and restore nodes between recursive calls by reusing the data structures mentioned in Sect. 3, e.g., traversing the vector for $N(S) \setminus X$ so that deletion of nodes is modeled by a flexible starting index.

Below we adopt the notation CAGE-1, CAGE-2, and CAGE-3, with the intent of distinguishing among the base cases $|S| = k - 1$, $|S| = k - 2$, and $|S| = k - 3$, respectively. Indeed, CAGE-3 is actually CAGE, and the others can be easily derived from by simplifying the pseudocode in Algorithm 1.

We compared our algorithms against several different algorithms from the literature, including recent and well known approaches from [22, Table 2], i.e., PGD [1], Kavosh [11], FaSE [19] and Escape [20] (although we exclude the matrix-based approach [16], unsuitable for large graphs), as well as the more recent KS-Simple [12]. The aforementioned methods adopt different strategies to exactly count the number of k-graphlets, but they all share an enumerative core, either via implicit counting (i.e., PGD and Escape) or explicit construction of the graphlets (the others). While we do not discuss their implementations for

---

[4] https://github.com/DavideR95/CAGE.

lack of space, we refer to their papers for the available sources. Since the available code of KS-Simple [12] is in Python, we reimplemented it in C++ to avoid penalizing it, following the same guidelines as for CAGE. We then conducted the following experiments:

(i) Analysis of cache efficiency and hotspots in the code using VTune, setting a timeout of 15 min on a subset of the dataset built to represent differently sized graphs,

(ii) Analysis of the recursion tree of KS-Simple for $k = 4, 5, 7, 9$ on the entire dataset with a time limit of 30 min, showing that the number of failure leaves is typically less than 1% of the total number of leaves, and never more than 9% (details omitted for space reasons),

(iii) A stress test for all the algorithms and competitors, for $k \in [4, 10]$ with a timeout of 12 h.

Table 2. VTune Profiler cache access statistics for our algorithms and competitors with $k = 7$. †: execution stopped after 15 min. *: execution stopped due to a memory allocation problem.

Graph	Algorithm	Time (s)	#Graphlets Found ($k = 7$)	L3 Misses	L1 Bound	L2 Bound	L3 Bound	Loads	Stores
ca-GrQc	KS-Simple †	†	8,577,821,416	0	7.9%	0.7%	0%	3,531 E+9	1,394 E+9
	Kavosh	†	884,849,128	65,553,660	8.2%	0.3%	0.1%	3,075 E+9	1,782 E+9
	FaSE	†	2,448,373,561	5,429,820	10.2%	0.7%	0%	1,912 E+9	276 E+9
	CAGE-1	76	**15,186,322,814**	0	8.8%	0.6%	0%	303 E+9	117 E+9
	CAGE-2	39	**15,186,322,814**	0	11.3%	1.5%	0%	116 E+9	7 E+9
	CAGE-3	**34**	**15,186,322,814**	0	15.4%	2.8%	0%	95 E+9	2 E+9
roadnet-TX	KS-Simple	14	**203,059,778**	0	14.9%	0%	0%	43 E+9	23 E+9
	Kavosh	*	–	–	–	–	–	–	–
	FaSE	46	**203,059,778**	0	14.2%	0.5%	0%	128 E+9	37 E+9
	CAGE-1	5	**203,059,778**	0	16.2%	0%	0.6%	14 E+9	8 E+9
	CAGE-2	4	**203,059,778**	0	22.3%	0%	0%	6 E+9	2 E+9
	CAGE-3	**3**	**203,059,778**	0	28.2%	0%	1.3%	5 E+9	1 E+9
auto	KS-Simple	†	4,775,173,331	0	14.2%	0.3%	0%	2,360 E+9	1,051 E+9
	Kavosh	*	–	–	–	–	–	–	–
	FaSE	†	3,032,335,810	27,202,000	12.5%	0.7%	0.1%	2,027 E+9	365 E+9
	CAGE-1	†	80,580,776,005	0	14.2%	0.3%	0%	2,176 E+9	1,014 E+9
	CAGE-2	†	135,096,265,408	0	21.9%	0.5%	0.1%	1,480 E+9	138 E+9
	CAGE-3	†	**152,621,021,219**	0	21.1%	1.4%	0.1%	2,080 E+9	53 E+9
arabic-2005	KS-Simple	†	4,001,309,731	27,157,305	6%	0.7%	5.1%	3,626 E+9	593 E+9
	Kavosh	*	–	–	–	–	–	–	–
	FaSE	*	–	–	–	–	–	–	–
	CAGE-1	†	22,084,290,111,889	48,516,858	5.9%	0.8%	4.9%	2,976 E+9	552 E+9
	CAGE-2	†	27,208,214,120,342	21,697,928	7.8%	1%	4.2%	1,820 E+9	11 E+9
	CAGE-3	†	**47,718,156,097,277**	173,529,344	1%	0.3%	0.2%	3,329 E+9	23 E+9

4.3 Cache Analysis

During our implementation of CAGE, we took our design decisions based upon the performance metrics given by Intel VTune Profiler, tweaking the code according to the statistics of cache accesses and misses. The results of the VTune analysis for the optimized versions of KS-Simple, CAGE-1, CAGE-2, and CAGE-3 are summarized in Table 2, along with the same data for our competitors Kavosh [11] and FaSE [19]. We chose not to include PGD [1] and ESCAPE [20] in this analysis since they only work for $k \leq 5$, while our focus is on larger values of k, in order to show the performance scaling to larger working sets (i.e. larger S and $N(S)$). The macro rows correspond to four graphs, whose names are given in the first column. We report the cache performance by taking the best out of 5 non-consecutive executions of VTune, where the computation was stopped after 15 min for two large graphs[5]. The other columns report: the run time in seconds, number of graphlets found for $k = 7$, number of L3 cache misses (i.e. accesses to the RAM), how much L1, L2, L3 cache affected on the percentage of clock ticks where the CPU was stalled waiting on that level of cache (the lower, the better), and the total number of load and store operations.

From the results on our algorithms, it clearly emerges that L2 and L3 bounds are very small, whereas L1 is larger for small graphs. This is a sign of good cache usage, as the L2 and L3 cache misses are definitely more expensive than the L1 cache misses (both data and instructions). For the small graphs ca-GrQc and roadnet-TX and mid-size graph auto, we have zero L3 misses (as they probably fit in the L3); for large graph arabic-2005, the number of L3 cache misses increases going through the rows for KS-Simple, CAGE-1, CAGE-2, and CAGE-3. This may not be obvious as there is a timeout of 15 min for this large graph: consequently, KS-Simple produces very few solutions compared to the others, and CAGE-3 more solutions (the apparent anomaly for CAGE-3 on arabic-2005 can be explained by dividing the numbers by 2 in its last row in the table, so that we roughly get the same number of solutions as CAGE-1 and CAGE-2, we have also similar numbers in the other columns, except for L3 misses, which we discuss in a while). The number of loads/stores follows a similar pattern to that for L3 misses. The number of L3 cache misses is negligible with respect to the total number of load and store instructions issued, allowing the percentage of L2- and L3-bound to stay always within 5%. On the other hand, our competitors are able to achieve similar results in terms of L1, L2, and L3 bound, but the number of L3 misses is higher even on ca-GrQc. FaSE is able to compute all the solutions with zero cache misses on roadnet-TX, but with a much higher time requirement compared to CAGE. Additionally, the Kavosh algorithm started having bad memory allocation[6] issues already with roadNet-TX, while FaSE fails later on the largest graph.

We also evaluated the size of the data structures for $N(S)$, $N(u)$, and $N(z)$, examined during the for loop of Algorithm 1 along with the average degree of the networks, as shown in Table 3, for $k = 7$. They clearly fit into the cache most of the times, recalling that each node identifier is a 32-bit integer and that the L2 cache in each core is 1.25MB on our machine. For arabic-2005, even if its maximum degree Δ is large, its average degree is small, as well as the median

---

[5] This timeout is due to the size of the data gathered by VTune, which grows quickly over time.

[6] i.e. the runtime raised a `bad_alloc` exception or a segmentation fault while reading the input graph.

values for $N(\cdot)$. Lemma 1 indicates that CAGE-3 is cache-friendly under this condition on the average.

Table 3. Median size of $N(S)$, $N(u)$ and $N(z)$ for a subset of the dataset ($k = 7$).

Graph	Δ	avg. degree	$N(S)$	$N(u)$	$N(z)$
ca-GrQc	81	5.5	52	13	9
roadnet-TX	12	2.8	3	3	8
auto	37	14.7	35	15	15
arabic-2005	575,628	48.7	115	36	49

According to these results, we believe that CAGE-3 is preferable to CAGE-1 and CAGE-2 as it finds more solutions for the given timeout, even though in some cases CAGE-2 could perform as well as CAGE-3.

Finally, we remark that CAGE-4 (i.e., adopting $|S| = k - 4$ as base case) might not be faster than CAGE-3, as a recursive call needs to address 8 scenarios individually and explore $N^4(S)$ instead of $N^3(S)$: this increases code complexity, requires extensive fine-tuning and could reduce cache friendliness (adding nested for-loops and more load operations); as such we leave this study for future work.

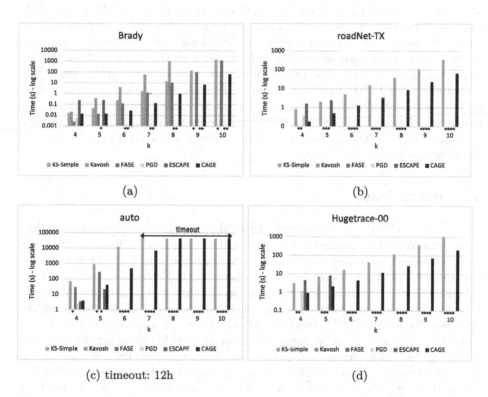

(a)   (b)

(c) timeout: 12h   (d)

Fig. 3. Running time of our competitors vs. CAGE on four significant input graphs. *: an execution issue occurred.

4.4   Running Time Analysis

Armed with our fine-tuned implementation of CAGE (i.e. CAGE-3), we compared it to the other methods mentioned in Sect. 4.2. We performed an extensive validation phase, measuring both the running time and the number of graphlets found, with a timeout of 12 h. Some methods could not terminate the execution for some issues, e.g. they could not handle a graph so large, ran into memory issues, or were killed by the operating system.

Figure 3 shows a summary of the results obtained in four graphs of increasing size, Brady, roadNet-TX, auto, and Hugetrace-0000 (see Table 1 for their characteristics). On the y-axis, we reported the running time in seconds, in logarithmic scale; on the x-axis, we report increasing values of $k \in [4, 10]$. For each value of k, each bar corresponds to an algorithm as specified in the legend, where lower bars mean better performances. The executions that had the issues mentioned above, have an asterisk '*' in place of the corresponding bar. For graph auto, the executions for $k = 7, 8, 9, 10$ were in timeout (indicated with a bar).

We can observe that KS-Simple and CAGE perform quite well, with CAGE running faster than KS-Simple by roughly an order of magnitude (for $k = 7$ on auto, KS-Simple was in timeout whereas CAGE was the only one to terminate its execution). Both compare significantly better with the other methods, as the latter ones are either too slow or cannot execute that instance of the graph. An exception is the PGD algorithm [1], as it achieves faster runnning time than CAGE on Brady and auto, but it does not scale well for the other two graphs; moreover, PGD is explicitly designed to work only with $k = 3, 4$. These results, combined with the code profiling statistics of Sect. 4.3, confirm the benefits of our design principles for CAGE.

5   Conclusions

We presented CAGE, a cache-aware algorithm for k-graphlet enumeration that is an evolution of the classical binary partition technique. CAGE builds upon our empirical observation that binary partition produces only few failure leaves in the computational recursion tree. We developed and analyzed several variants of CAGE, iteratively cutting more recursion levels, that were fine-tuned during an extensive experimental phase based on cache analysis and number of graphlets reported. We believe CAGE can be used as the core for more sophisticated enumeration tasks, such as orbit enumeration or isomorphic graphlet classification [15,21] as well as arbitrary user-defined queries, boosting analysis possibilities for large networks. Future work will consider applying the analysis capabilities of CAGE to graph analysis tasks, as well as attempting to increase the levels of recursion-tree pruning beyond 3.

Acknowledgements. Work partially supported by MIUR project n. 20174LF3T8 Algorithms for Harnessing networked Data (AHeAD).

References

1. Ahmed, N.K., Neville, J., Rossi, R.A., Duffield, N.: Efficient graphlet counting for large networks. In: 2015 IEEE International Conference on Data Mining, pp. 1–10. Atlantic City, NJ, USA, IEEE (2015)
2. Aparício, D., Ribeiro, P., Silva, F.: Graphlet-orbit transitions (GoT): a fingerprint for temporal network comparison. PLoS ONE **13**(10), e0205497 (2018)
3. Aparício, D., Ribeiro, P., Silva, F., Silva, J.: Finding dominant nodes using graphlets. In: Cherifi, H., Gaito, S., Mendes, J.F., Moro, E., Rocha, L.M. (eds.) COMPLEX NETWORKS 2019. SCI, vol. 881, pp. 77–89. Springer, Cham (2020). https://doi.org/10.1007/978-3-030-36687-2_7
4. Bhuiyan, M.A., Rahman, M., Rahman, M., Al Hasan, M.: Guise: uniform sampling of graphlets for large graph analysis. In: 2012 IEEE 12th International Conference on Data Mining, pp. 91–100. IEEE (2012)
5. Bressan, M., Chierichetti, F., Kumar, R., Leucci, S., Panconesi, A.: Motif counting beyond five nodes. ACM TKDD **12**(4), 1–25 (2018)
6. Dutta, A., Riba, P., Lladós, J., Fornés, A.: Hierarchical stochastic graphlet embedding for graph-based pattern recognition. Neural Comput. Appl. **32**(15), 11579–11596 (2020)
7. Elenberg, E.R., Shanmugam, K., Borokhovich, M., Dimakis, A.G.: Beyond triangles: a distributed framework for estimating 3-profiles of large graphs. In: ACM SIGKDD, pp. 229–238 (2015)
8. Frigo, M., Leiserson, C.E., Prokop, H., Ramachandran, S.: Cache-oblivious algorithms. ACM Trans. Algorithms **8**(1), 4:1–4:22 (2012)
9. Harris, S.L., Harris, D.: 8 - memory systems. In: Harris, S.L., Harris, D. (eds.) Digital Design and Computer Architecture, pp. 498–541. Morgan Kaufmann, Burlington (2022)
10. Jazayeri, A., Yang, C.C.: Motif discovery algorithms in static and temporal networks: a survey. J. Complex Netw. **8**(4), cnaa031 (2020)
11. Kashani, Z.R.M., et al.: Kavosh: a new algorithm for finding network motifs. BMC Bioinform. **10**(1), 318 (2009)
12. Komusiewicz, C., Sommer, F.: Enumerating connected induced subgraphs: improved delay and experimental comparison. Discret. Appl. Math. **303**, 262–282 (2021)
13. Leskovec, J., Krevl, A.: SNAP Datasets: Stanford large network dataset collection (2014). http://snap.stanford.edu/data
14. Marino, A. , Crescenzi, P.: LASAGNE Networks: Laboratory of Algorithms, modelS, and Analysis of Graphs and NEtworks (2015). http://www.pilucrescenzi.it/lasagne/content/networks.html
15. Melckenbeeck, I., Audenaert, P., Colle, D., Pickavet, M.: Efficiently counting all orbits of graphlets of any order in a graph using autogenerated equations. Bioinformatics **34**(8), 1372–1380 (2017)
16. Melckenbeeck, I., Audenaert, P., Van Parys, T., Van De Peer, Y., Colle, D., Pickavet, M.: Optimising orbit counting of arbitrary order by equation selection. BMC Bioinform. **20**(1), 1–13 (2019)
17. Milo, R., Shen-Orr, S., Itzkovitz, S., Kashtan, N., Chklovskii, D., Alon, U.: Network motifs: simple building blocks of complex networks. Science **298**(5594), 824–827 (2002)
18. Pagh, R., Rodler, F.F.: Cuckoo hashing. J. Algorithms **51**(2), 122–144 (2004)

19. Paredes, P., Ribeiro, P.: Towards a faster network-centric subgraph census. In: IEEE/ACM ASONAM, pp. 264–271, New York, NY, USA, ACM (2013)
20. Pinar, A., Seshadhri, C., Vishal, V.: Escape: efficiently counting all 5-vertex sub-graphs. In: The Web Conference (WWW), pp. 1431–1440 (2017)
21. Pržulj, N.: Biological network comparison using graphlet degree distribution. Bioinformatics **23**(2), e177–e183 (2007)
22. Ribeiro, P., Paredes, P., Silva, M.E., Aparicio, D., Silva, F.: A survey on subgraph counting: concepts, algorithms, and applications to network motifs and graphlets. ACM Comput. Surv. **54**(2), 1–36 (2021)
23. Rossi, R., Ahmed, N.: The network data repository with interactive graph analytics and visualization. In: AAAI (2015)
24. Ruskey, F.: Combinatorial generation. Preliminary Working Draft, vol. 11, pp. 20. University of Victoria, Victoria, BC, Canada (2003)
25. Shannon, C.E.: A mathematical theory of communication. Bell Syst. Tech. J. **27**(3), 379–423 (1948)
26. Shervashidze, N., Vishwanathan, S.V.N., Petri, T., Mehlhorn, K., Borgwardt, K.: Efficient graphlet kernels for large graph comparison. In: van Dyk, D., Welling, M. (eds.) AISTATS, vol. 5, pp. 488–495 (2009). PMLR 16–18
27. Shioura, A., Tamura, A., Uno, T.: An optimal algorithm for scanning all spanning trees of undirected graphs. SIAM J. Comput. **26**(3), 678–692 (1997)
28. Tomita, E., Tanaka, A., Takahashi, H.: The worst-case time complexity for generating all maximal cliques and computational experiments. Theoret. Comput. Sci. **363**(1), 28–42 (2006)
29. Wang, P., Lui, J.C., Ribeiro, B., Towsley, D., Zhao, J., Guan, X.: Efficiently estimating motif statistics of large networks. ACM TKDD **9**(2), 1–27 (2014)
30. Wang, P., et al.: Moss-5: a fast method of approximating counts of 5-node graphlets in large graphs. IEEE TKDE **30**(1), 73–86 (2017)
31. Wernicke, S.: Efficient detection of network motifs. IEEE/ACM Trans. Comput. Biol. Bioinf. **3**(4), 347–359 (2006)
32. Windels, S.F., Malod-Dognin, N., Pržulj, N.: Graphlet eigencentralities capture novel central roles of genes in pathways. PLoS ONE **17**(1), e0261676 (2022)

Space-Time Trade-Offs for the LCP Array of Wheeler DFAs

Nicola Cotumaccio[1,2]($\boxtimes$) (iD), Travis Gagie[2] (iD), Dominik Köppl[3] (iD),
and Nicola Prezza[4] (iD)

[1] GSSI, L'Aquila, Italy
nicola.cotumaccio@gssi.it
[2] Dalhousie University, Halifax, Canada
{nicola.cotumaccio,travis.gagie}@dal.ca
[3] University of Münster, Münster, Germany
dominik.koeppl@uni-muenster.de
[4] University Ca' Foscari, Venice, Italy
nicola.prezza@unive.it

Abstract. Recently, Conte et al. generalized the longest-common prefix (LCP) array from strings to Wheeler DFAs, and they showed that it can be used to efficiently determine matching statistics on a Wheeler DFA [DCC 2023]. However, storing the LCP array requires $O(n \log n)$ bits, n being the number of states, while the compact representation of Wheeler DFAs often requires much less space. In particular, the BOSS representation of a de Bruijn graph only requires a linear number of bits, if the size of alphabet is constant.

In this paper, we propose a sampling technique that allows to access an entry of the LCP array in logarithmic time by only storing a linear number of bits. We use our technique to provide a space-time trade-off to compute matching statistics on a Wheeler DFA. In addition, we show that by augmenting the BOSS representation of a k-th order de Bruijn graph with a linear number of bits we can navigate the underlying variable-order de Bruijn graph in time logarithmic in k, thus improving a previous bound by Boucher et al. which was linear in k [DCC 2015].

Keywords: Wheeler graphs · LCP array · de Bruijn graphs ·
Matching statistics · Variable-order de Bruijn graphs

1 Introduction

In 1973, Weiner invented the *suffix tree* of a string [28], a versatile data structure which allows to efficiently handle a variety of problems, including solving pattern matching queries, determining matching statistics, identifying combinatorial properties of the string and computing its Lempel-Ziv decomposition. However, the space consumption of a suffix tree can be too high for some applications (including bioinformatics), so over the past 30 years a number of *compressed* data structures simulating the behavior of a suffix tree have been designed, thus

© The Author(s), under exclusive license to Springer Nature Switzerland AG 2023
F. M. Nardini et al. (Eds.): SPIRE 2023, LNCS 14240, pp. 143–156, 2023.
https://doi.org/10.1007/978-3-031-43980-3_12

leading to compressed suffix trees [26]. In many applications, one does not need the full functionality of a suffix tree, so it may be sufficient to store only some of these data structures. Among the most popular data structures, we have the suffix array [21], the longest common prefix (LCP) array [21], the Burrows-Wheeler transform (BWT) [6] and the FM-index [13].

In the past 20 years, the ideas behind the suffix array, the BWT and the FM-index have been generalized to trees [12,14], de Bruijn graphs [5], Wheeler graphs [1,17] and arbitrary graphs and automata [8,9]. Broadly speaking, Wheeler graphs concisely capture the intuition behind these data structures in a graph setting; thus, they can be regarded as a benchmark for extending suffix tree functionality to graphs. In particular, the LCP array of a string remarkably extends the functionality of the suffix array, and a recent paper [7] shows that the LCP array can also be generalized to Wheeler DFAs, which represents a remarkable step toward fully simulating suffix-tree functionality in a graph setting. However, the solution in [7] is not space efficient: storing the LCP array of a Wheeler DFA requires $O(n \log n)$ bits, n being the number of states. If the size σ of the alphabet is small, this space can be considerably larger than the space required to store the Wheeler DFA itself. As we will see, if $\sigma \log \sigma = o(\log n)$, then the space required to store the Wheeler DFA is $o(n \log n)$, and if $\sigma = O(1)$, then the space required to store the Wheeler DFA is $O(n)$. The latter case is especially relevant in practice, because de Bruijn graphs are the prototypes of Wheeler graphs, and in bioinformatics de Bruijn graphs are defined over the constant-size alphabet $\Sigma = \{A, C, G, T\}$.

In this paper, we show that we can *sample* entries of the LCP array in such a way that, by storing only a linear number of additional bits on top of the Wheeler graph, we can compute each entry of the LCP array in logarithmic time, thus providing a space-time trade-off. More precisely:

Theorem 1. *We can augment the compact representation of a Wheeler DFA $\mathcal{A}$ with $O(n)$ bits ($O(n \log \log \sigma)$ bits, respectively), where n is the number of states and σ is the size of the alphabet, in such a way that we can compute each entry of the LCP array of $\mathcal{A}$ in $O(\log n \log \log \sigma)$ time ($O(\log n)$ time, respectively).*

We present two applications of our result: computing matching statistics on Wheeler DFAs and navigating variable-order de Bruijn graphs.

Matching Statistics on Wheeler DFAs. The problem of computing matching statistics on a Wheeler DFA is defined as follows: given a pattern of length m and a Wheeler DFA with n states, determine the longest suffix of each prefix of the pattern that occurs in the graph (that is, that can be read by following some edges on the graph and concatenating the labels). This problem is a natural generalization of the problem of computing matching statistics on strings. Conte et al. [7] proved the following result:

Theorem 2. *We can augment the compact representation of a Wheeler DFA $\mathcal{A}$ with $O(n \log n)$ bits, where n is the number of states and σ is the size of the* *alphabet, in such a way that we can compute the matching statistics of a pattern of length m with respect to the Wheeler DFA in $O(m \log n)$ time.*

We will show that if we only want to use linear space, then we can use Theorem 1 to obtain the following trade-off.

Theorem 3. *We can augment the compact representation of a Wheeler DFA $\mathcal{A}$ with $O(n \log \log \sigma)$ bits, where n is the number of states and σ is the size of the alphabet, in such a way that we can compute the matching statistics of a pattern of length m with respect to the Wheeler DFA in $O(m \log^2 n)$ time.*

Variable-Order de Bruijn Graphs. Wheeler graphs are a generalization of de Bruijn graphs; in particular, the compact representation of a Wheeler graph is a generalization of the BOSS representation of a de Bruijn graph [5], and our results on the LCP array also apply to a de Bruijn graph. Many assemblers [3,19,24,27] consider all k-mers occurring in a set of reads and build a k-th order de Bruijn graph (on the alphabet $\Sigma = \{A, C, G, T\}$) to perform Eulerian sequence assembly [18,25]. However, the choice of the parameter k impacts the assembly quality, so some assemblers try several choices for k [3,24], which slows down the process because several de Bruijn graphs need to be built. In [4] it was shown that the k-order de Bruijn graph of $\mathcal{S}$ can be used to *implicitly* store the k'-th order de Bruijn graph of $\mathcal{S}$ for *every* $k' \leq k$, thus leading to a *variable-order de Bruijn graph*. The challenge is to navigate this implicit representation (that is, how to follow edges in a forward or backward fashion). In [4], it was shown that the navigation is possible by storing or by simulating an array $\overline{\mathsf{LCP}}_G$ which can be seen as a simplification of the LCP array of the Wheeler graph G. More precisely, we have the following result (see [4]; we assume $\sigma = O(1)$).

Theorem 4. 1. *We can augment the BOSS representation of a k-th order de Bruijn graph with $O(n \log k)$ bits, where n is the number of nodes, so that the underlying variable-order de Bruijn graph can be navigated in $O(\log k)$ time per visited node.*

2. *We can augment the BOSS representation of a k-th order de Bruijn graph with $O(n)$ bits, where n is the number of nodes, so that the underlying variable-order de Bruijn graph can be navigated in $O(k \log n)$ time per visited node.*

Essentially, the first solution in Theorem 4 explicitly stores $\overline{\mathsf{LCP}}_G$, while the second solution in Theorem 4 computes the entries of $\overline{\mathsf{LCP}}_G$ by exploiting the BOSS representation. In general, a big k (close to the size of the reads) allows to retrieve the expressive power on an overlap graph [11], so in Theorem 4 we cannot assume that k is small. On the one hand, the *space* required for the first solution can be too large, because a de Bruijn graph can be stored by using only $O(n)$ bits. On the other hand, the *time* bound in the second solution increases substantially. We can now improve the second solution by providing a data structure that achieves the best of both worlds. As we did in Theorem 1, we can conveniently sample some entries of $\overline{\mathsf{LCP}}_G$. We will prove the following result.

Theorem 5. *We can augment the BOSS representation of a k-th order de Bruijn graph with $O(n)$ bits, where n is the number of nodes, so that the underlying variable-order de Bruijn graph can be navigated in $O(\log k \log n)$ time per visited node.*

2 Definitions

Sets and Relations. Let V be a set. A total order on V is a binary relation $\leq$ which is reflexive, antisymmetric and transitive. We say that U is a $\leq$-interval (or simply an interval) if for all $v_1, v_2, v_3 \in V$, if $v_1, v_3 \in U$ and $v_1 < v_2 < v_3$, then $v_2 \in U$. If $u, v \in V$, with $u \leq v$, we denote by $[u, v]$ the smallest interval containing u and v, that is $[u, v] = \{z \in V \mid u \leq z \leq v\}$. In particular, if V is the set of integers, then we assume that $\leq$ is the standard total order, hence $[u, v] = \{u, u+1, \dots, v-1, v\}$.

Strings. Let Σ be a finite alphabet, with $\sigma = |\Sigma|$. Let Σ^* be the set of all finite strings on Σ and let Σ^ω be the set of all (countably) infinite strings on Σ. If $\alpha \in \Sigma^*$, then α^R is the reverse string of α. If $\alpha, \beta \in \Sigma^* \cup \Sigma^\omega$, we denote by $\mathsf{lcp}(\alpha, \beta)$ the length of longest common prefix between α and β. In particular, if $\alpha \in \Sigma^*$, then $\mathsf{lcp}(\alpha, \beta) \leq |\alpha|$ and if $\alpha, \beta \in \Sigma^\omega$ with $\alpha = \beta$, then $\mathsf{lcp}(\alpha, \beta) = \infty$. Let $\preceq$ be a fixed total order on Σ. We extend the total order $\preceq$ from Σ to $\Sigma^* \cup \Sigma^\omega$ lexicographically.

DFAs. Throughout the paper, let $\mathcal{A} = (Q, E, s_0, F)$ be a deterministic finite automaton (DFA), where Q is the set of states, $E \subseteq Q \times Q \times \Sigma$ is the set of labeled edges, $s_0 \in Q$ is the initial state and $F \subseteq Q$ is the set of final states. The alphabet Σ is *effective*, that is, every $c \in \Sigma$ labels some edge. Since $\mathcal{A}$ is deterministic, for every $u \in Q$ and for every $a \in \Sigma$ there exists at most one edge labeled a leaving u. Following [1], we assume that (i) s_0 has no incoming edges, (ii) every state is reachable from the initial state and (iii) all edges entering the same state have the same label (*input-consistency*). For every $u \in Q \setminus \{s_0\}$, let $\lambda(u) \in \Sigma$ be the label of all edges entering u. We define $\lambda(s_0) = \#$, where $\# \notin \Sigma$ is a special character such that $\# \prec a$ for every $a \in \Sigma$ (the character $\#$ plays the same role as the termination character $\$$ in suffix arrays, suffix trees and Burrows-Wheeler transforms). As a consequence, an edge (u', u, a) can be simply written as (u', u), because it must be $a = \lambda(u)$.

Compact Data Structures. Let A be an array of length n containing elements from a finite totally-ordered set. A *range minimum query* on A is defined as follows: given $1 \leq i \leq j \leq n$, return one of the indices k with $1 \leq k \leq n$ such that (i) $i \leq k \leq j$ and $A[k] = \min\{A[i], A[i+1], \dots, A[j-1], A[j]\}$. We write $k = RMQ_A(i, j)$. Then, there exists a data structure of $2n + o(n)$ such that in $O(1)$ time we can compute $RMQ_A(i, j)$ for every $1 \leq i \leq n$, *without the need to* *access* A [15, 16]. This result is essentially optimal, because every data structure solving range minimum queries on A requires at least $2n - \Theta(\log n)$ bits [16, 20].

Let A be a bitvector of length n. Let $rank(A, i) = |\{j \in \{1, 2, \ldots, i - 1, i\} \mid A[j] = 1\}|$ be the number of 1's among the first i bits of A. Then, there exists a data structure of $n + o(n)$ bits such that in $O(1)$ time we can compute $rank(A, i)$ for $1 \le i \le n$ [23].

3   Wheeler DFAs

Let us recall the definition of Wheeler DFA [7].

Definition 1. *Let* $\mathcal{A} = (Q, E, s_0, F)$ *be a DFA. A* Wheeler order *on* $\mathcal{A}$ *is a total order* $\le$ *on* Q *such that* $s_0 \le u$ *for every* $u \in Q$ *and:*

1. *(Axiom 1) If* $u, v \in Q$ *and* $u < v$, *then* $\lambda(u) \preceq \lambda(v)$.
2. *(Axiom 2) If* $(u', u), (v', v) \in E$, $\lambda(u) = \lambda(v)$ *and* $u < v$, *then* $u' < v'$.

A DFA $\mathcal{A}$ *is* Wheeler *if it admits a Wheeler order.*

Every DFA admits at most one Wheeler order [1], so the total order $\le$ in Definition 1 is *the* Wheeler order on $\mathcal{A}$. In the following, we fix a Wheeler DFA $\mathcal{A} = (Q, E, s_0, F)$, with $n = |Q|$ and $e = |E|$, and we write $Q = \{u_1, \ldots, u_n\}$, with $u_1 < u_2 < \cdots < u_n$ in the Wheeler order. In particular, $u_1 = s_0$. Following [7], we assume that s_0 has a self-loop labeled #, which is consistent with Axiom 1, because $\# \prec a$ for every $a \in \Sigma$). This implies that every state has at least one incoming edge, so for every state u_i there exists at least one infinite string $\alpha \in \Sigma^\omega$ that can be read starting from u_i and following edges in a backward fashion. We denote by I_{u_i} the nonempty set of all such strings. Formally:

Definition 2. *Let* $1 \le i \le n$. *Define:*

$$I_{u_i} = \{\alpha \in \Sigma^\omega \mid \text{there exist integers } f_1, f_2, \ldots \text{ in } [1, n] \text{ such that } (i)f_1 = i,$$
$$(ii) \ (u_{f_{k+1}}, u_{f_k}) \in E \text{ for every } k \ge 1 \text{ and } (iii)\alpha = \lambda(u_{f_1})\lambda(u_{f_2})\ldots \}.$$

For every $1 \le i \le n$, let $p_{\min}(i)$ be the smallest $1 \le i' \le n$ such that $(u_{i'}, u_i) \in E$ and let $p_{\max}(i)$ be the largest $1 \le i'' \le n$ such that $(u_{i''}, u_i) \in E$. Both $p_{\min}(i)$ and $p_{\max}(i)$ are well-defined because every state has at least one incoming edge. For every $1 \le i \le n$, define $p_{\min}^1(i) = p_{\min}(i)$ and recursively, for $j \ge 2$, let $p_{\min}^j(i) = p_{\min}(p_{\min}^{j-1}(i))$. Then, $\lambda(u_i)\lambda(p_{\min}(i))\lambda(p_{\min}^2(i))\lambda(p_{\min}^3(i))\ldots$ is the lexicographically *smallest* string in I_{u_i}, which we denote by $\min_i$ [7]. Analogously, one can define the lexicographically *largest* string in I_{u_i} by using $p_{\max}$. Moreover, in [7] it was shown that:

$$\min_1 \preceq \max_1 \preceq \min_2 \preceq \max_2 \preceq \cdots \preceq \max_{n-1} \preceq \min_n \preceq \max_n.$$

Intuitively, the previous equation shows that the permutation of the set of all states of $\mathcal{A}$ induced by the Wheeler order can be seen as a generalization of the permutation of positions induced by the prefix array of a string α (or equivalently, the suffix array of the reverse string of α^R). Indeed, a string α can also be seen as a DFA $\mathcal{A}' = (Q', E', s_0', F')$, where $Q' = \{q_0', q_1' \ldots, q_{|\alpha|}'\}$, $s_0' = q_0'$, $F' = \{q_{|\alpha|}'\}$ (the set F plays no role here), $\lambda(q_i')$ is the i-th character of α for $1 \leq i \leq n$ and $E' = \{(q_{i-1}', q_i') \mid 1 \leq i \leq n\}$ (every state is reached by exactly one string so the minimum and the maximum string reaching each state are equal).

Let $1 \leq r \leq s \leq n$ and let $c \in \Sigma$. Let $E_{r,s,c}$ be the set of all states that can be reached from a state in $[r, s]$ by following edges labeled c; formally, $E_{r,s,c} = \{1 \leq j \leq n \mid \lambda(u_j) = c$ and $(u_i, u_j) \in E$ for some $i \in [r, s]\}$. Then, $E_{r,s,c}$ is again an interval, that is, there exist $1 \leq r' \leq s' \leq n$ such that $E_{r,s,c} = [r', s']$ [17]. This property enables a compression mechanism that generalizes the Burrows-Wheeler transform [6] and the FM-index [13] to Wheeler DFAs. The Wheeler DFA $\mathcal{A}$ can be stored by using only $2e + 4n + e \log \sigma + \sigma \log e$ bits (up to lower order terms), including n bits to mark the set F of final states and n bits to mark all $1 \leq i \leq n$ such that $\lambda(u_i) \neq \lambda(u_{i-1})$, which allows us to retrieve each $\lambda(u_i)$ in $O(1)$ time by using a rank query [17] (recall that n is the number of states and e is the number of edges). Since $\mathcal{A}$ is a DFA, we have $e \leq n\sigma$, so the required space is $O(n\sigma \log \sigma)$. If the alphabet is small — that is, if $\sigma \log \sigma = o(\log n)$ — then the number of required bits is $o(n \log n)$; if $\sigma = O(1)$, then the number of required bits is $O(n)$. This compact representation supports efficient navigation of the graph and it allows to solve pattern matching queries. More precisely, by resorting to state-of-the art select queries [23] in $O(\log \log \sigma)$ time (i) for $1 \leq i \leq n$, we can compute $p_{\min}(i)$ and $p_{\max}(i)$ and (ii) given $1 \leq r \leq s \leq n$ and $c \in \Sigma$, we can compute $[r', s'] = E_{r,s,c}$ [17]. In particular, query (ii) is the so-called *forward-search*, which generalizes the analogous mechanism of the FM-index, thus allowing to solve pattern matching queries on the graph.

The Wheeler order generalizes the notion of suffix array from strings to DFA. It is also possible to generalize LCP-arrays from strings to graph [7].

Definition 3. *The* LCP-array *of the Wheeler DFA $\mathcal{A}$ is the array* $\mathsf{LCP}_{\mathcal{A}} = \mathsf{LCP}_{\mathcal{A}}[2, 2n]$ *which contains the following $2n - 1$ values in the following order:* $\mathsf{lcp}(\min_1, \max_1)$, $\mathsf{lcp}(\max_1, \min_2)$, $\mathsf{lcp}(\min_2, \max_2)$, $\ldots$, $\mathsf{lcp}(\max_{n-1}, \min_n)$, $\mathsf{lcp}(\min_n, \max_n)$. *In other words,* $\mathsf{LCP}_{\mathcal{A}}[2i] = \mathsf{lcp}(\min_i, \max_i)$ *for* $1 \leq i \leq n$ *and* $\mathsf{LCP}_{\mathcal{A}}[2i - 1] = \mathsf{lcp}(\max_{i-1}, \min_i)$ *for* $2 \leq i \leq n$.

It can be proved that for every $2 \leq i \leq n$, if $\mathsf{LCP}_{\mathcal{A}}[i]$ is finite, then $\mathsf{LCP}_{\mathcal{A}}[i] < 3n$ [7]. As a consequence, $\mathsf{LCP}_{\mathcal{A}}$ can be stored by using $O(n \log n)$ bits.

4   A Space-Time Trade-Off for the LCP Array

By storing an LCP array on top of the compact representation of a Wheeler graph, we have additional information that we can use to efficiently solve problems such as computing the matching statistics; however, we need to store $O(n \log n)$ bits. As we have seen, $O(n \log n)$ dominates the number of bits required to store $\mathcal{A}$ itself, if the alphabet is small. In this section, we show

Algorithm 1. Building $V(h)$

$V(h) \leftarrow \emptyset$
$U \leftarrow \emptyset$
while there exists $v \in V$ such that (a) $v(i)$ is defined for $0 \leq i \leq h-1$, (b) $v(i) \neq v(j)$ for $0 \leq j < i \leq h-1$, (c) $v(i) \notin U$ for $0 \leq i \leq h-1$ **do**
    Pick such a v, add $v(h-1)$ to $V(h)$ and add $v(i)$ to U for every $0 \leq i \leq h-1$
end while

Algorithm 2. Input: $h \in [2, 2n]$. Output: $\mathsf{LCP}_{\mathcal{A}}[h]$.

---
procedure MAIN_FUNCTION(h)
    Initialize a global bit array $D[2, 2n]$ to zero   ▷ $D[2, 2n]$ marks the entries already considered
    return LCP(h)
end procedure

procedure LCP(h)
    $D[h] \leftarrow 1$
    if $C[h] = 1$ **then**   ▷ The desired value has been sampled
        return $\mathsf{LCP}_{\mathcal{A}}^*[rank(C, h)]$
    else if h is odd **then**
        $i \leftarrow \lceil h/2 \rceil$
        if $\lambda(u_{i-1}) \neq \lambda(u_i)$ **then**
            return 0
        else
            $k \leftarrow p_{\max}(i-1)$
            $k' \leftarrow p_{\min}(i)$
            $j \leftarrow RMQ_{\mathsf{LCP}_{\mathcal{A}}}(2k+1, 2k'-1)$
            if $D[j] = 1$ **then**   ▷ We have already considered this entry before, so there is a cycle
                return ∞
            else
                return $1 + \mathsf{LCP}(j)$
            end if
        end if
    else
        $i \leftarrow h/2$
        $k \leftarrow p_{\min}(i)$
        $k' \leftarrow p_{\max}(i)$
        $j \leftarrow RMQ_{\mathsf{LCP}_{\mathcal{A}}}(2k, 2k')$
        if $D[j] = 1$ **then**   ▷ We have already considered this entry before, so there is a cycle
            return ∞
        else
            return $1 + \mathsf{LCP}(j)$
        end if
    end if
end procedure

---

that we can store a data structure of only $O(n \log \log \sigma)$ bits which allows to compute every entry $\mathsf{LCP}_{\mathcal{A}}[i]$ in $O(\log n)$ time, thus proving Theorem 1. This will be possible by sampling some entries of $\mathsf{LCP}_{\mathcal{A}}$. The sampling mechanism is obtained by conveniently defining an auxiliary graph from the entries of the LCP array. We will immediately describe our technique, our sampling mechanism being general-purpose.

Sampling. Let $G = (V, H)$ be a finite (unlabeled) directed graph such that every node has at most one incoming edge. For every $v \in V$ and for every $i \geq 0$, there exists at most one node $v' \in V$ such that there exists a directed path from v' to v having i edges; if v' exists, we denote it by $v(i)$. Fix a parameter $h \geq 1$.

Let us prove that there exists $V(h) \subseteq V$ such that (i) $|V(h)| \leq \frac{|V|}{h}$ and (ii) for every $v \in V$ there exists $0 \leq i \leq 2h - 2$ such that $v(i)$ is defined and either $v(i) \in V(h)$ or $v(i)$ has no incoming edges or $v(i) = v(j)$ for some $0 \leq j < i$. We build $V(h)$ incrementally following Algorithm 1. Let us prove that, at the end of the algorithm, properties (i) and (ii) are true. For every $v \in V(h)$, define $S_v = \{v, v(1), v(2), \ldots, v(h-1)\}$, which is possible because by construction if $v \in V(h)$, then $v(i)$ is defined for every $0 \leq i \leq h - 1$. It must be $v(i) \neq v(j)$ for $0 \leq i < j \leq h-1$, so $|S_v| = h$. If $v, v' \in V(h)$ and $v \neq v'$, then by construction S_v and $S_{v'}$ are disjoint. As a consequence, $|V| \geq \sum_{v \in V(h)} |S_v| = \sum_{v \in V(h)} h = h|V_h|$ and so $|V_h| \leq \frac{|V|}{h}$, which proves property (i). Let us prove property (ii). Pick $v \in V$; we must prove that there exists $0 \leq i \leq 2h - 2$ such that $v(i)$ is defined and either $v(i) \in V(h)$ or $v(i)$ has no incoming edges or $v(i) = v(j)$ for some $0 \leq j < i$. We distinguish three cases:

1. there exists i with $1 \leq i \leq h - 1$ such that $v(i - 1)$ is defined but $v(i)$ is not defined. Then, $v(i-1)$ has no incoming edges.
2. there exist i, j with $0 \leq j < i \leq h - 1$ such that $v(j)$ and $v(i)$ are defined and $v(i) = v(j)$. In this case, the conclusion is immediate.
3. $v(i)$ is defined for every $0 \leq i \leq h$ and $v(i) \neq v(j)$ for $0 \leq j < i \leq h - 1$. Since Algorithm 1 has terminated, then there exists $0 \leq j \leq h - 1$ such that $v(j) \in U$. The construction of U implies that there exists $v' \in V$ and $0 \leq j' \leq h-1$ such that $v(j) = v'(j')$ and $v'(h-1) \in V(h)$. As a consequence $v(h - 1 + j - j') = v(j)(h - 1 - j') = (v'(j'))(h - 1 - j') = v'(h-1) \in V(h)$. Since $j \leq h - 1$ and $j' \geq 0$, we conclude $h - 1 + j - j' \leq 2h - 2$ and we are done.

Computing the LCP Array Using a Linear Number of Bits. First, let us store a data structure of $O(n)$ bits which in $O(1)$ time determines $RMQ_{\mathsf{LCP}_\mathcal{A}}(i, j)$ for every $2 \leq i \leq j \leq 2n$.

Notice that $\mathsf{LCP}_\mathcal{A}[2i] \geq 1$ for $1 \leq i \leq n$ because the first character of $\min_i$ and the first character of $\max_i$ are equal to $\lambda(u_i)$. Moreover, we have $\mathsf{LCP}_\mathcal{A}[2i-1] \geq 1$ if and only if $\lambda(u_{i-1}) = \lambda(u_i)$, for $2 \leq i \leq n$.

Consider the entry $\mathsf{LCP}_\mathcal{A}[2i - 1] = \mathsf{lcp}(\max_{i-1}, \min_i)$, for $2 \leq i \leq n$, and assume that $\mathsf{LCP}_\mathcal{A}[2i - 1] \geq 1$. Let $k = p_{\max}(i - 1)$ and $k' = p_{\min}(i)$. Since $\mathsf{LCP}_\mathcal{A}[2i - 1] \geq 1$, then there exists $a \in \Sigma$ such that $\max_{i-1} = a\max_k$ and $\min_{i-1} = a\min_{k'}$. In particular, $(u_k, u_{i-1}, a) \in E$ and $(u_{k'}, u_i, a) \in E$, so from Axiom 2 we obtain $k < k'$. Moreover, we have $\mathsf{LCP}_\mathcal{A}[2i - 1] = \mathsf{lcp}(\max_{i-1}, \min_i) = \mathsf{lcp}(a\max_k, a\min_{k'}) = 1 + \mathsf{lcp}(\max_k, \min_{k'})$. Notice that:

$$\mathsf{lcp}(\max_k, \min_{k'}) = \min\{\mathsf{lcp}(\max_k, \min_{k+1}), \mathsf{lcp}(\min_{k+1}, \max_{k+1}), \ldots,$$
$$= \mathsf{lcp}(\min_{k'-1}, \max_{k'-1}), \mathsf{lcp}(\max_{k'-1}, \min_{k'})\} =$$
$$= \min\{\mathsf{LCP}_\mathcal{A}[2k + 1], \mathsf{LCP}_\mathcal{A}[2k + 2], \ldots, \mathsf{LCP}_\mathcal{A}[2k' - 2], \mathsf{LCP}_\mathcal{A}[2k' - 1]\}.$$

Let $j = RMQ_{\mathsf{LCP}_\mathcal{A}}(2k + 1, 2k' - 1)$. Then, $\mathsf{LCP}_\mathcal{A}[j] = \min\{\mathsf{LCP}_\mathcal{A}[2k + 1], \mathsf{LCP}_\mathcal{A}[2k+2], \ldots, \mathsf{LCP}_\mathcal{A}[2k'-2], \mathsf{LCP}_\mathcal{A}[2k'-1]\}$, so $\mathsf{LCP}_\mathcal{A}[2i-1] = 1 + \mathsf{LCP}_\mathcal{A}[j]$

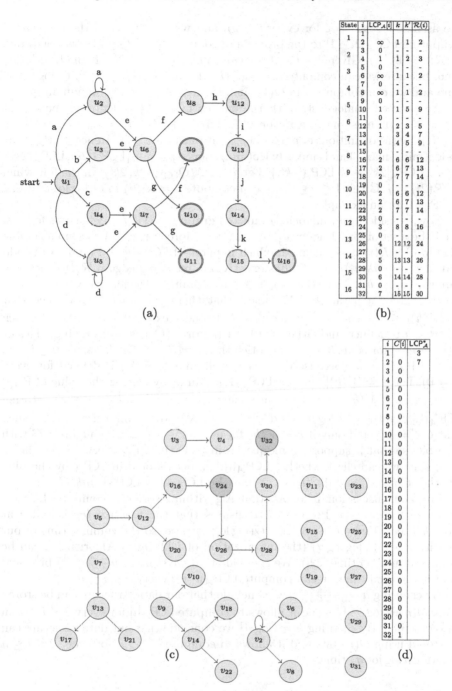

Fig. 1. (a) A Wheeler DFA. States are numbered according to the Wheeler order. (b) The array $\mathsf{LCP}_{\mathcal{A}}$, and the values needed to compute $G = (V, H)$. We assume that a range minimum query returns the *largest* position of a minimum value. (c) The graph $G = (V, H)$, with $V(\lceil \log n \rceil) = V(4) = \{v_{24}, v_{32}\}$ (yellow states). (d) The data structures that we store.

(we assume $t + \infty = \infty$ for every $t \geq 0$), and we have reduced the problem of computing $\mathsf{LCP}_\mathcal{A}[2i-1]$ to the problem of computing $\mathsf{LCP}_\mathcal{A}[j]$. In the following, let $\mathcal{R}(2i-1) = j$. Given $2 \leq i \leq n$, we can compute $j = \mathcal{R}(2i-1)$ in $O(\log \log \sigma)$ time, because we can compute $k = p_{\max}(i-1)$ and $k' = p_{\min}(i)$ in $O(\log \log \sigma)$ time and we can compute j in $O(1)$ time by means of a range minimum query.

We proceed analogously with the entries $\mathsf{LCP}_\mathcal{A}[2i] = \mathsf{lcp}(\min_i, \max_i)$, for $1 \leq i \leq n$ (it must necessarily be $\mathsf{LCP}_\mathcal{A}[2i] \geq 1$). Let $k = p_{\min}(i)$ and $k' = p_{\max}(i)$; by the definitions of $p_{\min}$ and $p_{\max}$ it must be $k \leq k'$. Hence, $\mathsf{LCP}_\mathcal{A}[2i] = 1 + \mathsf{lcp}(\min_k, \max_{k'})$ and similarly $\mathsf{lcp}(\min_k, \max_{k'}) = \min\{\mathsf{LCP}_\mathcal{A}[2k], \mathsf{LCP}_\mathcal{A}[2k+1], \ldots, \mathsf{LCP}_\mathcal{A}[2k'-1], \mathsf{LCP}_\mathcal{A}[2k']\}$. Let $j = RMQ_{\mathsf{LCP}_\mathcal{A}}(2k, 2k')$. In the following, let $\mathcal{R}(2i) = j$. Given $1 \leq i \leq n$, we can compute $j = \mathcal{R}(2i)$ in $O(\log \log \sigma)$ time. See Fig. 1 for an example.

Now, consider the (unlabeled) directed graph $G = (V, H)$ defined as follows. Let V be a set of $2n-1$ nodes $v_2, v_3, \ldots, v_{2n}$. Moreover, $v_i \in V$ has no incoming edge in G if $\mathcal{R}(i)$ is not defined, which happens if $\mathsf{LCP}_\mathcal{A}[i] = 0$ (and so i is odd and $\lambda(u_{i-1}) \neq \lambda(u_i)$); $v_i \in V$ has exactly one incoming edge if $\mathcal{R}(i)$ is defined, namely, $(v_{\mathcal{R}(i)}, v_i)$. Note that v_{2i} has an incoming edge for every $1 \leq i \leq n$. Let $h \geq 1$ be a parameter. We know that there exists $V(h) \subseteq V$ such that (i) $|V(h)| \leq \frac{|V|}{h}$ and (ii) for every $v_i \in V$ there exists $0 \leq k \leq 2h - 2$ such that $v_i(k)$ is defined and either $v_i(k) \in V(h)$ or $v_i(k)$ has no incoming edges or $v_i(k) = v_i(l)$ for some $0 \leq l < k$. Notice that if $v_i(k) = v_i(l)$ for some $0 \leq l < k$, then $\mathsf{LCP}_\mathcal{A}[i] = \infty$ (because there is a cycle and so $v_i(k')$ is defined for every $k' \geq 0$). Let $n' = |V(h)|$, and let $\mathsf{LCP}_\mathcal{A}^*[1, n']$ an array storing the value $\mathsf{LCP}_\mathcal{A}[i]$ for each $v_i \in V(h)$, sorted by increasing i. Since $n' \leq \frac{|V|}{h} = \frac{2n-1}{h}$, storing $\mathsf{LCP}_\mathcal{A}^*[1, n']$ takes $n'O(\log n) = O(\frac{n \log n}{h})$ bits. We store a bitvector $C[2, 2n]$ such that $C[i] = 1$ if and only if $v_i \in V(h)$ for every $2 \leq i \leq 2n$; we augment C with $o(n)$ bits so that it supports rank queries in $O(1)$ time. For every $2 \leq i \leq 2n$, in $O(1)$ time we can check whether $\mathsf{LCP}_\mathcal{A}[i]$ has been stored in $\mathsf{LCP}_\mathcal{A}^*$ by checking whether $C[i] = 1$, and if $C[i] = 1$ it must be $\mathsf{LCP}_\mathcal{A}[i] = \mathsf{LCP}_\mathcal{A}^*[rank(C, i)]$.

From our discussion, it follows that Algorithm 2 correctly computes $\mathsf{LCP}_\mathcal{A}[i]$ for every $2 \leq i \leq n$. Property (ii) ensures that the function lcp is called at most h times. Every call requires $O(\log \log \sigma)$ time, so the running time of our algorithm is $O(h \log \log \sigma)$ (the initialization of $D[2, 2n]$ in Algorithm 2 can be simulated in $O(1)$ time [22]). We conclude that we store $O(n + \frac{n \log n}{h})$ bits, and in $O(h \log \log \sigma)$ time we can compute $\mathsf{LCP}_\mathcal{A}[i]$ for every $2 \leq i \leq n$.

By choosing $h = \lceil \frac{\log n}{\log \log \sigma} \rceil$, we conclude that our data structure can be stored using $O(n \log \log \sigma)$ bits and it allows to compute $\mathsf{LCP}_\mathcal{A}[i]$ for every $2 \leq i \leq n$ in $O(\log n)$ time. By choosing $h = \lceil \log n \rceil$ we conclude that our data structure can be stored using $O(n)$ bits and it allows to compute $\mathsf{LCP}_\mathcal{A}[i]$ for every $2 \leq i \leq n$ in $O(\log n \log \log \sigma)$ time.

5 Applications

Matching Statistics. Let us recall how the bounds in Theorem 2 are obtained. The space bound is $O(n \log n)$ bits because we need to store $\mathsf{LCP}_\mathcal{A}$. We also store a data structure to solve range minimum queries on LCP_A, which only takes $O(n)$ bits. The time bound $O(m \log n)$ follows from performing $O(m)$ steps to compute all matching statistics. In each of these $O(m)$ steps, we may need to perform a binary search on LCP_A. In each step of the binary search, we need to solve a range minimum query once and we need to access LCP_A once, so the binary search takes $O(\log n)$ time per step. By Theorem 1, if we store only $O(n \log \log \sigma)$ bits, we can access LCP_A in $O(\log n)$ time, so the time for the binary search becomes $O(\log^2 n)$ per step and Theorem 3 follows.

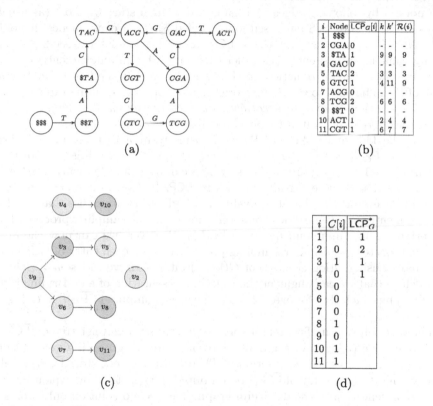

i	Node	$\overline{\mathsf{LCP}}_G[i]$	k	k'	$\mathcal{R}(i)$
1	$$$				
2	CGA	0	-	-	-
3	$TA	1	9	9	9
4	GAC	0	-	-	-
5	TAC	2	3	3	3
6	GTC	1	4	11	9
7	ACG	0	-	-	-
8	TCG	2	6	6	6
9	$$T	0	-	-	-
10	ACT	1	2	4	4
11	CGT	1	6	7	7

(a)   (b)

i	$C[i]$	LCP^*_G
1		1
2	0	2
3	1	1
4	0	1
5	0	
6	0	
7	0	
8	1	
9	0	
10	1	
11	1	

(c)   (d)

Fig. 2. The 3-rd order de Bruijn graph for the set $\mathcal{S} = \{CGAC, GACG, GACT, TACG, GTCG, ACGA, ACGT, TCGA, CGTC\}$ from [4]. We proceed like in Fig. 1 (now we only consider odd entries of LCP_G, and $h = \lceil \log k \rceil = 2$).

Variable-Order de Bruijn Graphs. Let $k \geq 0$ be a parameter, and let $\mathcal{S}$ be a set of strings on the alphabet $\Sigma = \{A, C, G, T\}$ (in this application we always assume $\sigma = O(1)$). The k-th order de Bruijn graph of $\mathcal{S}$ is defined as follows. The set of nodes is the set of all strings of Σ of length k that occur as a substring of some string in $\mathcal{S}$. There is an edge from node α to node β labeled $c \in \Sigma$ if and only if (i) the suffix of α of length $k-1$ is equal to the prefix of β of length $k-1$ and (ii) the last character of β is c. If some node

α has no incoming edges, then we add nodes $\$^i\alpha_{k-i}$ for $1 \le i \le k$, where α_j is the prefix of α of length j and $\$$ is a special character, and we add edges as above; see Fig. 2 for an example. Wheeler DFAs are a generalization of de Bruijn graphs (we do not need to define an initial state and a set of final states, because here we are not interested in studying the applications of de Bruijn graphs and Wheeler automata to automata theory [2,10]); the Wheeler order is the one such that node α comes before node β if and only if the string α^R is lexicographically smaller than the string β^R [17].

Notice that, in a k-th order de Bruijn graph G, all strings that can be read from node α by following edges in a backward fashion start with α^R (as usual, we assume that node $\$\$\$$ has a self-loop labeled $\$$). As a consequence, it holds $\mathsf{LCP}_G[2i] \ge k$ for every $1 \le i \le n$ and $\mathsf{LCP}_G[2i-1] \le k-1$ for every $2 \le i \le n$ (so any value in an odd entry is smaller than any value in an even entry).

As we saw in the introduction, in [4] it was shown that the k-order de Bruijn graph of $\mathcal{S}$ can be used to *implicitly* store the k'-th order de Bruijn graph of $\mathcal{S}$ for *every* $k' \le k$, thus leading to a *variable-order de Bruijn graph*. The navigation of a variable-order de Bruijn graph is possible by storing or by simulating the values in the odd entries of the LCP array. Formally, in order to avoid confusion, we define $\overline{\mathsf{LCP}}_G[i] = \mathsf{LCP}_G[2i-1]$ for every $2 \le i \le n$; see Fig. 2. Note that $\overline{\mathsf{LCP}}_G[i] \le k-1$ for every $2 \le i \le n$, so $\overline{\mathsf{LCP}}_G$ can be stored by using $O(n \log k)$ bits. Notice that Theorem 1 also applies to $\overline{\mathsf{LCP}}_G[i]$ (we do not need to store values in the even entries because a value in an odd entry is smaller than a value in an even entry, so even entries are never selected in the sampling process when answering a range minimum query on LCP_G). However, we can now choose a better parameter $h \ge 1$ in our sampling process. Indeed, each entry of $\overline{\mathsf{LCP}}_G$ can be stored by using $O(\log k)$ bits (not $O(\log n)$ bits), so if we choose $h = \lceil \log k \rceil$, we conclude that we can augment the BOSS representation of a de Bruijn graph with $O(n)$ bits such that for every $2 \le i \le n$ we can compute $\overline{\mathsf{LCP}}_G[i]$ in $O(\log k)$ time.

The first solution in Theorem 4 consists in storing a wavelet tree on $\overline{\mathsf{LCP}}_G$, which requires $O(n \log k)$ bits and allows to navigate the graph in $O(\log k)$ time per visited node. The second solution in Theorem 4 does not store $\overline{\mathsf{LCP}}_G$ at all; whenever needed, an entry of $\overline{\mathsf{LCP}}_G$ is computed in $O(k)$ time by exploiting the BOSS representation of the de Bruijn graph. The second solution only stores a data structures of $O(n)$ bits to solve range minimum queries. The details can be found in [4]. Essentially, the time bound $O(k \log n)$ comes from performing binary searches on $\overline{\mathsf{LCP}}_G$ while explicitly computing an entry of $\overline{\mathsf{LCP}}_G$ at each step in $O(k)$ time. However, we have seen that, while staying within the $O(n)$ space bound, we can augment the BOSS representation so that we can compute the entries of $\overline{\mathsf{LCP}}_G$ in $O(\log k)$ time, so the time bound $O(k \log n)$ becomes $O(\log k \log n)$, which implies Theorem 5.

Acknowledgements. *Travis Gagie*: funded by National Institutes of Health (NIH) NIAID (grant no. HG011392), the National Science Foundation NSF IIBR (grant no. 2029552) and a Natural Science and Engineering Research Council (NSERC) Discovery Grant (grant no. RGPIN-07185-2020). *Dominik Köppl*: supported by JSPS KAKENHI with No. JP21K17701 and JP23H04378. *Nicola Prezza*: funded by the European Union (ERC, REGINDEX, 101039208). Views and opinions expressed are however those of the author(s) only and do not necessarily reflect those of the European Union or the European Research Council. Neither the European Union nor the granting authority can be held responsible for them.

References

1. Alanko, J., D'Agostino, G., Policriti, A., Prezza, N.: Regular languages meet prefix sorting. In: Proceedings of the 31st Symposium on Discrete Algorithms, (SODA 2020), pp. 911–930. SIAM (2020). https://doi.org/10.1137/1.9781611975994.55
2. Alanko, J., D'Agostino, G., Policriti, A., Prezza, N.: Wheeler languages. Inf. Comput. **281**, 104820 (2021)
3. Bankevich, A., et al.: SPAdes: a new genome assembly algorithm and its applications to single-cell sequencing. J. Comput. Biol. **19**(5), 455–477 (2012). https://doi.org/10.1089/cmb.2012.0021, pMID: 22506599
4. Boucher, C., Bowe, A., Gagie, T., Puglisi, S.J., Sadakane, K.: Variable-order de Bruijn graphs. In: 2015 Data Compression Conference, pp. 383–392 (2015). https://doi.org/10.1109/DCC.2015.70
5. Bowe, A., Onodera, T., Sadakane, K., Shibuya, T.: Succinct de bruijn graphs. In: Raphael, B., Tang, J. (eds.) WABI 2012. LNCS, vol. 7534, pp. 225–235. Springer, Heidelberg (2012). https://doi.org/10.1007/978-3-642-33122-0_18
6. Burrows, M., Wheeler, D.J.: A block-sorting lossless data compression algorithm. Technical report 124, Digital Equipment Corporation (1994)
7. Conte, A., Cotumaccio, N., Gagie, T., Manzini, G., Prezza, N., Sciortino, M.: Computing matching statistics on Wheeler DFAs. In: 2023 Data Compression Conference (DCC), pp. 150–159 (2023). https://doi.org/10.1109/DCC55655.2023.00023
8. Cotumaccio, N.: Graphs can be succinctly indexed for pattern matching in $O(|E|^2 + |V|^{5/2})$ time. In: 2022 Data Compression Conference (DCC), pp. 272–281 (2022). https://doi.org/10.1109/DCC52660.2022.00035
9. Cotumaccio, N., Prezza, N.: On indexing and compressing finite automata. In: Proceedings of the 32nd Symposium on Discrete Algorithms, (SODA 2021), pp. 2585–2599. SIAM (2021). https://doi.org/10.1137/1.9781611976465.153
10. Cotumaccio, N., D'Agostino, G., Policriti, A., Prezza, N.: Co-lexicographically ordering automata and regular languages - part I. J. ACM **70**, 1–73 (2023). https://doi.org/10.1145/3607471
11. Díaz-Domínguez, D., Gagie, T., Navarro, G.: Simulating the DNA overlap graph in succinct space. In: Pisanti, N., Pissis, S.P. (eds.) 30th Annual Symposium on Combinatorial Pattern Matching (CPM 2019). Leibniz International Proceedings in Informatics (LIPIcs), vol. 128, pp. 26:1–26:20. Schloss Dagstuhl-Leibniz-Zentrum fuer Informatik, Dagstuhl, Germany (2019). https://doi.org/10.4230/LIPIcs.CPM.2019.26, http://drops.dagstuhl.de/opus/volltexte/2019/10497
12. Ferragina, P., Luccio, F., Manzini, G., Muthukrishnan, S.: Structuring labeled trees for optimal succinctness, and beyond. In: proceedings of the 46th Annual IEEE Symposium on Foundations of Computer Science (FOCS 2005), pp. 184–193 (2005). https://doi.org/10.1109/SFCS.2005.69
13. Ferragina, P., Manzini, G.: Opportunistic data structures with applications. In: Proceedings of the 41st Annual Symposium on Foundations of Computer Science (FOCS 2000), pp. 390–398 (2000). https://doi.org/10.1109/SFCS.2000.892127

14. Ferragina, P., Luccio, F., Manzini, G., Muthukrishnan, S.: Compressing and indexing labeled trees, with applications. J. ACM **57**(1) (2009). https://doi.org/10.1145/1613676.1613680
15. Fischer, J.: Optimal succinctness for range minimum queries. In: López-Ortiz, A. (ed.) LATIN 2010. LNCS, vol. 6034, pp. 158–169. Springer, Heidelberg (2010). https://doi.org/10.1007/978-3-642-12200-2_16
16. Fischer, J., Heun, V.: Space-efficient preprocessing schemes for range minimum queries on static arrays. SIAM J. Comput. **40**(2), 465–492 (2011). https://doi.org/10.1137/090779759
17. Gagie, T., Manzini, G., Sirén, J.: Wheeler graphs: a framework for BWT-based data structures. Theor. Comput. Sci. **698**, 67–78 (2017). https://doi.org/10.1016/j.tcs.2017.06.016, https://www.sciencedirect.com/science/article/pii/S0304397517305285, algorithms, Strings and Theoretical Approaches in the Big Data Era (In Honor of the 60th Birthday of Professor Raffaele Giancarlo)
18. Idury, R.M., Waterman, M.S.: A new algorithm for DNA sequence assembly. J. Comput. Biol. **2**(2), 291–306 (1995)
19. Li, R., et al.: De novo assembly of human genomes with massively parallel short read sequencing. Genome Res. **20**, 265–72 (2009). https://doi.org/10.1101/gr.097261.109
20. Liu, M., Yu, H.: Lower bound for succinct range minimum query. In: Proceedings of the 52nd Annual ACM SIGACT Symposium on Theory of Computing, pp. 1402–1415. STOC 2020, Association for Computing Machinery, New York, NY, USA (2020). https://doi.org/10.1145/3357713.3384260
21. Manber, U., Myers, G.: Suffix arrays: a new method for on-line string searches. SIAM J. Comput. **22**(5), 935–948 (1993). https://doi.org/10.1137/0222058
22. Navarro, G.: Spaces, trees, and colors: the algorithmic landscape of document retrieval on sequences. ACM Comput. Surv. **46**(4) (2014). https://doi.org/10.1145/2535933
23. Navarro, G.: Compact Data Structures - A Practical Approach. Cambridge University Press (2016). http://www.cambridge.org/de/academic/subjects/computer-science/algorithmics-complexity-computer-algebra-and-computational-g/compact-data-structures-practical-approach?format=HB
24. Peng, Yu., Leung, H.C.M., Yiu, S.M., Chin, F.Y.L.: IDBA – a practical iterative de bruijn graph de novo assembler. In: Berger, B. (ed.) RECOMB 2010. LNCS, vol. 6044, pp. 426–440. Springer, Heidelberg (2010). https://doi.org/10.1007/978-3-642-12683-3_28
25. Pevzner, P.A., Tang, H., Waterman, M.S.: An Eulerian path approach to DNA fragment assembly. Proc. Nat. Acad. Sci. **98**(17), 9748–9753 (2001). https://doi.org/10.1073/pnas.171285098
26. Sadakane, K.: Compressed suffix trees with full functionality. Theor. Comp. Sys. **41**(4), 589–607 (2007). https://doi.org/10.1007/s00224-006-1198-x
27. Simpson, J., Wong, K., Jackman, S., Schein, J., Jones, S., Birol, I.: ABySS: a parallel assembler for short read sequence data. Genome Res. **19**, 1117–1123 (2009). https://doi.org/10.1101/gr.089532.108
28. Weiner, P.: Linear pattern matching algorithms. In: Proceedings of the 14th IEEE Annual Symposium on Switching and Automata Theory, pp. 1–11 (1973). https://doi.org/10.1109/SWAT.1973.13

Computing All-vs-All MEMs
in Grammar-Compressed Text

Diego Díaz-Domínguez[(✉)] and Leena Salmela

Department of Computer Science, University of Helsinki, Helsinki, Finland
{diego.diaz,leena.salmela}@helsinki.fi

Abstract. We describe a compression-aware method to find all-vs-all maximal exact matches (MEM) among strings of a repetitive collection $\mathcal{T}$. The key concept in our work is the construction of a fully-balanced grammar $\mathcal{G}$ from $\mathcal{T}$ that meets a property that we call *fix-free*: the expansions of the nonterminals that have the same height in the parse tree form a fix-free set (i.e., prefix-free and suffix-free). The fix-free property allows us to compute the MEMs of $\mathcal{T}$ incrementally over $\mathcal{G}$ using a standard suffix-tree-based MEM algorithm, which runs on a subset of grammar rules at a time and does not decompress nonterminals. By modifying the locally-consistent grammar of Christiansen et al. [7], we show how we can build $\mathcal{G}$ from $\mathcal{T}$ in linear time and space. We also demonstrate that our MEM algorithm runs on top of $\mathcal{G}$ in $O(G + occ)$ time and uses $O((G + occ) \log G)$ bits, where G is the grammar size, and occ is the number of MEMs in $\mathcal{T}$. In the conclusions, we discuss how to modify our idea to perform approximate pattern matching in compressed space.

Keywords: MEMs · Text Compression · Context-free grammars

1 Introduction

A *maximal exact match* (MEM) between two strings is a match that cannot be extended without introducing mismatches or reaching an end in one of the strings. This notion plays an important role in biological sequence analyses [18, 20,21] as they simplify finding long stretches of identical sequences. However, with the rapid development of DNA sequencing technologies, genomic collections have grown to hundreds of GBs or even TBs in size, and computing MEMs in such volumes of data has become impractical using state-of-the-art techniques.

Seed-and-extend heuristics is a popular solution to scale the problem of approximate pattern matching in large collections [2,16,20–22]. The efficiency of these heuristics largely depends on the seeding mechanism they employ. In this regard, using MEMs to seed alignments of near-identical sequences, such as genomes or proteins, offers two important benefits. Firstly, it improves alignment

Supported by Academy of Finland (Grants 323233 and 339070), and by Basal Funds FB0001, Chile (first author).

accuracy and reduces the cost of the heuristic's extension phase as approximate alignments of highly-similar strings often consist of long MEMs separated by small edits. Secondly, computing MEMs takes linear time [25,34], and in small collections, it does not impose a substantial overhead. The problem, as mentioned before, is that genomic data have grown to a point where fitting the necessary data structures to detect MEMs into main memory is difficult.

State-of-the-art methods [4,26,29–32] find MEMs in large inputs using compressed text indexes [9,13] and matching statistics [5]. These approaches have demonstrated that compression greatly reduces the overhead of computing MEMs in repetitive collections, but they are limited to situations where one needs to compare one string against many. This setting falls short for computationally-expensive genomic tasks such as genome assembly or multiple genome alignment that require all-vs-all approximate alignments of multiple strings. By extending the use of compression to find all-vs-all MEMs in collections, we could scale the execution of these genomic applications to TBs of data, which in turn could have important implications for Bioinformatics.

Our Contribution. We present a method to compute all-vs-all MEMs in a collection $\mathcal{T}$ of repetitive strings. Our idea consists of building a context-free grammar $\mathcal{G}$ from $\mathcal{T}$, from which we compute the MEMs. Our grammar algorithm ensures a property that we call *fix-free*, which means that the expansions of nonterminals with the same height form a set that is prefix-free and suffix-free. This idea enables the fast computation of MEMs by incrementally indexing parts of the grammar with simple data structures. Section 5 shows how we can build a fix-free grammar in linear time and space using a variant of the locally-consistent grammar of Christiansen et al. [7]. In Sect. 6, we describe how to compute a list $\mathcal{L}$ of primary MEMs (the precursors of the real MEMs) from $\mathcal{G}$ in $O(G+|\mathcal{L}|)$ time and $O((G+|\mathcal{L}|)\log G)$ bits of space, where G is the grammar size. In Sect. 6, we show how to enumerate all the occ all-vs-all MEMs from $\mathcal{L}$ and $\mathcal{G}$, yielding thus an algorithm that runs in $O(G + occ)$ time and uses $O((G + occ)\log G)$ bits.

2   Preliminaries

String Data Structures. Consider a string $T[1..n]$ over the alphabet Σ whose smallest symbol $\$ \in \Sigma$ is a sentinel that only occurs at $T[n]$. The *suffix array* [24] of T is a permutation $SA[1..n]$ that enumerates the suffixes of T in increasing lexicographic order, $T[SA[j]..n] < T[SA[j+1]..n]$, for $j \in [1..n-1]$. The *longest common prefix* array $LCP[1..n]$ [24] stores in $LCP[j]$ the longest common prefix between $T[SA[j-1]..n]$ and $T[SA[j]..n]$. Given an array $V[1..n]$ of integers, a *range minimum query* $rmq(V, j, j')$, with $j < j'$, returns the minimum value in the range $V[j..j']$. By augmenting LCP with support for rmq [12], one can obtain the length of the longest prefix shared by two arbitrary suffixes $T[SA[j]..n]$ and $T[SA[j']..n]$ by performing $rmq(LCP, j, j')$.

Locally-Consistent Parsing. *Parsing* consists in breaking a text $T[1..n]$ into a sequence of phrases. The parsing is *locally-consistent* [10] if, for any pattern P, its occurrences in T are largely partitioned in the same way. There is more than one way to make a parsing locally consistent (see [3,7,33] for more details), but this work focuses on those using *local minima*. A position j is a local minimum if $T[j-1] > T[j] < T[j+1]$. A method that uses this idea scans T, and for each pair of consecutive local minima j and j', it defines the phrase $T[j..j'-1]$. The procedure to compare adjacent positions in T can vary. For instance, Christiansen et al. [7] first create a new string $\hat{T}$ where they replace equal-symbol runs by metasymbols. Then, they define a random permutation π for the alphabet of $\hat{T}$, and compute the breaks as $\pi(\hat{T}[j-1]) > \pi(\hat{T}[i]) < \pi(\hat{T}[i+1])$. On the other hand, Nogueira et al. [28] compare consecutive suffixes rather than positions. Concretely, j is a local minimum if the suffix $T[j..n]$ is lexicographically smaller than the suffixes $T[j-1..n]$ and $T[j+1..n]$. Nong et al. [27] proposed this suffix-based local minimum in their suffix array algorithm SAIS. They refer to it as an LMS-type position and to the phrases covering consecutive LMS-type positions as LMS-substrings.

Grammar Compression. Grammar compression [6,17] is a form of lossless compression that encodes a string $T[1..n]$ as a context-free grammar $\mathcal{G}$ that generates only T. Formally, $\mathcal{G} = \{\Sigma, V, \mathcal{R}, S\}$ is a tuple of four elements, where Σ is the set of terminals, V is the set of nonterminals, $\mathcal{R}$ is a set of derivation rules in the form $X \to F$, with $X \in V$ being a nonterminal and $F \in (\Sigma \cup V)^*$ being its replacement, and $S \in V$ is the start symbol. In grammar compression, each nonterminal $X \in V$ appears only once on the left-hand sides of $\mathcal{R}$, which ensures the unambiguous decompression of T.

The graphical sequence of derivations that transforms S into T is called the parse tree. The root of this tree is labelled S and has $|A|$ children, with $S \to A$. The root's children are labelled from left to right according to A's sequence. The subtrees for the root's children are recursively defined in the same way. The height of a nonterminal X is the longest path in the parse subtree induced by X's recursive expansion. The grammar is said to be fully-balanced if, for each nonterminal $X \in V$ at height i, the symbols in the right-hand side of $X \to A_1 A_2 \cdots A_x \in \mathcal{R}$ are at height $i-1$.

The *grammar tree* is a pruned version of the parse tree that, for each $X \in V$, keeps only the leftmost internal node labelled $X \in V$. The remaining internal nodes labelled X are transformed into leaves. The leaves of the grammar tree induce a partition over T: for each grammar tree leaf u, its corresponding phrase is the substring in T mapping the terminal symbols under the parse tree node from which u was originated. One can classify the occurrences in T of a pattern $P \in \Sigma^*$ into primary and secondary. A *primary occurrence* of P crosses two or more phrases in the partition induced by the grammar tree. On the other hand, a *secondary occurrence* of P is fully contained within a phrase.

Locally-Consistent Grammar. A grammar $\mathcal{G}$ generating $T[1..n]$ is locally consistent if the occurrences of the same pattern are largely compressed in the same way [7,14]. A mechanism to build $\mathcal{G}$ is by transforming T in successive rounds of locally-consistent parsing [7,11,28]. In every round i, the construction algorithm receives as input a string $T^i[1..n^i]$ over an alphabet Σ^i, which represents a partially-compressed version of T (when $i = 1$, $T^1 = T$ and $\Sigma^1 = \Sigma$). Then, it breaks T^i using its local minima and creates a set $\mathcal{S}^i$ with the distinct phrases of the parsing. For every phrase $F \in \mathcal{S}^i$, the algorithm defines a new metasymbol X and appends the new rule $X \to F$ into $\mathcal{R}$. The value of X is the number of symbols in $\Sigma \cup V$ before the parsing round i plus the rank of F in an arbitrary order of $\mathcal{S}^i$. After creating the new rules, the algorithm replaces the phrases in T^i with their corresponding metasymbols to produce another string $T^{i+1}[1..n^{i+1}]$, which is the input for the next round $i+1$. Notice that the alphabet $\Sigma^{i+1} \subset V$ for T^{i+1} is the subset of metasymbols that the algorithm assigned to the phrases in T^i. When T^{i+1} does not have local minima, the algorithm creates the final rule $S \to T^{i+1}$ for the start symbol S of $\mathcal{G}$ and finishes. The whole process runs in $O(n)$ time and produces a fully-balanced grammar. Christiansen at al. [7] showed that it is possible to obtain a locally-consistent grammar of size $O(\gamma \log n/\gamma)$ in $O(n)$ expected time, where γ is the size of the smallest string attractor for T [15].

3   Definitions

Let us consider a collection $\mathcal{T} = \{T_1, T_2, \ldots, T_m\}$ of m strings over the alphabet Σ. We start by defining the concept of MEM between elements of $\mathcal{T}$.

Definition 1. *MEM: let $T_x[1..n_x]$ and $T_y[1..n_y]$ be two strings in $\mathcal{T}$. A maximal exact match $T_x[a..b] = T_y[a'..b']$ of length $\ell = b - a + 1 = b' - a' + 1$, denoted $M(x, y, a, a', \ell)$, has the following properties: (i) $a = 1$ or $a' = 1$, or $a, a' > 1$ and $T_x[a-1] \neq T_y[a'-1]$. (ii) $b = n_x$ or $b' = n_y$, or $b < n_x$, $b' < n_y$, and $T_x[b+1] \neq T_y[b'+1]$.*

Now we formulate the problem we address in this work as follows:

AvAMem
Input: a string collection $\mathcal{T} = \{T_1, T_2, \ldots, T_m\}$ and an input integer τ.
Output: every possible $M(x, y, a, a', \ell)$ with $T_x, T_y \in \mathcal{T}$ and $\ell \geq \tau$.

We will refer to the parsing of Sect. 2 as LCPar, and the grammar algorithm of Sect. 2 that relies on this technique as LCGram. Let $\mathcal{G} = (\Sigma, V, \mathcal{R}, S)$ be the fully-balanced grammar constructed by LCGram from $\mathcal{T}$ in h rounds of parsing. We assume for the moment that the definition of local minima is arbitrary, but sequence-based. We denote the sum of the right-hand side lengths of $\mathcal{R}$ as G and the number of nonterminals as $g = |V|$. Additionally, we consider the partition $V = \{V^1, \ldots, V^h\}$ such that every V^i, with $i \in [1..h]$, has the nonterminals generated during the parsing round i. Similarly, the partition $\mathcal{R} = \{\mathcal{R}^1, \ldots, R^h\}$ groups in $\mathcal{R}^i$ the rules generated during the parsing round i. We denote as G^i the sum of the right-hand side lengths of $\mathcal{R}^i$ and $g^i = |V^i|$. We will refer to the sequence $[1..h]$ as the *levels* of the grammar, which is read bottom-up in the parse tree. $\mathcal{T}$ is at level 0.

We assume LCGRAM compresses the strings of T independently but collapses all the rules in one single grammar $\mathcal{G}$. Thus, the rule $S \to A_1..A_{m'}$ encodes the compressed strings of T concatenated in the string $A_1 \ldots A_{m'}$.

The operator $exp(X)$ returns the string in Σ^* resulting from the recursive expansion of $X \in V$. The function exp can also receive as input a sequence $X_1 \cdots X_b \in V^*$, in which case it returns the concatenation $exp(X_1) \cdots exp(X_b)$. The operation $lcp^i(X, Y)$ receives two nonterminals $(X, Y) \in V^i$ and returns the longest common prefix between $exp(X)$ and $exp(Y)$. Similarly, the operator $lcs^i(X, Y)$ returns the longest common suffix between $exp(X)$ and $exp(Y)$.

Definition 2. *A set $\mathcal{S}$ of strings is fix-free iff, for any pair $F, Q \in \mathcal{S}$, the string F is not a suffix nor a prefix of Q, and vice-versa.*

Definition 3. *A grammar $\mathcal{G}$ is fix-free iff it is fully balanced, and for any level i, the set $\mathcal{S}^i = \{exp(X_1), \ldots, exp(X_{g^i})\}$, with $X_1, \ldots, X_{g^i} \subset V^i$, is fix-free.*

Definition 4. *Primary MEM (prMEM): let $M(x, y, a, a', \ell)$, with $T_x, T_y \in T$, be a MEM. $M(x, y, a, a', \ell)$ is primary if both $T_x[a..a+\ell-1]$ and $T_y[a'..a'+\ell-1]$ are primary occurrences of the pattern $T_x[a..a+\ell-1] = T_y[a'..a'+\ell-1]$.*

4   Overview of Our Algorithm

We solve AVAMEM(T, τ) in three steps: (i) we build a fix-free grammar from T using a variant of LCGRAM, (ii) we compute a list $\mathcal{L}$ storing the prMEMs of T, and (iii) we use $\mathcal{L}$ and $\mathcal{G}$ to report the positions in T of all the MEMs.

The advantage of $\mathcal{G}$ being fix-free is that we can use the following lemma:

Lemma 1. *Let $\mathcal{G}$ be a fix-free grammar. Two rules $X \to AZB$, $Y \to CZD \in \mathcal{R}^i$, with $Z \in V^{i-1*}$ and $A, B, C, D \in V^{i-1}$ yield a prMEM if $\ell = lcs^{i-1}(A, C) + |exp(Z)| + lcp^{i-1}(B, D) \geq \tau$.*

Proof. Both $exp(X)$ and $exp(Y)$ contain $exp(Z)$ as a substring as the rules for X and Y have occurrences of Z, and there is only one string $exp(Z)$ the grammar can produce. Still, the strings $exp(A)$ and $exp(C)$ (respectively, $exp(B)$ and $exp(D)$) could share a suffix (respectively, a prefix), meaning that the prMEM extends to the left of Z (respectively, the right of Z). The fix-free property guarantees that the left boundary of the prMEM lies at some index in the right-to-left comparison of $exp(A)$ and $exp(C)$, and that the right boundary lies at some position within the left-to-right comparison of $exp(B)$ and $exp(D)$.

Lemma 1 offers a simple solution to detect prMEMs as we do not have to look into other parts of $\mathcal{G}$'s parse tree to check the boundaries of the prMEM in (X, Y). The only remaining aspects to consider are, first, how to get a fix-free grammar, and then, how to compute $lcs(A, C)$, $lcp(B, D)$, and $|exp(Z)|$ efficiently. Once we solve these problems, finding prMEMs reduces to run a suffix-tree-based MEM algorithm over the right-hand sides of each $\mathcal{R}^i$. On the other hand, getting the positions in T of the MEMs (step three of in the overview) requires traversing the rules of $\mathcal{G}$, but it is not necessary to perform any string comparison.

5   Building the Fix-Free Grammar

We first describe the parsing we will use in our variant of LCGRAM. We refer to this procedure as FFPAR. The input is a string collection T' over the alphabet $\Sigma' = \{\$\} \cup \Sigma \cup \{\#\}$, where each string $T_x = \$T\$\# \in T'$ is flanked by the symbols $\{\$, \#\} \notin \Sigma$ that do not occur in the internal substring $T \in \Sigma^*$.

We choose a function $h : \Sigma' \rightarrow [0..p+1]$ that maps symbols in Σ to integers in the range $[1..p]$ uniformly at random, where $p > |\Sigma|$ is a large prime number. When $c \in \Sigma$, the function returns $h(c) = (ac + b) \bmod p$, where $a, b \in [1..p]$ are chosen uniformly at random. If $c = \$ \notin \Sigma$, $h(c) = 0$, and if $c = \#$, $h(c) = p + 1$.

We use h to define the local minima in T'. The idea is to combine h with a scheme to classify symbols similar to that of SAIS [27]. Let $T_x[1..n_x]$ be a string in T'. A position $j \in [2..n_x - 1]$ has three possible classifications:

- L-type : $h(T_x[j]) > h(T_x[j+1])$ or $T_x[j] = T_x[j+1]$ and $T_x[j+1]$ is L-type.
- S-type : $h(T_x[j]) < h(T_x[j+1])$ or $T_x[j] = T_x[j+1]$ and $T_x[j+1]$ is S-type.
- LMS-type : $T_x[j]$ is S-type and $T_x[j-1]$ is L-type.

The LMS-type positions are the local minima of T'. Randomising the local minimum definition aims to protect us against adversarial inputs. Notice that h is not a random permutation π like Christiansen et al. [7], but it still defines a total order over Σ as it does not assign the same random value to two symbols, which is enough for our purposes.

Lemma 2. FFPAR: *let $T_x[1..n_x] \in T'$ be a string. For each pair of consecutive local minima $1 < j < j' < n_x$, we create the phrase $T_x[j-1..j'+1]$. For the leftmost local minimum j in T_x, we create $T_x[1..j+1]$, and for its rightmost local minimum j', we create $T_x[j'-1..n_x]$. The set of parsing phrases is fix-free.*

Proof. Let us first consider the set S created by LCPAR. We will say that a phrase $T_x[j..j'-1]$ in LCPAR is an *instance* of $F \in S$ if $T_x[j..j'-1] = F$. When F occurs in T' as a proper suffix or prefix of another phrase in S, it is not an instance, only an occurrence of F.

Let $W \subset S$ be a subset of phrases and let $F \in S \setminus W$ be a phrase occurring as a proper prefix in each element of W. Consider any pair of instances $F = T_x[j..j'-1]$ and $Q = T_y[l..l'-1] \in W$ such that $T_x, T_y \in T'$, and (l, l') (respectively, (j, j')) are consecutive local minima. The substring $T_x[j'..j'+1]$ following F's instance cannot be equal to the substring $T_y[l+|F|..l+|F|+1]$ following F's occurrence within Q because j' is a local minimum while $l + |F|$ is not. We know that $T_y[l+|F|]$ is within Q because $T_y[l..l+|F|-1]$ is an occurrence of F that is a proper prefix of Q. Therefore, if $l + |F|$ were a local minimum, Q would also be an instance of F. In conclusion, running FFPAR will right-extend every instance of F in T' such that none of the resulting phrases is a proper prefix in the right-extended phrases of W.

We now develop a similar argument for the left extension. Suppose, in this case, $F \in S \setminus W$ is a proper suffix in $W \subset S$. As before, we assume S was built using LCPAR and that $F = T_x[j..j'-1]$ and $Q = T_y[l..l'-1] \in W$ are phrase

Fig. 1. Execution of FFGRAM on $gtaatagtagtacc$#. The left side shows the parse tree up to level 2, and the right side shows the corresponding rules for those levels. The underlined symbols are local minima.

instances. The symbol $T_x[j-1]$ preceding F's instance differs from the symbol $T_y[l'-|F|-1]$ preceding the occurrence $F = T_y[l'-|F|..l'-1]$ within Q. The reason is because the position j in $F[1] = T_x[j]$ is a local minimum, while the position $l'-|F|$ of $Q[|Q|-|F|+1] = T_y[l'-|F|] = F[1]$ is not. Hence, FFPAR left-extends each occurrence of F such that none of the resulting phrases is a proper suffix in the left-extended elements of $\mathcal{W}$.

The FFGram Algorithm. FFGRAM is our LCGRAM variant that builds a fix-free grammar by applying successive rounds of FFPAR (Definition 2). The input of FFGRAM is the collection T^1 built from T by adding the special flanking symbols T_x# in every $T_x \in T$. The alphabet of T^1 is $\Sigma^1 = \{\$\} \cup \Sigma \cup \{\#\}$.

In every round i, we create a random hash function $h^i : \Sigma^i \to [0..p+1]$, with $p > |\Sigma^i|$, to define the local minima of T^i. Then, we run FFPAR using h^i and sort the resulting set $\mathcal{S}^i$ of phrases in lexicographical order starting from the leftmost proper suffix of each string. The relative order of strings of $\mathcal{S}^i$ differing only in the leftmost symbol is arbitrary. Let c be the number of symbols in $\Sigma \cup V$ before parsing round i and let r the rank of $F \in \mathcal{S}^i$ in the ordering we just defined. We assign the metasymbol $X = c+r \in V^i$ to F and append the new rule $X \to F \in \mathcal{R}^i$. The last step in the round is to create the collection T^{i+1} by replacing the phrases with their corresponding metasymbols. Additionally, we define two special new symbols $\{\$, \#\}$, which we append at the ends of the strings in T^{i+1}. Thus, the alphabet of T^i becomes $\Sigma^i = \{\$\} \cup V^i \cup \{\#\}$. We assume the occurrences of the special symbols $\#, \$$ in the right-hand sides of $\mathcal{R}$ expand to the empty string. FFGRAM stops after h parsing rounds, when the input T^i does not have local minima. Figure 1 shows an example.

Lemma 3. *The grammar $\mathcal{G}$ resulting from running FFGRAM over T is fix-free.*

Proof. Consider the execution of FFPAR over T^1 in the first round of FFGRAM. The output set $\mathcal{S}^1$ is over the alphabet $\{\$\} \cup \Sigma \cup \{\#\}$ and is, by Lemma 2, fix-free. Thus, the symbols of V^1 meet the fix-free property of Definition 3. Now consider the parsing rounds $i-1$ with $i \geq 2$. Assume without loss of generality that the nonterminals in V^{i-1} meet Definition 3. The recursive definition of FFGRAM implies that the phrases in $\mathcal{S}^i$ are over the alphabet V^{i-1}. Thus, for any pair of different strings $F, Q \in \mathcal{S}^i$ sharing a prefix $F[1..q] = Q[1..q]$, the symbols $F[q+1] \neq Q[q+1]$ expand to different sequences $exp(F[q+1]) \neq exp(Q[q+1])$ that are not prefix one of the other as $F[q+1], Q[q+1]$ belong to V^{i-1}. Therefore,

$exp(F)$ and $exp(Q)$ do not prefix one to the other either. The same argument applies when F and Q share a suffix. We conclude then that the metasymbols of V^i also meet the fix-free property of Definition 3.

Overlaps in the Fix-Free Grammar. A relevant feature of FFGRAM is that consecutive nonterminals in the right-hand sides of $\mathcal{R}$ cover overlapping substrings of $\mathcal{T}$. This property allows us to compute prMEMs as described in Lemma 1. The downside, however, is that expanding substrings of $\mathcal{T}$ from $\mathcal{G}$ is now more difficult. However, the number of symbols overlapping in every grammar level is constant (one to the left and two to the right). Depending on the situation, we might want to decompress nonterminals considering (or not) skipping overlaps. Thus, we define the operations $efexp(X)$, $lexp(X)$, and $rexp(X)$ that return different types of nonterminal expansions. These functions only differ in the edges they skip in X's subtree during the decompression.

- $efexp(X)$: recursively skips the leftmost and two rightmost edges.
- $lexp(X)$: recursively skips the two rightmost edges, and the leftmost edges when the parent node does not belong to the leftmost branch.
- $rexp(X)$ recursively skips the leftmost edges, and the two rightmost edges when the parent node does not belong to the rightmost branch.

Figure 2 shows examples of $efexp(X)$, $lexp(X)$ and $rexp(X)$. We also modify the functions lcs^i and lcp^i described in Definitions 3. Let $X, Y \in V^i$ be two nonterminals at level i. The function $lcs^i(X, Y)$ now returns the longest common suffix between $lexp(X)$ and $lexp(Y)$; and $lcp^i(X, Y)$ returns the longest common prefix between $rexp(X)$ and $rexp(Y)$.

We remark that $efexp$, $lexp$, $rexp$, lcs^i, and lcp^i are virtual as our algorithm to find prMEMs never calls them directly. Instead, it incrementally produces satellite data structures with precomputed answers to solve them in $O(1)$ time.

6   Computing prMEMs in the Fix-Free Grammar

Our MEM algorithm uses the grammar $\mathcal{G} = \{\Sigma, V, \mathcal{R}, S\}$ resulted from running FFGRAM with $\mathcal{T}'$, and produces the list $\mathcal{L}$ of prMEMs in $\mathcal{T}$. We consider a set $\mathcal{O}$ of $g = |\mathcal{R}|$ rules storing the cumulative lengths of the $efexp$ expansions for the right-hand sides of $\mathcal{R}$. We also define a logical partition for $\mathcal{O}$ according to the grammar levels. Let $X \to A_1 A_2 \cdots A_x \in \mathcal{R}^i$ be a rule at level i. The rule $X \to c_1 \cdots c_x \in \mathcal{O}^i$ stores in c_j, with $j \in [3..x-1]$, the value $c_j = efexp(A_2) + \cdots + efexp(A_{j-1})$. To avoid recursive overlaps, we set $c_1 = 0, c_2 = 0$, and $c_{x-1} = c_{x-2}, c_x = c_{x-2}$. The leftmost tree of Fig. 2 shows an example of a rule in $\mathcal{O}$. We assume FFGRAM already constructed $\mathcal{O}$.

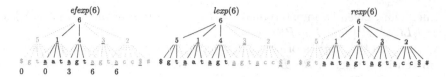

Fig. 2. Expansions for nonterminal 6 of Fig. 1. Dashed lines are skipped branches. The sequence of numbers below $efexp(6)$ corresponds to the rule $6 \rightarrow 0\ 0\ 3\ 6\ 6 \in \mathcal{O}$.

prMEM Encoding. Every element of $\mathcal{L}$ is a tuple (X, Y, o_X, o_Y, ℓ) of five elements. X and Y are the nonterminals labelling the lowest nodes in $\mathcal{G}$'s grammar tree that encode the primary occurrences for the MEM's sequence. The fields o_X and o_Y are the number of terminal symbols preceding the prMEM within $efexp(X)$ and $efexp(Y)$ (respectively), and ℓ is the length of the prMEM.

Grammar Encoding. We encode every subset $\mathcal{R}^i \subset \mathcal{R}$ as an individual string collection concatenated in one single array $R^i[1..G^i]$. We create an equivalent array O^i for the rules of $\mathcal{O}^i \subset \mathcal{O}$. We also consider a function $map(X) = j$ that indicates that the right-hand side $F = R^i[a..a']$ of $X \rightarrow F \in \mathcal{R}^i$ is the jth string of R^i from left to right. Additionally, we define the function $parent(b) = X$ that returns X for each position $b \in [a..a']$. We assume map and $parent$ are implemented in $O(1)$ time and $G_i + o(G_i)$ bits using bit vectors.

Our prMEM algorithm is an iterative process that, each step i, searches for prMEMs in the rules of the grammar level i. Still, there is a slight difference between the iteration $i = 1$ and the others $i > 1$, so we explain them separately.

First Iteration. The first step in iteration $i = 1$ is to create a sparse suffix array SA for R^i that discards each position $SA[j]$ meeting one of the following conditions: (i) $R^i[SA[j]]$ is the start of a phrase, (ii) $R^i[SA[j] + 1]$ is the end of a phrase, or (iii) $R^i[SA[j]]$ is the end of a phrase. We refer to the resulting sparse suffix array as A. The next step is to produce the LCP array for A, which we name $GLCP$ for convenience. We then run the suffix-tree-based MEM algorithm [34] using A and $GLCP$. We implement this step by simulating with $GLCP$ a traversal over the compact trie induced by the suffixes of R^i in A (see [1,19] for the traversal). Every time the MEM algorithm reports a triplet $(A[u], A[u'], l)$ for a MEM $R^i[A[u]..A[u] + l - 1] = R^i[A[u']..A[u'] + l - 1]$, we append the tuple $(X = parent(A[u]), Y = parent(A[u']), o_X = O^i[A[u]], o_Y = O^i[A[u']], \ell = l)$ into $\mathcal{L}$. Once we finish running the MEM algorithm, we obtain satellite data structures for the next iteration $i + 1$:

- A vector $P^1[1..g^1]$ encoding the permutation of R^1 resulted from sorting the $lexp$ expansions of the strings in colexicographical order.
- A vector $LCS^1[1..g^1]$ storing LCS (longest common suffix) values between the $lexp$ expansions of strings in R^i that are consecutive in the permutation P^1. We encode LCS^1 with support for range minimum queries in $O(1)$

time. Thus, given two nonterminals $X, Y \in V^1$, we implement $lcs^1(X, Y)$ as $rmq(LCS^1, P^1[map(X)], P^1[map(Y)])$.

- A vector $LCP^1[1..g^1]$ storing LCP values between the $rexp$ expansions of consecutive strings of R^1. LCP^1 also supports $O(1)$-time rmq queries so we implement $lcs^1(X, Y)$ as $rmq(LCP^1, X, Y)$.

Next Iterations. For $i > 1$, we receive as input the tuple (R^i, O^i) and the vectors $P^{i-1}, LCS^{i-1}, LCP^{i-1}$. We assume LCS^{i-1} and LCP^{i-1} support rmq queries in $O(1)$ time so we can implement lcs^{i-1} and lcp^{i-1} in $O(1)$ time as well.

We compute A and $GLCP$ for R^i as in Sect. 6. However, we transform $GLCP$ to store the LCP values of the $rexp$ expansions of the suffixes of R^i in A. Let $R^i[A[j]..|R^i|]$ and $R^i[A[j+1]..|R^i|]$ be two consecutive suffixes of R^i in A sharing a prefix of length $GLCP[j+1] = l$. We update this value to $GLCP[j+1] = O^i[A[j+1]+l] + lcp^{i-1}(R^i[A[j]+l]), R^i[A[j+1]+l])$.

We use $GLCP$ and A to simulate a traversal over the compact trie induced by the $rexp$ expansions of the R^i suffixes in A. The purpose of the traversal is, again, to run the suffix-tree-based MEM algorithm. Every time this procedure reports a triplet $(A[u], A[u'], l)$ as a MEM, we compute $o = lcs(R^i[u-1], R^i[u'-1])$, and insert the tuple $(X = parent(A[u]), Y = parent(A[u'])), o_X = O^i[A[u]] - o + 1, o_Y = O^i[A[u']] - o + 1, \ell = o + l)$ into $\mathcal{L}$. The final step in the iteration is to produce the satellite data structures:

- $P^i[1..g^i]$: we sort the strings in R^i colexicographically using P^{i-1}. We define the relative order of any pair of strings $F, Q \in R^i$ by comparing their sequences $P^{i-1}[F[1]]\cdots P^{i-1}[F[|F|-2]]$ and $P^{i-1}[Q[1]]\cdots P^{i-1}[Q[|Q|-2]]$ from right to left. The resulting permutation P^i has the following property: let $X, X' \in V^i$ be two nonterminals. If $P^i[map(X)] < P^i[map(X')]$, it means $lexp(X)$ is colexicographically equal or smaller than $lexp(X')$.
- $LCS^i[1..g^i]$: we scan the strings of R^i in P^i order. Let $R^i[a..a']$ and $R^i[b..b']$ be two consecutive strings in the permutation of P^i. That is, $X = parent(a)$ and $X' = parent(b)$ such that $P^i[map(X)] = j$ and $P^i[map(X')] = j+1$. Assume their prefixes $R^i[a..a'-2]$ and $R^i[b..b'-2]$ share a suffix of length $l \geq 0$. We set $LCS^i[j+1] = O^i[a'] - O^i[a'-l-1] + lcs^{i-1}(R^i[a'-l-2], R^i[b'-l-2])$.
- $LCP^i[1..g^i]$: we scan the strings of R^i from left to right. Let $R^i[a..a']$ and $R^i[b..b']$ be two strings in R^i with $X = parent(a)$ and $X' = parent(b) = X+1$. Assume their suffixes $R^i[a+1..a']$ and $R^i[b+1..b']$ share a prefix of length $l \geq 0$. We set $LCS^i[map(X')] = O^i[b+l+1] + lcp^{i-1}(R^i[a+1+l], R^i[b+1+l])$.

Theorem 1. *Let $\mathcal{G} = \{\Sigma, V, \mathcal{R}, S\}$ be a fix-free grammar of size G built with* FFGRAM *using the collection $\mathcal{T} = \{T_1, \ldots, T_u\}$. It is possible to obtain from $\mathcal{G}$ the list $\mathcal{L}$ with the prMEMs of $\mathcal{T}$ in $O(G + |\mathcal{L}|)$ time and $O((G + |\mathcal{L}|)\log G)$ bits.*

Proof. In iteration $i = 1$, constructing A and $GLCP$ takes $O(G_1)$ time and $O(G^1 \log G^1)$ bits [19, 27]. Then, we implement the suffix-tree-based algorithm to report MEMs in R^i by combining the method of Abouelhoda et al. [1], that visits the nodes of the compact trie induced by $GLCP$ in $O(G^i)$ time, with Lemma

11.4 of Mäkinen et al. [23], which reports the MEMs of every internal node. These ideas combined take $O(G^1 + e^1)$ time and $O((G^1 + e^1) \log G^1)$ bits, where e^1 is the number of prMEMs in the grammar level 1. The final step is to build LCP^1, LCS^1, and P^1. FFGRAM sorted the strings of R^1 in lexicographical order according to their $rexp$ expansions, so building LCP^1 reduces to scan R^1 from left to right and compute LCP values between consecutive elements. This process takes $O(G^1)$ time and $O(G^1 \log G^1)$ bits. Then, we obtain P^1 by running SAIS in the reversed strings of R^1, which also takes $O(G^1)$ time. We use the reversed strings to build LCS^1 as we did with LCP^1. Finally, giving rmq support to LCP^1 and LCP^1 takes $O(G^1)$ time and $O(G^1 \log G^1)$ bits if we use the data structure of Johannes Fischer [12]. Summing up, the iteration i runs in $O(G^1 + e^1)$ time and uses $O((G^1 + e^1) \log G^1)$ bits. Each iteration $i > 1$ performs the same operations, but it also updates $GLCP, LCP^i$, and LCS^i. These updates require linear scans of the arrays as processing each position performs $O(1)$ access to O^i and $O(1)$ calls to lcs^{i-1} or lcp^{i-1}, which we implement in $O(1)$ time. Thus, the cost of iteration i is $O(G^i + e^i)$ time and $O((G^i + e^i) \log G^i)$ bits, where e^i is the number prMEMs in the grammar level i. Combining the h grammar levels, the cost to compute $\mathcal{L}$ is $O(G + |\mathcal{L}|)$ time and $O((G + |\mathcal{L}|) \log G)$ bits.

7   Positioning MEMs in the Text

The last aspect we cover to solve AvAMEM$(\mathcal{T}, \tau)$ is computing from $(\mathcal{L}, \mathcal{G})$ the positions in $\mathcal{T}$ of the MEMs. We assume that the collections $\{R^1, \ldots, R^h\}$ of Sect. 6 are concatenated in one single array $R[1..G]$, and that the collections $\{O^1, \ldots, O^h\}$ are concatenated in another array $O[1..G]$. We define the function $stringid$, which takes as input an index u within R with $S = parent(u)$ (start symbol in $\mathcal{G}$), and returns the identifier of the string of $\mathcal{T}$ where $R[u]$ lies.

We first simplify $\mathcal{G}$ to remove the unary paths in its parse tree This extra step is not strictly necessary, but it is convenient to avoid redundant work. After the simplification, we create an array $N[1..G]$, which we divide into g buckets. Each bucket $b \in [1..g]$ stores in an arbitrary order the positions in R for the occurrences of $b \in \Sigma \cup V$. Additionally, we create a vector $C[1..g]$ that stores in $C[b]$ the position in N where the bucket for $b \in \Sigma \cup V$ starts.

We report MEMs as follows: we insert the tuples of $\mathcal{L}$ into a stack. Then, we extract the tuple (X, Y, o_Y, o_X, ℓ) from the top of the stack and compute $s_X = C[X], e_X = C[X+1] - 1$ and $s_Y = C[Y], e_Y = C[Y+1] - 1$. For every pair of indexes $(u, u') \in [s_X, e_X] \times [s_Y, e_Y]$, we get $X' = parent(N[u])$, and if $X' = S$, we set $X' = stringid(N[u])$. We do the same with $N[u']$ and store the result in a variable Y'. Now we produce the new tuple $(X', Y', o_{X'} = O[N[u]] + o_X, o_{Y'} = O[N[u']] + o_Y, \ell)$. If both X' and Y' are strings identifiers, we report the tuple $M(X', Y', o_{X'}, o_{Y'}, \ell)$ as an output of AvAMEM, otherwise we insert the new tuple into the stack. If X' or Y' is a string identifier rather than a nonterminal, we flag the tuple to indicate that one of the elements is a string. Thus, when we visit the tuple again, we avoid recomputing its values. The report of MEMs ends when the stack becomes empty.

Theorem 2. *Let $\mathcal{G}$ be a fix-free grammar of size G constructed with* FFGRAM *using the collection $\mathcal{T} = \{T_1, \ldots, T_u\}$, and let $\mathcal{L}$ be the list of prMEMs in $\mathcal{G}$ with length $> \tau$, where τ is an input parameter. Given the simplified version of $\mathcal{G}$ and $\mathcal{L}$, it is possible to report the positions in $\mathcal{T}$ of the occ MEMs of length $> \tau$ in $O(G + occ)$ time and $O((G + occ) \log G)$ bits.*

Proof. The G term in the time complexity comes from the grammar simplification and the construction of N and C. Let $(X, Y, o_X, o_Y, \ell) \in \mathcal{L}$ be a prMEM whose sequence in $\mathcal{T}$ is L, with $|L| = \ell$. Let us assume aLb is the primary occurrence of L under X and xLz is the primary occurrence under Y. In our algorithm, the access pattern in N simulates a bottom-up traversal of $\mathcal{G}$'s grammar tree that visits every node labelled X' such that $exp(X')$ has aLb, and every node labelled Y' such that $exp(Y')$ has xLz. Our idea is similar to reporting secondary occurrences in the grammar self-index of Claude and Navarro [8]. They showed that the cost of traversing the grammar to enumerate the occ_P occurrences of a pattern P amortizes to $O(occ_P)$ time. The argument is that, in a simplified $\mathcal{G}$, each node we visit in the grammar tree yields at least one occurrence in $\mathcal{T}$. However, our traversal processes two patterns simultaneously (aLb and yLz), pairing each occurrence of one with each occurrence of the other. Thus, our amortized time to process a prMEM tuple is $O(occ_X \times occ_Y)$, where occ_X and occ_Y are the numbers of occurrences in $\mathcal{T}$ for aLb and yLz, respectively. Summing up, the cost of processing all the tuples in $\mathcal{L}$ is $O(occ)$ time.

Corollary 1. *Let $\mathcal{T}$ be a string collection of n symbols containing occ MEMs of length $\geq \tau$, τ being a parameter. We can solve* AVAMEM$(\mathcal{T}, \tau)$ *by building a fix-free grammar $\mathcal{G}$ of size G in $O(n)$ time and $O(G \log G)$ bits, and then computing the occ MEMs of $\mathcal{T}$ over $\mathcal{G}$ in $O(G + occ)$ time and $O((G + occ) \log G)$ bits.*

8   Concluding Remarks

We have presented a method to compute all-vs-all MEMs that rely on grammar compression to reduce memory overhead and save redundant calculations. In particular, given a collection $\mathcal{T}$ of n symbols and occ MEMs, we can get a grammar $\mathcal{G}$ of size G in $O(n)$ time and $O(G \log G)$ bits of space, and find the MEMs of $\mathcal{T}$ on top of $\mathcal{G}$ in $O(G + occ)$ time and using $O((G + occ) \log G)$ bits. We believe our framework is of practical interest as it uses mostly plain data structures that store satellite data about the grammar. Besides, we can choose to compute the MEMs at the same time we construct the grammar or do it later. However, it remains open to check how far is G from $O(\gamma \log \frac{n}{\gamma})$. The comparison is reasonable as FFGRAM, our grammar algorithm, resembles the locally-consistent grammar of Christiansen et al. [7], which achieves that bound. Still, it is unclear to us how the overlap produced by FFGRAM affects the bound. Overlapping phrases has an exponential effect on the grammar size but also affects how the text is parsed. An interesting idea would be to find a way to chain prMEMs as approximate matches and then report those in the last step of our algorithm instead of the MEMs. An efficient implementation of such a procedure could significantly reduce the cost of biological sequence analyses in massive collections.

References

1. Abouelhoda, M.I., Kurtz, S., Ohlebusch, E.: Replacing suffix trees with enhanced suffix arrays. J. Discrete Algorithms **2**(1), 53–86 (2004)
2. Altschul, S.F., Gish, W., Miller, W., Myers, E.W., Lipman, D.J.: Basic local alignment search tool. J. Mol. Biol. **215**(3), 403–410 (1990)
3. Batu, T., Ergun, F., Sahinalp, C.: Oblivious string embeddings and edit distance approximations. In: Proceedings of the 17th Symposium on Discrete Algorithms (SODA), pp. 792–801 (2006)
4. Boucher, C., et al.: PHONI: streamed matching statistics with multi-genome references. In: Proceedings of the 21st Data Compression Conference (DCC), pp. 193–202 (2021)
5. Chang, W.I., Lawler, E.L.: Sublinear approximate string matching and biological applications. Algorithmica **12**(4), 327–344 (1994)
6. Charikar, M., et al.: The smallest grammar problem. IEEE Trans. Inf. Theory **51**(7), 2554–2576 (2005)
7. Christiansen, A.R., Ettienne, M.B., Kociumaka, T., Navarro, G., Prezza, N.: Optimal-time dictionary-compressed indexes. ACM Trans. Algorithms **17**(1), 1–39 (2020)
8. Claude, F., Navarro, G.: Improved grammar-based compressed indexes. In: Calderón-Benavides, L., González-Caro, C., Chávez, E., Ziviani, N. (eds.) SPIRE 2012. LNCS, vol. 7608, pp. 180–192. Springer, Heidelberg (2012). https://doi.org/10.1007/978-3-642-34109-0_19
9. Claude, F., Navarro, G., Pacheco, A.: Grammar-compressed indexes with logarithmic search time. J. Comput. Syst. Sci. **118**, 53–74 (2021)
10. Cole, R., Vishkin, U.: Deterministic coin tossing and accelerating cascades: micro and macro techniques for designing parallel algorithms. In: Proceedings of the 18th Annual Symposium on Theory of Computing (STOC), pp. 206–219 (1986)
11. Díaz-Domínguez, D., Navarro, G.: A grammar compressor for collections of reads with applications to the construction of the BWT. In: Proceedings of the 31st Data Compression Conference (DCC), pp. 83–92 (2021)
12. Fischer, J.: Optimal succinctness for range minimum queries. In: López-Ortiz, A. (ed.) LATIN 2010. LNCS, vol. 6034, pp. 158–169. Springer, Heidelberg (2010). https://doi.org/10.1007/978-3-642-12200-2_16
13. Gagie, T., Navarro, G., Prezza, N.: Fully-functional suffix trees and optimal text searching in BWT-runs bounded space. J. ACM **67**(1) (2020). Article 2
14. Jeż, A.: Approximation of grammar-based compression via recompression. Theor. Comput. Sci. **592**, 115–134 (2015)
15. Kempa, D., Prezza, N.: At the roots of dictionary compression: string attractors. In: Proceedings of the 50th Annual ACM SIGACT Symposium on Theory of Computing (STOC), pp. 827–840 (2018)
16. Kent, W.J.: BLAT-the BLAST-like alignment tool. Genome Res. **12**(4), 656–664 (2002)
17. Kieffer, J., Yang, E.-H.: Grammar-based codes: a new class of universal lossless source codes. IEEE Trans. Inf. Theory **46**(3), 737–754 (2000)
18. Kurtz, S., et al.: Versatile and open software for comparing large genomes. Genome Biol. **5**, 1–9 (2004)
19. Kasai, T., Lee, G., Arimura, H., Arikawa, S., Park, K.: Linear-time longest-common-prefix computation in suffix arrays and its applications. In: Amir, A. (ed.) CPM 2001. LNCS, vol. 2089, pp. 181–192. Springer, Heidelberg (2001). https://doi.org/10.1007/3-540-48194-X_17

20. Langmead, B., Salzberg, S.L.: Fast gapped-read alignment with bowtie 2. Nat. Methods **9**(4), 357–359 (2012)
21. Li, H.: Aligning sequence reads, clone sequences and assembly contigs with BWA-MEM. arXiv preprint arXiv:1303.3997 (2013)
22. Li, H.: Minimap2: pairwise alignment for nucleotide sequences. Bioinformatics **34**(18), 3094–3100 (2018)
23. Mäkinen, V., Belazzougui, D., Cunial, F., Tomescu, A.I.: Genome-Scale Algorithm Design. Cambridge University Press, Cambridge (2015)
24. Manber, U., Myers, G.: Suffix arrays: a new method for on-line string searches. SIAM J. Comput. **22**(5), 935–948 (1993)
25. McCreight, E.M.: A space-economical suffix tree construction algorithm. J. ACM **23**(2), 262–272 (1976)
26. Navarro, G.: Computing MEMs on repetitive text collections. In: Proceedings of the 34th Annual Symposium on Combinatorial Pattern Matching (CPM), pp. article 22 (2023)
27. Nong, G., Zhang, S., Chan, W.H.; Linear suffix array construction by almost pure induced-sorting. In; Proceedings of the 19th Data Compression Conference (DCC), pp. 193–202 (2009)
28. Nunes, D.S.N., Louza, F., Gog, S., Ayala-Rincón, M., Navarro, G.: A grammar compression algorithm based on induced suffix sorting. In: Proceedings of the 28th Data Compression Conference (DCC), pp. 42–51 (2018)
29. Ohlebusch, E., Fischer, J., Gog, S.: CST++. In: Chavez, E., Lonardi, S. (eds.) SPIRE 2010. LNCS, vol. 6393, pp. 322–333. Springer, Heidelberg (2010). https://doi.org/10.1007/978-3-642-16321-0_34
30. Rossi, M., Oliva, M., Bonizzoni, P., Langmead, B., Gagie, T., Boucher, C.: Finding maximal exact matches using the r-index. J. Comput. Biol. **29**(2), 188–194 (2022)
31. Rossi, M., Oliva, M., Langmead, B., Gagie, T., Boucher, C.: MONI: a pangenomic index for finding maximal exact matches. J. Comput. Biol. **29**(2), 169–187 (2022)
32. Sadakane, K.: Compressed suffix trees with full functionality. Theory Comput. Syst. **41**(4), 589–607 (2007)
33. Sahinalp, S.C., Vishkin, U.: On a parallel-algorithms method for string matching problems (overview). In: Bonuccelli, M., Crescenzi, P., Petreschi, R. (eds.) CIAC 1994. LNCS, vol. 778, pp. 22–32. Springer, Heidelberg (1994). https://doi.org/10.1007/3-540-57811-0_3
34. Weiner, P.: Linear pattern matching algorithms. In: Proceedings of the 14th Annual Symposium on Switching and Automata Theory (SWAT), pp. 1–11 (1973)

Sublinear Time Lempel-Ziv (LZ77) Factorization

Jonas Ellert$^{(\boxtimes)}$ [iD]

Technical University of Dortmund, Dortmund, Germany
jonas.ellert@tu-dortmund.de

Abstract. The Lempel-Ziv (LZ77) factorization of a string is a widely-used algorithmic tool that plays a central role in data compression and indexing. For a length-n string over integer alphabet $[0, \sigma)$ with $\sigma = n^{\mathcal{O}(1)}$, and on a word RAM of width $w = \Theta(\log n)$, it can be computed in $\mathcal{O}(n)$ time. However, the packed representation of the string occupies only $\Theta(n \log \sigma)$ bits or equivalently $\Theta(n/\log_\sigma n)$ words of space, and hence we can hope for algorithms that run in $\mathcal{O}(n/\log_\sigma n)$ time and words of space. Kempa showed how to compute the LZ77 factorization with overlaps in $\mathcal{O}(n/\log_\sigma n + z \log^{11} n)$ time and $\mathcal{O}(n/\log_\sigma n + z \log^{10} n)$ words of space, where z is the number of phrases in the LZ77 factorization (SODA 2019). We significantly improve this result by achieving $\mathcal{O}(n/\log_\sigma n + z \log^{3+\epsilon} z)$ time with overlaps, and $\mathcal{O}(n/\log_\sigma n + z \log^{23/5+\epsilon} z)$ without overlaps (for any constant $\epsilon \in \mathbb{R}^+$). In both cases, we require only $\mathcal{O}(n/\log_\sigma n)$ words of space. One ingredient of the solution is a novel approximation algorithm that computes an LZ-like parsing of at most $3z$ phrases in $\mathcal{O}(n/\log_\sigma n)$ time and words of space. All algorithms are deterministic.

Keywords: Lempel-Ziv · LZ77 · LZ-like · lossless compression · word-packing · sublinear time · string algorithms · approximation algorithms

1 Introduction

The Lempel-Ziv (LZ) factorization [45] of a string decomposes it into a series of phrases. Each phrase is either the leftmost occurrence of a symbol (a *literal phrase*), or the longest substring that can be read at an earlier position in the string (a *referencing phrase*). Each referencing phrase can be replaced by an integer pair consisting of the length of the phrase and the (absolute or relative) position of an earlier occurrence of the phrase. This way, one can store the string in $\mathcal{O}(z \log n)$ bits of space, where n is the length of the string and z is the number of phrases. Further compression can be achieved by encoding the integers, e.g., by applying a universal code. In an LZ-*like* factorization, referencing phrases do not need to be of maximal length.

Background and Related Work. The LZ factorization was first introduced in 1976, when Lempel and Ziv proposed the number z of phrases in the factorization

© The Author(s), under exclusive license to Springer Nature Switzerland AG 2023
F. M. Nardini et al. (Eds.): SPIRE 2023, LNCS 14240, pp. 171–187, 2023.
https://doi.org/10.1007/978-3-031-43980-3_14

as a complexity measure for strings (aimed at evaluating the "randomness" of a string) [45]. Over 45 years later, it is still a standard measure for dictionary-based compression (see, e.g., [24]). This is because z can be computed in linear time [56], and because it lower-bounds other measures like the size of the smallest grammar that generates the string [9,57]. Many compressibility measures are within polylogarithmic factors of z, e.g., the size of the smallest bidirectional macro scheme [24], the number of runs in the BWT [34], the size of the smallest string attractor [35], and the normalized substring complexity [36,55].

Apart from introducing z as a measure, Ziv and Lempel also used their factorization to derive the compression scheme now commonly known as LZ77 [67]. (Nowadays, the LZ factorization – despite being introduced in 1976 – is often referred to as the LZ77 factorization.) Since then, the LZ factorization has become a cornerstone of practical compression; LZ-based techniques are a crucial ingredient of the most commonly used compressors (e.g., gzip, 7zip, rar, brotli) and compressed formats (e.g., PDF, PNG). There has been extensive work aimed at computing (versions of) the LZ factorization (we list only a few examples). This includes parallel algorithms [13,15,48,59,60], online and streaming algorithms [5,53,54,61,66], external memory algorithms [3,43], and approximation algorithms [18,40]. Another line of research improves the compression rate by optimizing the encoding of phrases [1,4,10,12,14,17,38,44]. There are several text indices that rely on LZ compression [6,7,16,22,23,31,42,50,63,64]. Despite this plethora of results, the ever-increasing relevance of compression still drives the development of new ways to compute LZ(-like) factorizations [21,28,37,51,58,65].

In this work, we consider sequential algorithms on a word RAM of word-width $\Theta(\log n)$. The string is over integer alphabet $\{0, \ldots, \sigma - 1\}$, such that a symbol can be stored in $\lceil \log_2 \sigma \rceil$ bits. Hence the string occupies $n \lceil \log_2 \sigma \rceil$ bits or $\Theta(n/\log_\sigma n)$ words of memory. This makes $\Omega(n/\log_\sigma n)$ a natural lower bound for the time and words of space needed to compute the LZ factorization. Many algorithms take $\mathcal{O}(n)$ time (see, e.g., [19,25,26,30]) or $\mathcal{O}(n/\log_\sigma n)$ words of working space (see, e.g., [4,29,39,41,52,53,61,66]), and at least one algorithm achieves both [20]. Kempa [32] introduced an algorithm that takes $\mathcal{O}(n/\log_\sigma n + r \log^9 n + z \log^9 n)$ time and $\mathcal{O}(n/\log_\sigma n + r \log^8 n)$ words of space, where $r = \mathcal{O}(z \log^2 n)$ [34] is the number of BWT runs. However, it appears that there is no algorithm with both space (in words) and time in $\mathcal{O}(n/\log_\sigma n)$.

Contributions. We propose new deterministic algorithms for computing LZ(-like) factorizations, summarized by the theorems below. The space asymptotically matches the space needed for storing the string. The time for the exact factorization is optimal if $z = \mathcal{O}(n \log \sigma / \log^{4+\epsilon} n)$, i.e., for compressible strings. We adapt Theorem 2 to the non-overlapping version of LZ in Sect. 5.

Theorem 1. *Let $T \in [0, \sigma)^n$ be a string, and let z be the number of phrases in the LZ factorization of T. An LZ-like factorization of T that consists of at most $3z$ phrases can be computed in $\mathcal{O}(n/\log_\sigma n)$ time and $\mathcal{O}(n \log \sigma)$ bits of space.*

Theorem 2. *Let $T \in [0, \sigma)^n$ be a string, and let $\epsilon \in \mathbb{R}^+$ be an arbitrarily small positive constant. The LZ factorization $T = f_1 \ldots f_z$ can be computed in $\mathcal{O}(n/\log_\sigma n + z \log^{3+\epsilon} z)$ time and $\mathcal{O}(n \log \sigma)$ bits of space.*

2   Preliminaries

Strings and Computational Model. For $i, j \in \mathbb{Z}$, we write $[i, j] = [i, j+1)$ instead of $\{k \in \mathbb{Z} \mid i \le k \le j\}$. Let $n \in \mathbb{N}$ and $\sigma \in [1, n]$. A string $T \in [0, \sigma)^n$ of length $|T| = n$ is a sequence $T = T[1]T[2] \ldots T[n]$ of symbols from alphabet $[0, \sigma)$. For $i, j \in [1, n]$, substring $T[i..j] = T[i..j+1)$ is the sequence $T[i]T[i+1] \ldots T[j]$ (or the empty string ε if $j < i$). A substring shorter than T is *proper*. Substrings $T[1..i]$ and $T[i..n]$ are respectively called prefix and suffix of T. The reversal of T is $\mathrm{rev}(T) = T[n]T[n-1] \ldots T[1]$. Two (sub-)strings S_1 and S_2 are equal, written $S_1 = S_2$, if $|S_1| = |S_2|$ and $\forall i \in [1, |S_1|] : S_1[i] = S_2[i]$. We write $S_1 \prec S_2$ and say that S_1 is lexicographically smaller than S_2 if and only if either S_1 equals a proper prefix of S_2, or $\exists \ell \in [1, \min(|S_1|, |S_2|)]$ such that $S_1[1..\ell] = S_2[1..\ell]$ and $S_1[\ell] < S_2[\ell]$. We write $S_2 \preceq S_1$ instead of $\neg(S_1 \prec S_2)$. We say that S_1 is co-lexicographically smaller than S_2 if and only if $\mathrm{rev}(S_1) \prec \mathrm{rev}(S_2)$. The concatenation of S_1 and S_2 is denoted by $S_1 \cdot S_2$ or $S_1 S_2$.

All algorithms, lemmas, and intermediate results assume the following model of computation. String $T \in [0, \sigma)^n$ is processed on a word RAM (see, e.g., [27]) of word-width $w \ge \log n$. Each symbol is stored in $\lceil \log_2 \sigma \rceil$ bits, and the entire string occupies $n \lceil \log_2 \sigma \rceil$ bits (or $\mathcal{O}(n/\log_\sigma n)$ words) of consecutive memory.

Lempel-Ziv and Longest Common Extensions. The uniquely defined Lempel-Ziv (LZ) factorization of a string T decomposes it into a series of z phrases $T = f_1 f_2 \ldots f_z$. Each phrase $f_{i'} = T[i..i + |f_{i'}|)$ with $i' \in [1, z]$ and $i = 1 + \sum_{k=1}^{i'-1} |f_k|$ is either a single symbol $T[i]$ that does not occur in $T[1..i)$ (a *literal phrase*), or otherwise it is the longest prefix of $T[i..n]$ that has a previous occurrence $T[j..j + |f_{i'}|) = f_{i'}$ with $j \in [1, i)$ (a *referencing phrase*). Position i is *the destination* of $f_{i'}$. If $f_{i'}$ is a referencing phrase, then j is *a source* of $f_{i'}$. This is Storer and Szymanski's version of the factorization [62]. All presented algorithms can be trivially modified to compute Lempel and Ziv's original version instead, in which each phrase is a combination of a lengthwise maximal (possibly empty) reference and a literal symbol. It holds $z = \mathcal{O}(n/\log_\sigma n)$ [45, Theorem 2]. An LZ-like factorization is defined exactly like the LZ factorization, but without the requirement that referencing phrases are of maximal length.

Given $i, j \in [1, n]$ with $i \le j$, their longest common extension (LCE) is $\mathrm{LCE}(i, j) = \mathrm{LCE}(j, i) = \max(\{\ell \in [0, n - j + 1] \mid T[i..i + \ell) = T[j..j + \ell)\})$. This is closely related to LZ because a referencing phrase $f_{i'}$ with source j and destination i is of length $|f_{i'}| = \mathrm{LCE}(i, j) = \max_{j' \in [1, i)} (\mathrm{LCE}(i, j'))$. LCEs also reveal the lexicographical order of substrings. For any substrings $T[i..i + \ell_i)$ and $T[i'..i' + \ell_{i'})$, it holds $T[i..i + \ell_i) \prec T[i'..i' + \ell_{i'})$ if and only if either $\mathrm{LCE}(i, i') \ge \ell_i$ and $\ell_i < \ell_{i'}$, or $\mathrm{LCE}(i, i') < \min(\ell_i, \ell_{i'})$ and $T[i + \mathrm{LCE}(i, i')] < T[i' + \mathrm{LCE}(i, i')]$. A data structure by Kempa and Kociumaka provides constant time LCE queries, and thus also constant time lexicographical order testing of substrings.

Lemma 1 ([33, Theorem 5.4]). *For a string $T \in [0, \sigma)^n$, a data structure that supports constant time LCE queries (given $i, j \in [1, n]$, output $\mathrm{LCE}(i, j)$) and lexicographical order testing (given $i, j \in [1, n]$, output if $T[i..n] \prec T[j..n]$) can be computed in $\mathcal{O}(n/\log_\sigma n)$ time and $\mathcal{O}(n \log \sigma)$ bits of working space.*

3 Algorithm for 3-Approximate LZ-Like Factorization

We accelerate the computation with precomputed lookup tables. We access the tables with short substrings of T. A (sub-)string $P \in [0, \sigma)^m$ is a bitstring of length $m \cdot \lceil \log_2 \sigma \rceil$. Hence we can interpret P as an integer $\mathrm{int}(P) \in [1, 2^{m \cdot \lceil \log_2 \sigma \rceil}]$. If $m \leq \log_2 n / \lceil \log_2 \sigma \rceil$, then P fits in a word of memory and can be extracted from T in constant time. We can then obtain $\mathrm{int}(P)$ and use it to access a lookup table in constant time. As a warm-up result (and for later usage), we describe a lookup table that detects periodicities (Lemma 2) and a set of tables for leftmost pattern matching queries (Lemma 3). A string P is of period $p \in \mathbb{N}^+$ if and only if $P[1..\,|P|-p] = P[1+p..\,|P|]$ (or equivalently if $\forall i \in [1, |P|-p] : P[i] = P[i+p]$). We say that p is *the* period of P, if it is the minimal period of P.

Lemma 2. *Let $n \in [1, 2^w]$. There is a data structure that, given pattern $P \in [0, \sigma)^m$ with $m \leq \log_2 n / (2 \lceil \log_2 \sigma \rceil)$, outputs the shortest period of P in constant time. It can be computed in $\mathcal{O}(\sqrt{n}\,\mathrm{polylog}(n))$ time and words of space.*

Proof. Let $P \in [0, \sigma)^m$, then $\mathrm{int}(P) \in [1, n']$ with $n' \leq 2^{\log_2 n / 2} = \mathcal{O}(\sqrt{n})$. For each $P \in [0, \sigma)^m$, we naively compute its period in $\mathcal{O}(m^2) \subseteq \mathcal{O}(\log^2 n)$ time, and store it in entry $Q_m[\mathrm{int}(P)]$ of a lookup table Q_m. There are $\mathcal{O}(\log n)$ tables (one per possible value of m), and each table has $\mathcal{O}(n')$ entries. Hence the total time and words of space are bounded by $\mathcal{O}(n' \log^3 n) = \mathcal{O}(\sqrt{n}\,\mathrm{polylog}(n))$.   □

Lemma 3. *Let $T \in [0, \sigma)^n$. Let $\epsilon \in \mathbb{R}^+$ be constant. There is a data structure that, given a query pattern $P \in [0, \sigma)^m$ with $m \leq \log_2 n / ((2 + \epsilon) \lceil \log_2 \sigma \rceil)$, outputs the leftmost occurrence of P in T in constant time. It can be computed in $\mathcal{O}(n / \log_\sigma n)$ time and $o(n / \log n)$ bits of space.*

Proof. Let $k = \lfloor \log_2 n / ((2 + \epsilon) \lceil \log_2 \sigma \rceil) \rfloor$ be the maximal allowed pattern length. Let $S \in [0, \sigma)^{2k}$, then $\mathrm{int}(S) \in [1, n']$ with $n' \in \mathcal{O}(n^{1-\hat{\epsilon}})$, where $\hat{\epsilon} = \epsilon / (2 + \epsilon) > 0$. In a table $M[1..n']$, we compute for every $S \in [0, \sigma)^{2k}$ the value $M[\mathrm{int}(S)] = ik + 1$, where $i \in [0, \frac{n}{k} - 2]$ is the minimal value with $S = T[ik + 1..ik + 2k + 1)$. If no such i exists, we store $M[\mathrm{int}(S)] = n + 1$. Computing the table takes $\mathcal{O}(n / \log_\sigma n)$ time. We simply iterate over the $\mathcal{O}(n / \log_\sigma n)$ possible values of $ik + 1$ in decreasing order. For each of them, we take constant time to assign $M[\mathrm{int}(T[ik + 1..ik + 2k + 1))] = ik + 1$.

Now we use M to compute the leftmost occurrence of each possible pattern of length at most k. We create k lookup tables $L_1, L_2, \ldots, L_k$. For $P \in [0, \sigma)^m$, entry $L_m[\mathrm{int}(P)]$ will contain the leftmost occurrence of P in T. We compute L_m as follows. Initially, all entries are set to $n + 1$. Now we consider each string $S \in [0, \sigma)^{2k}$. For every $j \in [0, 2k - m]$, we let $P = S[1 + j..1 + j + m)$ and assign $L_m[\mathrm{int}(P)] = \min(L_m[\mathrm{int}(P)], M[\mathrm{int}(S)] + j)$. The leftmost occurrence of any length-m pattern is fully contained in a length-$2k$ substring at some position $ik + 1$. Hence the computed values are correct. For each of the n' possible $S \in [0, \sigma)^{2k}$, we have to consider $\mathcal{O}(k^2)$ substrings, and for each of them we spend constant time to update some table L_m. The time is $\mathcal{O}(n' \cdot k^2) \subset \mathcal{O}(n^{1-\hat{\epsilon}}\,\mathrm{polylog}(n))$. There are $k + 1$ lookup tables, and each has at most n' entries. Hence $\mathcal{O}(n^{1-\hat{\epsilon}}\,\mathrm{polylog}(n)) \subset o(n/\log n)$ bits of space are sufficient. □

String Synchronizing Sets. We will work with a small subset of sample positions that has convenient synchronizing properties.

Definition 1 ([33]). *Let $T \in [0,\sigma)^n$ and $\tau \in [1, \lfloor\frac{n}{2}\rfloor]$. A set $S \subseteq [1, n - 2\tau + 1]$ is τ-synchronizing (with respect to T) if and only if the following conditions hold.*

- Synchronizing condition: *For any $i, j \in [1, n - 2\tau + 1]$ with $T[i..i + 2\tau) = T[j..j + 2\tau)$, it holds $i \in S$ if and only if $j \in S$.*
- Density condition: *For any $i \in [1, n - 3\tau + 2]$, it holds $S \cap [i, i + \tau) = \emptyset$ if and only if the period of $T[i..i + 3\tau - 2]$ is at most $\frac{\tau}{3}$.*

Lemma 4 ([33, **Theorems 4.3 and 8.11**]). *Let $T \in [0,\sigma)^n$. There is a $\lfloor\log_2 n/(8\lceil\log_2\sigma\rceil)\rfloor$-synchronizing set of size $\mathcal{O}(n/\log_\sigma n)$. It takes $\mathcal{O}(n/\log_\sigma n)$ time and $\mathcal{O}(n\log\sigma)$ bits of working space to compute the set, and to lexicographically sort all the suffixes that start at positions in the set.*

3.1 Computing Longest Previous Factors of Sample Positions

Let $\tau = \lfloor\log_2 n/(8\lceil\log_2\sigma\rceil)\rfloor$. We use a τ-synchronizing set of sample positions. We start by computing for each sample position the longest referencing phrase that could hypothetically start at that position. This is similar to computing longest previous factors [11] in the sequential setting without word-packing.

We obtain a τ-synchronizing set $\{d_1, d_2, \ldots, d_N\}$ with $\forall x \in [1, N)$: $d_x < d_{x+1}$ and $N = \mathcal{O}(\frac{n}{\tau}) = \mathcal{O}(n/\log_\sigma n)$. We lexicographically sort the suffixes at synchronizing positions and obtain their sparse suffix array, which is the unique permutation suf of $[1, N]$ with $\forall x \in [1, N) : T[d_{\mathsf{suf}[x]}..n] \prec T[d_{\mathsf{suf}[x+1]}..n]$. This takes $\mathcal{O}(n/\log_\sigma n)$ time with Lemma 4. Next, we compute an array $\mathsf{LPF}[1..N]$ (for longest previous factor), where entry $\mathsf{LPF}[x]$ is a position from $[1, d_x)$ that maximizes $\mathrm{LCE}(d_x, \mathsf{LPF}[x])$ (this position may not be unique, and it is not necessarily a sample position). We first use Lemma 3 to find the minimal j with $T[j..j + 2\tau) = T[d_x..d_x + 2\tau)$ in constant time. If $j = d_x$, then $T[d_x..d_x + 2\tau)$ has no previous occurrence. In this case, we issue at most $\mathcal{O}(\tau)$ queries to Lemma 3 and find the maximal $\ell \in [0, 2\tau)$ such that $T[d_x..d_x + \ell)$ has a previous occurrence. This also reveals $\mathsf{LPF}[x]$ (we can choose any position from $[1, d_x)$ if $\ell = 0$), but it takes $\mathcal{O}(\tau)$ time. However, this can only happen once per distinct length-2τ substring, which limits the total time to $\mathcal{O}(2^{2\tau\cdot\lceil\log_2\sigma\rceil}\tau) \subset \mathcal{O}(n/\log n)$. If $j < d_x$, then $T[d_x..d_x + 2\tau)$ has a previous occurrence, and the synchronizing property of Definition 1 guarantees that *all* previous occurrences of $T[d_x..d_x + 2\tau)$ start at sample positions. Thus, we can compute LPF in the same way as it is usually done for the entire suffix array. A detailed description can be found, e.g., in [11], and we only give a brief summary. (Our LPF corresponds to PrevOcc in [11].) For each entry $\mathsf{suf}[x]$, we find

$$\mathsf{prev}[x] = \max(\{y \in [1, x) \quad | \; \mathsf{suf}[y] < \mathsf{suf}[x]\}) \quad \text{and}$$
$$\mathsf{next}[x] = \min(\{y \in [x + 1, m] \mid \mathsf{suf}[y] < \mathsf{suf}[x]\}),$$

which takes $\mathcal{O}(N)$ time with an algorithm for nearest smaller values (see, e.g., [2, Lemma 1]). We then use Lemma 1 to compute $\ell_1 = \text{LCE}(d_{\text{suf}[x]}, d_{\text{suf}[\text{prev}[x]]})$ and $\ell_2 = \text{LCE}(d_{\text{suf}[x]}, d_{\text{suf}[\text{next}[x]]})$, which are the respective maximal phrase lengths at destination $d_{\text{suf}[x]}$ that can be achieved with a lexicographically smaller and a lexicographically larger suffix starting at an earlier sample position. If $\max(\ell_1, \ell_2) < 2\tau$, then we have already assigned $\text{LPF}[\text{suf}[x]]$ with Lemma 3 as described above. Otherwise, if $\ell_1 > \ell_2$, then we assign $\text{LPF}[\text{suf}[x]] = \text{suf}[\text{prev}[x]]$. If, however, $\ell_2 \geq \ell_1$, then we assign $\text{LPF}[\text{suf}[x]] = \text{suf}[\text{next}[x]]$. (It is possible that $\text{prev}[x]$ and/or $\text{next}[x]$ are undefined, but treating this is trivial.) The correctness follows from the synchronizing property and the correctness of the same technique for the full suffix array [11]. The total time and space in words are $\mathcal{O}(N) = \mathcal{O}(n/\log_\sigma n)$.

3.2 Computing a Gapped Factorization

Now we compute a *gapped* LZ factorization $T = f_1 g_1 r_1 f_2 g_2 r_2 \ldots f_{z'} g_{z'} r_{z'}$, where:

- Each $f_{i'}$ is a *perfect phrase* at destination $i = 1 + \sum_{h=1}^{i'-1} |f_h g_h r_h|$ defined just like in the exact factorization. It is either the leftmost occurrence of a symbol (a literal phrase), or the longest prefix of $T[i..n]$ with an earlier occurrence $T[j..j+|f_{i'}|) = f_{i'}$ at some source $j \in [1, i)$ (a referencing phrase).
- Each $g_{i'}$ is a (possibly empty) *gap* at destination $i = 1 + |f_{i'}| + \sum_{h=1}^{i'-1} |f_h g_h r_h|$. A gap can be any string and does not necessarily have a previous occurrence.
- Each $r_{i'}$ is a *reference* at destination $i = 1 + |f_{i'} g_{i'}| + \sum_{h=1}^{i'-1} |f_h g_h r_h|$, which is either empty or it has an earlier occurrence $T[j..j+|r_{i'}|) = r_{i'}$ at source $j \in [1, i)$ (with no requirement of maximal length).

Lemma 5. *Any gapped LZ factorization $T = f_1 g_1 r_1 f_2 g_2 r_2 \ldots f_{z'} g_{z'} r_{z'}$ satisfies $z' \leq z$, where z is the number of phrases in the exact LZ factorization of T.*

Proof. A suffix $T[j..i+\ell)$ of an exact LZ phrase $T[i..i+\ell)$ at destination i has an earlier occurrence. Hence, if j is the destination of a perfect phrase $f_{j'}$ in the gapped factorization, then this phrase is of length at least $\ell - j$. This means that a phrase of the exact LZ factorization contains the destination of at most one perfect phrase of the gapped LZ factorization, which implies $z' \leq z$. □

Computing *any* gapped factorization is trivial (e.g., $T = fgr$ with $f = T[1]$, $g = [T_2..n]$, $r = \varepsilon$ is a gapped factorization). We will compute a gapped factorization with the additional property that none of the gaps contain a position from the synchronizing set, which makes it easy to eliminate the gaps in a post-processing. We compute the factorization from left to right using LPF.

The first perfect phrase is literal phrase $f_1 = T[1]$. After creating some perfect phrase $f_{i'}$ at destination i, we iterate over the upcoming sample positions until we reach the first $d_x \geq i + |f_{i'}|$. The next gap is $g_{i'} = T[i+|f_{i'}|..d_x)$, and the next reference $r_{i'}$ is empty. The next perfect phrase $f_{i'+1}$ at destination d_x is a literal phrase if $\text{LCE}(d_x, \text{LPF}[x]) = 0$. Otherwise, it is a referencing phrase with source $\text{LPF}[x]$ and length $\text{LCE}(d_x, \text{LPF}[x])$. As soon as we create a perfect phrase $f_{i'}$ at destination i with $i + |f_{i'}| > d_N$, we complete the factorization with gap $g_{i'} = T[i + |f_{i'}| .. n]$ and empty reference $r_{i'}$. We spend constant time per sample position, and hence the time is $\mathcal{O}(N) = \mathcal{O}(n/\log_\sigma n)$.

Eliminating Long Gaps. Now we eliminate the gaps by replacing them with references. We distinguish between short gaps of length at most 3τ, and long gaps of length more than 3τ. A long gap $g_{i'}$ at destination i is of length more than 3τ, and due to our method of computing the factorization it does not contain any of the synchronizing positions. By the density condition of Definition 1, $g_{i'}$ has period $p \le \tau/3$. The reference $r_{i'}$ is empty (because all references in the initial gapped factorization are empty, and we only replace them with non-empty references when eliminating long gaps). We replace $g_{i'}r_{i'}$ with $g'_{i'}r'_{i'}$, where $g'_{i'} = g_{i'}[1..3\tau]$ and $r'_{i'} = g_{i'}[3\tau+1..|g_{i'}|]$. Since $g_{i'}$ has period p, the new reference $r'_{i'}$ at destination $i + 3\tau$ has an earlier occurrence at source $j = i + 3\tau - p$. Hence the replacement retains the properties of a gapped factorization.

If $p \le \frac{\tau}{3}$ is the shortest period of $g_{i'}$ of length at least 3τ, then it is easy to see that also $g_{i'}[1..3\tau]$ has shortest period p (because $g_{i'}[1..3\tau]$ contains all the length $2p$ substrings of $g_{i'}$, and thus a shorter period would directly translate to the entire $g_{i'}$). Hence we can simply use Lemma 2 with query pattern $g_{i'}[1..3\tau]$ to lookup p in constant time. This way, replacing a long gap takes constant time, and the total time needed for all long gaps is $\mathcal{O}(z) = \mathcal{O}(n/\log_\sigma n)$.

Eliminating Short Gaps and Finalizing the Factorization. We have eliminated all long gaps, and from now on we simply say gap rather than short gap. A non-empty gap $g_{i'}$ at destination i is *referencing* if there is some $j < i$ with $T[j..j + |g_{i'}|) = g_{i'}$ (we could replace the gap with a reference). We first identify all the non-referencing non-empty gaps. For each non-empty gap, we use Lemma 3 to find the minimal j with $T[j..j + |g_{i'}|) = g_{i'}$ in constant time. If and only if $j = i$, then $g_{i'}$ is non-referencing. The total time needed is $\mathcal{O}(z) \subseteq \mathcal{O}(n/\log_\sigma n)$.

We process each non-referencing non-empty gap $g_{i'}$ separately. We find the maximal $\ell \in [0, |g_{i'}|)$ such that the prefix $T[i..i + \ell)$ of $g_{i'}$ has a previous occurrence. We do so by issuing $\mathcal{O}(|g_{i'}|)$ queries to Lemma 3, which takes $\mathcal{O}(\tau)$ time. This also reveals the source position $j \in [1, i)$ of the previous occurrence. We adjust the gapped factorization by re-factorizing the gap as $g_{i'} = grfg'$, where g is a new empty gap, r is a new empty reference, $f = T[i..i + \max(1, \ell))$ is a new perfect phrase (with source j if $\ell > 0$), and $g' = T[i+\ell..i+|g_{i'}|)$ is the remainder of the gap. Note that this replacement retains the properties of a gapped factorization, and by Lemma 5 there are still at most z factors of each type. If the new gap g' is still non-empty and non-referencing (we check this in the same way as before), then we replace g' by applying the same re-factorization procedure again, and we keep doing so until the remainder of the gap is either empty or referencing. Each application takes $\mathcal{O}(\tau)$ time and decreases the length of the remainder. Hence the total time for processing $g_{i'}$ is $\mathcal{O}(\tau^2)$.

A non-referencing gap $g_{i'}$ at destination i implies that $T[i..i + |g_{i'}|)$ is the leftmost occurrence of a substring of length at most 3τ. There are fewer than

$2^{3\tau \cdot \lceil \log_2 \sigma \rceil} \cdot 3\tau$ distinct substrings of this length, and hence re-factorizing all the non-referencing gaps takes $\mathcal{O}(2^{3\tau \cdot \lceil \log_2 \sigma \rceil} \cdot \tau^3) \subseteq \mathcal{O}(n^{3/8} \log^3 n)$ time.

Now all non-empty gaps are referencing. We obtain an LZ-like parsing by discarding all empty factors. This leaves at most z perfect phrases and $2z$ referencing phrases. The total time is $\mathcal{O}(n/\log_\sigma n)$. Lemmas 1 to 4 require $\mathcal{O}(n \log \sigma)$ bits of memory. Apart from that, we only use arrays of size N, which also require $\mathcal{O}(N \log n) = \mathcal{O}(n \log \sigma)$ bits of memory. Hence we have shown Theorem 1. It is easy to see that, instead of first computing a gapped factorization and then closing the gaps, we could just as well directly compute the approximate factorization from left to right. This may result in a faster practical implementation.

4   Algorithm for Exact LZ Factorization

In the approximate algorithm, we create perfect phrases for which both source and destination are samples. For the exact LZ factorization, we have to admit arbitrary sources and destinations. We will define a new set of sample positions such that, if a phrase $f_{i'}$ has source j, there will be at least one sample position $j' \in [j, j + \min(\delta, |f_{i'}|))$ for some parameter δ. We can conceptually divide the phrase into a head $T[j..j']$ and a tail $T[j'..j + |f_{i'}|)$. Computing a phrase means finding a sample position with matching head, and with tail of maximal length. If we co-lexicographically sort the prefixes that end at sample positions, then we group together samples that admit the same head. Similarly, if we lexicographically sort the suffixes that start at sample positions, then we group together samples that admit the same tail. This motivates a geometric interpretation of sample positions, in which each sample is represented by the lexicographical rank of its suffix and the co-lexicographical rank of its prefix. (This technique is similar to what was done in [4, Section 6.2].) Ultimately, we use geometric data structures for insertion-only orthogonal range one-reporting to handle most of the computational effort. (We could also use static data structures with an extra dimension or weighted points; however, there are few such data structures with known construction times.)

Definition 2. *Let $N \in [1, 2^w]$ and let π be a permutation of $[1, N]$. The task of insertion-only orthogonal range one-reporting is to maintain a set of points $\mathcal{P} \subseteq \{(i, \pi(i)) \mid i \in [1, N]\}$ (initially empty) with the following operations:*

- *insert $p \in \{(i, \pi(i)) \mid i \in [1, N]\}$ into $\mathcal{P}$*
- *given $\mathcal{Q} = [a_1, a_2] \times [b_1, b_2]$, output any point from $\mathcal{Q} \cap \mathcal{P}$, or report $\mathcal{Q} \cap \mathcal{P} = \emptyset$*

Now we show how to find previous occurrences of substrings by using orthogonal range reporting and an arbitrary set of sample positions.

Lemma 6. *Let $T \in [0, \sigma)^n$ be a string, and let $\mathcal{A}[1..N]$ be an array of N distinct samples from $[1, n]$ in increasing order. Let $u_\mathcal{A}$ and $q_\mathcal{A}$ be respectively the insertion and query time of a data structure for insertion-only orthogonal range one-reporting, and let $s_\mathcal{A}$ be the maximum number of words occupied by this data structure after N insertions. After an $\mathcal{O}(n/\log_\sigma n + N \log N)$ time preprocessing, and in $\mathcal{O}(n/\log_\sigma n + N + s_\mathcal{A})$ words of space, a subset $\mathcal{X}$ of sample positions can be maintained with the following operations.*

- given $h \in [1, N]$, insert $\mathcal{A}[h]$ into $\mathcal{X}$ in $\mathcal{O}(u_{\mathcal{A}})$ time
- given $i \in [1, n]$ and $k \in [0, n-i]$, find the (possibly not unique) $x \in \mathcal{X} \cap (k, n]$ with $T[x - k..x] = T[i..i + k]$ and maximal $\ell = \text{LCE}(x, i+k)$ in $\mathcal{O}(\log \ell \cdot (\log N + q_{\mathcal{A}}))$ time, or report that j does not exist in $\mathcal{O}(\log N + q_{\mathcal{A}})$ time

Proof. We start by preprocessing the sample positions. Let suf be the unique permutation of $[1, N]$ that lexicographically sorts the suffixes of T that start at sample positions, i.e., $\forall h \in [1, N) : T[\mathcal{A}[\text{suf}[h]]..n] \prec T[\mathcal{A}[\text{suf}[h+1]]..n]$ (a sparse suffix array). We use comparison sorting with Lemma 1 for constant time lexicographical suffix comparisons and obtain suf in $\mathcal{O}(n/\log_\sigma n + N \log N)$ time and $\mathcal{O}(n/\log_\sigma n + N)$ words of space. Analogously, we obtain the unique permutation pref of $[1, N]$ that co-lexicographically sorts the prefixes of T that end at sample positions, i.e., $\forall h \in [1, N) : \text{rev}(T[1..\mathcal{A}[\text{pref}[h]]]) \prec \text{rev}(T[1..\mathcal{A}[\text{pref}[h+1]]])$. We compute $\text{rev}(T)$ in $\mathcal{O}(n/\log_\sigma n)$ time and words of space with universal lookup tables (see, e.g., [4, Section 6.2]). By comparison sorting with the data structure from Lemma 1 constructed for $\text{rev}(T)$, we obtain pref in $\mathcal{O}(n/\log_\sigma n + N \log N)$ time and $\mathcal{O}(n/\log_\sigma n + N)$ words of space. It is trivial to compute the respective inverse permutations suf-rank and pref-rank of suf and pref in $\mathcal{O}(N)$ time and words of space. This concludes the preprocessing.

Insertions. In order to insert $\mathcal{A}[h]$ into $\mathcal{X}$, we insert the two-dimensional point $(\text{suf-rank}(h), \text{pref-rank}(h))$ into the geometric data structure for orthogonal range reporting, which leads to the claimed insertion time and space complexity.

Queries. We first show a fast way to answer a slightly simpler type of query. Given suffix $T[i..n]$, offset k, and length estimate ℓ, we want to find some $\mathcal{A}[h] \in \mathcal{X}$ such that $T[\mathcal{A}[h] - k..\mathcal{A}[h] + \ell) = T[i..i + k + \ell)$ (if it exists). The lexicographical order groups together suffixes of T that share a long prefix. Thus, there is an interval $\text{suf}[a_1..a_2]$ that contains exactly the $h \in [1, N]$ with $T[\mathcal{A}[h]..\mathcal{A}[h] + \ell) = T[i + k..i + k + \ell)$. We compute a_1 by binary searching in suf for the lexicographically minimal suffix that starts at a sample position and has prefix $T[i + k..i + k + \ell)$. This works similarly to pattern matching with the suffix array [46]. If $\text{suf}[h']$ is the center of the search interval, then we compute $\ell' = \text{LCE}(\mathcal{A}[\text{suf}[h']], i + k)$. If $\ell' \geq \ell$, or if $T[i + k..n] \preceq T[\mathcal{A}[\text{suf}[h']]..n]$, then we proceed in the left half of the search interval (including $\text{suf}[h']$). Otherwise, it holds $\ell' < \ell$ and $T[i+k..n] \succ T[\mathcal{A}[\text{suf}[h']]..n]$, and we continue in the right half of the interval (excluding $\text{suf}[h]$). Computing LCEs and performing lexicographical comparisons takes constant time with Lemma 1, and hence the binary search takes $\mathcal{O}(\log N)$ time. Analogously, we compute a_2 in $\mathcal{O}(\log N)$ time. The co-lexicographical order groups together prefixes that share a long suffix. There is an interval $\text{pref}[b_1..b_2]$ that contains exactly the $h \in [1, N]$ with $T[\mathcal{A}[h] - k..\mathcal{A}[h]] = T[i..i + k]$, and we can compute the interval borders in $\mathcal{O}(\log N)$ time (analogously to the computation of a_1 and a_2, but with the LCE data structure for $\text{rev}(T)$). A sample $\mathcal{A}[h]$ satisfies $T[\mathcal{A}[h] - k..\mathcal{A}[h] + \ell) = T[i..i + k + \ell)$ if and only if $(\text{suf-rank}(h), \text{pref-rank}(h)) \in [a_1, a_2] \times [b_1, b_2]$. If we have already inserted a sample position that satisfies this condition, then a query to the geometric data structure returns a matching point $(\text{suf-rank}(h), \text{pref-rank}(h))$ in $q_{\mathcal{A}}$ time.

Obtaining $\mathcal{A}[h]$ from the point takes constant time due to $h = \mathsf{suf}(\mathsf{suf\text{-}rank}(h))$. Otherwise, the data structure returns that the point does not exist. Thus, we can find some $\mathcal{A}[h] \in \mathcal{X}$ such that $T[\mathcal{A}[h] - k..\mathcal{A}[h] + \ell) = T[i..i + k + \ell)$, or report that such a sample does not exist, in $\mathcal{O}(\log N + q_A)$ time.

Finally, in order to answer a query of the type stated in the lemma, we use exponential search to find the maximal $\ell \in \mathbb{N}^+$ such that there is some $\mathcal{A}[h] \in \mathcal{X}$ with $T[\mathcal{A}[h] - k..\mathcal{A}[h] + \ell) = T[i..i + k + \ell)$. This way, we obtain $\mathcal{A}[h]$ in $\mathcal{O}(\log \ell \cdot (\log N + q_A))$ time (or the first query with $\ell = 1$ to the geometric data structure comes back negative, and we report that no matching sample position exists in $\mathcal{O}(\log N + q_A)$ time). $\qquad\square$

4.1   Computing the Exact LZ Factorization

Now we show how to compute the LZ factorization $T = f_1 f_2 \ldots f_z$. We distinguish between short phrases of length less than δ and long phrases of length at least δ, where $\delta \in [1, n]$ is a parameter to be fixed later (it will be polylogarithmic in n). We compute the factorization one phrase at a time and in left-to-right order. When computing $f_{i'}$ at destination $i \in [1, \delta)$, we compute $\mathrm{LCE}(j, i)$ for each $j \in [1, i)$ with Lemma 1, which reveals the length and source of the phrase in $\mathcal{O}(\delta)$ time (or all LCEs are zero and $f_{i'}$ is a literal phrase). When computing a phrase $f_{i'}$ at destination $i \geq \delta$, we use three different methods depending on the leftmost source j of $f_{i'}$. We do not know j in advance, and thus we try each of the methods and then choose the result that yields the longest phrase.

Method 1: Close Sources. If $j \in [i - \delta, i)$, then we obtain the phrase by computing $\mathrm{LCE}(j', i)$ for each $j' \in [i - \delta, i)$ and keeping track of the longest LCE.

Methods 2 and 3: Far Sources. If $j \in [1, i - \delta)$, then we use two instances of the data structure from Lemma 6. The first one maintains a subset $\mathcal{X}$ of evenly spaced samples from an array $\mathcal{A}[1..N]$ with $N = \lfloor \frac{n}{\delta} \rfloor$ and $\forall h \in [1, N] : \mathcal{A}[h] = h\delta$. The second one maintains a subset $\mathcal{Y}$ of samples from an array $\mathcal{B}[1..M]$ of size $M = \mathcal{O}(z)$ that is sorted in increasing order. The samples are chosen such that, if $T[j'..j' + \ell)$ is the leftmost occurrence of any substring (e.g., the leftmost source occurrence of an LZ phrase), then $[j', j' + \ell)$ contains at least one sample position. This is achieved with the LZ-like factorization from Theorem 1, which we explain later. The space usage is $\mathcal{O}(n/\log_\sigma n + N + M + s_A + s_B)$, and the preprocessing time is $\mathcal{O}(n/\log_\sigma n + N \log N + M \log M)$. For both instances, we maintain the following invariant. At the time at which we compute phrase $f_{i'}$ at destination i, we have inserted exactly the samples that satisfy $\mathcal{A}[h] < i$ into $\mathcal{X}$, and the samples that satisfy $\mathcal{B}[h] < i$ into $\mathcal{Y}$. Since we compute the phrases from left to right, we also insert the samples of each array in left to right order. Thus, there is no time overhead for finding the next sample to insert. The total insertion time is $\mathcal{O}(N \cdot u_A + M \cdot u_B)$. Now we use $\mathcal{X}$ and $\mathcal{Y}$ to compute phrases.

Method 2: Long Phrases. If $|f_{i'}| \geq \delta$, then $[j, j + \delta)$ contains the sample position $\mathcal{A}[h] = h\delta < j + \delta < i$ with $h = \lceil j/\delta \rceil$ (where $j + \delta < i$ due to the assumption that $j \in [1, i - \delta))$. By the invariant on $\mathcal{X}$, we have already inserted $\mathcal{A}[h]$ into $\mathcal{X}$. Let $k = \mathcal{A}[h] - j \in [0, \delta)$, then it holds $T[j..j+k] = T[\mathcal{A}[h] - k..\mathcal{A}[h]] = T[i..i+k]$. Thus, if we query $\mathcal{X}$ with position i and offset k, then we obtain either $\mathcal{A}[h]$ or another sample position $\mathcal{A}[h'] < i$ with $\text{LCE}(\mathcal{A}[h'] - k, i) = \text{LCE}(\mathcal{A}[h] - k, i) = \text{LCE}(j, i) = |f_{i'}|$. This way, we find both a source and the length of $f_{i'}$. Since we do not know k in advance, we issue one query for each possible value of k and keep track of the maximal LCE, which takes $\mathcal{O}(\delta \cdot (\log N + q_{\mathcal{A}}) \cdot \log |f_{i'}|)$ time.

Method 3: Short Phrases. This works analogously to the method for long phrases (but with $\mathcal{Y}$ instead of $\mathcal{X}$). If $|f_{i'}| < \delta$, then $[j, j + |f_{i'}|)$ contains at least one sample position $\mathcal{B}[h] < j + |f_{i'}| < j + \delta < i$ (where $j + \delta < i$ due to the assumption that $j \in [1, i - \delta)$, and a sample position is present because $T[j..j + |f_{i'}|)$ is the leftmost occurrence of a substring). By the invariant on $\mathcal{Y}$, we have already inserted $\mathcal{B}[h]$ into $\mathcal{Y}$. Let $k = \mathcal{B}[h] - j \in [0, |f_{i'}|)$, then it holds $T[j..j + k] = T[\mathcal{B}[h] - k..\mathcal{B}[h]] = T[i..i + k]$. Thus, if we query $\mathcal{Y}$ with position $T[i..n]$ and offset k, we obtain either $\mathcal{B}[h]$ or another sample position $\mathcal{B}[h'] < i$ for which it holds $\text{LCE}(\mathcal{B}[h'] - k, i) = \text{LCE}(\mathcal{B}[h] - k, i) = \text{LCE}(j, i) = |f_{i'}|$. This way, we find both a source and the length of $f_{i'}$. Since we do not know k in advance, we issue one query for each possible value of k and keep track of the maximal LCE, which takes $\mathcal{O}(\delta \cdot (\log N + q_{\mathcal{B}}) \cdot \log |f_{i'}|)$ time.

Analyzing the Procedure. Each method requires that j and possibly $|f_{i'}|$ satisfies some condition. It is easy to see that every scenario is covered by one of the conditions. Thus, there is always at least one method that computes a phrase of maximal length. If we run any of the methods even though the respective condition is not satisfied, then the result is still an LCE between i and a smaller position. Thus, we will never overestimate $|f_{i'}|$, and it is indeed correct to always run all three methods and produce the longest phrase admitted by any of them.

Now we analyze time and space complexity. We use $\delta = \Theta(\log^2 n)$ and two different geometric data structures (with amortized time bounds). The first one [8,47] has space complexity $s_{\mathcal{A}} = \mathcal{O}(N \log N) \subseteq \mathcal{O}(n / \log n)$ and performs insertions and queries in $u_{\mathcal{A}} = \mathcal{O}(\log n)$ and $q_{\mathcal{A}} = \mathcal{O}(\log n)$ time (the precise bounds are better, but we do not need them for our purposes). The second one [49] uses linear space $s_{\mathcal{B}} = \mathcal{O}(M) = \mathcal{O}(z) \subseteq \mathcal{O}(n / \log_\sigma n)$ and performs insertions and queries in $u_{\mathcal{B}} = \mathcal{O}(\log^{3+\epsilon} n)$ and $q_{\mathcal{B}} = \mathcal{O}(\log n)$ time (where $\epsilon \in \mathbb{R}^+$ is an arbitrarily small constant). The total space usage is $\mathcal{O}(n / \log_\sigma n)$ words.

The preprocessing time is $\mathcal{O}(n / \log_\sigma n + N \log N + M \log M) \subseteq \mathcal{O}(n / \log_\sigma n + z \log z)$, and the total time for insertions is $\mathcal{O}(N \log n + M \log^{3+\epsilon} n) \subseteq \mathcal{O}(n / \log n + z \log^{3+\epsilon} n)$. The time for applying all three methods is $\mathcal{O}(\log^3 n \cdot \log |f_{i'}|)$ for phrase $f_{i'}$. If $|f_{i'}| = \Omega(\log^5 n)$, then the time amortizes to $\mathcal{O}(1 / \log n)$ per symbol in $f_{i'}$, which results in $\mathcal{O}(n / \log n)$ time in total. If $|f_{i'}| = \mathcal{O}(\log^5 n)$, then the time is $\mathcal{O}(\log^3 n \cdot \log \log n)$, or $\mathcal{O}(z \log^{3+\epsilon} n)$ for all phrases. Thus, the overall time is $\mathcal{O}(n / \log_\sigma n + z \log^{3+\epsilon} n)$, and the space is $\mathcal{O}(n / \log_\sigma n)$ words or $\mathcal{O}(n \log \sigma)$ bits. For a purely cosmetic improvement, assume that $(z \log^{3+\epsilon} n) > (n / \log_\sigma n)$. Then $z > (n / \log^5 n)$ and $\log z = \Omega(\log n)$. Thus, the time bound is equivalent to $\mathcal{O}(n/\log_\sigma n + z\log^{3+\epsilon} z)$, and the complexities match the ones in Theorem 2.

Computing $\mathcal{B}$. We conclude the proof of Theorem 2 by showing how to compute $\mathcal{B}$ in $\mathcal{O}(n/\log_\sigma n)$ time and words of space. We compute an LZ-like factorization $T = f_1' f_2' \dots f_M'$ of $M = \Theta(z) \subseteq \mathcal{O}(n/\log_\sigma n)$ phrases with Theorem 1. Now it is easy to obtain $\mathcal{B}[1..M]$ with $\forall h \in [1,M] = \sum_{i'=1}^{h} |f_{i'}'|$ (the end positions of all phrases). Every leftmost occurrence of a symbol is a literal phrase in any LZ-like parsing. Thus, the leftmost occurrence of any symbol is a sample position. Now assume that $T[j'..j'+\ell)$ with $\ell > 1$ is the leftmost occurrence of a substring. If $[j', j'+\ell)$ contains no sample position, then $T[j'..j'+\ell)$ is fully contained within a referencing LZ-like phrase. However, every such phrase has a previous occurrence, which contradicts the fact that $T[j'..j'+\ell)$ has no previous occurrence. Thus $\mathcal{B}$ functions as required by the algorithm.

5   Computing the Non-overlapping LZ Factorization

A common variation of the LZ factorization (also proposed by Storer and Szymanski [62]) requires that the leftmost source j of a referencing phrase $f_{i'}$ at destination i satisfies $j + |f_{i'}| \le i$. Theorem 2 can be adapted to compute this *non-overlapping* factorization. When using the first method, we avoid overlaps by simply truncating each LCE to $\min(\text{LCE}(j,i), i-j)$.

For the second method, we repeatedly use the geometric data structure to decide if, for position i, offset k, and length estimate ℓ, there is any $\mathcal{A}[h] \in \mathcal{X}$ with $T[\mathcal{A}[h] - k..\mathcal{A}[h] + \ell) = T[i..i + k + \ell)$. To avoid overlaps, we have to ensure $\mathcal{A}[h] \le i - \ell$. We add a dimension and represent each $\mathcal{A}[h]$ as a point $(\text{suf-rank}(h), \text{pref-rank}(h), \mathcal{A}[h])$. The query interval becomes a three-dimensional hyper-rectangle $[a_1, a_2] \times [b_1, b_2] \times [1, i - \ell]$. There is a three-dimensional geometric data structure with amortized times $u_{\mathcal{A}} = \mathcal{O}(\log^{8/5+\epsilon} N)$ and $q_{\mathcal{A}} = \mathcal{O}((\log N/\log\log N)^2)$ [8]. Its space complexity is $s_{\mathcal{A}} = \mathcal{O}(N\log^{8/5+\epsilon} N)$ words. To accommodate for the more expensive update time and higher space complexity, we use a larger $\delta = \Theta(\log_\sigma n \cdot \log^{8/5+\epsilon} n)$. This increases the time for computing a phrase to $\mathcal{O}(\log^{23/5+\epsilon} n \cdot \log|f_{i'}|/\log\sigma)$, and the total time to $\mathcal{O}(n/\log_\sigma n + z\log^{23/5+2\epsilon} z/\log\sigma)$. The space remains $\mathcal{O}(n/\log_\sigma n)$ words.

For avoiding overlaps in the third method, we adjust the first method such that it considers leftmost sources in $[i - 2\delta, i)$ rather than $[i - \delta, i)$ (which does not increase the complexity). Now we can change the invariant for $\mathcal{Y}$ such that, at the time at which we compute $f_{i'}$, we have inserted exactly the samples with $\mathcal{B}[h] < i - \delta$ into $\mathcal{Y}$. We do not miss any sources this way due to our previous adjustment of the first method. Now any source reported by the third method is from $[1, i - \delta)$. The third method is only used for phrases of length less than δ, and thus an overlap is impossible. (If the third method reports a phrase of length at least δ, then we simply ignore it; it will be computed by the second method already.) The increased value of δ increases the number of queries for the third method, but the time is still dominated by $\mathcal{O}(z\log^{23/5+\epsilon} z/\log\sigma)$.

Corollary 1. *Let $T \in [0, \sigma)^n$ be a string, and let $\epsilon \in \mathbb{R}^+$ be an arbitrarily small constant. The non-overlapping LZ factorization $T = f_1 \ldots f_z$ can be computed in $\mathcal{O}(n/\log_\sigma n + z \log^{23/5+\epsilon} z / \log \sigma)$ time and $\mathcal{O}(n \log \sigma)$ bits of working space.*

References

1. Amir, A., Landau, G.M., Ukkonen, E.: Online timestamped text indexing. Inf. Process. Lett. **82**(5), 253–259 (2002). https://doi.org/10.1016/S0020-0190(01)00275-7
2. Barbay, J., Fischer, J., Navarro, G.: LRM-trees: compressed indices, adaptive sorting, and compressed permutations. In: Giancarlo, R., Manzini, G. (eds.) CPM 2011. LNCS, vol. 6661, pp. 285–298. Springer, Heidelberg (2011). https://doi.org/10.1007/978-3-642-21458-5_25
3. Belazzougui, D., Kärkkäinen, J., Kempa, D., Puglisi, S.J.: Lempel-Ziv decoding in external memory. In: Goldberg, A.V., Kulikov, A.S. (eds.) SEA 2016. LNCS, vol. 9685, pp. 63–74. Springer, Cham (2016). https://doi.org/10.1007/978-3-319-38851-9_5
4. Belazzougui, D., Puglisi, S.J.: Range predecessor and Lempel-Ziv parsing. In: Proceedings of the 27th Annual Symposium on Discrete Algorithms (SODA 2016), Arlington, VA, USA, pp. 2053–2071 (2016). https://doi.org/10.1137/1.9781611974331.ch143
5. Bille, P., Cording, P.H., Fischer, J., Gørtz, I.L.: Lempel-Ziv compression in a sliding window. In: Proceedings of the 28th Annual Symposium on Combinatorial Pattern Matching (CPM 2017), Warsaw, Poland, pp. 15:1–15:11 (2017). https://doi.org/10.4230/LIPIcs.CPM.2017.15
6. Bille, P., Ettienne, M.B., Gørtz, I.L., Vildhøj, H.W.: Time-space trade-offs for Lempel-Ziv compressed indexing. Theor. Comput. Sci. **713**, 66–77 (2018). https://doi.org/10.1016/j.tcs.2017.12.021
7. Bille, P., Gørtz, I.L., Steiner, T.A.: String indexing with compressed patterns. In: Proceedings of the 37th International Symposium on Theoretical Aspects of Computer Science (STACS 2020), Montpellier, France, pp. 10:1–10:13 (2020). https://doi.org/10.4230/LIPIcs.STACS.2020.10
8. Chan, T.M., Tsakalidis, K.: Dynamic orthogonal range searching on the ram, revisited. J. Comput. Geom. **9**(2), 45–66 (2018). https://doi.org/10.20382/jocg.v9i2a5
9. Charikar, M., et al.: The smallest grammar problem. IEEE Trans. Inf. Theory **51**(7), 2554–2576 (2005). https://doi.org/10.1109/TIT.2005.850116
10. Crochemore, M., Giambruno, L., Langiu, A., Mignosi, F., Restivo, A.: Dictionary-symbolwise flexible parsing. J. Discret. Algorithms **14**, 74–90 (2012). https://doi.org/10.1016/j.jda.2011.12.021
11. Crochemore, M., Ilie, L.: Computing longest previous factor in linear time and applications. Inf. Process. Lett. **106**(2), 75–80 (2008). https://doi.org/10.1016/j.ipl.2007.10.006
12. Crochemore, M., Langiu, A., Mignosi, F.: The rightmost equal-cost position problem. In: Proceedings of the 2013 Data Compression Conference (DCC 2013), Snowbird, UT, USA, pp. 421–430 (2013). https://doi.org/10.1109/DCC.2013.50
13. Crochemore, M., Rytter, W.: Efficient parallel algorithms to test square-freeness and factorize strings. Inf. Process. Lett. **38**(2), 57–60 (1991). https://doi.org/10.1016/0020-0190(91)90223-5
14. Ellert, J., Fischer, J., Pedersen, M.R.: New advances in rightmost Lempel-Ziv. In: Proceedings of the 30th International Symposium on String Processing and Information Retrieval (SPIRE 2023), Pisa, Italy (2023)

15. Farach, M., Muthukrishnan, S.: Optimal parallel dictionary matching and compression (extended abstract). In: Proceedings of the 7th Annual Symposium on Parallel Algorithms and Architectures (SPAA 1995), Santa Barbara, CA, USA, pp. 244–253 (1995). https://doi.org/10.1145/215399.215451
16. Ferrada, H., Gagie, T., Hirvola, T., Puglisi, S.J.: Hybrid indexes for repetitive datasets. Philos. Trans. R. Soc. A **372**(2016) (2014). https://doi.org/10.1098/rsta.2013.0137
17. Ferragina, P., Nitto, I., Venturini, R.: On the bit-complexity of Lempel-Ziv compression. SIAM J. Comput. **42**(4), 1521–1541 (2013). https://doi.org/10.1137/120869511
18. Fischer, J., Gagie, T., Gawrychowski, P., Kociumaka, T.: Approximating LZ77 via small-space multiple-pattern matching. In: Bansal, N., Finocchi, I. (eds.) ESA 2015. LNCS, vol. 9294, pp. 533–544. Springer, Heidelberg (2015). https://doi.org/10.1007/978-3-662-48350-3_45
19. Fischer, J., I, T., Köppl, D.: Lempel Ziv computation in small space (LZ-CISS). In: Cicalese, F., Porat, E., Vaccaro, U. (eds.) CPM 2015. LNCS, vol. 9133, pp. 172–184. Springer, Cham (2015). https://doi.org/10.1007/978-3-319-19929-0_15
20. Fischer, J., I, T., Köppl, D., Sadakane, K.: Lempel–Ziv factorization powered by space efficient suffix trees. Algorithmica **80**(7), 2048–2081 (2017). https://doi.org/10.1007/s00453-017-0333-1
21. Gagie, T.: Space-efficient RLZ-to-LZ77 conversion. CoRR abs/2211.13254 (2022). https://doi.org/10.48550/arXiv.2211.13254
22. Gagie, T., Gawrychowski, P., Kärkkäinen, J., Nekrich, Y., Puglisi, S.J.: LZ77-based self-indexing with faster pattern matching. In: Pardo, A., Viola, A. (eds.) LATIN 2014. LNCS, vol. 8392, pp. 731–742. Springer, Heidelberg (2014). https://doi.org/10.1007/978-3-642-54423-1_63
23. Gagie, T., Gawrychowski, P., Puglisi, S.J.: Approximate pattern matching in LZ77-compressed texts. J. Discret. Algorithms **32**, 64–68 (2015). https://doi.org/10.1016/j.jda.2014.10.003
24. Gagie, T., Navarro, G., Prezza, N.: On the approximation ratio of Lempel-Ziv parsing. In: Bender, M.A., Farach-Colton, M., Mosteiro, M.A. (eds.) LATIN 2018. LNCS, vol. 10807, pp. 490–503. Springer, Cham (2018). https://doi.org/10.1007/978-3-319-77404-6_36
25. Goto, K., Bannai, H.: Simpler and faster Lempel Ziv factorization. In: Proceedings of the 2013 Data Compression Conference (DCC 2013), Snowbird, UT, USA, pp. 133–142 (2013). https://doi.org/10.1109/DCC.2013.21
26. Goto, K., Bannai, H.: Space efficient linear time Lempel-Ziv factorization for small alphabets. In: Proceedings of the 2014 Data Compression Conference (DCC 2014), Snowbird, UT, USA, pp. 163–172 (2014). https://doi.org/10.1109/DCC.2014.62
27. Hagerup, T.: Sorting and searching on the word RAM. In: Morvan, M., Meinel, C., Krob, D. (eds.) STACS 1998. LNCS, vol. 1373, pp. 366–398. Springer, Heidelberg (1998). https://doi.org/10.1007/BFb0028575
28. Hong, A., Rossi, M., Boucher, C.: LZ77 via prefix-free parsing. In: Proceedings of the Symposium on Algorithm Engineering and Experiments (ALENEX 2023), Florence, Italy, pp. 123–134 (2023). https://doi.org/10.1137/1.9781611977561.ch11
29. Kärkkäinen, J., Kempa, D., Puglisi, S.J.: Lightweight Lempel-Ziv parsing. In: Bonifaci, V., Demetrescu, C., Marchetti-Spaccamela, A. (eds.) SEA 2013. LNCS, vol. 7933, pp. 139–150. Springer, Heidelberg (2013). https://doi.org/10.1007/978-3-642-38527-8_14

30. Kärkkäinen, J., Kempa, D., Puglisi, S.J.: Linear time Lempel-Ziv factorization: simple, fast, small. In: Fischer, J., Sanders, P. (eds.) CPM 2013. LNCS, vol. 7922, pp. 189–200. Springer, Heidelberg (2013). https://doi.org/10.1007/978-3-642-38905-4_19

31. Kärkkäinen, J., Sutinen, E.: Lempel-Ziv index for q-grams. Algorithmica 21(1), 137–154 (1998). https://doi.org/10.1007/PL00009205

32. Kempa, D.: Optimal construction of compressed indexes for highly repetitive texts. In: Proceedings of the 30th Annual Symposium on Discrete Algorithms (SODA 2019), San Diego, CA, USA, pp. 1344–1357 (2019). https://doi.org/10.1137/1.9781611975482.82

33. Kempa, D., Kociumaka, T.: String synchronizing sets: sublinear-time BWT construction and optimal LCE data structure. In: Proceedings of the 51st Annual Symposium on Theory of Computing (STOC 2019), Phoenix, AZ, USA, pp. 756–767 (2019). https://doi.org/10.1145/3313276.3316368

34. Kempa, D., Kociumaka, T.: Resolution of the burrows-wheeler transform conjecture. Commun. ACM 65(6), 91–98 (2022). https://doi.org/10.1145/3531445

35. Kempa, D., Prezza, N.: At the roots of dictionary compression: string attractors. In: Proceedings of the 50th Annual Symposium on Theory of Computing (STOC 2018), Los Angeles, CA, USA, pp. 827–840 (2018). https://doi.org/10.1145/3188745.3188814

36. Kociumaka, T., Navarro, G., Prezza, N.: Towards a definitive measure of repetitiveness. In: Kohayakawa, Y., Miyazawa, F.K. (eds.) LATIN 2021. LNCS, vol. 12118, pp. 207–219. Springer, Cham (2020). https://doi.org/10.1007/978-3-030-61792-9_17

37. Köppl, D.: Non-overlapping LZ77 factorization and LZ78 substring compression queries with suffix trees. Algorithms 14(2), 44 (2021). https://doi.org/10.3390/a14020044

38. Köppl, D., Navarro, G., Prezza, N.: HOLZ: high-order entropy encoding of Lempel-Ziv factor distances. In: Proceedings of the 2022 Data Compression Conference (DCC 2022), Snowbird, UT, USA, pp. 83–92 (2022). https://doi.org/10.1109/DCC52660.2022.00016

39. Kosolobov, D.: Faster lightweight Lempel-Ziv parsing. In: Italiano, G.F., Pighizzini, G., Sannella, D.T. (eds.) MFCS 2015. LNCS, vol. 9235, pp. 432–444. Springer, Heidelberg (2015). https://doi.org/10.1007/978-3-662-48054-0_36

40. Kosolobov, D., Valenzuela, D., Navarro, G., Puglisi, S.J.: Lempel–Ziv-like parsing in small space. Algorithmica 82(11), 3195–3215 (2020). https://doi.org/10.1007/s00453-020-00722-6

41. Köppl, D., Sadakane, K.: Lempel-Ziv computation in compressed space (LZ-CICS). In: Proceedings of the 2016 Data Compression Conference (DCC 2016), Snowbird, UT, USA, pp. 3–12 (2016). https://doi.org/10.1109/DCC.2016.38

42. Kreft, S., Navarro, G.: On compressing and indexing repetitive sequences. Theor. Comput. Sci. 483, 115–133 (2013). https://doi.org/10.1016/j.tcs.2012.02.006

43. Kärkkäinen, J., Kempa, D., Puglisi, S.J.: Lempel-Ziv parsing in external memory. In: Proceedings of the 2014 Data Compression Conference (DCC 2014), Snowbird, UT, USA, pp. 153–162 (2014). https://doi.org/10.1109/DCC.2014.78

44. Larsson, N.J.: Most recent match queries in on-line suffix trees. In: Kulikov, A.S., Kuznetsov, S.O., Pevzner, P. (eds.) CPM 2014. LNCS, vol. 8486, pp. 252–261. Springer, Cham (2014). https://doi.org/10.1007/978-3-319-07566-2_26

45. Lempel, A., Ziv, J.: On the complexity of finite sequences. IEEE Trans. Inf. Theory 22(1), 75–81 (1976). https://doi.org/10.1109/TIT.1976.1055501

46. Manber, U., Myers, E.W.: Suffix arrays: a new method for on-line string searches. SIAM J. Comput. **22**(5), 935–948 (1993). https://doi.org/10.1137/0222058
47. Mortensen, C.W.: Fully dynamic orthogonal range reporting on RAM. SIAM J. Comput. **35**(6), 1494–1525 (2006). https://doi.org/10.1137/S0097539703436722
48. Naor, M.: String matching with preprocessing of text and pattern. In: Albert, J.L., Monien, B., Artalejo, M.R. (eds.) ICALP 1991. LNCS, vol. 510, pp. 739–750. Springer, Heidelberg (1991). https://doi.org/10.1007/3-540-54233-7_179
49. Nekrich, Y.: Orthogonal range searching in linear and almost-linear space. Comput. Geom. **42**(4), 342–351 (2009). https://doi.org/10.1016/j.comgeo.2008.09.001
50. Nishimoto, T., I, T., Inenaga, S., Bannai, H., Takeda, M.: Dynamic index and LZ factorization in compressed space. Discret. Appl. Math. **274**, 116–129 (2020). https://doi.org/10.1016/j.dam.2019.01.014
51. Nishimoto, T., Tabei, Y.: LZRR: LZ77 parsing with right reference. Inf. Comput. **285** (2022). https://doi.org/10.1016/j.ic.2021.104859
52. Ohlebusch, E., Gog, S.: Lempel-Ziv factorization revisited. In: Giancarlo, R., Manzini, G. (eds.) CPM 2011. LNCS, vol. 6661, pp. 15–26. Springer, Heidelberg (2011). https://doi.org/10.1007/978-3-642-21458-5_4
53. Okanohara, D., Sadakane, K.: An online algorithm for finding the longest previous factors. In: Halperin, D., Mehlhorn, K. (eds.) ESA 2008. LNCS, vol. 5193, pp. 696–707. Springer, Heidelberg (2008). https://doi.org/10.1007/978-3-540-87744-8_58
54. Policriti, A., Prezza, N.: Fast online Lempel-Ziv factorization in compressed space. In: Iliopoulos, C., Puglisi, S., Yilmaz, E. (eds.) SPIRE 2015. LNCS, vol. 9309, pp. 13–20. Springer, Cham (2015). https://doi.org/10.1007/978-3-319-23826-5_2
55. Raskhodnikova, S., Ron, D., Rubinfeld, R., Smith, A.: Sublinear algorithms for approximating string compressibility. Algorithmica **65**, 685–709 (2013). https://doi.org/10.1007/s00453-012-9618-6
56. Rodeh, M., Pratt, V.R., Even, S.: Linear algorithm for data compression via string matching. J. ACM **28**(1), 16–24 (1981). https://doi.org/10.1145/322234.322237
57. Rytter, W.: Application of Lempel-Ziv factorization to the approximation of grammar-based compression. Theor. Comput. Sci. **302**(1), 211–222 (2003). https://doi.org/10.1016/S0304-3975(02)00777-6
58. Shigekuni, M., I, T.: Converting RLBWT to LZ77 in smaller space. In: Proceedings of the 2022 Data Compression Conference (DCC 2022), Snowbird, UT, USA, pp. 242–251 (2022). https://doi.org/10.1109/DCC52660.2022.00032
59. Shun, J.: Parallel Lempel-Ziv Factorization, chap. 13. Association for Computing Machinery and Morgan & Claypool (2018). https://doi.org/10.1145/3018787.3018801
60. Shun, J., Zhao, F.: Practical parallel Lempel-Ziv factorization. In: Proceedings of the 2013 Data Compression Conference (DCC 2013), Snowbird, UT, USA, pp. 123–132 (2013). https://doi.org/10.1109/DCC.2013.20
61. Starikovskaya, T.: Computing Lempel-Ziv factorization online. In: Rovan, B., Sassone, V., Widmayer, P. (eds.) MFCS 2012. LNCS, vol. 7464, pp. 789–799. Springer, Heidelberg (2012). https://doi.org/10.1007/978-3-642-32589-2_68
62. Storer, J.A., Szymanski, T.G.: Data compression via textual substitution. J. ACM **29**(4), 928–951 (1982). https://doi.org/10.1145/322344.322346
63. Sun, X., Wu, D., Mo, D., Cui, J., Zhong, H.: Accelerating Knuth-Morris-Pratt string matching over LZ77 compressed text. In: Proceedings of the 2021 Data Compression Conference (DCC 2021), Snowbird, UT, USA, p. 372 (2021). https://doi.org/10.1109/DCC50243.2021.00070

64. Valenzuela, D.: CHICO: a compressed hybrid index for repetitive collections. In: Goldberg, A.V., Kulikov, A.S. (eds.) SEA 2016. LNCS, vol. 9685, pp. 326–338. Springer, Cham (2016). https://doi.org/10.1007/978-3-319-38851-9_22
65. Wu, C.Y.: Improved LZ77 compression. In: Proceedings of the 2021 Data Compression Conference (DCC 2021), Snowbird, UT, USA, p. 377 (2021). https://doi.org/10.1109/DCC50243.2021.00066
66. Yamamoto, J., I, T., Bannai, H., Inenaga, S., Takeda, M.: Faster compact on-line Lempel-Ziv factorization. In: Proceedings of the 31st International Symposium on Theoretical Aspects of Computer Science (STACS 2014), Lyon, France, pp. 675–686 (2014). https://doi.org/10.4230/LIPIcs.STACS.2014.675
67. Ziv, J., Lempel, A.: A universal algorithm for sequential data compression. IEEE Trans. Inf. Theory **23**(3), 337–343 (1977). https://doi.org/10.1109/TIT.1977.1055714

New Advances in Rightmost Lempel-Ziv

Jonas Ellert[1]([✉])(iD), Johannes Fischer[1](iD), and Max Rishøj Pedersen[2](iD)

[1] Technical University of Dortmund, Dortmund, Germany
jonas.ellert@tu-dortmund.de, johannes.fischer@cs.tu-dortmund.de
[2] Technical University of Denmark, DTU Compute, Lyngby, Denmark
mhrpe@dtu.dk

Abstract. The Lempel-Ziv (LZ) 77 factorization of a string is a widely-used algorithmic tool that plays a central role in compression and indexing. For a length-n string over a linearly-sortable alphabet, e.g., $\Sigma = \{1, \ldots, \sigma\}$ with $\sigma = n^{\mathcal{O}(1)}$, it can be computed in $\mathcal{O}(n)$ time. It is unknown whether this time can be achieved for the *rightmost* LZ parsing, where each referencing phrase points to its rightmost previous occurrence. The currently best solution takes $\mathcal{O}(n(1 + \log \sigma / \sqrt{\log n}))$ time (Belazzougui & Puglisi SODA2016). We show that this problem is much easier to solve for the LZ-End factorization (Kreft & Navarro DCC2010), where the rightmost factorization can be obtained in $\mathcal{O}(n)$ time for the greedy parsing (with phrases of maximal length), and in $\mathcal{O}(n + z\sqrt{\log z})$ time for any LZ-End parsing of z phrases. We also make advances towards a linear time solution for the general case. We show how to solve multiple non-trivial subsets of the phrases of any LZ-like parsing in $\mathcal{O}(n)$ time. As a prime example, we can find the rightmost occurrence of all phrases of length $\Omega(\log^{6.66} n / \log^2 \sigma)$ in $\mathcal{O}(n / \log_\sigma n)$ time and space.

Keywords: Lempel-Ziv · LZ77 · rightmost LZ · LZ-End · lossless compression · string algorithms · linear time algorithms · word-packing

1 Introduction

The Lempel-Ziv (LZ) 77 factorization [28] of a string S decomposes it into a series of phrases $S = f_1 f_2 \ldots f_z$. Each phrase is either the leftmost occurrence of an alphabet symbol (a *literal phrase*), or the longest substring that can be read at an earlier position in the string (a *referencing phrase*). Compression can be achieved by replacing each referencing phrase with an integer pair consisting of the length and the distance to an earlier occurrence of the phrase. Further compression is possible by encoding the integers, e.g., by applying a universal code. Variable length codes often assign longer codewords to larger integers, and thus it is beneficial if every referencing phrase knows not only any of its previous occurrences, but the rightmost one (at the smallest distance).

In an LZ-*like* factorization, referencing phrases do not need to be of maximal length. The encoding works in the same way as for the exact LZ factorization.

Supported by Danish Research Council grant DFF-8021-002498.

Related Work. LZ(-like) parsings are well-studied, and there are fast factorization algorithms in multiple settings (we only list a few examples for each) including parallel [7,9,30,34,35], online [33,36,39] and external memory algorithms [25]. In the sequential setting, there are several linear-time solutions [11,14,15,18], and some that compute the parsing in small space [3,8,17,22,26,32,33,36,39], with the overall best using only $\mathcal{O}(n \log \sigma)$ bits and running in $\mathcal{O}(n)$ time [12] for a string of length n over integer alphabet $[0, \sigma)$.

LZ-End, introduced by Kreft and Navarro [23,24], is a family of LZ-like parsings where each referencing phrase must have a previous occurrence aligned with the end of a phrase, i.e., for f_k there must be $k' < k$ such that f_k is a suffix of $f_1 f_2 \ldots f_{k'}$. This has beneficial properties that lead to efficient compressed text indices (e.g., [21]). The uniquely defined *greedy* LZ-End parsing, in which each referencing phrase is of maximal length, can be computed in linear time [20], and the number of phrases is within an $\mathcal{O}(\log^2 n)$ factor of the exact LZ factorization [21]. Bannai et al. [2] proved that computing the optimal LZ-End parsing (with minimal number of phrases) is NP-hard and gave a lower bound of 2 for the approximation ratio of optimal LZ-End to greedy LZ-End.

The first theoretical result on computing the rightmost LZ parsing is by Amir et al. [1] and uses $\mathcal{O}(n \log n)$ time and working space. Larsson et al. [27] presented an online algorithm in the same time and space. Crochemore et al. [6] gave the first approximation algorithm, which runs in $\mathcal{O}(n \log n)$ time and $\mathcal{O}(n)$ space and finds the rightmost *equal-cost* position for each phrase, meaning it takes the same number of bits to encode as the rightmost position. Later, Bille et al. [4] gave an $(1 + \epsilon)$-approximation algorithm of the rightmost parsing in $\mathcal{O}(n(\log z + \log \log n))$ time and linear working space. The first exact algorithm to achieve $o(n \log n)$ time is by Ferragina et al. [10] and runs in $\mathcal{O}(n(1 + \log \sigma / \log \log n))$ time and $\mathcal{O}(n)$ words of space. This was improved by Belazzougui and Puglisi [3] with an algorithm using only $\mathcal{O}(n \log \sigma)$ bits of space and achieving $\mathcal{O}(n(\log \log \sigma + \log \sigma / \sqrt{\log n}))$ deterministic time or $\mathcal{O}(n(1 + \log \sigma / \sqrt{\log n}))$ time with randomization, which is the current state of the art.

Our Contributions. We present time-efficient deterministic algorithms for rightmost LZ parsings, summarized by Theorems 1 and 2 below.

Theorem 1. *Let $S \in [0, \sigma)^n$. Given an LZ-End factorization $S = f_1 \ldots f_z$, we can compute its rightmost LZ-End parsing in $\mathcal{O}(n + z\sqrt{\log z})$ time and $\mathcal{O}(n)$ words of space. For the greedy LZ-End factorization, we achieve $\mathcal{O}(n)$ time.*

Theorem 2. *Let $S \in [0, \sigma)^n$. Unless explicitly stated otherwise, the space complexity is $\mathcal{O}(n)$ words. Given any LZ-like factorization $S = f_1 \ldots f_z$, we can compute the rightmost previous occurrence of all referencing phrases*

(a) of length $\Omega(\log^{6.66} n / \log^2 \sigma)$ in $\mathcal{O}(n / \log_\sigma n)$ time and words of space
(b) f_k with $k \in F \subseteq [1, z]$ in $\mathcal{O}(n + |F| d^\epsilon)$ time, where $d = |\{f_{k'} \mid k' \in F\}| \leq |F|$
(c) f_k with $|\{k' \in [1, z] \mid f_{k'} = f_k\}| = \mathcal{O}(\log n)$ in $\mathcal{O}(n)$ time
(d) with rightmost previous occurrence at distance $\mathcal{O}(\log n)$ in $\mathcal{O}(n)$ time

We provide the solution for rightmost parsings of LZ-End factorizations (Theorem 1) in Sect. 3. The algorithms for subsolutions of general rightmost LZ-like parsings (Theorem 2) are presented in Sect. 4.

2   Preliminaries

Strings and Model of Computation. For $i, j \in \mathbb{N}$, we write $[i, j] = [i, j + 1)$ rather than $\{k \in \mathbb{N}^+ \mid i \leq k \leq j\}$. A string $S = S[1..n] = S[1]S[2]\ldots S[n]$ of length $|S| = n$ is a sequence of n symbols from an alphabet Σ. For $i, j \in [1, n]$, the substring $S[i..j] = S[i..j + 1)$ is the sequence $S[i]S[i + 1]\ldots S[j]$ (or the empty string ε if $j < i$). A substring shorter than S is *proper*. Substrings $S[1..i]$ and $S[i..n]$ are respectively called prefix and suffix of S. The reversal of S is $\mathsf{rev}(S) = S[n]S[n - 1]\ldots S[1]$. The concatenation of two string S_1 and S_2 is $S_1 S_2$. We only consider alphabets Σ that are totally ordered, which induces a lexicographical order over the set of all strings in the usual way. We write $S_1 \prec S_2$ to denote that S_1 is lexicographically smaller than S_2. We say that S_1 is *co*-lexicographically smaller than S_2 if $\mathsf{rev}(S_1) \prec \mathsf{rev}(S_2)$. For strings S and P, an *occurrence* of P in S is a position i such that P is a prefix of $S[i..|S|]$. For the occurrence i of substring $S[i..i + \ell)$ in S, a *previous occurrence* is an occurrence j of $S[i..i + \ell)$ in S with $j < i$. We assume that the string $S[1..n]$ is over integer alphabet $[0, \sigma)$ with $\sigma = n^{\mathcal{O}(1)}$, and we use a word RAM of width $w = \Theta(\log n)$ bits (see, e.g., [16]). Each symbol is stored in $\lceil \log \sigma \rceil$ bits, and thus the string occupies $\mathcal{O}(n/\log_\sigma n)$ words of space. From now on, space complexities are given in number of words.

We assume that the reader is familiar with tries [13]. The *suffix tree* [38] of S is the compact trie of all suffixes of $S\$$, where $\$ = -\infty$ is smaller than all symbols from the alphabet. Each leaf corresponds to a suffix of S and is labeled with the start position of this suffix. The outgoing edges of each node are arranged in increasing order of the first symbol of the respective edge label. Hence the leaves are ordered from left to right in lexicographical order of suffixes. In the present model of computation, the suffix tree can be computed in $\mathcal{O}(n)$ time and space [29]. The suffix array SA of S is the unique permutation of $[1, n]$ that lexicographically sorts the suffixes, i.e., $\forall i \in [1, n) : S[\mathsf{SA}[i]..n] \prec S[\mathsf{SA}[i + 1]..n]$. Equivalently, it consists of the leaf-labels of the suffix tree in left-to-right order and can therefore be constructed from the suffix tree in linear time.

Lempel-Ziv Parsings. The unique *LZ (77) factorization* $S = f_1 f_2 \ldots f_z$ decomposes S into z substrings called *phrases*. Each phrase f_k at *destination* $i = 1 + \sum_{j=1}^{k-1} |f_j|$ is either the leftmost occurrence of $S[i]$ (a *literal phrase*), or the longest prefix of $S[i..n]$ that has a previous occurrence (a *referencing phrase*). A previous occurrence $j \in [1, i)$ of a referencing phrase f_k is called a *source* of f_k. (This is Storer and Szymanski's version of the factorization [37].) An *LZ-like factorization* is defined exactly like the LZ factorization, but without the requirement that referencing phrases are of maximal length. The *rightmost parsing* of an LZ(-like) factorization annotates each referencing phrase with its rightmost source, i.e., f_k at destination i is annotated with the maximal $j \in [1, i)$ such that f_k is a prefix of $S[j..n]$.

A source j of some phrase f_k in an LZ-like factorization is *LZ-End aligned* if $S[1..j + |f_k|) = f_1 f_2 \ldots f_{k'}$ for some $k' \in [1, k)$ (i.e., f_k equals the suffix of $f_1 f_2 \ldots f_{k'}$ that starts at position j). An *LZ-End factorization* is an LZ-like factorization in which all referencing phrases have an LZ-End aligned source. (This is slightly different from [23] and leads to a simpler description; the presented results can be easily modified to work for the original definition.) The *greedy LZ-End factorization* is the unique LZ-End factorization in which each f_k at destination i is the longest prefix of $S[i..n]$ that is a suffix of $f_1 f_2 \ldots f_{k'}$ for some $k' \in [1, k)$. We could define the rightmost parsing for LZ-End in the same way as for arbitrary LZ-like factorizations (i.e., annotate each phrase with its rightmost source), but this is undesirable because the rightmost source might not be LZ-End aligned. Hence the rightmost parsing of an LZ-End factorization annotates each referencing phrase with its rightmost LZ-End aligned source.

From now on, we use z (commonly used to denote the number of phrases in the exact LZ 77 factorization) to denote the number of phrases in the factorization at hand, even if it is an LZ-like or LZ-End factorization. Instead of saying that we compute the rightmost source of f_k, we simply say that we resolve f_k.

3 Computing Rightmost LZ-End Parsings

In this section, we provide the solutions for Theorem 1. We exploit the fact that an LZ-End phrase only has to choose from less than z sources, while a general LZ-like phrase has to consider up to $\Omega(n)$ possible sources. This makes the computation significantly easier for LZ-End factorizations.

Rightmost Greedy LZ-End Parsing. We start by computing an arbitrary LZ-End aligned source for each referencing phrase f_k. We build the suffix array of the reversed text $\text{rev}(S)$, and use filtering and rank reduction to obtain in $\mathcal{O}(n)$ time the unique permutation co of $[1, z]$ that satisfies $\forall k' \in [1, z)$: $\text{rev}(f_1 f_2 \ldots f_{\text{co}(k')}) \prec \text{rev}(f_1 f_2 \ldots f_{\text{co}(k'+1)})$. (This permutation rearranges the prefixes that end at phrase boundaries in co-lexicographical order.) We also compute its inverse permutation co^{-1}. Any referencing phrase f_k has a previous occurrence as a suffix of $f_1 f_2 \ldots f_{k'}$, where k' and k are neighbors in co (because the co-lexicographical order groups together prefixes that share a long suffix). More precisely, if $\text{co}^{-1}(k) = 1$ then $k' = \text{co}(2)$. If $\text{co}^{-1}(k) = z$ then $k' = \text{co}(z-1)$. Otherwise, $k' \in \{k^-, k^+\}$ with $k^- = \text{co}(\text{co}^{-1}(k)-1)$ and $k^+ = \text{co}(\text{co}^{-1}(k)+1)$. In the latter case, we naively check if f_k is a suffix of $f_1 f_2 \ldots f_{k^-}$. If this is the case, then we use $k' = k^-$. Otherwise, we use $k' = k^+$. Hence we can compute a suitable k' for each referencing phrase f_k in total time $\mathcal{O}(n + z + \sum_{j=1}^{z} |f_j|) = \mathcal{O}(n)$. We then report $|f_1 f_2 \ldots f_{k'}| - |f_k| + 1$ as an LZ-End aligned source of f_k.

The computed sources are already rightmost for all phrases that only have a single LZ-End aligned source. It remains to correct the sources of phrases that have multiple LZ-End aligned sources, for which we observe the following.

Proposition 1. *Let f_k be a referencing phrase in the greedy LZ-End factorization, and let $k', k'' \in [1, k)$ with $k'' < k'$ be such that f_k is a suffix of both $f_1 f_2 \ldots f_{k'}$ and $f_1 f_2 \ldots f_{k''}$. Then f_k is a suffix of $f_{k'-1} f_{k'}$.*

Proof. If f_k is a suffix of $f_1 f_2 \ldots f_{k'}$ but not of $f_{k'-1} f_{k'}$, then $f_{k'-1} f_{k'}$ is a suffix of f_k. Since f_k is a suffix of $f_1 f_2 \ldots f_{k''}$, this implies that $f_{k'-1} f_{k'}$ is a suffix of $f_1 f_2 \ldots f_{k''}$. Hence $f_{k'-1} f_{k'}$ has a previous occurrence that satisfies the LZ-End property. Thus, $f_{k'-1}$ is not of maximal length, which contradicts the definition of the greedy LZ-End factorization. □

We compute a compacted trie that contains for each $k' \in [2, z]$ the string $\mathrm{rev}(f_{k'-1} f_{k'})$. Note that the total length of the strings is less than $2n$. We make the respective nodes that spell $\mathrm{rev}(f_{k'})$ and $\mathrm{rev}(f_{k'-1} f_{k'})$ explicit (if they are not explicit already), and store pointers to these nodes. We will not need fast navigation on the trie; in fact, we only need the parent operation. Hence we can construct the trie in $\mathcal{O}(n)$ deterministic time using standard techniques (e.g., from the suffix array of $\mathrm{rev}(f_1 f_2 \# f_2 f_3 \# \ldots \# f_{z-1} f_z)$ where $\#$ is a special separator symbol). Now we process the phrase pairs $f_{k'-1} f_{k'}$ with $k' \in [2, z]$ from right to left. Whenever we finish processing a pair, we annotate the node that spells $\mathrm{rev}(f_{k'})$ with k' (indicating that the rightmost LZ-End aligned source of $f_{k'}$ has not been found yet). Before adding this annotation, we first check if $f_{k'-1} f_{k'}$ resolves other phrases. For this purpose, we traverse the path from the leaf that spells $\mathrm{rev}(f_{k'-1} f_{k'})$ to the root of the trie. For each node on the path, we check if it has been annotated with some value k. If we find such an annotation, then the corresponding node spells $\mathrm{rev}(f_k)$, and f_k is a suffix of $f_{k'-1} f_{k'}$. Hence we store $|f_1 f_2 \ldots f_{k'}| - |f_k| + 1$ as the maximal LZ-End aligned source of f_k, and remove the annotation of the node. By Proposition 1 and the right-to-left order of processing, we correctly find the rightmost LZ-End aligned source of any phrase that has multiple LZ-End aligned sources.

A node might spell the reversal of a phrase that has multiple occurrences in the parsing. Nevertheless, each node has at most one annotation at any given point in time. This is because we annotate the node that spells $\mathrm{rev}(f_{k'})$ only after we finish processing pair $f_{k'-1} f_{k'}$. If the node is already annotated with some $k > k'$ (because $f_k = f_{k'}$), then we also find the source $|f_1 f_2 \ldots f_{k'}| - |f_k| + 1$ of f_k while processing pair $f_{k'-1} f_{k'}$, and hence we remove annotation k before adding annotation k'.

We need $\mathcal{O}(n)$ time for computing the trie. Processing a pair $f_{k'-1} f_{k'}$ takes time linear in the depth of the node that spells $\mathrm{rev}(f_{k'-1} f_{k'})$. This is limited by $\mathcal{O}(|f_{k'-1} f_{k'}|)$, which sums to $\mathcal{O}(n)$ over all phrase pairs. The space for the trie is $\mathcal{O}(n)$. Hence we have shown Theorem 1 for the greedy LZ-End factorization.

Rightmost (Arbitrary) LZ-End Parsing. If the given LZ-End factorization does not satisfy the greedy property, then Proposition 1 no longer holds. However, each referencing phrase f_k is still a suffix of some $f_1 f_2 \ldots f_{k'}$ with $k' \in [1, k)$, which limits the number of possible sources. We will again exploit properties of the co-lexicographical order of prefixes.

We compute a compacted trie that contains for each $k' \in [1, z]$ the reversed prefix $\mathrm{rev}(f_1 f_2 \ldots f_{k'})$ of the text. We make the respective nodes that spell $\mathrm{rev}(f_{k'})$ and $\mathrm{rev}(f_1 f_2 \ldots f_{k'})$ explicit (if they are not explicit already), and store pointers to these nodes. We annotate the node that spells $\mathrm{rev}(f_1 f_2 \ldots f_{k'})$ with its co-lexicographical rank $\mathrm{co}^{-1}(k')$ (defined as before). Additionally, we annotate the node that spells $\mathrm{rev}(f_{k'})$ with its co-lexicographical range, which is given by the respectively smallest and largest co-lexicographical ranks $c_{k'}^{\min}$ and $c_{k'}^{\max}$ that were used to annotate any of its descendants (or itself). Again, we do not need fast navigation on the trie; for writing the annotations, it suffices if we can perform a preorder traversal in linear time. Hence we can construct the trie and its annotations in $\mathcal{O}(n)$ deterministic time using standard techniques (e.g., from the suffix array of $\mathrm{rev}(S)$).

Now we show how to find the rightmost LZ-End aligned source of referencing phrase f_k. We have annotated the node that spells $\mathrm{rev}(f_k)$ with the co-lexicographical range $[c_k^{\min}, c_k^{\max}]$. We store the permutation co (defined as before) in an array. Note that, by design of the trie, the range $\mathrm{co}[c_k^{\min}, c_k^{\max}]$ contains exactly all the k' for which f_k is a suffix of $f_1 f_2 \ldots f_{k'}$. Hence finding the rightmost LZ-End aligned source of f_k is equivalent to answering the following so-called *range predecessor query*. Given the range $[c_k^{\min}, c_k^{\max}] \subseteq [1, z]$ and the threshold k, find the largest value $k' < k$ in $\mathrm{co}[c^{\min}, c^{\max}]$. Then, the rightmost LZ-End aligned source of f_k is $|f_1 f_2 \ldots f_{k'}| - |f_k| + 1$.

Belazzougui and Puglisi show how to compute a data structure in $\mathcal{O}(z \sqrt{\log z})$ time and $\mathcal{O}(z)$ space that answers range predecessor queries on a permutation of $[1, z]$ in $\mathcal{O}(\log^\epsilon z)$ time (for any constant $0 < \epsilon < 1$). We issue less than z queries, and thus the total construction and query time is $\mathcal{O}(z \sqrt{\log z})$. The total time for computing the rightmost parsing (including the construction of the trie) is $\mathcal{O}(n + z \sqrt{\log z})$, and the total space is $\mathcal{O}(n)$. Hence we have shown Theorem 1 for an arbitrary LZ-End factorization.

4   Partially Solving Rightmost LZ-Like Parsings

In this section, we show how to efficiently compute the rightmost sources for some subsets of the phrases of an LZ-like factorization (Theorem 2).

4.1   Long Phrases

Belazzougui and Puglisi [3] find the rightmost sources of all phrases of length $\Omega(\log^5 n)$ in $\mathcal{O}(n)$ time and $\mathcal{O}(n/\log_\sigma n)$ space. We show a similar result for resolving all phrases of length $\Omega(\log^{33/5+\epsilon} n/\log^2 \sigma)$ in $\mathcal{O}(n/\log_\sigma n)$ time and space. The main contribution here is that we achieve sublinear time. The solution works for an arbitrary LZ-like factorization $S = f_1 f_2 \ldots f_z$.

Let $\delta = \Omega(\log^2 n/\log \sigma)$ be a parameter to be fixed later. We start by performing a preprocessing as follows. In $\mathcal{O}(n/\log_\sigma n)$ time, we compute the reversed text $\mathrm{rev}(S)$ as described in [3, Section 6.2] (essentially, we use a precomputed lookup table to reverse the text one half-word rather than one symbol at a time). We consider a set $\mathcal{D} = \{d \in [1, n] \mid d \equiv 0 \pmod{\delta}\}$ of

$m = |D| = \mathcal{O}(\frac{n}{\delta})$ regularly sampled positions. We construct the respectively unique permutations pref and suf of $[1, m]$ such that for every $h \in [1, m)$ it holds $S[\mathsf{suf}(h)\delta..n] \prec S[\mathsf{suf}(h+1)\delta..n]$ and $\mathsf{rev}(S[1..\mathsf{pref}(h)\delta]) \prec \mathsf{rev}(S[1..\mathsf{pref}(h+1)\delta])$ (these are sparse suffix arrays of the string and its reversal). We use comparison sorting and obtain the permutations with $\mathcal{O}(m \log m) \subseteq \mathcal{O}(\frac{n}{\delta} \log n) \subset \mathcal{O}(n/\log_\sigma n)$ lexicographical comparisons between suffixes of either S or $\mathsf{rev}(S)$. With an LCE data structure by Kempa and Kociumaka [19] (constructed for both S and $\mathsf{rev}(S)$), each lexicographical comparison takes constant time. The data structure can be constructed in $\mathcal{O}(n/\log_\sigma n)$ time and space. We use $\mathcal{O}(m \log m) \subseteq \mathcal{O}(\frac{n}{\delta} \log n) \subset \mathcal{O}(n \log \sigma)$ bits of space to store pref, suf, and their respective inverse permutations pref-rank and suf-rank.

A long phrase is of length at least $\gamma > \delta$, where γ is another parameter. When resolving a long phrase f_k with rightmost source j and destination i, we will use the fact that $j+q$ with $q = (\delta - (j \bmod \delta)) \in [1, \delta]$ is a sample position. For now, assume that we know the value of q in advance (we will later simply try all the possible values of q). Finding the rightmost source of f_k means that we have to find the rightmost sample position $h\delta < i+q$ with $S[h\delta - q..h\delta] = S[i..i+q]$ and $S[h\delta..h\delta - q + |f_k|) = S[i + q..i + |f_k|)$. Note that the co-lexicographical order groups together prefixes that share a long suffix, and hence all the values of h for which $S[i..i+q]$ is a suffix of $S[1....h\delta]$ form a consecutive interval $\mathsf{pref}[p_1..p_2]$ (we treat the permutations like arrays). We can find the boundaries p_1 and p_2 by binary searching in pref for the respectively co-lexicographically minimal and maximal prefixes of S that have suffix $S[i..i+q]$. This takes $\mathcal{O}(\log m)$ time because we can perform each LCE computation and lexicographical comparison in constant time using the same LCE data structure as before. Similarly, it takes $\mathcal{O}(\log m)$ time to compute the interval $\mathsf{suf}[s_1..s_2]$ that contains exactly the values of h for which $S[i + q..i + |f_k|)$ is a prefix of $S[h\delta..n]$.

We associate a three-dimensional point $(\mathsf{pref\text{-}rank}(h), \mathsf{suf\text{-}rank}(h), h)$ with each sample position. For resolving the phrase, we have to find the point $(p, s, \hat{h})$ with $p \in [p_1, p_2]$, $s \in [s_1, s_2]$, and maximal value $\hat{h}\delta < i + q$ (or equivalently $h < \frac{i+q}{\delta}$). Given this point, it is easy to compute the rightmost source $\hat{h}\delta - q$ of f_k. For solving the geometric query, we use a data structure for three-dimensional orthogonal range searching [5, Theorem 4]. For our m points from $[1, m]^3$, it can be constructed in $\mathcal{O}(m \log^{8/5+\epsilon} m)$ time and space (for any constant $\epsilon \in \mathbb{R}^+$). Given a three-dimensional six-sided orthogonal query range, it returns a point in the range or reports that it is empty in $\mathcal{O}(\log^2 m)$ time (the precise bound is slightly better, but not needed for our purposes). For our queries, we have to find the point with maximal coordinate in the third dimension. Thus, we binary search for this point with $\mathcal{O}(\log n)$ queries to the geometric data structure, which increases the query time to $\mathcal{O}(\log^3 n)$. Note that this dominates the $\mathcal{O}(\log m)$ time needed to compute the query range. Finally, we do not actually know the value of q in advance. Hence we try all the possible $q \in [1, \delta]$. For each of them, we compute the query range and find the rightmost admitted source in $\mathcal{O}(\log^3 n)$ time. Thus, the time needed per phrase is $\mathcal{O}(\delta \cdot \log^3 n)$.

We need $\mathcal{O}(n/\log_\sigma n)$ time for computing the (co-)lexicographically sorted permutations of samples, $\mathcal{O}(\frac{n}{\delta}\log^{8/5+\epsilon} n)$ time for computing the geometric data structure, and $\mathcal{O}(\frac{n\delta}{\gamma} \cdot \log^3 n)$ time for actually resolving the phrases. We want δ to be small in order to minimize the time for resolving phrases. On the other hand, the time needed for computing the geometric data structure should become $\mathcal{O}(n/\log_\sigma n)$. Hence we use $\delta = \Theta(\log^{13/5+\epsilon} n/\log\sigma)$, which achieves the desired construction time and implies that we take $\mathcal{O}(\frac{n}{\gamma} \cdot \log^{28/5+\epsilon} n/\log\sigma)$ time for resolving phrases. Thus, in order to achieve $\mathcal{O}(n/\log_\sigma n)$ time, long phrases have to be of length at least $\gamma = \Omega(\log^{33/5+\epsilon} n/\log^2 \sigma) \subset \Omega(\log^{6.66} n/\log^2 \sigma)$. For all steps (including the geometric data structure), the space is linear in the time spent, and hence it is $\mathcal{O}(n/\log_\sigma n)$. This concludes the proof of Theorem 2(a).

4.2   Arbitrary Subsets of Phrases

Now we show how to solve an arbitrary subset of phrases of any LZ-like factorization $S = f_1 f_2 \ldots f_z$. The subset is given by $F \subseteq [1, z]$, and the time complexity depends on $d = |\{f_k \mid k \in F\}| \leq F$, i.e., on the number of distinct phrases in the subset. In a slight abuse of terminology, we will say that f_k is a phrase from F if $k \in F$. We show how to resolve all phrases from F in $\mathcal{O}(\frac{n}{\epsilon} + |F| d^\epsilon)$ time and $\mathcal{O}(\frac{n}{\epsilon})$ space for arbitrary $\epsilon \in \mathbb{R}^+$ with $\epsilon \leq \frac{1}{2}$, or $\mathcal{O}(n + |F| d^\epsilon)$ time and $\mathcal{O}(n)$ space for constant ϵ. If the string is highly compressible, say, $z = \mathcal{O}(n^{1-\epsilon})$, then the time is $\mathcal{O}(n)$. The idea is to use range maximum data structures to find the rightmost sources. We note that this solution is very similar to [10], and mostly differs in the choice of the range maximum data structure.

We start with the following preprocessing. We arrange the distinct phrases of F into a tree of $d+1$ nodes, and we start using the terms node and phrase interchangeably (even though multiple phrases may refer to the same node). The parent of phrase f_k is the longest phrase $f_{k'}$ from F that is a proper prefix of f_k (or the artificial root node ε if $f_{k'}$ does not exist), and we call this tree the *phrase trie*. This is a slight abuse of terminology, since the tree is only similar to a trie. An example is provided in Fig. 1. We annotate f_k with its *preorder number* p_k, which is the rank of f_k in a preorder traversal of the phrase trie, as well as the maximal preorder number q_k of a descendant of f_k. We also annotate each text position i with the preorder number of the longest phrase from F that is a prefix of $S[i..n]$, if any. This concludes the preprocessing.

In order to resolve the phrases, we traverse S from left to right and track in an array $A[1..d]$ the last position at which we encountered each preorder number as an annotation. When we reach the destination i of some phrase f_k from F, the rightmost previous occurrence will be at position $\max_{p\in[p_k,q_k]} A[p]$ (the solution of a *range maximum query*), as any occurrence of f_k is annotated with either p_k or the preorder number $p_{k'}$ of a phrase $f_{k'}$ that is a descendant of f_k in the phrase trie. Hence, if we have a dynamic data structure for range maximum queries, then we can compute each rightmost occurrence with one query.

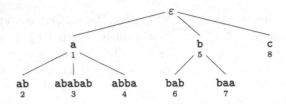

Fig. 1. The phrase trie for the LZ factorization a│b│b│a│ab│ababab│bab│c│abba│baa where F is all the distinct phrases. Below each node is the preorder number.

The phrase trie can be obtained as follows. We compute the suffix tree for the string $S' = S\#_0 f_1 \#_1 f_2 \#_2 \dots \#_{z-1} f_z \#_z$, where each $\#_k$ is a unique seperator symbol. This takes $\mathcal{O}(n)$ time. For any f_k from F, the parent of the leaf that spells suffix $f_k \#_k \dots$ is exactly the node that spells f_k. Thus, we can mark the d nodes that spell phrases from F in $\mathcal{O}(|F|)$ time. It is then easy to compute the nearest marked ancestor of each node in $\mathcal{O}(n)$ time. The phrase trie is obtained by creating a new tree that contains only the marked nodes and an artificial root. The new parent of a marked node is its nearest marked ancestor (or the artificial root node if it does not exist). Finally, we compute the preorder numbers in the phrase trie, and also annotate the corresponding marked nodes in the suffix tree with these numbers. Then, the annotation of text position i is the annotation of the nearest marked ancestor of the leaf that corresponds to text position i in the suffix tree. Hence we obtain the annotations in $\mathcal{O}(n)$ time.

Finally, we solve dynamic range maximum queries (RMQ) for A. The updates are incremental in the sense that every update is the new global maximum (i.e., the rightmost text position processed so far). Therefore, we can maintain a dynamic RMQ data structure for A with $\mathcal{O}(\frac{1}{\epsilon})$ time updates and $\mathcal{O}(d^\epsilon)$ time queries using the standard technique of square-root decomposition, generalized to arbitrary ϵ. For $\epsilon = \frac{1}{2}$, we split A into blocks of size $\Theta(\sqrt{d})$ and maintain the maximum of each block, which we can update in constant time whenever we update an entry of A. To answer queries we need to scan at most $\mathcal{O}(\sqrt{d})$ elements in A that are in blocks that are only partially overlapped by the query range. Then, we also scan the $\mathcal{O}(\sqrt{d})$ maxima of blocks that are fully contained in the query range. Thus, we take $\mathcal{O}(\sqrt{d})$ time. This generalizes to smaller ϵ by recursively subdividing the blocks into $\frac{1}{\epsilon}$ layers, leading to $\mathcal{O}(\frac{1}{\epsilon})$ update time and $\mathcal{O}(d^\epsilon)$ query time. Each phrase in F incurs a range query and each text position an update. We perform $|F|$ range queries and n updates in $\mathcal{O}(\frac{n}{\epsilon} + |F| d^\epsilon)$ time. This concludes the proof of Theorem 2(b).

4.3   Infrequent Phrases

Given an LZ-like parsing $S = f_1 \dots f_z$, we say that a phrase f_k is *infrequent* if $|\{k' \in [1, z] \mid f_{k'} = f_k\}| = \mathcal{O}(\log n)$, i.e., if it occurs at most $\mathcal{O}(\log n)$ times in the parsing. We now show how to resolve all infrequent phrases in $\mathcal{O}(n)$ time, and we begin by establishing a data structure that is crucial for our solution.

Lemma 1. *Let* $m, n \in [1, 2^w]$. *For a tree of* m *nodes, labeled with preorder numbers from* $[1, m]$, *after an* $\mathcal{O}(m) + o(n)$ *time preprocessing, and in* $\mathcal{O}(m) + o(n)$ *space, we can maintain a data structure for nearest marked ancestor queries with the following operations.*

- *mark/unmark a node* $i \in [1, m]$ *with* d_i *descendants in* $\mathcal{O}(1 + d_i/\log n)$ *time*
- *check if a node* $i \in [1, m]$ *is marked in* $\mathcal{O}(1)$ *time*
- *check if a node* $i \in [1, m]$ *has a marked ancestor in* $\mathcal{O}(1)$ *time*
- *output the nearest marked ancestor* j *of a node* $i \in [1, m]$ *in* $\mathcal{O}(1 + d_j/\log n)$ *time, where* d_j *is the number of descendants of* j.

Proof. We compute the balanced parenthesis sequence [31, Chapter 7] (BPS) $B[1..2m]$ of the tree by re-running the traversal used to obtain preorder numbers (with an artificial parent edge for the root to start the traversal). When we walk down the edge to node i, we append i's opening parenthesis to B, when we walk up the edge from node i we append its closing one. The ith opening parenthesis (in left to right order) belongs to node i, and between i's opening and closing parentheses there are exactly all the parentheses corresponding to descendants of i. We preprocess B such that given node $i \in [1, m]$ we can lookup the positions open(i) and close(i) of its respective opening and closing parentheses in B in constant time. This is possible with a simply linear scan in $\mathcal{O}(m)$ time and space. For open, we also compute the inverse mapping prenum(open(i)) $= i$.

We use two additional bitvectors $A[1..2m]$ and $R[1..2m]$, both initialized with zeroes. When asked to mark node i, we set the bits $A[\text{open}(i)]$ and $A[\text{close}(i)]$ (marking the respective parentheses in B as *active*), and additionally we set the entire range $R[\text{open}(i) + 1..\text{close}(i)]$ one word at a time (indicating that nodes whose opening parentheses lie in this region have a marked ancestor). If i has d_i descendants, then it holds close(i) − open(i) = $1 + 2d_i$, and thus the procedure takes $\mathcal{O}(1 + d_i/w)$ time. A node i is marked if and only if $A[\text{open}(i)]$ is set, and it has a marked ancestor if and only if $R[\text{open}(i)]$ is set (we do not consider a node to be its own ancestor). Both can be tested in constant time. Finding the nearest marked ancestor of i is more involved, and we explain it later.

When unmarking a node i, we unset the bits $A[\text{open}(i)]$ and $A[\text{close}(i)]$. If i currently has a marked ancestor, then there is no need to unset the range in R associated with i. Otherwise, we cannot simply unset the entire range $R[\text{open}(i) + 1..\text{close}(i)]$ because it may have also been set by descendants of i. Hence we have to leave segments corresponding to marked nodes untouched. Starting at position $k = \text{open}(i) + 1$, we scan $A[k..\text{close}(i)]$ from left to right and keep track of the excess of opening active parentheses, which is initially $e = 0$. We perform the scan in blocks of size $w' = \lfloor \log n/7 \rfloor$. Processing $A[k..k + w']$ works as follows. We scan the block from left to right. For each position $A[j]$ in the block, we first check if currently $e = 0$. If yes, then we unset bit $R[j]$. Afterwards, if $A[j] = 1$, we increment e if $B[j]$ is an opening parenthesis, and decrement e otherwise. Once we reach the end of the block, we increase k by w' and continue with the next block, until we reach position close(i). This way, we avoid unsetting parts of R that have to remain active. However, the procedure takes $\mathcal{O}(d_i)$ time, or $\mathcal{O}(w')$ time per block.

The processing of block $A[k..k + w')$ depends only on $A[k..k + w')$, $B[k..k + w')$, $R[k..k + w')$ and $\min(e, w')$ (if $e > w'$, then the excess cannot reach 0 while processing the block). Thus it depends on $3w' + \log w' \leq \log n/2$ bits of information, and in principle there are fewer than $2^{\log n/2} = \sqrt{n}$ distinct instances of the procedure. In a lookup table, we precompute for each possible $A[k..k+w')$, $B[k..k+w')$, $R[k..k+w')$, and $\min(e, w')$ the result of the procedure, i.e., the total increment or decrement that we have to apply to e, and the new value of $R[k..k + w')$. The lookup table has $\mathcal{O}(\sqrt{n})$ entries, and each of them can be computed naively in $\mathcal{O}(\text{polylog}(n))$ time. Using the table, an entire block $A[k..k + w')$ can be processed in constant time (and handling the last block that is possibly shorter than w' can be solved with additional lookup tables for each shorter block length). Thus, we can unmark a node in $\mathcal{O}(1 + d_i/w') = \mathcal{O}(1 + d_i/\log n)$ time.

We have already shown how to check if i has a marked ancestor in constant time. If we also want to output the nearest marked ancestor, then we start at position $o = \text{open}(i)$. Similarly to the technique for unmarking nodes, we now scan $A[1..o]$ and $B[1..o]$ from *right to left* and keep track of the excess of active *closing* parentheses. As soon as the excess becomes negative, we have found the opening parenthesis of the nearest marked ancestor. If this parenthesis is at position o', then the ancestor is $j = \text{prenum}(o')$. We can implement this procedure with lookup tables (similar to unmarking nodes), and thus it takes $\mathcal{O}(1 + d_j/\log n)$ time, where d_j is the number of descendants of j.   □

Resolving the Phrases. Now we are ready to resolve the infrequent phrases. We first build the phrase trie including only the infrequent phrases, and compute the mapping from phrases to preorder numbers. We also annotate each text position i with the preorder number corresponding to the longest infrequent phrase that is a prefix of $S[i..n]$ (this works just like in Sect. 4.2). We prepare the phrase trie for nearest marked ancestor queries with Lemma 1.

Now we scan S from right to left. For each text position i, we first try to resolve phrases, which we explain in a moment. After that, if i is the destination of a phrase f_k with preorder number p_k, we mark node p_k in the phrase trie (indicating that the phrase needs to be resolved). We also store $P[p_k] = k$ in an array of size at most z. This is necessary because the preorder numbers correspond to the *distinct* infrequent phrases, and thus the mapping from preorder numbers to phrases is not necessarily injective. Later, we resolve f_k by discovering that node p_k is marked, and we will then need to be able to lookup $k = P[p_k]$. Note that we never try to resolve two phrases with the same preorder number at the same time, since the one further to the left would have already resolved the other one.

For every text position i, if its annotation is q_i, we check if q_i has a marked ancestor. If this is the case, then we obtain the nearest marked ancestor p of q_i, which corresponds to phrase $f_{P[p]}$. By the construction of the phrase trie and the annotations of text positions, $f_{P[p]}$ is a prefix of $T[i..n]$. Since we have not unmarked the node yet, and due to the right-to-left processing order, it follows that i is the rightmost source of $f_{P[p]}$. We unmark node p.

Analyzing the Complexity. The preprocessing for the nearest marked ancestor structure takes $\mathcal{O}(z) + o(n)$ time and space. For each text position, annotated with q_i, we check if q_i has a marked ancestor in overall $\mathcal{O}(n)$ time. Whenever this is the case, we also find its nearest marked ancestor. However, we will then also immediately unmark the nearest marked ancestor, and thus the total time for finding marked ancestors is the same as the time for unmarking nodes, which is bounded by the time for marking them.

Now we analyze the total time for marking nodes. Let m be the number of nodes in the phrase trie (or equivalently the number of distinct infrequent phrases). We mark nodes $\mathcal{O}(z)$ times, and thus the total time is $\mathcal{O}(z)$ plus the sum of all the $\mathcal{O}(d_i / \log n)$ terms. For now, we assume that each node gets marked exactly once. Then the time is $\mathcal{O}(\frac{1}{\log n} \cdot \sum_{i=1}^m d_i)$. Let a_i denote the number of ancestors of a node i, and observe that $\sum_{i=1}^m d_i = \sum_{i=1}^m a_i$ (because in both sums each combination of descendant and ancestor contributes value 1 to the sum). If node i corresponds to a phrase f_k, then the number of ancestors of i is bounded by $a_i < |f_k|$, since each ancestor represents a phrase that is a proper prefix of f_k. Hence the time is $\mathcal{O}(\frac{1}{\log n} \cdot \sum_{i=1}^z |f_k|) = \mathcal{O}(n / \log n)$. We assumed that each node gets marked exactly once. Since we only consider infrequent phrases, each node gets marked $\mathcal{O}(\log n)$ times, and thus the time is $\mathcal{O}(n)$. This concludes the proof of Theorem 2(c).

4.4 Close Phrases

Given an LZ-like parsing $S = f_1 \ldots f_z$, we say that a phrase f_k with destination i is *close* if its rightmost source is j and $i - j = \mathcal{O}(\log n)$. We now show how to resolve all close phrases in $\mathcal{O}(n)$ time. Let $\gamma = \Theta(\log n)$. If a phrase at destination i is of length at least γ, then we can afford $\mathcal{O}(\log n)$ time to resolve it. We consider each $j \in [i-r, i)$ with $r = \mathcal{O}(\log n)$ as a potential source. Checking if j is a source of i takes constant time with an LCE data structure (e.g., [19]). Thus we can resolve all close phrases of length at least γ in $\mathcal{O}(n)$ time.

For the phrases of length less than γ, we extract copies of overlapping segments $s_0, \ldots, s_{\lfloor n/2\gamma \rfloor}$ where $\forall i \in [1, \lfloor n/2\gamma \rfloor] : s_i = S[1 + 2(i - 1)\gamma \ldots \min(2(i + 1)\gamma, n)]$. We modify each segment s_i by rank-reducing the alphabet of s_i to (a subset of) $[1, 4\gamma]$, which takes $\mathcal{O}(n)$ total time by radix sorting all segments in batch. Then, we offset the alphabets such that s_i is over alphabet $[1 + 4(i - 1)\gamma, 4i\gamma]$. We concatenate all segments s_i into $S' = s_0 s_1 \ldots s_{\lfloor n/2\gamma \rfloor}$.

Each phrase of length less than γ is fully contained in the right half of at least one segment (apart from possible phrases with destination in the first 2γ position of S, which we solve with the LCE data structure in $\mathcal{O}(\text{polylog}(n))$ time). We map each phrase of length less than γ to a corresponding destination in S' such that if the destination is within some segment s_j then the phrase is fully contained in the right half of s_j. This results in a subset of an LZ-like factorization of S'. Since the segments have disjoint alphabets, all phrases in the subset are infrequent an can be solved with Theorem 2(c). We only have to map the sources back to original text positions, which is easily done in linear time. Hence we have shown Theorem 2(d).

References

1. Amir, A., Landau, G.M., Ukkonen, E.: Online timestamped text indexing. Inf. Process. Lett. **82**(5), 253–259 (2002). https://doi.org/10.1016/S0020-0190(01)00275-7
2. Bannai, H., Funakoshi, M., Kurita, K., Nakashima, Y., Seto, K., Uno, T.: Optimal LZ-end parsing is hard. In: Proceedings of the 34th Annual Symposium on Combinatorial Pattern Matching (CPM 2023) (2023). https://doi.org/10.4230/LIPIcs.CPM.2023.3
3. Belazzougui, D., Puglisi, S.J.: Range predecessor and Lempel-Ziv parsing. In: Proceedings of the 27th Annual Symposium on Discrete Algorithms (SODA 2016), pp. 2053–2071. Arlington, VA, USA (2016). https://doi.org/10.1137/1.9781611974331.ch143
4. Bille, P., Cording, P.H., Fischer, J., Gørtz, I.L.: Lempel-Ziv compression in a sliding window. In: Proceedings of the 28th Annual Symposium on Combinatorial Pattern Matching (CPM 2017), pp. 15:1–15:11. Warsaw, Poland (2017). https://doi.org/10.4230/LIPIcs.CPM.2017.15
5. Chan, T.M., Tsakalidis, K.: Dynamic orthogonal range searching on the ram, revisited. J. Comput. Geom. **9**(2), 45–66 (2018). https://doi.org/10.20382/jocg.v9i2a5
6. Crochemore, M., Langiu, A., Mignosi, F.: The rightmost equal-cost position problem. In: Proceedings of the 2013 Data Compression Conference (DCC 2013), pp. 421–430. Snowbird, UT, USA (2013). https://doi.org/10.1109/DCC.2013.50
7. Crochemore, M., Rytter, W.: Efficient parallel algorithms to test square-freeness and factorize strings. Inf. Process. Lett. **38**(2), 57–60 (1991). https://doi.org/10.1016/0020-0190(91)90223-5
8. Ellert, J.: Sublinear time Lempel-Ziv (LZ77) factorization. In: Proceedings of the 30th International Symposium on String Processing and Information Retrieval (SPIRE 2023). Pisa, Italy (2023)
9. Farach, M., Muthukrishnan, S.: Optimal parallel dictionary matching and compression (extended abstract). In: Proceedings of the 7th Annual Symposium on Parallel Algorithms and Architectures (SPAA 1995), pp. 244–253. Santa Barbara, California, USA (1995). https://doi.org/10.1145/215399.215451
10. Ferragina, P., Nitto, I., Venturini, R.: On the bit-complexity of Lempel-Ziv compression. SIAM J. Comput. **42**(4), 1521–1541 (2013). https://doi.org/10.1137/120869511
11. Fischer, J., I, T., Köppl, D.: Lempel Ziv computation in small space (LZ-CISS). In: Cicalese, F., Porat, E., Vaccaro, U. (eds.) CPM 2015. LNCS, vol. 9133, pp. 172–184. Springer, Cham (2015). https://doi.org/10.1007/978-3-319-19929-0_15
12. Fischer, J., Tomohiro, I., Köppl, D., Sadakane, K.: Lempel-Ziv factorization powered by space efficient suffix trees. Algorithmica **80**(7), 2048–2081 (2018). https://doi.org/10.1007/s00453-017-0333-1
13. Fredkin, E.: Trie memory. Commun. ACM **3**(9), 490–499 (1960). https://doi.org/10.1145/367390.367400
14. Goto, K., Bannai, H.: Simpler and faster Lempel Ziv factorization. In: Proceedings of the 2013 Data Compression Conference (DCC 2013), pp. 133–142. Snowbird, UT, USA (2013). https://doi.org/10.1109/DCC.2013.21
15. Goto, K., Bannai, H.: Space efficient linear time Lempel-Ziv factorization for small alphabets. In: Proceedings of the 2014 Data Compression Conference (DCC 2014), pp. 163–172. Snowbird, UT, USA (2014). https://doi.org/10.1109/DCC.2014.62

16. Hagerup, T.: Sorting and searching on the word RAM. In: Morvan, M., Meinel, C., Krob, D. (eds.) STACS 1998. LNCS, vol. 1373, pp. 366–398. Springer, Heidelberg (1998). https://doi.org/10.1007/BFb0028575
17. Kärkkäinen, J., Kempa, D., Puglisi, S.J.: Lightweight Lempel-Ziv parsing. In: Bonifaci, V., Demetrescu, C., Marchetti-Spaccamela, A. (eds.) SEA 2013. LNCS, vol. 7933, pp. 139–150. Springer, Heidelberg (2013). https://doi.org/10.1007/978-3-642-38527-8_14
18. Kärkkäinen, J., Kempa, D., Puglisi, S.J.: Linear time Lempel-Ziv factorization: simple, fast, small. In: Fischer, J., Sanders, P. (eds.) CPM 2013. LNCS, vol. 7922, pp. 189–200. Springer, Heidelberg (2013). https://doi.org/10.1007/978-3-642-38905-4_19
19. Kempa, D., Kociumaka, T.: String synchronizing sets: sublinear-time BWT construction and optimal LCE data structure. In: Proceedings of the 51st Annual Symposium on Theory of Computing (STOC 2019), pp. 756–767. Phoenix, AZ, USA (2019). https://doi.org/10.1145/3313276.3316368
20. Kempa, D., Kosolobov, D.: LZ-end parsing in linear time. In: Proceedings of the 25th Annual European Symposium on Algorithms (ESA 2017), pp. 53:1–53:14. Vienna, Austria (2017). https://doi.org/10.4230/LIPIcs.ESA.2017.53
21. Kempa, D., Saha, B.: An upper bound and linear-space queries on the LZ-end parsing. In: Proceedings of the 33rd Annual Symposium on Discrete Algorithms (SODA 2022), pp. 2847–2866. Alexandria, VA, USA (Virtual Conference) (2022). https://doi.org/10.1137/1.9781611977073.111
22. Kosolobov, D.: Faster lightweight Lempel-Ziv parsing. In: Italiano, G.F., Pighizzini, G., Sannella, D.T. (eds.) MFCS 2015. LNCS, vol. 9235, pp. 432–444. Springer, Heidelberg (2015). https://doi.org/10.1007/978-3-662-48054-0_36
23. Kreft, S., Navarro, G.: LZ77-like compression with fast random access. In: Proceedings of the 2010 Data Compression Conference (DCC 2010), pp. 239–248. Snowbird, UT, USA (2010). https://doi.org/10.1109/DCC.2010.29
24. Kreft, S., Navarro, G.: On compressing and indexing repetitive sequences. Theor. Comput. Sci. **483**, 115–133 (2013). https://doi.org/10.1016/j.tcs.2012.02.006
25. Kärkkäinen, J., Kempa, D., Puglisi, S.J.: Lempel-Ziv parsing in external memory. In: Proceedings of the 2014 Data Compression Conference (DCC 2014), pp. 153–162. Snowbird, UT, USA (2014). https://doi.org/10.1109/DCC.2014.78
26. Köppl, D., Sadakane, K.: Lempel-Ziv computation in compressed space (LZ-CICS). In: Proceedings of the 2016 Data Compression Conference (DCC 2016), pp. 3–12. Snowbird, UT, USA (2016). https://doi.org/10.1109/DCC.2016.38
27. Larsson, N.J.: Most recent match queries in on-line suffix trees. In: Kulikov, A.S., Kuznetsov, S.O., Pevzner, P. (eds.) CPM 2014. LNCS, vol. 8486, pp. 252–261. Springer, Cham (2014). https://doi.org/10.1007/978-3-319-07566-2_26
28. Lempel, A., Ziv, J.: On the complexity of finite sequences. IEEE Trans. Inf. Theory **22**(1), 75–81 (1976). https://doi.org/10.1109/TIT.1976.1055501
29. Manber, U., Myers, E.W.: Suffix arrays: a new method for on-line string searches. SIAM J. Comput. **22**(5), 935–948 (1993). https://doi.org/10.1137/0222058
30. Naor, M.: String matching with preprocessing of text and pattern. In: Albert, J.L., Monien, B., Artalejo, M.R. (eds.) ICALP 1991. LNCS, vol. 510, pp. 739–750. Springer, Heidelberg (1991). https://doi.org/10.1007/3-540-54233-7_179
31. Navarro, G.: Compact Data Structures: A Practical Approach. Cambridge University Press, Cambridge (2016). https://doi.org/10.1017/CBO9781316588284
32. Ohlebusch, E., Gog, S.: Lempel-Ziv factorization revisited. In: Giancarlo, R., Manzini, G. (eds.) CPM 2011. LNCS, vol. 6661, pp. 15–26. Springer, Heidelberg (2011). https://doi.org/10.1007/978-3-642-21458-5_4

33. Okanohara, D., Sadakane, K.: An online algorithm for finding the longest previous factors. In: Halperin, D., Mehlhorn, K. (eds.) ESA 2008. LNCS, vol. 5193, pp. 696–707. Springer, Heidelberg (2008). https://doi.org/10.1007/978-3-540-87744-8_58

34. Shun, J.: Parallel Lempel-Ziv factorization, chap. 13. Association for Computing Machinery and Morgan & Claypool (2018). https://doi.org/10.1145/3018787.3018801

35. Shun, J., Zhao, F.: Practical parallel Lempel-Ziv factorization. In: Proceedings of the 2013 Data Compression Conference (DCC 2013). pp. 123–132. Snowbird, UT, USA (2013). https://doi.org/10.1109/DCC.2013.20

36. Starikovskaya, T.: Computing Lempel-Ziv factorization online. In: Rovan, B., Sassone, V., Widmayer, P. (eds.) MFCS 2012. LNCS, vol. 7464, pp. 789–799. Springer, Heidelberg (2012). https://doi.org/10.1007/978-3-642-32589-2_68

37. Storer, J.A., Szymanski, T.G.: Data compression via textual substitution. J. ACM **29**(4), 928–951 (1982). https://doi.org/10.1145/322344.322346

38. Weiner, P.: Linear pattern matching algorithms. In: Proceedings of the 14th Annual Symposium on Switching and Automata Theory (SWAT 1973), pp. 1–11. Iowa City, IA, USA (1973). https://doi.org/10.1109/SWAT.1973.13

39. Yamamoto, J., I, T., Bannai, H., Inenaga, S., Takeda, M.: Faster compact on-line Lempel-Ziv factorization. In: Proceedings of the 31st International Symposium on Theoretical Aspects of Computer Science (STACS 2014), pp. 675–686. Lyon, France (2014). https://doi.org/10.4230/LIPIcs.STACS.2014.675

Engineering a Textbook Approach to Index Massive String Dictionaries

Paolo Ferragina⬤, Mariagiovanna Rotundo⬤, and Giorgio Vinciguerra(✉)⬤

Department of Computer Science, University of Pisa, Pisa, Italy
{paolo.ferragina,giorgio.vinciguerra}@unipi.it,
m.rotundo1@studenti.unipi.it

Abstract. We study the problem of engineering space-time efficient indexes that support membership and lexicographic (rank) queries on *very* large static dictionaries of strings.

Our solution is based on a very simple approach that consists of decoupling string storage and string indexing by means of a blockwise compression of the sorted dictionary strings (to be stored in external memory) and a succinct implementation of a Patricia trie (to be stored in internal memory) built on the first string of each block.

Our experimental evaluation on two new datasets, which are at least one order of magnitude larger than the ones used in the literature, shows that (i) the state-of-the-art compressed string dictionaries (such as FST, PDT, CoCo-trie) do not provide significant benefits if used in an indexing setting compared to Patricia tries, and (ii) our two-level approach enables the indexing of 3.5 billion strings taking 273 GB in less than 200 MB of internal memory, which is available on any commodity machine, while still guaranteeing comparable or faster query performance than those offered by array-based solutions used in modern storage systems, such as RocksDB, thus possibly influencing their future designs.

Keywords: String dictionary problem · Trie data structure · String compression · Algorithm engineering · Key-value store

1 Introduction

The string dictionary problem is a classic one in the string-matching field. It is defined on a set S of n strings of variable length, drawn from an alphabet Σ. The goal is to build an indexing data structure on S that efficiently answers a *membership query* on any query string $q \in \Sigma^+$, namely: "does $q \in S$?" Sometimes, the data structure is required to answer a more powerful query, which finds the *lexicographic position* of q within the sorted set S (aka the *rank* of q in S). The attention to this operation is motivated by the fact that the implementation of several other operations on S—such as the *prefix search*, which finds all the strings in S prefixed by q, and the *range search*, which finds all the strings in S that fall in a given query range—boil down to solving it.

© The Author(s), under exclusive license to Springer Nature Switzerland AG 2023
F. M. Nardini et al. (Eds.): SPIRE 2023, LNCS 14240, pp. 203–217, 2023.
https://doi.org/10.1007/978-3-031-43980-3_16

In this paper, we assume that S is static, and thus it cannot be updated, but its total length N and number n of strings is so large that it has to be stored in slow storage, such as HDDs or SSDs. In fact, the recent explosion in the availability of massive string dictionaries in several applications—such as databases [33, 43], bioinformatic tools [11], search engines [28], code repositories [13], and string embeddings (see e.g. [27, 44]), just to name a few—has revitalised the interest in solving the problem in efficient time and space by taking into account the hierarchy of memory levels that are involved in their processing.

To solve the string dictionary problem, different approaches were proposed over the years in the literature. A trivial one consists of using an array of string pointers and deploying a binary search to answer queries, which causes random memory accesses and possibly I/Os. The classic one is the trie [22], a multiway tree that stores each string in S as a root-to-leaf path, and whose edges are labelled with either one character from Σ (the so-called uncompacted trie) or a substring from the strings in S (the so-called compacted trie). This historical solution has undergone over the years many significant developments that improved its query or space efficiency (see also [6] and refs therein) such as compacting subtries [6, 40], using adaptive representations for its nodes [2, 4, 31], succinct representations of its topology [25, 43], cache-aware or disk-based layouts [18, 21], and even replacing it with learned models [16].

Among the most recent and performing variants of tries, which are pertinent to our discussion, we mention: ART [31], CART [42], Path Decomposed Trie (PDT) [24], Fast Succinct Trie (FST) [43], ctrie++ [40], and CoCo-trie [6]. According to the experimental results published in [6], we know that ART, CART, and ctrie++ are space inefficient and offer query times on par with the other data structures, which is a strong limitation in the massive-dictionary context we consider in this paper. The other three proposals—namely, FST, PDT, and CoCo-trie—stand out as the most interesting ones because they offer the best space-time trade-offs. Nevertheless, they incur three main "limitations": they are very complex to be implemented; their code is highly engineered, and thus difficult to be maintained or adapted to different scenarios (e.g., rank operations, adding satellite information); and, finally, they are designed to compress and index the string dictionary entirely in internal memory. In this paper, we ask ourselves whether this "sophistication" is really needed in practice to achieve efficient time and space performance on massive string dictionaries.

Inspired by the theoretical proposals of [12, 17, 19, 21], our solution consists of decoupling string indexing and string storage, via a two-level approach [15]. The on-disk storage level compresses the sorted strings in S via rear coding [18] and partitions them into blocks of fixed size. The indexing level exploits a succinctly-encoded Patricia trie built on the first string of each block, so that it plays the role of a *router* for determining the block that possibly contains the query string q. Then, that block is fetched from the storage level and eventually scanned to search for the (lexicographic position of the) string q. Now, as long as the indexing level is small enough to fit in internal memory, we can solve the query in at most two disk I/Os without resorting to more complicated solutions [17,

21]. Additionally, as for LSM-trees [33,38], decoupling indexing from storage allows us to support some dictionary updates, thus making our proposed solution suitable to manage datasets with high insertion rates too.

To perform our massive-scale experiments, we first notice that datasets from previous evaluations [6,24,43] are inadequate because their size is at most about 7 GB and the number of strings is at most 114 million. We, therefore, increase these sizes by at least an order of magnitude via two new datasets, one consisting of URLs from various Web crawls (272 GB, 3.5 billion strings) [8], the other consisting of filenames of source code files from the Software Heritage initiative (69 GB, 2 billion strings) [32].

Our first experimental finding is that sophisticated compressed string dictionaries (i.e., FST, PDT, CoCo-trie) are too complex for the indexing level, and they do not provide substantial space-time performance advantage compared to our well-engineered succinct Patricia trie, which is also much faster to construct.

Then, we show that our overall two-level approach based on succinct Patricia tries enables the indexing of the largest dataset with only at most 195 MB of internal memory (at least $\approx 1400\times$ smaller than the dataset size). This small memory footprint allows dedicating much more memory to caching disk pages and this, in turn, determines a query efficiency that is comparable to or faster than the one offered by array-based solutions (which however take 5.2× more internal memory).

For these reasons, our two-level approach is a robust candidate for indexing massive string dictionaries, and it paves the way for further investigations and engineering, as we elaborate upon in the conclusions.

2 Background

A Patricia trie (PT) [36] for a string set S is derived from the trie of S by compacting each unary path into a single edge labelled with its first character, and by storing at each node the length of the (uncompacted) root-to-node path. Figure 1 shows an example of a PT built on a set of 8 strings.

Even if the PT strips out some information from the compacted trie, it is still able to support the search for the lexicographic position of a pattern $P[1,p]$ among a sorted sequence of strings, with the significant advantage (discussed below) that this search needs to access only one single string, and hence execute typically 1 I/O instead of the p I/Os potentially incurred by the traversal of the compacted trie due to accessing its (possibly long) edge labels. This algorithm is called *blind search* in the literature [15,17]. It is a little bit more complicated than prefix searching in classic tries, because of the presence of only one character per edge label. Technically speaking, blind search consists of three stages.

Stage 1: Downward traversal. Trace a downward path in the PT to locate a leaf l which points to one of the indexed strings sharing the longest common prefix (LCP) with P (see [17] for the proof). The traversal compares the characters of P with the single characters which label the traversed edges

until either a leaf is reached or no further branching is possible. In this last
case, we can choose l as any descendant leaf from the last traversed node; in
our implementation, we will take the leftmost one.

Stage 2: LCP computation. Compare P against the string s pointed to by
leaf l, in order to determine their LCP $\ell \geq 0$.

Stage 3: Upward traversal. Traverse upward the PT from l to determine the
edge $e = (u, v)$ where the mismatched character $s[\ell + 1]$ lies. If $s[\ell + 1]$ is a
branching character (and recall that $s[\ell + 1] \neq P[\ell + 1]$), then we determine
the lexicographic position of $P[\ell + 1]$ among the branching characters of u.
Say this is the ith child of u, the lexicographic position of P is therefore to
the immediate left of the subtree descending from this ith child. Otherwise,
the character $s[\ell + 1]$ lies within the edge e and after its first character, so the
lexicographic position of P is to the immediate right of the subtree descending
from edge e, if $P[\ell + 1] > s[\ell + 1]$, otherwise it is to the immediate left of that
subtree.

The topology of the PT can be represented in several different ways, like, for
example, using pointers or succinct encodings. Since we aim for space savings,
we will use the latter and, in particular, the Level-Order Unary Degree Sequence
(LOUDS) [25] and the Depth-First Unary Degree Sequence (DFUDS) [5]. Both
encode the trie topology with a bitvector in which a node of degree d is represented by the binary string $1^d 0$. The difference is the order in which the nodes
are visited and the corresponding binary strings are written in the bitvector: in
level-wise left-to-right for LOUDS, and in preorder for DFUDS. For our implementation of DFUDS, we follow [37] and prepend 110 to the representation. For
our implementation of LOUDS, we follow [43] and prepend no bits. See Fig. 1
for an example of LOUDS and DFUDS representation.

Regarding compressing a lexicographically-sorted set of strings, two simple
techniques are front coding [15,18] and rear coding [18]. Front coding represents each string with two values: an integer denoting the length of the LCP
between the considered string and the previous one, and the remaining suffix of
the considered string obtained by removing that LCP. If the string has not a
predecessor, the LCP length is set to 0. In rear coding, the suffix is obtained in
the same way as in front coding, but the integer represents the number of characters to remove from the previous string to obtain the longest common prefix.
Rear coding may be more efficient than front coding since it does not encode
the length of repeated prefixes [18,21].

3   Our Two-Level Approach

As anticipated in the Introduction, our string dictionary consists of two levels: a
storage level (residing on disk), which consists of a sequence of fixed-size blocks
where strings are stored in lexicographic order and compressed; and an indexing level (residing in internal memory), which consists of a succinctly-encoded
Patricia trie (PT) that indexes the first string of every block.

3.1   Storage Level

For the on-disk storage level, let us consider the sequence of lexicographically-sorted strings, and disk blocks of size 4, 8, 16, and 32 KiB. The first string of each block is stored explicitly (i.e., not compressed), whereas the subsequent strings are compressed with rear coding until the block is (almost) full, that is, it cannot host the subsequent rear-coded string s. In this case, the current block is padded with zeroes, and a new block is started by setting its first string to s. The lengths in rear coding are stored with a variable-byte encoder to keep byte alignment, and thus speed up string decompression.

Since the blocks are of fixed size, the indexing level just needs to return the rank of the block containing the query string, which is then multiplied by the block size to get the byte offset of that block on disk.

To efficiently compute the rank of the query string q in S, we store for each block b an integer indicating how many dictionary strings appear before it in the lexicographic order, denoted with $c(b)$. This way, let $\hat{b}$ be the disk block containing the lexicographic position of the query string q: the rank of q is then computed by summing $c(\hat{b})$ with the *relative* rank of q among the strings in $\hat{b}$. The latter value is obtained via a linear scan and decompression of the block $\hat{b}$, which takes advantage of rear coding and LCP length information to possibly skip some characters, as detailed in [34, §6]. For simplicity, we store the integers $c(b)$ in an in-memory packed array that allocates a number of bits per element sufficient to contain the largest one. It goes without saying that, since these integers are increasing, one could save some further space by using a randomly-accessible compressed integer dictionary (see e.g. [7,20] and references therein), but this is deferred to subsequent studies.

Clearly, one can apply other compression techniques on top of or in place of rear coding, such as entropy coding, grammar compression, and dictionary compression. These techniques have been shown to be useful to reduce the space of in-memory string dictionaries [3,9,10,30,34], but since we are dealing with strings kept in (the much cheaper, but slower) secondary storage, we opt for the simplicity of rear coding, which is shown next to be already very effective in our context. In fact, even for datasets of billions of strings, the number of created blocks (and thus "first strings" to be indexed in memory) is sufficiently small that the indexing level (i.e., the succinct PT) fits in a few MBs (e.g., up to 195 MB for a dictionary of 273 GB, see Sect. 4). We finally mention that, compared to the approach of creating variable-sized blocks with a fixed number of (front- or rear-coded) strings [30,34], our use of fixed-size blocks allows for better compression because it may take more advantage of runs of consecutive strings sharing long common prefixes, which thus result highly compressible in one single block.

The storage level is accessed by memory-mapping the corresponding file (via the mmap system call), which compared to explicit reads of disk blocks allows a simpler implementation and often faster performance [39].

3.2   Indexing Level

We succinctly encode the Patricia Trie (PT), forming the indexing level, by considering one of two succinct representations of its topology, i.e. LOUDS or DFUDS, and using two additional sequences: one for the single characters labelling the edges of the PT, and the other for the root-to-node path lengths. Both sequences are stored as packed arrays whose elements are ordered according to the topology representation, thus in level-wise order for LOUDS and in preorder for DFUDS. To reduce the number of bits needed to store the lengths, we consider the length of the edge that leads to a node and not the one of the whole root-to-node path, which can be easily recovered by summing the lengths of the visited nodes during the downward traversal (see Sect. 2).

If LOUDS is used, we need one more sequence that maps each leaf in the level-wise ordering to the lexicographic rank of the corresponding string, which we need to jump to the corresponding block in the storage level. If DFUDS is used, such a sequence is not needed since the leaves are ordered according to the lexicographic rank of the corresponding strings. Figure 1 shows an example of the sequences created for the encoding of a PT.

Downward Traversal with LOUDS. To downward traverse the PT encoded with LOUDS, rank and select primitives are used: $rank_b(i)$ counts the number of bits equal to b up to position i, while $select_b(i)$ finds the position of the ith bit equal to b. Assuming that the nodes, their children, and the bits of the binary sequences are counted starting from 0, it is well known [25,37,43] that we can traverse the trie downwards by computing the position of the kth child

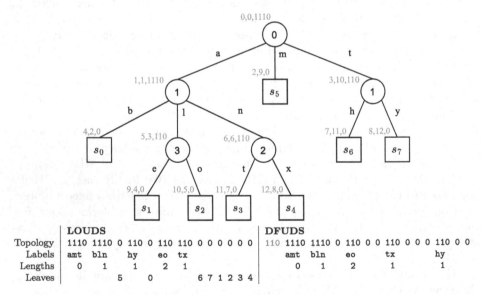

Fig. 1. At the top, the Patricia Trie on the strings {*abduct, algebra, algorithm, ant, anxiety, machine, three, typo*} corresponding to the leaves $s_0, \ldots, s_7$. Outside each node, we denote its position in the LOUDS order, in the DFUDS order, and its degree in unary, respectively. At the bottom, the corresponding succinct representations.

of the node that starts at position p with the formula $select_0(rank_1(p+k))+1$. Actually, it is not hard to show that we do not need $rank_1$, because its result can be computed with proper arithmetic operations during the traversal. This fact allows in practice to save space, because we discard the auxiliary data structure needed for constant-time $rank_1$ operations, and to save time, because several CPU cycles and possibly cache misses are needed for $rank_1$.

Fact 1. *The downward traversal of a Patricia trie encoded with LOUDS can be executed with just $select_0$ operations.*

When a leaf is reached, we compute its rank in the leaf sequence by counting how many leaves appear before its position x in the LOUDS representation of the PT. This rank is given by $rank_0(x) - rank_{10}(x)$, where the first value denotes the number of nodes (internal and leaves) that appear in LOUDS before the considered one, and the second value denotes the number of internal nodes (not leaves) that appear before position x. Now we notice that the value $rank_0(x) = x - rank_1(x) + 1$ can be computed by substituting $rank_1(x)$ with the value returned by the arithmetic operations executed during the downward traversal.

Thus, we build overall just the $select_0$ and $rank_{10}$ data structures on the LOUDS sequence (due to their time efficiency [29], we use the **sux** library [41] for the former, and the **sdsl** library [23] for the latter).

Downward Traversal with DFUDS. To downward traverse the PT encoded with DFUDS, we compute the position of the kth child of the node whose encoding starts at position p with the formula $close(succ_0(p) - (k+1)) + 1$ [37]. Here, $succ_0(p)$ returns the position of the first 0 that follows p in the DFUDS sequence, and it is implemented by using a linear scan starting from the position p until a 0 is found. Since DFUDS can be seen as a sequence of balanced parenthesis, we have that if i is the position of an open parenthesis, $close(i)$ returns the position of the corresponding close one. For $close$ we adopt the **sdsl::bp_support_sada** implementation of balanced parenthesis.

When a leaf is reached, we compute its rank among the leaves with a $rank_1$ and $rank_{10}$ operation. By knowing the position where the leaf starts, the $rank_1$ allows us to derive the number of nodes that appear in the sequence before it, while the $rank_{10}$, as for LOUDS above, allows us to compute how many of these nodes are internal nodes, thus by exploiting the results of these operations we get the rank of the leaf. Therefore, in our implementation of DFUDS, we exploit data structures that allow us to execute in constant time operations of $rank_{10}$, $close$, and $rank_1$ (these last two ones are included in **sdsl::bp_support_sada**).

Upward Traversal in LOUDS and DFUDS. For the upward traversal of a PT (either encoded with LOUDS or DFUDS), we need to scan back the nodes accessed during the downward traversal. But, instead of executing any of the bit-operations above (as typically done for the upward traversal of trees [25]), we adopt a much simpler and time-efficient approach that pushes in a stack the LOUDS/DFUDS positions of the nodes visited during the downward traversal, and then it pops them from the stack during the upward traversal.

4   Experiments

Experimental Setting. We use a machine with a KIOXIA KPM61RUG960G SSD and two NUMA nodes, each with a 1.80 GHz Intel Xeon E5-2650L v3 CPU and 30 GB local DDR4 RAM. The machine runs Ubuntu 20.04.4 LTS with Linux 5.4.0, and the compiler is GCC 9.4.0. We schedule experiments on a single node via numactl. For the mmap in the storage level, we tested both the MAP_SHARED and MAP_PRIVATE flags and noticed no significant performance difference (indeed, the storage level is read-only), so we choose the former. The MAP_POPULATE flag too did not impact the query performance, so we do not set it. We alternate datasets given to mmap to try to prevent caching by the operating system. Our source code is available at https://github.com/MariagiovannaRotundo/Two-level-indexing.

Datasets. Datasets used in previous experimental evaluations of state-of-the-art solutions (i.e., FST [43], PDT [24], and CoCo-trie [6]) are quite small. Their size is indeed no more than 0.5 GB and 25M strings for FST, 2.7 GB and 40.5M strings for the CoCo-trie, and 7.1 GB and 114.3M strings for PDT.

   Since we want to evaluate our solution on big datasets, we introduce two new ones. The first, *URLs*, combines web page addresses from various crawls [8], has a size of 272.7 GB, and contains 3.7 billion strings. The second, *Filenames*, consists of the name of source code files collected by Software Heritage [1,13,14,32], has a size of around 68.9 GB, and contains 2.3 billion strings. So our datasets are larger than the ones used in previous evaluations by up to 32.0× in number of strings and up to 38.4× in size. Also, we point out that our datasets are up to one order of magnitude larger than the internal memory of our machine, described above.

   About the features of the new datasets, we briefly report that URLs contains long strings (avg. 73.6, max. 2083) with long LCPs among them (avg. 53.7), on a medium-size alphabet (88 characters); whereas Filenames offers the opposite features, namely shorter string (avg. 29.1, max. 16051) with even shorter LCPs among them (15.4), on a large alphabet (241 characters).

Competitors. For the indexing level, we consider the set S', composed of the first string of every block truncated at its minimum distinguishing prefix, to construct an in-memory index, and then we discard S'. As the index, other than our PT-LOUDS and PT-DFUDS implementations, we consider FST [43], PDT [24], CoCo-trie [6], and a simple and commonly-used solution [34,35]—that we name Array—which stores S' contiguously in an array and binary searches on it via an auxiliary packed array of offsets to the beginning of the strings. Notice that, for all solutions, the truncation of strings in S' saves space in the resulting index and still allows identifying the correct block in the storage level (actually, upon accessing the first string of a block we might find that the sought string is in the preceding block, which nonetheless is likely to be loaded quickly thanks to disk prefetching). On the other hand, PT does not store the distinguishing prefixes but only $\Theta(|S'|)$ characters/edges/nodes, thus occupying a space that is independent of the string lengths. We also anticipate that all these implementations of the indexing level allow us to fit it in the internal memory of our machine and thus solve a query with at most two random I/Os to the storage level.

In what follows, we first evaluate in Sect. 4.1 the different data structures for the indexing level in isolation, i.e. without considering the access to the storage level that concludes the query. Then, in Sect. 4.2 we evaluate the performance of the overall two-level approach.

4.1 Indexing Level Evaluation

Construction Time. Figure 2 shows the time to construct the various data structures from the set S' loaded in memory. CoCo-trie is constructed only on URLs because the current implementation [6] supports only ASCII alphabets. Moreover, we point out that its construction time for blocks of 4 and 8 KiB is not shown due to its high-memory consumption that required a machine with a much larger internal memory and thus different performance (still, we constructed these CoCo-tries because we test their search time in Fig. 3).

Unsurprisingly, Array has the fastest construction because it involves just strings and offsets storage. Our PT-LOUDS and PT-DFUDS implementations have the second-fastest construction, which is based on scanning prefixes at increasing lengths of (ranges) of strings, determining sub-ranges corresponding to deeper levels of the PT, and handling these sub-ranges recursively in LOUDS order or DFUDS order. Finally, we notice that FST, PDT, and CoCo-trie are significantly slower to construct than our PT, up to 7×, 5×, 42×, respectively.

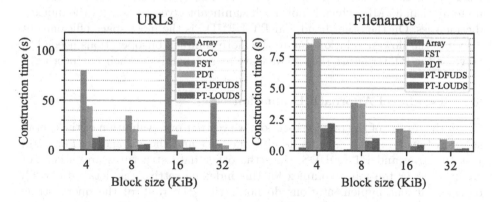

Fig. 2. Times needed to construct each data structure in the indexing level.

Space-Time Performance. Figure 3 shows the performance of data structures for the indexing level. The query time refers to the average time needed to perform a membership query on a sample of 10% strings drawn from the set of distinguishing prefixes S', without any access to the storage level. In particular,

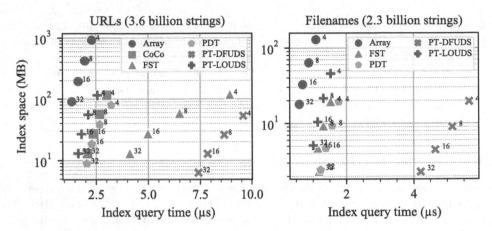

Fig. 3. Space and average query time of different data structures for the indexing level.

for PT, since such access is needed for Stage 2 of the blind search (cf. Sect. 2), the time is evaluated by executing a downward and an upward traversal.

The results show that Array is the fastest but also the most space-hungry solution. FST is competitive only for the Filenames dataset due to its shorter strings. Our PT approaches, despite their simplicity, are very competitive and on the Pareto space-time frontier of both experimented datasets. In particular, PT-LOUDS is the second-fastest data structure with a space occupancy that is competitive with that of the most sophisticated solutions such as CoCo and PDT. We notice in fact that the difference in space with those data structures is no more than 35 MB, which is not much significant given the size of the indexed dictionaries. On the other hand, our PT-DFUDS is the most space efficient but also it is the slowest solution due to the more complex bit-operations needed to traverse the PT structure (hence, we leave as an open issue their engineering).

4.2   Two-Level Approach Evaluation

Given the results of the previous section, we restrict our evaluation of the overall solution (involving the indexing level in memory and the storage level on disk) just to Array and PT-LOUDS, since the other data structures are either not competitive or too much complex for this indexing setting (as detailed above), or their current implementations do not return the rank of the query string among the indexed ones, being this a crucial information to jump to the correct disk block. We mention here that returning the rank of the query string in the LOUDS-based FST requires adding an integer for each leaf (as we did with our PT-LOUDS, cf. Fig. 1), thus increasing the space of FST, or it requires switching to the much slower DFUDS representation, thus increasing the query time. On the other hand, returning the rank of a query string in PDT requires more complex trie traversals thus increasing the query time. So Fig. 3 underestimates the space-time performance of FST or PDT when they are used in the two-level setting, which justifies our choice of experimenting below just with Array and PT-LOUDS (henceforth referred to simply as PT).

The following paragraphs discuss the experimental results reported in Fig. 4. Note that, as stated in Sect. 3.1, we need to keep in memory the array of integers $c(b)$ to answer rank queries on the indexed strings (which is why the index space in Fig. 4 is larger than the one reported in Fig. 3).

Storage Level Size. We begin by reporting that our storage level with blocks of size 4–32 KiB compresses the URLs dataset to 80.5–82.2 GB, and the Filenames dataset to 35.9–36.1 GB. Therefore, our approach to the blocked-compressed storage of dictionary strings achieves a compression factor of up to 3.4× on URLs, and up to 1.9× on Filenames, which is an interesting achievement given the simplicity of rear coding.

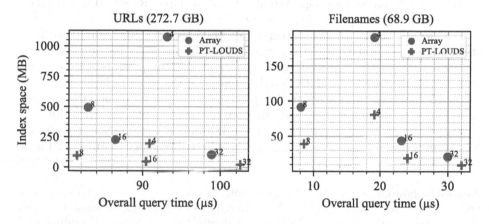

Fig. 4. Space and average query time of our two-level approach.

Space-Time Performance. Figure 4 shows that the PT and Array configurations with 8 KiB blocks are the fastest solutions overall. In particular, PT is faster on URLs and Array on Filenames (although PT is very close), but PT takes 5.2× less memory than Array on URLs, and 2.3× less on Filenames.

For increasing block sizes from 8 to 32 KiB, both solutions with PT and Array get from 1.3× to 3.7× slower, because of the larger block to scan and decompress, but more space efficient. Notably, as the block size halves, PT scales better in memory consumption compared to Array, because its space does not depend on the length of the strings but just on their number (as already observed above).

Interestingly enough, the PT and Array configurations with 4 KiB blocks are dominated by the corresponding ones with 8 KiB blocks. This occurs because the indexing level takes more space and thus there is less memory available for caching disk pages, hence making page faults more frequent, as we have verified with the `mincore` system call. The more space available for caching explains also why PT is not slowed down by the execution of one more random I/O compared to Array because of Stage 2 of the blind search (c.f. Sect. 2).

5   Conclusions and Future Work

Our two-level approach based on a succinct Patricia trie is a robust candidate for indexing massive string dictionaries. As we showed above, it enables indexing up to 272.7 GB with less than 195 MB of internal memory (a space at least 1396.3× smaller than the dictionary's size). This small memory footprint allows dedicating much more memory to caching disk pages and this, in turn, determines a query efficiency that is comparable to or faster than the one offered by Array-based solutions (which take 5.2× more memory). We believe these findings are significant not only for static dictionaries but also for dynamic ones that occur in the design of modern storage systems. As an example, RocksDB [35] is based on (static) runs of strings with in-memory Array-based indexes.

As future work, other than investigating the impact of our findings on these storage systems we suggest: for the indexing level, combining Patricia tries with dynamic succinct tree representations [26] or proper compressors for node fanouts (similarly to FST and CoCo-trie); and, for the storage level, designing solutions that take into account the query distribution to reduce the average time for block decompression/scan, or that use more sophisticated techniques on top of rear coding (such as dictionary and grammar compression) to improve block compression thus further reducing the internal-memory footprint of Patricia tries.

Acknowledgements. We thank Antonio Boffa for executing some tests on the CoCo-trie, and the Green Data Centre at the University of Pisa for machines and technical support. We also thank Roberto Di Cosmo, Valentin Lorentz, Stefano Zacchiroli, and the Software Heritage team for providing us with the Filenames dataset. This work was made possible by Software Heritage, the great library of source code: https://www.softwareheritage.org.

This work has been supported by the European Union – Horizon 2020 Program under the scheme "INFRAIA-01-2018-2019 – Integrating Activities for Advanced Communities", Grant Agreement n. 871042, "SoBigData++: European Integrated Infrastructure for Social Mining and Big Data Analytics" http://www.sobigdata.eu, by the NextGenerationEU – National Recovery and Resilience Plan (Piano Nazionale di Ripresa e Resilienza, PNRR) – Project: "SoBigData.it - Strengthening the Italian RI for Social Mining and Big Data Analytics" – Prot. IR0000013 – Avviso n. 3264 del 28/12/2021, by the spoke "FutureHPC & BigData" of the ICSC – Centro Nazionale di Ricerca in High-Performance Computing, Big Data and Quantum Computing funded by European Union – NextGenerationEU – PNRR, by the Italian Ministry of University and Research "Progetti di Rilevante Interesse Nazionale" project: "Multicriteria data structures and algorithms" (grant n. 2017WR7SHH).

References

1. Abramatic, J., Di Cosmo, R., Zacchiroli, S.: Building the universal archive of source code. Commun. ACM **61**(10), 29–31 (2018). https://doi.org/10.1145/3183558
2. Acharya, A., Zhu, H., Shen, K.: Adaptive algorithms for cache-efficient trie search. In: Goodrich, M.T., McGeoch, C.C. (eds.) ALENEX 1999. LNCS, vol. 1619, pp. 300–315. Springer, Heidelberg (1999). https://doi.org/10.1007/3-540-48518-X_18
3. Arz, J., Fischer, J.: LZ-compressed string dictionaries. In: Proceedings of the 24th Data Compression Conference (DCC), pp. 322–331 (2014). https://doi.org/10.1109/DCC.2014.36
4. Baskins, D.: A 10-minute description of how Judy arrays work and why they are so fast (2002). http://judy.sourceforge.net/doc/10minutes.htm
5. Benoit, D., Demaine, E.D., Munro, J.I., Raman, R., Raman, V., Rao, S.S.: Representing trees of higher degree. Algorithmica **43**(4), 275–292 (2005). https://doi.org/10.1007/s00453-004-1146-6
6. Boffa, A., Ferragina, P., Tosoni, F., Vinciguerra, G.: Compressed string dictionaries via data-aware subtrie compaction. In: Arroyuelo, D., Poblete, B. (eds.) SPIRE 2022. LNCS, vol. 13617, pp. 233–249. Springer, Cham (2022). https://doi.org/10.1007/978-3-031-20643-6_17. Implementation available at https://github.com/aboffa/CoCo-trie
7. Boffa, A., Ferragina, P., Vinciguerra, G.: A learned approach to design compressed rank/select data structures. ACM Trans. Algorithms **18**(3) (2022). https://doi.org/10.1145/3524060
8. Boldi, P., Marino, A., Santini, M., Vigna, S.: BUbiNG: massive crawling for the masses. ACM Trans. Web **12**(2), 12:1–12:26 (2018). https://doi.org/10.1145/3160017. Datasets of URLs available at https://law.di.unimi.it/datasets.php
9. Boncz, P., Neumann, T., Leis, V.: FSST: fast random access string compression. PVLDB **13**(12), 2649–2661 (2020). https://doi.org/10.14778/3407790.3407851
10. Brisaboa, N.R., Cerdeira-Pena, A., de Bernardo, G., Navarro, G.: Improved compressed string dictionaries. In: Proceedings of the 28th ACM International Conference on Information and Knowledge Management (CIKM), pp. 29–38 (2019). https://doi.org/10.1145/3357384.3357972
11. Chikhi, R., Holub, J., Medvedev, P.: Data structures to represent a set of k-long DNA sequences. ACM Comput. Surv. **54**(1) (2021). https://doi.org/10.1145/3445967
12. Clark, J.L.: PATRICIA-II. Two-level overlaid indexes for large libraries. Int. J. Parallel Program. **2**(4), 269–292 (1973). https://doi.org/10.1007/BF00985662
13. Di Cosmo, R.: Should we preserve the world's software history, and can we? In: Silvello, G., et al. (eds.) TPDL 2022. LNCS, vol. 13541, pp. 3–7. Springer, Cham (2022). https://doi.org/10.1007/978-3-031-16802-4_1
14. Di Cosmo, R., Zacchiroli, S.: Software Heritage: why and how to preserve software source code. In: Proceedings of the 14th International Conference on Digital Preservation (iPRES) (2017). https://hdl.handle.net/11353/10.931064
15. Ferragina, P.: Pearls of Algorithm Engineering. Cambridge University Press (2023). https://doi.org/10.1017/9781009128933
16. Ferragina, P., Frasca, M., Marinò, G.C., Vinciguerra, G.: On nonlinear learned string indexing. IEEE Access **11**, 74021–74034 (2023). https://doi.org/10.1109/ACCESS.2023.3295434
17. Ferragina, P., Grossi, R.: The string B-tree: a new data structure for string search in external memory and its applications. J. ACM **46**(2), 236–280 (1999). https://doi.org/10.1145/301970.301973

18. Ferragina, P., Grossi, R., Gupta, A., Shah, R., Vitter, J.S.: On searching compressed string collections cache-obliviously. In: Proceedings of the 27th ACM Symposium on Principles of Database Systems (PODS), pp. 181–190 (2008). https://doi.org/10.1145/1376916.1376943
19. Ferragina, P., Luccio, F.: String search in coarse-grained parallel computers. Algorithmica **24**(3–4), 177–194 (1999). https://doi.org/10.1007/PL00008259
20. Ferragina, P., Manzini, G., Vinciguerra, G.: Compressing and querying integer dictionaries under linearities and repetitions. IEEE Access **10**, 118831–118848 (2022). https://doi.org/10.1109/ACCESS.2022.3221520
21. Ferragina, P., Venturini, R.: Compressed cache-oblivious string B-tree. ACM Trans. Algorithms **12**(4), 52:1–52:17 (2016). https://doi.org/10.1145/2903141
22. Fredkin, E.: Trie memory. Commun. ACM **3**(9), 490–499 (1960). https://doi.org/10.1145/367390.367400
23. Gog, S., Beller, T., Moffat, A., Petri, M.: From theory to practice: plug and play with succinct data structures. In: Gudmundsson, J., Katajainen, J. (eds.) SEA 2014. LNCS, vol. 8504, pp. 326–337. Springer, Cham (2014). https://doi.org/10.1007/978-3-319-07959-2_28
24. Grossi, R., Ottaviano, G.: Fast compressed tries through path decompositions. ACM J. Exp. Algorithmics **19** (2015). https://doi.org/10.1145/2656332. Implementation available at https://github.com/ot/path_decomposed_tries
25. Jacobson, G.: Space-efficient static trees and graphs. In: Proceedings of the 30th IEEE Symposium on Foundations of Computer Science (FOCS), pp. 549–554 (1989). https://doi.org/10.1109/SFCS.1989.63533
26. Joannou, S., Raman, R.: Dynamizing succinct tree representations. In: Klasing, R. (ed.) SEA 2012. LNCS, vol. 7276, pp. 224–235. Springer, Heidelberg (2012). https://doi.org/10.1007/978-3-642-30850-5_20
27. Joulin, A., Grave, E., Bojanowski, P., Douze, M., Jégou, H., Mikolov, T.: Fasttext.zip: compressing text classification models. CoRR abs/1612.03651 (2016). http://arxiv.org/abs/1612.03651
28. Krishnan, U., Moffat, A., Zobel, J.: A taxonomy of query auto completion modes. In: Proceedings of the 22nd Australasian Document Computing Symposium (ADCS) (2017). https://doi.org/10.1145/3166072.3166081
29. Kurpicz, F.: Engineering compact data structures for rank and select queries on bit vectors. In: Arroyuelo, D., Poblete, B. (eds.) SPIRE 2022. LNCS, vol. 13617, pp. 257–272. Springer, Cham (2022). https://doi.org/10.1007/978-3-031-20643-6_19
30. Lasch, R., Oukid, I., Dementiev, R., May, N., Demirsoy, S.S., Sattler, K.: Fast & strong: the case of compressed string dictionaries on modern CPUs. In: Proceedings of the 15th International Workshop on Data Management on New Hardware (DaMoN), pp. 4:1–4:10 (2019). https://doi.org/10.1145/3329785.3329924
31. Leis, V., Kemper, A., Neumann, T.: The adaptive radix tree: ARTful indexing for main-memory databases. In: Proceedings of the 29th IEEE International Conference on Data Engineering (ICDE), pp. 38–49 (2013). https://doi.org/10.1109/ICDE.2013.6544812
32. Lorentz, V., Di Cosmo, R., Zacchiroli, S.: The popular content filenames dataset: deriving most likely filenames from the Software Heritage archive. Technical report (2023). https://inria.hal.science/hal-04171177, preprint
33. Luo, C., Carey, M.J.: LSM-based storage techniques: a survey. VLDB J. **29**(1), 393–418 (2019). https://doi.org/10.1007/s00778-019-00555-y
34. Martínez-Prieto, M.A., Brisaboa, N.R., Cánovas, R., Claude, F., Navarro, G.: Practical compressed string dictionaries. Inf. Syst. **56**, 73–108 (2016). https://doi.org/10.1016/j.is.2015.08.008

35. Meta Platforms Inc.: RocksDB. https://rocksdb.org/
36. Morrison, D.R.: PATRICIA—practical algorithm to retrieve information coded in alphanumeric. J. ACM **15**(4), 514–534 (1968). https://doi.org/10.1145/321479. 321481
37. Navarro, G.: Compact Data Structures: A Practical Approach. Cambridge University Press (2016). https://doi.org/10.1017/CBO9781316588284
38. O'Neil, P.E., Cheng, E., Gawlick, D., O'Neil, E.J.: The log-structured merge-tree (LSM-tree). Acta Informatica **33**(4), 351–385 (1996). https://doi.org/10.1007/s002360050048
39. Silberschatz, A., Galvin, P.B., Gagne, G.: Operating System Concepts, 10th edn. Wiley, Hoboken (2018)
40. Tsuruta, K., et al.: C-trie++: a dynamic trie tailored for fast prefix searches. Inf. Comput. **285**, 104794 (2022). https://doi.org/10.1016/j.ic.2021.104794
41. Vigna, S.: Broadword implementation of rank/select queries. In: McGeoch, C.C. (ed.) WEA 2008. LNCS, vol. 5038, pp. 154–168. Springer, Heidelberg (2008). https://doi.org/10.1007/978-3-540-68552-4_12
42. Zhang, H., Andersen, D.G., Pavlo, A., Kaminsky, M., Ma, L., Shen, R.: Reducing the storage overhead of main-memory OLTP databases with hybrid indexes. In: Proceedings of the ACM International Conference on Management of Data (SIGMOD), pp. 1567–1581 (2016). https://doi.org/10.1145/2882903.2915222
43. Zhang, H., et al.: Succinct range filters. ACM Trans. Database Syst. **45**(2) (2020). https://doi.org/10.1145/3375660. Fork of the implementation available at https://github.com/kampersanda/fast_succinct_trie
44. Zhang, W., et al.: TernaryBERT: distillation-aware ultra-low bit BERT. In: Proceedings of the 2020 Conference on Empirical Methods in Natural Language Processing (EMNLP), pp. 509–521 (2020). https://doi.org/10.18653/v1/2020.emnlp-main.37

Count-Min Sketch with Variable Number of Hash Functions: An Experimental Study

Éric Fusy and Gregory Kucherov[(✉)][iD]

LIGM, CNRS, Univ. Gustave Eiffel, Marne-la-Vallée, France
{Eric.Fusy,Gregory.Kucherov}@univ-eiffel.fr

Abstract. Conservative Count-Min, a stronger version of the popular Count-Min sketch [Cormode, Muthukrishnan 2005], is an online-maintained hashing-based sketch summarizing element frequency information of a stream. Although several works attempted to analyze the error of conservative Count-Min, its behavior remains poorly understood. In [Fusy, Kucherov 2022], we demonstrated that under the uniform distribution of input elements, the error of conservative Count-Min follows two distinct regimes depending on its load factor.

In this work, we present a series of results providing new insights into the behavior of conservative Count-Min. Our contribution is twofold. On one hand, we provide a detailed experimental analysis of Count-Min sketch in different regimes and under several representative probability distributions of input elements. On the other hand, we demonstrate improvements that can be made by assigning a variable number of hash functions to different elements. This includes, in particular, reduced space of the data structure while still supporting a small error.

1 Introduction

In most general terms, *Count-Min sketch* is a data structure for representing an associative array of numbers indexed by elements (keys) drawn from a large universe, where the array is provided through a stream of (key, value) updates so that the current value associated to a key is the sum of all previous updates of this key. Perhaps the most common setting for applying Count-Min, that we focus on in this paper, is the *counting* setting where all update values are +1. In this case, the value of a key is its *count* telling how many times this key has appeared in the stream. In other words, Count-Min can be seen as representing a *multiset*, that is a mapping of a subset of keys to non-negative integers. With this latter interpretation in mind, each update will be called *insertion*. The main supported query of Count-Min is retrieving the count of a given key, and the returned estimate may not be exact, but can only overestimate the true count.

The counting version of Count-Min is applied to different practical problems related to data stream mining and data summarization. One example is tracking frequent items (*heavy hitters*) in streams [7,11,23]. It occurs in network traffic monitoring [17], optimization of cache usage [16]. It also occurs in non-streaming big data applications, e.g. in bioinformatics [1,25,29].

F. M. Nardini et al. (Eds.): SPIRE 2023, LNCS 14240, pp. 218–232, 2023.
https://doi.org/10.1007/978-3-031-43980-3_17

Count-Min relies on hash functions but, unlike classic hash tables, does not store elements but only count information (hence the term *sketch*). It was proposed in [12], however a very similar data structure was proposed earlier in [9] under the name *Spectral Bloom filter*. The latter, in turn, is closely related to *Counting Bloom filters* [19]. In this work, we adopt the definition of [9] but still call it Count-Min to be consistent with the name commonly adopted in the literature. A survey on Count-Min can be found e.g. in [10].

In this paper, we study a stronger version of Count-Min called *conservative*. This modification of Count-Min was introduced in [17] under the name *conservative update*, see [10]. It was also discussed in [9] under the name *minimal increase*. Conservative Count-Min provides strictly tighter count estimates using the same memory and thus strictly outperforms the original version. The price to pay is the impossibility to deal with deletions (negative updates), whereas the original Count-Min can handle deletions as well, provided that the cumulative counts remain non-negative (condition known as *strict turnstile model* [23]).

Analysis of error of conservative Count-Min is a difficult problem having direct consequences on practical applications. Below in Sect. 2.2 we survey known related results in more details. In our previous work [21], we approached this problem through the relationship with *random hypergraphs*. We proved, in particular, that if the elements represented in the data structure are uniformly distributed in the input, the error follows two different regimes depending on the *peelability* property of the underlying *hash hypergraph*. While properties of random hypergraphs have been known to be crucially related to some data structures (see Sect. 2.3), this had not been known for Count-Min.

Starting out from these results, in this paper we extend and strengthen this analysis in several ways, providing experimental demonstrations in support of our claims. Our first goal is to provide a fine analysis of the "anatomy" of conservative Count-Min, describing its behavior in different regimes. Our main novel contribution is the demonstration that assigning different number of hash functions to different elements can significantly improve the error, and, as a consequence, lead to memory saving. Another major extension concerns the probability distribution of input elements: here we study non-uniform distributions as well, in particular step distribution and Zipf's distribution, and analyze the behavior of Count-Min for these distributions. This analysis is important not only because non-uniform distributions commonly occur in practice, but also because this provides important insights for the *heavy hitters* problem [7,11,23]). In particular, we consider the "small memory regime" (*supercritical*, in our terminology) when the number of distinct represented elements is considerably larger than the size of the data structure, and analyse conditions under which most frequent elements are evaluated with negligible error. This has direct applications to the frequent elements problem.

2   Background and Related Work

2.1   Conservative Count-Min: Definitions

A Count-Min sketch is a counter array A of size n together with a set of hash functions mapping elements (keys) of a given universe U to $[1..n]$. In this work, each element $e \in U$ can in general be assigned a different number k_e of hash functions. Hash functions are assumed fully random, therefore we assume w.l.o.g. that an element e is assigned hash functions $h_1, \ldots, h_{k_e}$.

At initialization, counters $A[i]$ are set to 0. When processing an insertion of an input element e, basic Count-Min increments by 1 each counter $A[h_i(e)]$, $1 \le i \le k_e$. The conservative version of Count-Min increments by 1 only the smallest of all $A[h_i(e)]$. That is, $A[h_i(e)]$ is incremented by 1 if and only if $A[h_i(e)] = \min_{1 \le j \le k_e}\{A[h_j(e)]\}$ and is left unchanged otherwise.

In both versions, the *estimate* of the number of occurrences of a queried element e is computed by $c(e) = \min_{1 \le i \le k_e}\{A[h_i(e)]\}$. It is easily seen that for any input sequence of elements, the estimate computed by original Count-Min is greater than or equal to the one computed by the conservative version.

In this work, we study the conservative version of Count-Min. Let H denote a selection of hash functions $H = \{h_1, h_2, \ldots\}$. Consider an input sequence I of N insertions and let E be the set of distinct elements in I. The *relative error* of an element e is defined by $err(e) = (c(e) - occ(e))/occ(e)$, where $occ(e)$ is the number of occurrences of e in the input. The *combined error* is an average error over all elements in I weighted by the number of occurrences, i.e.

$$err = \frac{1}{N}\sum_{e \in E} occ(e) \cdot err(e) = \frac{1}{N}\sum_{e \in E}(c(e) - occ(e)).$$

We assume that I is an i.i.d. random sequence drawn from a probability distribution on a set of elements $E \subseteq U$. A key parameter is the size of E relative to the size n of A. By analogy to hash tables, $\lambda = |E|/n$ is called the *load factor*, or simply the *load*.

2.2   Analysis of Conservative Count-Min: Prior Works

Motivated by applications to traffic monitoring, [5] was probably the first work devoted to the analysis of conservative Count-Min in the counting setting. Their model assumed that all $\binom{n}{k}$ counter combinations are equally likely, where k hash functions are applied to each element. This implies the regime when $|E| \gg n$. The focus of [5] was on the analysis of the *growth rate* of counters, i.e. the average number of counter increments per insertion, using a technique based on Markov chains and differential equations. Another approach proposed in [16] simulates a conservative Count-Min sketch by a hierarchy of ordinary Bloom filters. Obtained error bounds are expressed via a recursive relation based on false positive rates of corresponding Bloom filters.

Recent works [2,3] propose an analytical approach for computing error bounds depending on element probabilities assumed independent but not necessarily uniform, in particular leading to improved precision bounds for detecting heavy hitters. However the efficiency of this technique is more limited when all element probabilities are small. In particular, if the input distribution is uniform, their approach does not bring out any improvement over the general bounds known for original Count-Min.

In our recent work [21], we proposed an analysis of conservative Count-Min based on its relationship with random hypergraphs. We summarize the main results of this work below in Sect. 2.4.

2.3 Hash Hypergraph

Many hashing-based data structures are naturally associated with hash hypergraphs so that hypergraph properties are directly related to the proper functioning of the data structure. This is the case with Cuckoo hashing [27] and Cuckoo filters [18], Minimal Perfect Hash Functions and Static Functions [24], Invertible Bloom Lookup Tables [22], and some others. [30] provides an extended study of relationships between hash hypergraphs and some of those data structures.

A Count-Min sketch is associated with a *hash hypergraph* $H = (V, E)$ where $V = \{1..n\}$ and $E = \{\{h_1(e), ...h_{k_e}(e)\}\}$ over all distinct input elements e. We use notation $\mathcal{H}_{n,m}$ for hypergraphs with n vertices and m edges, and $\mathcal{H}_{n,m}^k$ for k-uniform such hypergraphs, where all edges have cardinality k. In the latter case, since our hash functions are assumed fully random, a hash hypergraph is a k-uniform Erdős-Rényi random hypergraph.

As inserted elements are assumed to be drawn from a random distribution, it is convenient to look at the functioning of a Count-Min sketch as a stochastic process on the associated hash hypergraph [21]. Each vertex holds a counter initially set to zero, and therefore each edge is associated with a set of counters held by corresponding vertices. Inserting an element consists in incrementing the minimal counters of the corresponding edge, and retrieving the estimate of an element returns the minimum value among the counters of the corresponding edge. From now on in our presentation, we will interchangeably speak of distinct elements and edges of the associated hash hypergraph, as well as of counters and vertices. Thus, we will call the *vertex value* the value of the corresponding counter, and the *edge value* the estimate of the corresponding element. Also, we will speak about the *load* of a hypergraph understood as the density $|E|/|V|$.

2.4 Hypergraph Peelability and Phase Transition of Error

A hypergraph $H = (V, E)$ is called *peelable* if iterating the following step starting from H results in the empty graph: if the graph has a vertex of degree 1 or 0, delete this vertex together with the incident edge (if any). As many other properties of random hypergraphs, peelability undergoes a phase transition. Consider the Erdős-Rényi k-uniform hypergraph model where graphs are drawn from $\mathcal{H}_{n,m}^k$ uniformly at random. It is shown in [26] that a phase transition occurs at a (computable) peelability threshold λ_k: a random graph from $\mathcal{H}^k_{n,\lambda n}$ is with high probability (w.h.p.) peelable if $\lambda < \lambda_k$, and w.h.p. non-peelable if $\lambda > \lambda_k$. The first values are $\lambda_2 = 0.5$, $\lambda_3 \approx 0.818$, $\lambda_4 \approx 0.772$, etc., λ_3 being the largest. Note that the case $k = 2$ makes an exception to peelability: for $\lambda < \lambda_2$, a negligible fraction of vertices remain after peeling.

Peelability is known to be directly relevant to certain constructions of Minimal Perfect Hash Functions [24] as well as to the proper functioning of Invertible Bloom filters [22]. In [21], we proved that it is relevant to Count-Min as well.

Theorem 1 ([21]). *Consider a conservative Count-Min where each element is hashed using k random hash functions. Assume that the input I of length N is drawn from a uniform distribution on a set $E \subseteq U$ of elements and let $\lambda = |E|/n$, where n is the number of counters. If $\lambda < \lambda_k$, then for a randomly chosen element e, the relative error $err(e)$ is $o(1)$ w.h.p. when both n and N/n grow.*

In the complementary regime $\lambda > \lambda_k$, we showed in [21], under some additional assumptions, that err is $\Theta(1)$. Thus, the peelability threshold for random hash hypergraphs corresponds to phase transition in the error produced by conservative Count-Min for uniform distribution of input. We call regimes $\lambda < \lambda_k$ and $\lambda > \lambda_k$ *subcritical* and *supercritical*, respectively.

2.5 Variable Number of Hash Functions: Mixed Hypergraphs

The best peelability threshold $\lambda_3 \approx 0.818$ can be improved in at least two different ways. One way is to use a carefully defined class of hash functions which replace uniform sampling of k-edges by a specific non-uniform sampling. Thus, [15] showed that the peelability threshold can be increased to ≈ 0.918 for $k = 3$ and up to ≈ 0.999 for larger k's if a special class of hypergraphs is used.

Another somewhat surprising idea, that we apply in this paper, is to apply a different number of hash functions to differents elements, that is to consider non-uniform hypergraphs. Following [14], [28] showed that non-uniform hypergraphs may have a larger peelability threshold than uniform ones. More precisely, [28] showed that *mixed hypergraphs* with two types of edges of different cardinalities, each constituting a constant fraction of all edges, may have a larger peelability threshold: for example, hypergraphs with a fraction of ≈ 0.887 of edges of cardinality 3 and the remaining edges of cardinality 21 have the peelability threshold ≈ 0.920, larger than the best threshold 0.818 achieved by uniform hypergraphs. We adopt the notation of [28] for mixed hypergraphs: by writing $k = (k_1, k_2)$ we express that the hypergraph contains edges of cardinality k_1 and k_2, and $k = (k_1, k_2; \alpha)$ specifies in addition that the fraction of k_1-edges is α.

The idea of using different number of hash functions for different elements has also appeared in data structures design. [6] proposed *weighted Bloom filters* which apply a different number of hash functions depending on the frequency with which elements are queried and on probabilities for elements to belong to the set. It is shown that this leads to a reduced false positive probability, where the latter is defined to be weighted by query frequencies. This idea was further refined in [31], and then further in [4], under the name *Daisy Bloom filter*.

3   Results

3.1   Uniform Distribution

We start with the case where input elements are uniformly distributed, i.e. edges of the associated hash hypergraph have equal probabilities to be processed for updates.

Subcritical Regime. Theorem 1 in conjunction with the results of Sect. 2.5 leads to the assumption that using a different number of hash functions for different elements one could "extend" the regime of $o(1)$ error of Count-Min sketch, which can be made into a rigorous statement (for simplicity we only give it with two different edge cardinalities).

Theorem 2. *Consider a conservative Count-Min with n counters. Assume that the input of length N is drawn from a uniform distribution on $E \subseteq U$ and let $\lambda < \lambda_k$. Assume further that elements of E are hashed according to a mixed hypergraph model $k = (k_1, k_2; \alpha)$. Let c_k be the peelability ratio associated to k. Then, when $\lambda < c_k$, the relative error $err(e)$ of a randomly chosen key e is $o(1)$ w.h.p., as both n and N/n grow.*

The proof can be found in the full version [20].

Figure 1 shows the average relative error as a function of the load factor for three types of hypergraphs: 2-uniform, 3-uniform and mixed hypergraph where a 0.885 fraction of edges are of cardinality 3 and the remaining ones are of cardinality 14. 2-uniform and 3-uniform hypergraphs illustrate phase transitions at load factors approaching respectively 0.5 and $\approx$0.818, peelability thresholds for 2-uniform and 3-uniform hypergraphs respectively. It is clearly seen that the phase transition for the mixed hypergraphs occurs at a larger value approaching $\approx$0.898 which is the peelability threshold for this class of hypergraphs [28].

While this result follows by combining results of [28] and [21], it has not been observed earlier and has an important practical consequence: *using a variable number of hash functions in Count-Min sketch allows one to increase the load factor while keeping negligibly small error.* In particular, for the same input, this leads to space saving compared to the uniform case.

Note that parameters $k = (3, 14; 0.885)$ are borrowed from [28] in order to make sure that the phase transition corresponds to the peelability threshold obtained in [28]. In practice, "simpler" parameters can be chosen, for example we found that $k = (2, 5; 0.5)$ produces essentially the same curve as $k = (3, 14; 0.885)$ (data not shown).

Supercritical Regime. When the load factor becomes large (supercritical regime), the situation changes drastically. When the load factor just surpasses the threshold, some edges are still evaluated with small or zero error, whereas for the other edges, the error becomes large. This "intermediate regime" has been illustrated in [21]. When the load factor goes even larger, the multi-level pattern of edge values disappears and all edge values become concentrated around the

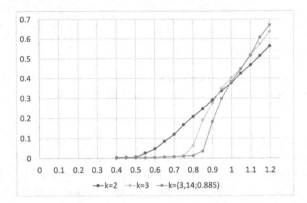

Fig. 1. *err* for small $\lambda = m/n$, for uniform distribution and different types of hypergraphs: 2-uniform, 3-uniform and (3,14)-mixed with a fraction of 0.885 of 3-edges (parameters borrowed from [28]). Data obtained for $n = 1000$. The input size in each experiment is 5,000 times the number of edges. Each average is taken over 10 random hypergraphs.

same value. We call this phenomenon *saturation*. For example, for $k = 3$ saturation occurs at around $\lambda = 6$ (data not shown). Under this regime, the hash hypergraph is dense enough so that its specific topology is likely to be irrelevant and the largest counter level "percolates" into all vertex counters. In other words, all counters grow at the same rate, without any of them "lagging behind" because of particular graph structural patterns (such as edges containing leaf vertices).

3.2 Step Distribution

In this section, we focus on the simplest non-uniform distribution – *step distribution* – in order to examine the behavior of Count-Min sketch in presence of elements with different frequencies. Our model is as follows. We assume that input elements are classified into two groups that we call *hot* and *cold*, where a hot element has a larger appearance probability than a cold one. Note that we assume that we have a prior knowledge on whether a given element belongs to hot or cold ones. This setting is similar to the one studied for Bloom filters augmented with prior membership and query probabilities [4]. Note that our definition of *err* assumes that the query probability of an element and its appearance probability in the input are equal.

We assume that the load factors of hot and cold elements are λ_h and λ_c respectively. That is, there are $\lambda_h n$ hot and $\lambda_c n$ cold edges in the hash hypergraph. $G > 1$, called *gap factor*, denotes the ratio between probabilities of a hot and a cold element respectively. Let p_h (resp. p_c) denote the probability for an input element to be hot (resp. cold). Then $p_h/p_c = G\lambda_h/\lambda_c$, and since $p_h + p_c = 1$, we have

$$p_h = \frac{G\lambda_h}{\lambda_c + G\lambda_h}, \quad p_c = \frac{\lambda_c}{\lambda_c + G\lambda_h}.$$

For example, if there are 10 times more distinct cold elements than hot ones ($\lambda_h/\lambda_c = 0.1$) but each hot element is 10 times more frequent than a cold one ($G = 10$), than we have about the same fraction of hot and cold elements in the input ($p_h = p_c = 0.5$).

In the rest of this section, we will be interested in the combined error of hot elements alone, denoted *errhot*. If $E_h \subseteq E$ is the subset of hot elements, and N_h is the total number of occurrences of hot elements in the input, then *errhot* is defined by

$$errhot = \frac{1}{N_h} \sum_{e \in E_h} occ(e) \cdot err(e) = \frac{1}{N_h} \sum_{e \in E_h} (c(e) - occ(e)).$$

"Interaction" of Hot and Cold Elements. A partition of elements into hot and cold induces the partition of the underlying hash hypergraph into two subgraphs that we call *hot* and *cold subgraphs* respectively. Since hot elements have larger counts, one might speculate that counters associated with hot edges are larger than counts of cold elements and therefore are not incremented by those. Then, *errhot* is entirely defined by the hot subgraph, considered under the uniform distribution of elements. In particular, *errhot* as a function of λ_h should behave the same way as *err* for the uniform distribution (see Sect. 3.1).

This conjecture, however, is not true in general. One reason is that there is a positive probability that all nodes of a cold edge are incident to hot edges as well. As a consequence, "hot counters" (i.e. those incident to hot edges) gain an additional increment due to cold edges, and the latter contribute to the overestimate of hot edge counts. Fig. 2a illustrates this point. It shows, for $k = 3$, *errhot* as a function of λ_h in presence of cold elements with $\lambda_c = 5$, for the gap value $G = 20$. For the purpose of comparison, the orange curve shows the error for the uniform distribution (as in Fig. 1), that is the error that hot elements would have if cold elements were not there. We clearly observe the contribution of cold elements to the error, even in the load interval below the peelability threshold.

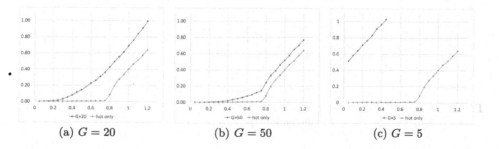

(a) $G = 20$ (b) $G = 50$ (c) $G = 5$

Fig. 2. *errhot* for $k = 3$ depending on λ_h, in presence of cold elements with $\lambda_c = 5$ (blue curves) and without any cold elements (orange curve). (Color figure online)

226   É. Fusy and G. Kucherov

Figure 2b illustrates that when the gap becomes larger (here, $G = 50$), the contribution of cold elements diminishes and the curve approaches the one of the uniform distribution. A larger gap leads to larger values of hot elements and, as a consequence, to a smaller relative impact of cold ones.

Another reason for which the above conjecture may not hold is the following: even if the number of hot elements is very small but the gap factor is not large enough, the cold edges may cause the counters to become large if λ_c is large enough, in particular in the saturation regime described in Sect. 3. As a consequence, the "background level" of counters created by cold edges may be larger than true counts of hot edges, causing their overestimates. As an example, consider again the configuration with $k = 3$ and $\lambda_c = 5$. The cold elements taken alone would have an error of about 6 on average (≈ 6.25, to be precise, data not shown) which means an about 7× overestimate. Since the graph is saturated in this regime (see Sect. 3), this means that most of the counters will be about 7 times larger than counts of cold edges. Now, if a hot element is only 5 times more frequent than a cold one, those will be about 1.4× overestimated, i.e. will have an error of about 0.4, This situation is illustrated in Fig. 2c.

Mixed Hypergraphs. The analysis above shows that in presence of a "background" formed by large number of cold elements, the error of hot elements starts growing for much smaller load factors than without cold elements, even if the latter are much less frequent than the former. Inspired by results of Sect. 3, one may ask if the interval of negligible error can be extended by employing the idea of variable number of hash functions. Note that here this idea applies more naturally by assigning a different number of hash functions to hot and cold elements.

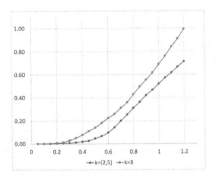

Fig. 3. *errhot* as a function of λ_h for $k = 3$, $\lambda_c = 5$ and $G = 20$ (same as in Fig. 2a) vs. $k = (2,5)$ for hot and cold elements respectively

Figure 3 illustrates that this is indeed possible by assigning a smaller number of hash functions to hot elements and a larger number to cold ones. It is clearly seen that the interval supporting close-to-zero errors is extended. This happens because when the hot subgraph is not too dense, increasing the cardinality of cold edges leads to a higher probability that at least one of the vertices of such an edge is not incident to a hot edge. As a consequence, this element does not affect the error of hot edges. For the same reason, decreasing the cardinality of hot edges (here, from 3 to 2) improves the error, as this increases the fraction of vertices non-incident to hot edges.

Saturation in Supercritical Regime. In Sect. 3 we discussed the saturation regime occurring for large load values: when the load grows sufficiently large, i.e. the hash hypergraph becomes sufficiently dense, all counters reach the same level, erasing distinctions between edges. In this regime, assuming a fixed load (graph density) and the uniform distribution of input, the edge value depends only on input size and not on the graph structure (with high probability).

It is an interesting, natural and practically important question whether this saturation phenomenon holds for non-uniform distributions as well, as it is directly related to the capacity of distinguishing elements of different frequency. A full and precise answer to this question is not within the scope of this work. We believe that the answer is positive at least when the distribution is piecewise uniform, when edges are partitioned into several classes and are equiprobable within each class, provided that each class takes a linear fraction of all elements. Here we illustrate this thesis with the step distribution.

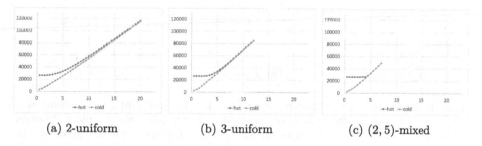

| (a) 2-uniform | (b) 3-uniform | (c) $(2,5)$-mixed |

Fig. 4. Convergence of average estimates of hot and cold elements for 2-uniform (4a), 3-uniform (4b) and $(2,5)$-mixed (4c) hypergraphs. x-axis shows the total load $\lambda = \lambda_h + \lambda_c$ with $\lambda_h = 0.1 \cdot \lambda$ and $\lambda_c = 0.9 \cdot \lambda$ and $G = 10$ in all cases.

Figure 4 illustrates the saturation phenomenon by showing average values of hot and cold edges ($G = 10$) with three different configurations: 2-uniform, 3-uniform, and (2,5)-mixed. Note that the x-axis shows here the total load $\lambda = \lambda_h + \lambda_c$, where $\lambda_h = 0.1 \cdot \lambda$ and $\lambda_c = 0.9 \cdot \lambda$. That is, the number of both hot and cold edges grows linearly when the total number of edges grows.

One can observe that in all configurations, values of hot and cold edges converge, which is a demonstration of the saturation phenomenon. Interestingly, the "convergence speed" heavily depends on the configuration: the convergence is "slower" for uniform configurations, whereas in the mixed configuration, it occurs right after the small error regime for hot edges.

3.3   Zipf's Distribution

Power law distributions are omnipresent in practical applications. The simplest of those is Zipf's distribution which is often used as a test case for different algorithms including Count-Min sketches [3,5,8,13,16]. Under Zipf's distribution, element probabilities in descending order are proportional to $1/i^\beta$, where i is the rank of the key and $\beta \geq 0$ is the *skewness* parameter. Note that for $\beta = 0$, Zipf's distribution reduces to the uniform one.

Zipf's distribution is an important test case for our study as well, as it forces several (few) most frequent elements to have very large counts and a large number of elements (*heavy tail*) to have small counts whose values decrease only polynomially on the element rank and are therefore of the same order of magnitude. Bianchi et al. [5, Fig. 1] observed that for Zipf's distribution in the supercritical regime, the estimates follow the "waterfall-type behavior": the most frequent elements have essentially exact estimates whereas the other elements have all about the same estimate regardless of their frequency. Figure 5 illustrates this phenomenon for different skewness values.

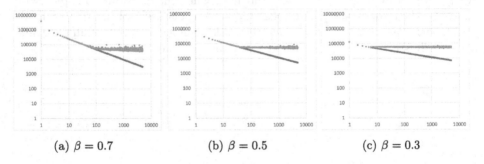

(a) $\beta = 0.7$ (b) $\beta = 0.5$ (c) $\beta = 0.3$

Fig. 5. Exact (blue) and estimated (orange) edge values for Zipf's distribution as a function on the element frequency rank, plotted in double log scale. All plots obtained for $n = 1000$, $\lambda = 5$, $k = 2$, and the input size $50 \cdot 10^6$. Estimates are averaged over 10 hash function draws. (Color figure online)

The waterfall-type behavior for Zipf's distribution is well explained by the analysis we developed in the previous sections. The "waterfall pool level" of values (called *error floor* in [5]) is the effect of saturation formed by heavy tail elements. The few "exceptionally frequent" elements are too few to affect the saturation level (their number is $\ll n$), they turn out to constitute "peaks" above the level and are thus estimated without error. Naturally, smaller skewness values make the distribution less steep and reduce the number of "exceptionally frequent" elements. For example, according to Fig. 5, for $\lambda = 5$ and $k = 2$, about 50 most frequent elements are evaluated without error for $\beta = 0.7$, about 40 for $\beta = 0.5$ and only 5 for $\beta = 0.3$.

Following our results from previous sections, we studied whether using a variable number of hash functions can extend the range of frequent elements estimated with small error. We found that for moderate loads λ, this is possible indeed. More specifically, using a variable number of hash functions can lead to a sharper "break point" compared to the constant number of hash functions, see Fig. 5. As a result, although the "waterfall pool level" may be higher, a larger range of most frequent elements are evaluated with small error. This observation matches the phenomenon illustrated earlier in Fig. 4. Due to space limitation, we refer to the full version [20] for the data illustrating this point.

4 Conclusions

In this paper, we presented a series of experimental results providing new insights into the behavior of conservative Count-Min sketch. Some of them have direct applications to practical usage of this data structure. Main results can be summarized as follows.

- For the uniform distribution of input elements, assigning a different number of hash functions to different elements extends the subcritical regime (range of load factors λ) that supports asymptotically vanishing relative error. This immediately implies space saving for Count-Min configurations verifying this regime. For non-uniform distributions, variable number of hash functions allows extending the regime of negligible error for most frequent elements,
- Under "sufficiently uniform distributions", including uniform and step distributions, a Count-Min sketch reaches a saturation regime when λ becomes sufficiently large. In this regime, counters become concentrated around the same value and elements with different frequency become indistinguishable,
- Frequent elements that can be estimated with small error can be seen as those which surpass the saturation level formed by the majority of other elements. For example, in case of Zipf's distribution, those elements are a few "exceptionally frequent elements", whereas the saturation is insured by the heavy-tail elements. Applying a variable number of hash functions can increase the number of those elements for moderate loads λ.

Many of those results lack a precise mathematical analysis. Perhaps the most relevant to practical usage of Count-Min is the question of saturation level ("waterfall pool level"), as it provides a lower bound to the frequency of elements that will be estimated with small error, which in turn is a fundamental information for heavy-hitter type of applications. Bianchi et al. [5] observed that in the case of non-uniform distribution of input elements, the "waterfall pool level" is upper-bounded by the saturation level for the uniform distribution of input. This latter is computed in [5] using a method based on Markov chains and differential equations. We believe that this method can be extended to the case of mixed graphs as well and leave it for future work. However, providing an analysis for more complex distributions including Zipf's distribution is an open problem.

References

1. Behera, S., Gayen, S., Deogun, J.S., Vinodchandran, N.: KmerEstimate: a streaming algorithm for estimating k-mer counts with optimal space usage. In: Proceedings of the 2018 ACM International Conference on Bioinformatics, Computational Biology, and Health Informatics, pp. 438–447 (2018)
2. Ben Mazziane, Y., Alouf, S., Neglia, G.: A formal analysis of the count-min sketch with conservative updates. In: IEEE INFOCOM WNA 2022 - The second Workshop on Networking Algorithms (WNA), New York, USA (2022). https://doi.org/10.1109/INFOCOMWKSHPS54753.2022.9798146
3. Ben Mazziane, Y., Alouf, S., Neglia, G.: Analyzing count min sketch with conservative updates. Comput. Netw. **217**, 109315 (2022). https://www.sciencedirect.com/science/article/pii/S1389128622003607
4. Bercea, I.O., Houen, J.B.T., Pagh, R.: Daisy Bloom filters. CoRR abs/2205.14894 (2022)
5. Bianchi, G., Duffy, K., Leith, D.J., Shneer, V.: Modeling conservative updates in multi-hash approximate count sketches. In: 24th International Teletraffic Congress, ITC 2012, Kraków, Poland, 4–7 September 2012, pp. 1–8. IEEE (2012). https://ieeexplore.ieee.org/document/6331813/
6. Bruck, J., Gao, J., Jiang, A.: Weighted Bloom filter. In: Proceedings 2006 IEEE International Symposium on Information Theory, ISIT 2006, The Westin Seattle, Seattle, Washington, USA, 9–14 July 2006, pp. 2304–2308. IEEE (2006). https://doi.org/10.1109/ISIT.2006.261978
7. Charikar, M., Chen, K., Farach-Colton, M.: Finding frequent items in data streams. Theor. Comput. Sci. **312**(1), 3–15 (2004)
8. Chen, P., Wu, Y., Yang, T., Jiang, J., Liu, Z.: Precise error estimation for sketch-based flow measurement. In: Proceedings of the 21st ACM Internet Measurement Conference, IMC 2021, pp. 113–121. Association for Computing Machinery, New York (2021). https://doi.org/10.1145/3487552.3487856
9. Cohen, S., Matias, Y.: Spectral Bloom filters. In: Halevy, A.Y., Ives, Z.G., Doan, A. (eds.) Proceedings of the 2003 ACM SIGMOD International Conference on Management of Data, San Diego, California, USA, 9–12 June 2003, pp. 241–252. ACM (2003). https://doi.org/10.1145/872757.872787
10. Cormode, G.: Count-min sketch. In: Liu, L., Özsu, M.T. (eds.) Encyclopedia of Database Systems, 2nd edn., pp. 653–659. Springer, New York (2018). https://doi.org/10.1007/978-1-4614-8265-9_87
11. Cormode, G., Hadjieleftheriou, M.: Finding frequent items in data streams. Proc. VLDB Endow. **1**(2), 1530–1541 (2008)
12. Cormode, G., Muthukrishnan, S.: An improved data stream summary: the count-min sketch and its applications. J. Algorithms **55**(1), 58–75 (2005)
13. Cormode, G., Muthukrishnan, S.: Summarizing and mining skewed data streams. In: Kargupta, H., Srivastava, J., Kamath, C., Goodman, A. (eds.) Proceedings of the 2005 SIAM International Conference on Data Mining, SDM 2005, Newport Beach, CA, USA, 21–23 April 2005, pp. 44–55. SIAM (2005). https://doi.org/10.1137/1.9781611972757.5
14. Dietzfelbinger, M., Goerdt, A., Mitzenmacher, M., Montanari, A., Pagh, R., Rink, M.: Tight thresholds for cuckoo hashing via XORSAT. In: Abramsky, S., Gavoille, C., Kirchner, C., Meyer auf der Heide, F., Spirakis, P.G. (eds.) ICALP 2010. LNCS, vol. 6198, pp. 213–225. Springer, Heidelberg (2010). https://doi.org/10.1007/978-3-642-14165-2_19

15. Dietzfelbinger, M., Walzer, S.: Dense peelable random uniform hypergraphs. In: Bender, M.A., Svensson, O., Herman, G. (eds.) 27th Annual European Symposium on Algorithms, ESA 2019, Munich/Garching, Germany, 9–11 September 2019. LIPIcs, vol. 144, pp. 38:1–38:16. Schloss Dagstuhl - Leibniz-Zentrum für Informatik (2019). https://doi.org/10.4230/LIPIcs.ESA.2019.38

16. Einziger, G., Friedman, R.: A formal analysis of conservative update based approximate counting. In: International Conference on Computing, Networking and Communications, ICNC 2015, Garden Grove, CA, USA, 16–19 February 2015, pp. 255–259. IEEE Computer Society (2015). https://doi.org/10.1109/ICCNC.2015.7069350

17. Estan, C., Varghese, G.: New directions in traffic measurement and accounting. In: Mathis, M., Steenkiste, P., Balakrishnan, H., Paxson, V. (eds.) Proceedings of the ACM SIGCOMM 2002 Conference on Applications, Technologies, Architectures, and Protocols for Computer Communication, Pittsburgh, PA, USA, 19–23 August 2002, pp. 323–336. ACM (2002). https://doi.org/10.1145/633025.633056

18. Fan, B., Andersen, D.G., Kaminsky, M., Mitzenmacher, M.D.: Cuckoo filter: practically better than Bloom. In: Proceedings of the 10th ACM International on Conference on Emerging Networking Experiments and Technologies, CoNEXT 2014, pp. 75–88. Association for Computing Machinery, New York (2014). https://doi.org/10.1145/2674005.2674994

19. Fan, L., Cao, P., Almeida, J., Broder, A.: Summary cache: a scalable wide-area web cache sharing protocol. IEEE/ACM Trans. Netw. 8(3), 281–293 (2000). https://doi.org/10.1109/90.851975

20. Fusy, É., Kucherov, G.: Count-min sketch with variable number of hash functions: an experimental study. CoRR abs/2302.05245 (2023). https://doi.org/10.48550/arXiv.2302.05245, to appear in SPIRE'23

21. Fusy, É., Kucherov, G.: Phase transition in count approximation by count-min sketch with conservative updates. In: Mavronicolas, M. (ed.) CIAC 2023. LNCS, vol. 13898, pp. 232–246. Springer, Cham (2023). https://doi.org/10.1007/978-3-031-30448-4_17. Full version in arxiv:2203.15496

22. Goodrich, M.T., Mitzenmacher, M.: Invertible Bloom lookup tables. In: 2011 49th Annual Allerton Conference on Communication, Control, and Computing (Allerton), pp. 792–799. IEEE (2011)

23. Liu, H., Lin, Y., Han, J.: Methods for mining frequent items in data streams: an overview. Knowl. Inf. Syst. 26(1), 1–30 (2011)

24. Majewski, B.S., Wormald, N.C., Havas, G., Czech, Z.J.: A family of perfect hashing methods. Comput. J. 39(6), 547–554 (1996)

25. Mohamadi, H., Khan, H., Birol, I.: ntCard: a streaming algorithm for cardinality estimation in genomics data. Bioinformatics 33(9), 1324–1330 (2017)

26. Molloy, M.: Cores in random hypergraphs and Boolean formulas. Random Struct. Algorithms 27(1), 124–135 (2005)

27. Pagh, R., Rodler, F.F.: Cuckoo hashing. J. Algorithms 51(2), 122–144 (2004)

28. Rink, M.: Mixed hypergraphs for linear-time construction of denser hashing-based data structures. In: van Emde Boas, P., Groen, F.C.A., Italiano, G.F., Nawrocki, J., Sack, H. (eds.) SOFSEM 2013. LNCS, vol. 7741, pp. 356–368. Springer, Heidelberg (2013). https://doi.org/10.1007/978-3-642-35843-2_31

29. Shibuya, Y., Kucherov, G.: Set-min sketch: a probabilistic map for power-law distributions with application to k-mer annotation. bioRxiv, p. 2020.11.14.382713 (2020). https://doi.org/10.1101/2020.11.14.382713

30. Walzer, S.: Random hypergraphs for hashing-based data structures. Ph.D. thesis, Technische Universität Ilmenau, Germany (2020). https://www.db-thueringen.de/receive/dbt_mods_00047127
31. Wang, X., Ji, Y., Dang, Z., Zheng, X., Zhao, B.: Improved weighted bloom filter and space lower bound analysis of algorithms for approximated membership querying. In: Renz, M., Shahabi, C., Zhou, X., Cheema, M.A. (eds.) DASFAA 2015. LNCS, vol. 9050, pp. 346–362. Springer, Cham (2015). https://doi.org/10.1007/978-3-319-18123-3_21

Dynamic Compact Planar Embeddings

Travis Gagie[ID], Meng He[ID], and Michael St Denis[(✉)][ID]

Dalhousie University, Halifax, NS B3H 4R2, Canada
{travis.gagie,michael.stdenis}@dal.ca, mhe@cs.dal.ca

Abstract. This paper presents a way to compactly represent dynamic connected planar embeddings, which may contain self loops and multi-edges, in $4m + o(m)$ bits, to support basic navigation in $O(\lg n)$ time and edge and vertex insertion and deletion in $O(\lg^{1+\epsilon} n)$ time, where n and m are respectively the number of vertices and edges currently in the graph and ϵ is an arbitrary positive constant. Previous works on dynamic succinct planar graphs either consider decremental settings only or are restricted to triangulations where the outer face must be a simple polygon and all inner faces must be triangles. To the best of our knowledge, this paper presents the first representation of dynamic compact connected planar embeddings that supports a full set of dynamic operations without restrictions on the sizes or shapes of the faces.

Keywords: Planar embedding · Dynamic planar embedding ·
Dynamic compact data structures

1 Introduction

A particular type of graphs, planar graphs, may be used to model the famous initial graph problem known as the seven bridges of Königsberg [21]. Aside from this application, planar graphs are also applicable to some maps in general, VLSI circuits [11], chemical molecules [1], and spatial partitions in geographical information systems (GIS) [11].

A more contemporary problem concerns the dramatic growth of problem sizes with respect to the growth in computer memory [7,16]. Although computer memories are growing, and our ability to store data in secondary or even tertiary storage is still sufficient, being able to process this data in main memory is becoming more cumbersome. Secondary to this concern is the size of the data structure built on the data which is used to perform queries and updates. These data structures often occupy much more space than the data itself. Hence, Jacobson proposed to study succinct data structures [12].

This paper exists at the intersection of graph theory and compact data structures, examining a way to represent a connected planar graph embedding using compact data structures. Much research has been conducted on static planar

This work was supported by NSERC of Canada.

F. M. Nardini et al. (Eds.): SPIRE 2023, LNCS 14240, pp. 233–245, 2023.
https://doi.org/10.1007/978-3-031-43980-3_18

graphs where $O(n)$ bits are used to support common navigational operations such as adjacency testing and listing a vertex's neighbors [3–5,7,12,14,19].

Static data structures offer easy access to the objects they store. However, objects either cannot be added or deleted from the structure or doing so would require rebuilding the entire structure itself. None of the prior works cited support the insertion or deletion of an edge or a vertex. Prior work on dynamic succinct planar graphs is restricted to triangulations where the outer face must be a simple polygon and all inner faces must be triangles [2] or consider the decremental setting under which the graph is updated by edge contractions and vertex deletions only [13]. This paper presents a way to dynamize a compact representation of a connected planar embedding.

The operations we aim to support include the following:

- Given a vertex v, or an edge (v, u), list the edges incident to v in clockwise or counterclockwise order, starting from (v, u) when given.
- Given an edge (v, u) and a face F that (v, u) is incident to, list the edges incident to F in clockwise or counterclockwise order starting from (v, u).
- Given two corners of a face, insert an edge between their apexes, bisecting these corners. Here, a *corner* is defined as the space between consecutive edges incident to a vertex [10], and this vertex is the *apex* of the corner.
- Delete a given edge from G so long as G remains connected.
- Given a corner with apex v, insert a degree-1 vertex u in the corner as a new neighbor of v.
- Given a degree-1 vertex v, delete v and the edge e it is incident to from G.

These operations allow for the transformation from one connected planar embedding to another connected planar embedding.

1.1 Our Contribution

Our contribution is summarized by the following theorem:

Theorem 1. *Given a connected planar embedding G, possibly containing multi-edges and self loops, on n vertices and m edges, there is a compact representation of G occupying $4m + o(m)$ bits that can list the edges incident to a given vertex in clockwise or counterclockwise order in $O(\lg n)$ time per edge, list the edges incident to a face in $O(\lg n)$ time per edge, and support insertion or deletion of an edge or a vertex in $O(\lg^{1+\epsilon} n)$ time for any constant $\epsilon > 0$.*

To the best of our knowledge, this paper presents the first dynamic compact connected planar embedding that supports fast insertion and deletion of an edge or a vertex and has no restrictions on the sizes or shapes of the faces. Additionally, we present the marker model to support navigational operations within a given connected planar embedding. This model is similar to the finger-update model [6] where a finger, or marker, is maintained on a given vertex and updates to the structure are limited to the position of the finger, or marker. The difference between the marker model and the finger-update model is that a marker has an indicator that points to a specific corner in the given planar embedding.

2   Related Work

We survey previous results on succinct representations of planar embeddings [3,4,7,14,19,20]. Succinct planar graph representations that cannot encode an arbitrary embedding are not included; see [4] for a survey including those results. Tutte [20] enumerated rooted planar maps and his results implied that $m \lg 12 = 3.58m$ bits are required to encode an m-edge planar embedding. Turán [19] derived a simple succinct encoding of planar graphs that uses $4m$ bits. Keeler and Westbrook [14] then showed an encoding of planar maps that achieves Tutte's lower space bound, but no operations are supported. Additionally, their encoding and decoding algorithms run in linear time. Later, researchers designed succinct data structures for planar embeddings, and the operations supported by these structures include listing the neighbors of a given vertex in counterclockwise or clockwise order. Barbay et al. [3] showed how to represent a simple planar embedding in $18n+o(m)$ bits to support degree and adjacency queries in constant time and the listing of neighbors in constant time per neighbor. Blelloch and Farzan [4] developed a data structure that occupies $3.58m+o(m)$ bits and provide the same support for queries. Ferres et al. [7] modified Turán's [19] structure to occupy $o(m)$ additional bits and showed the following: the edges incident to vertex v can be listed in clockwise or counterclockwise order in constant time per edge, the edges limiting a face can be traversed in constant time per edge, a vertex's degree can be found in $O(f(m))$ time for any function $f(m) \in \omega(1)$ and testing if two given vertices are neighbors in $O(f(m))$ time for any function $f(m) \in \omega(\lg m)$. This data structure is simpler than previous solutions and can be constructed in parallel efficiently. Fuentes-Sepúlveda et al. [8] further showed that by using *half-edges* and condensing Ferres et al.'s [7] data structures, a full set of topological queries can be supported efficiently.

All the above-mentioned works concern succinct representations of planar graphs in the static case. With respect to the dynamic case, Aleardi et al. [2] designed a succinct representation of an n-vertex triangulated graph with fixed genus g and a simple polygon boundary that supports standard navigation in $O(1)$ time, vertex addition in $O(1)$ amortized time without supporting access to satellite data associated with each vertex, $O(\lg n)$ amortized time with data access, and vertex deletion or edge flip[1] in $O(\lg^2 n)$ amortized time. This data structure occupies $2.17m + o(m)$ bits and uses an additional $O(g \lg m)$ bits for representing triangulations on genus g surfaces. Kammer and Meintrup [13] provided a dynamic data structure, a modification of Blelloch and Farzan's [4], that encodes a planar graph in $\mathcal{H}(n) + o(n)$ bits to support an arbitrary sequence of edge contractions and vertex deletions in $O(n)$ time, where $\mathcal{H}(n)$ is the entropy of encoding an n-vertex planar graph. It can compute the degree of a vertex in $O(1)$ time and list the neighbors in $O(1)$ time per neighbor. Edge or vertex insertions are not supported, so their solution is for the decremental setting only.

---

[1] Edge flipping refers to removing the edge e and replacing it with the other diagonal of Q, where Q is the union of two triangles which create a quadrilateral.

3 Preliminaries

3.1 Notation

We assume the word-RAM model of computation. For the remainder of this paper, let G be a given connected planar embedding on n vertices, m edges and f faces that may contain multi-edges and self loops. All logarithms are written as lg and are base 2, unless otherwise specified.

3.2 Dynamic Bitvectors

A **dynamic bitvector** $B[1..n]$, supporting the following operations, is a key compact data structure ($b \in \{0, 1\}$ in the following definitions):

- access(B, i): return the bit in $B[i]$ for any i such that $1 \le i \le n$.
- rank$_b$(B, i): return the number of occurrences of b in $B[1..i]$ where $1 \le i \le n$.
- select$_b$(B, j): return the index of the jth occurrence of b in B.
- insert(B, i, b): inserts bit b between $B[i - 1]$ and $B[i]$ where $1 \le i \le n$.
- delete(B, i): deletes the bit in position $B[i]$ where $1 \le i \le n$.
- link(B, p, B'): attaches all the bits in B' between bits $B[p]$ and $B[p + 1]$ where $1 \le p \le n$ and $|B'|$ is another bit vector with $O(n)$ bits.
- cut(B, i, j): returns bitvector B' which contains all the detached bits in B from $B[i]$ up to and including $B[j]$ where $1 \le i < j \le n$. B is then the concatenation of the bit ranges $[1..(i - 1)]$ and $[(j + 1)..n]$.

Navarro and Sadakane [17] present the following:

Lemma 1 ([17]). *There exists a succinct dynamic bitvector structure that supports* access, rank, select, insert *and* delete *in* $O(\frac{\lg n}{\lg \lg n})$ *time and* link *and* cut *in* $O(\lg^{1+\epsilon} n)$ *time, where n is the number of bits currently in the bitvector and ϵ is an arbitrary positive constant.*

3.3 Planar Graph Traversal

The traversal of a planar embedding G developed by Turán [19], which is used by Ferres et al. [7] to generate a binary sequence A, is now described. From here forward, we will refer to this traversal as a *Turán traversal*. An arbitrary spanning tree T, which is rooted at some vertex v_0 on the outer face of G, is computed before the traversal. We note that a Turán traversal can be performed in counterclockwise or clockwise order. Without the loss of generality, we use counterclockwise order and design our data structures with respect to this ordering. We call an edge in G that is also in T a *primal edge* and an edge in G not in T a *dual edge*. In this Turán traversal, after we visit a vertex v and traverse or examine an edge, (v, u), incident to it, we view (v, u) as a directed edge by orienting it from v to u, even though the graph G is undirected.

The traversal of G begins from the vertex selected as the root v_0 of T and examines one of its incident edges (v_0, u) such that (v_0, u) is on the boundary

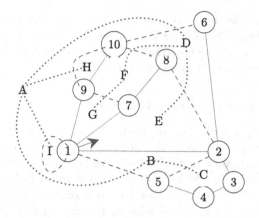

A: 0 1 0 1 1 1 0 0 1 1 1 1 0 1 0 1 1 1 0 0 1 0 1 1 0 1 0 0 0 1 0 1 0 0

Fig. 1. Planar embedding G where the red solid edges correspond to edges in T, the black dashed edges correspond to edges in $G \setminus T$, and blue dotted edges correspond to edges in T^*, the dual of G complementary to T. Each vertex of G is labeled by its preorder rank in T, while each face is labeled alphabetically in the order they are first visited in the Euler tour of T^*. The violet arrow on the vertex labeled 1 and directed at the face labeled E depicts a marker (to be discussed in Sect. 4.2). Bitvector A contains a Turán traversal beginning with edge $(1,5)$ and proceeding counterclockwise. The violet boldfaced 1 bit indicates the marker as represented by the violet arrow on vertex 1 which is marker 16. (Color figure online)

of the outer face and the outer face is to its right. The traversal is a modified depth first search (DFS) where we visit vertices in counterclockwise order. In a standard DFS, we examine an edge (v, u) and do not traverse from v to u if u has already been visited, unless we are returning to a parent. In the Turán traversal, we examine an edge (v, u), and if (v, u) is a primal edge, we traverse from v to u and record a 1 in a bitvector A. Afterwards, we examine the next edge incident to u after (u, v) in counterclockwise order. Otherwise, (v, u) is a dual edge, and we do not traverse it. Instead, we remain on v, record a 0 in A and examine the next edge in counterclockwise order. This process is repeated recursively until we have visited all vertices and returned to the root v_0 of T. All edges in G have been traversed or examined twice and $A[i]$ indicates whether the ith edge examined in the Turán traversal is a primal or dual edge.

Observe that a Turán traversal of G performs an Euler tour traversal of T and an Euler tour traversal of the spanning tree of the dual of G with respect to T, which we refer to as T^*. More specifically, to define T^*, consider some dual edge (v, u) in G not yet examined. Let f_r (f_l) be the face on the right (left) side of (v, u). Examining (v, u) advances the Euler tour of T^* from f_r to f_l, establishing f_r as the parent of f_l in T^*, and we refer to edge (u, v) as the *entry edge* of f_l. Thus, T^* encodes a spanning tree on the faces of G and each edge in G not in T is crossed by an edge in T^*. In this way, every connected planar embedding can be represented as interdigitating spanning trees of the primal and the dual [18]. As the Turán traversal in this paper is performed in counterclockwise order, the traversal of T^* is performed clockwise from its root, i.e., the outer face. Figure 1 gives an example.

3.4   Dynamic Succinct Euler-Tour Trees

Gagie and Wild [9] describe how to succinctly represent a set of Euler-Tour trees of an n-vertex forest in $2n+o(n)$ bits. An Euler-Tour tree contains directed edges (u, v) and (v, u) for every undirected edge in the given forest and preserves an encoding of the order in which edges are visited in an Euler tour. Each tree in the forest is unrooted, and its Euler tour determines the current parent-child relationship among its nodes; this relationship may change during updates. We use $\{a, b\}$ to refer to the corner of Euler-Tour tree T between the two edges traversed at the ath and bth steps of the Euler tour of T. In [9], the merge and split operations are implied, but we state them explicitly as operations 10 and 11. Throughout this paper, we refer to operations 3, 4 and 5 in the lemma below as vertex, entry and inverse, respectively. Note that Gagie and Wild did not state the support for entry, but their existing data structures can support it easily. The following lemma summerizes their results:

Lemma 2 ([9]). *Given a planar embedding of a forest F on n vertices, F can be encoded in a data structure occupying $2n + o(n)$ bits such that operations 1 through 5 below take constant time, operations 6 and 7 take $O(\lg n)$ time, and operations 8 through 11 take $O(\lg^{1+\epsilon} n)$ time for any constant $\epsilon > 0$.*

1. *return the predecessor and successor in the Euler tour of the tree containing the given directed edge e.*
2. *return the predecessor and successor in the counterclockwise order of the edges incident to u when given directed edge (u, v).*
3. *return vertex v such that v is the vertex arrived at after traversing the given directed edge e in the Euler tour of T.*
4. *return the directed edge e encountered in the Euler tour traversal of T such that, after traversing e, the Euler tour arrived at the given vertex v the first time, i.e., e links the parent of v in the Euler tour traversal to v;*
5. *return the edge e' such that, given a tree T and an edge e encountered in the Euler tour of T, edge e' corresponds to the inverse edge of e.*
6. *return the edge e' such that the distance from the given directed edge e to e' is the given distance t in the Euler tour of the tree containing e.*
7. *return the Euler tour distance between the given edges e and e' so long as the two edges are in the same tree.*
8. *delete the given edge e from the tree containing it and return the representations of the two resulting trees.*
9. *insert an edge between T and T' at the given corners, bisecting those corners, and return the representation of the resulting tree.*
10. *merge the adjacent vertices u and v to become one vertex and retain all other edges adjacent to u and v.*

11. *split v into two adjacent vertices, v_1 and v_2, where v_1 is a parent of v_2. Two incident boundary edges e_i and e_j are also given as parameters so that edges incident to v starting from e_i to e_j in counterclockwise order are to be incident to v_1, while the remaining edges are to be incident to v_2.*

Operation 9 supports the insertion of an edge between two arbitrary vertices in two trees. This operation cannot be supported by the dynamic succinct tree representation of Navarro and Sadakane [17] which is based on a balanced parenthesis representation, but it is needed in our dynamic planar graph representation. We also comment that, to support operations 1–5 in constant time, each edge is identified by a unique internal identifier. Operations 6 and 7 can perform mapping between this identifier and the rank of the edge in the Euler tour in $O(\lg n)$ time. Thus, when the context is clear, we may also pass or return the rank of an edge in the Euler tour when calling **vertex**, **entry** or **inverse**, and the increase in running time does not affect the complexity of our solution.

4   Data Structure and the Marker Model

4.1   Data Structure

Our representation of connected planar embedding G on n vertices, m edges, and f faces, contains the following components:

- A dynamic bitvector, A, which encodes a Turán traversal of G described in Sect. 3.3. It is represented by Lemma 1 in $2m + o(m)$ bits.
- A spanning tree, T, of G as defined in Sect. 3.3. It is represented by Lemma 2 in $2n + o(n)$ bits.
- A spanning tree, T^*, of the dual of G as defined in Sect. 3.3. It is represented by Lemma 2, in $2f + o(f)$ bits.

Observe that T and T^* represent succinct Euler-Tour trees on the vertices and faces of G, respectively. By Euler's formula [15], the total space cost of our data structure is $2m + o(m) + 2n + o(n) + 2(m - n + 2) = 4m + o(m)$ bits.

4.2   The Marker Model

The marker model provides a way to map an index in A to specific vertices in T and T^*. A marker, or a marker's value, is denoted by an index i in A, where $1 \leq i \leq 2m$. Recall that a Turán traversal of G induces an Euler tour traversal on T and T^*. Thus, we say that a marker stands on the vertex most recently visited in the Euler tour of T and points to the face most recently visited in the Euler tour of T^*. The number of 1's (0's) in A corresponds to the primal (dual) Euler-Tour tree edges just traversed in an Euler tour of T (T^*). Therefore, a marker with value i stands on the vertex of G that corresponds to node **vertex**$(T, \mathrm{rank}_1(A, i))$ in T, and the face it points to corresponds to node **vertex**$(T^*, \mathrm{rank}_0(A, i))$ in T^*. Figure 1 shows marker 16 as an example.

The following lemma shows how the face that a marker i points to relates to the edge $A[i]$ represents. Its proof is omitted due to space constraints.

Lemma 3. *Let (v, u) be the directed edge represented by $A[i]$. If (v, u) is a primal edge, then the marker i is standing on u and pointing to the face on the right side of (v, u). If (v, u) is a dual edge, then the marker is standing on v and pointing to the face on the left side of (v, u).*

Because we do not require G to be bi-connected, a vertex can be incident to multiple corners of the same face. This means multiple markers can stand on the same vertex and point to the same face, but each marker points to a different corner of the face. Therefore, we define the *orientation* of a marker as the vertex it stands on and the corner it points to. More formally, let marker i refer to directed edge (v, u). By Lemma 3, if (v, u) is primal, then marker i stands on u and points to the corner that has u as its apex and is on the right of directed edge (v, u). If (v, u) is a dual edge, then marker i stands on v and points to the corner that has v as its apex and is to the left of directed edge (v, u).

If more than one marker is maintained at a given time, then, after an update, all markers must be updated to preserve orientation. The index a marker refers to can be updated in $O(1)$ time via a constant number of comparisons and arithmetic operations due to the way we support updates. Thus, the time to update all markers is linear in the number of markers maintained.

4.3 Navigation

Now we define two rotation operations and a traverse operation. Either rotate operation changes the corner the marker is pointing to and the traverse operation changes the vertex the marker is standing on. These operations are necessary to move the marker to support queries and updates on our representation of G. For the operations described below, let v be the vertex marker i currently stands on and corner C of face F be the corner marker i currently points to; they will also be referred to when we discuss how to support these operations later.

- $\mathtt{rotate\_ccw}(i)$: Compute a new orientation of the marker such that the marker is still standing on v but is pointing to the corner next to C when listing all the corners incident to v in counterclockwise order.
- $\mathtt{rotate\_cw}(i)$: Compute a new orientation of the marker such that the marker is still standing on v but is pointing to the corner next to C when listing all the corners incident to v in clockwise order.
- $\mathtt{traverse}(i)$: Let the edge (v, w) be the $(i+1)$st edge examined in the Turán traversal. Compute a new orientation of the marker such that the marker is now standing on w but still pointing to F.

To support $\mathtt{rotate\_ccw}(i)$, first consider the ith edge examined in a Turán traversal. One of its endpoints is v, let u be the other endpoint, and let (v, w) be the next edge after (v, u) in counterclockwise order. If the ith edge is a primal edge, then, by Lemma 3, it is oriented from u to v. Furthermore, corner C is to the right of (u, v) and is thus to the left of (v, u). If this edge is a dual edge, then by Lemma 3, it is oriented from v to u, and corner C is again to the left of

(v, u). In either case, since (v, w) is the edge next to (v, u) in counterclockwise order, C is to the right of (v, w). Therefore, our goal is to compute a marker still standing on v but pointing to the corner, C', to the left of (v, w); C' is next to C when listing all the corners incident to v in counterclockwise order.

There are now two cases, depending on whether the $(i + 1)$st edge, (v, w), enumerated in a Turán traversal, is a dual or primal edge, i.e. whether $A[i+1]$ is 0 or 1. If $A[i+1] = 0$, then by Lemma 3 and the definition of marker orientation, marker $i + 1$ continues to stand on v but points to C'. Therefore, we return $i + 1$ as the answer. Otherwise, $A[i + 1] = 1$ and the answer is computed as the index in A when the Turán traversal returns to v from edge (w, v). This is computed by $j = \text{select}_1(A, \text{inverse}(T, \text{rank}_1(A, i + 1)))$. This follows from Lemma 3; since (w, v) is a primal edge, the marker j is standing on v and pointing to the corner to the right of (w, v). As the right side of (w, v) is the left side of (v, w), the marker computed is also pointing to C' in this case.

The details of how to support rotate_cw and traverse operations are omitted due to space constraints. In the worst case, at most two dynamic bitvector operations and one succinct Euler-Tour tree operation are performed to support each navigational operation. Combining this with Lemmas 1 and 2, we have

Lemma 4. *The structures in this section can support* rotate_ccw, rotate_cw, *and* traverse *in* $O(\lg n)$ *time.*

These rotation and traverse operations imply the support for listing the edges incident to a vertex or a face of G. For example, to list the edges incident to a vertex v in counterclockwise order, we start from a marker standing on v and call rotate_ccw repeatedly. Details are omitted due to space limitations. Thus, we have proved the support of navigational operations stated in Theorem 1.

5 Dynamization

We now prove the support of updates stated in Theorem 1. As the support for edge insertions and deletions (especially edge deletions) is more interesting, we discuss them here. Descriptions of how to insert or delete vertices are omitted due to space limitations.

5.1 Inserting an Edge

To minimize the changes to the Turán traversal, a new edge is always inserted as a dual edge. To insert an edge, we need two markers, i and j, pointing to two corners of the same face F. Let v be the vertex marker i stands on and let u be the vertex marker j stands on. These vertices, v and u, are the endpoints of the edge to be inserted. We assume, without the loss of generality, that $i < j$.

The rank_0 query on A, with parameters i and j, computes the Euler tour edges in T^* just processed at the ith and jth step in the Turán traversal. We denote these Euler tour edges as i' and j'. Recall that v is one endpoint of the ith edge examined in a Turán traversal of G, and let w be the other endpoint.

By Lemma 3, and similar to the reasoning from Sect. 4.3, no matter if this edge is a dual edge or a primal edge, the corner, C, that marker i points to is to the left of (v, w). If we rotate counterclockwise from (v, w), with v as the pivot, C is the first corner encountered and is encountered before any other edge incident to v. Therefore, when inserting an edge bisecting C, the new edge, (v, u), is the next edge examined in a Turán traversal. This means that (v, u) will be inserted as the dual edge examined in the $(i+1)$st step of the Turán traversal. Thus, we perform $\mathtt{insert}(A, i+1, 0)$. Due to the insertion of a new bit, we also increment j, so that marker j corresponds to the same edge after insertion. Then, by similar reasoning, we additionally perform $\mathtt{insert}(A, j+1, 0)$ to indicate that (u, v) is the dual edge examined in the $(j+1)$st step. To update T^*, we observe that drawing an edge across F splits the face. Therefore, we perform $\mathtt{split}(\mathtt{vertex}(T^*, i'), i', j')$. As inserting a dual edge only affects the faces and not the vertices, T is unaffected. Lastly, we increment m by 1. An example depicting this is omitted due to space constraints. Thus, we have the following lemma:

Lemma 5. *Given two corners of the same face, an edge connecting their apexes and bisecting these corners can be inserted into G in $O(\lg^{1+\epsilon} n)$ time.*

5.2   Deleting an Edge

We perform an edge deletion only if, after removing the edge, G remains connected. Deleting dual edges does not disconnect G, as the primal edges form the spanning tree T, connecting all vertices of G. However, G could become disconnected when deleting primal edges. Therefore, we allow primal edge deletions only if the deletion of that primal edge does not disconnect G. As the steps for dual edge deletion are symmetric to those for edge insertion, their descriptions are omitted while focusing on primal edge deletion in this section.

To discuss primal edge deletion, let edge (v, u) correspond to the edge enumerated at the ith step in a Turán traversal of G and be the primal edge we wish to delete. When deleting a primal edge from our representation of G, there are four items to consider. The first item to consider is how to determine if G would be disconnected after deleting (v, u). Second, if G remains connected after deleting (v, u), how do we choose a dual edge to promote to primal? The third item is, how are T and T^* affected by primal edge deletion? Lastly, recall from Sect. 3.3 that a Turán traversal follows primal edges, and thus, primal edge deletion changes the Turán traversal of G. We must then determine how we update A to reflect a valid Turán traversal after deleting (v, u).

To determine if G would be disconnected after deleting (v, u), we inspect the faces on either side of (v, u). If the faces are the same, then (v, u) cannot be deleted as it is the only edge linking two connected components of $G \setminus \{(v, u)\}$. If the faces are different, then (v, u) can be deleted while G remains connected.

Assuming the faces adjacent to (v, u) are different, we now discuss how to select a dual edge to promote to primal. Recall that the ith step in a Turán traversal corresponds to traversing (v, u) and let the jth step be the step traversing the same edge in the reverse direction, i.e. from u to v. We assume, without the loss of generality, that $i < j$. Observe that deleting a primal edge in G corresponds to deleting an edge in T and therefore disconnecting T into two trees. Our goal is to select a dual edge to promote to primal that reconnects these two trees. The following lemma will be useful when we select such an edge; when proving it, we define the *interval* of A corresponding to an edge of G (henceforth the interval of this edge for short) to be $[a, b]$ if this edge is examined in steps a and b of the Turán traversal with $a < b$, e.g., the interval of (v, u) is $[i, j]$.

Lemma 6. *Between the two faces incident to (v, u), at least one of them has the property that the interval of its entry edge does not enclose $[i, j]$.*

Proof. Let F_1 and F_2 be the two faces incident to (v, u). Let g_1 and g_2 be the indices in A corresponding to the entry edges of F_1 and F_2, respectively, and let k_1 and k_2 be the indices of the reverse of the edges corresponding to $A[g_1]$ and $A[g_2]$, respectively. We assume, without the loss of generality, that $g_1 < g_2$.

Assume to the contrary that both $[g_1, k_1]$ and $[g_2, k_2]$ enclose $[i, j]$. Then $[g_1, k_1]$ and $[g_2, k_2]$ must intersect. Furthermore, the endpoints of the interval of the entry edge of a face correspond to the first and the last time we visit the node of T^* representing this face in an Euler tour traversal of T^*. Therefore, if the entry edge intervals of two faces intersect, one must enclose the other. Since $g_1 < g_2$, we have $[g_2, k_2] \subset [g_1, k_1]$, and the node, f_2, of T^* representing F_2 is a descendant of the node, f_1, of T^* representing F_1. This means that between steps g_2 and k_2 of the Turán traversal, the induced Euler tour of T^* only visits nodes that are descendants of f_2 in T^*, including f_2 itself. Since f_1 is the parent of f_2, no marker between g_2 and k_2 can point to face F_1. However, either marker i or marker j points to F_1, and $[i, j] \subset [g_2, k_2]$, which is a contradiction. $\square$

Let F be a face incident to (v, u) such that the interval, $[g, k]$, of its entry edge, (w, x), does not enclose $[i, j]$; if both faces incident to (v, u) satisfy this condition, we choose F arbitrarily between them. Then marker g stands on w while marker k stands on x. We promote dual edge (w, x) to primal because:

Lemma 7. $(T \setminus \{(v, u)\}) \cup \{(w, x)\}$ *is a spanning tree of G.*

Proof. All the markers that point to F are in $[g, k]$. Since either marker i or marker j points to F, i or j must be strictly between g and k. Therefore, $[i, j]$ and $[g, k]$ must intersect, and $[g, k] \not\subseteq [i, j]$. As F is the face whose entry edge interval, $[g, k]$, does not enclose $[i, j]$, we observe the following two cases:

$$1 \leq g < i < k < j \leq 2m \tag{1}$$

$$1 \leq i < g < j < k \leq 2m \tag{2}$$

To prove our lemma in either case, observe that the removal of edge (v, u) disconnects T into two connected components. One of these two components is T_u, the subtree rooted at vertex u. By the definition of a Turán traversal, a marker stands on a vertex in T_u if and only if this marker is in $[i, j - 1]$. The inequalities for these two cases then guarantee that the vertex that g stands on (which is vertex w) and the vertex that k stands on (which is vertex x) are in different components of $T \setminus \{(u,v)\}$, and the lemma follows. □

We are now ready to describe our algorithm for primal edge deletion. To delete the primal edge (v,u) enumerated at the ith step of a Turán traversal of G, we first compute the step j where the Turán traversal traverses (v,u) in the reverse direction, i.e., from u to v. By Lemma 3, marker i points to the face on one side of (v,u) and marker j points to the face on the other side of (v,u). Hence, we perform $v_1 = \mathtt{vertex}(T^*, \mathtt{rank}_0(A,i))$ and $v_2 = \mathtt{vertex}(T^*, \mathtt{rank}_0(A,j))$ to compute the nodes of T^* that respectively represent the faces adjacent to (v,u). If these faces are the same, then deleting (v,u) would disconnect G, so we do not remove (v,u) and immediately return. Otherwise, the intervals of these faces are $[\mathtt{select}_0(A, \mathtt{entry}(T^*, v_1)), \mathtt{select}_0(A, \mathtt{inverse}(T^*, \mathtt{entry}(T^*, v_1)))]$ and $[\mathtt{select}_0(A, \mathtt{entry}(T^*, v_2)), \mathtt{select}_0(A, \mathtt{inverse}(T^*, \mathtt{entry}(T^*, v_2)))]$. We compare these intervals to $[i,j]$ to determine which of these two faces should be chosen to be F so that F is a face incident to (v,u) whose entry edge's interval, $[g,k]$, does not enclose $[i,j]$. Let F' be the other face incident to (v,u).

Next we update T and T^* to reflect the deletion of (v,u) and the promotion of (w,x). Deleting a primal edge from G corresponds to deleting an edge from T, thereby disconnecting T. By Lemma 7, after deleting (v,u) and promoting (w,x), T remains a spanning a tree of G. Let T_v be the tree containing v and T_u be the tree containing u, after the deletion of (v,u). By promoting (w,x), we are connecting T_v and T_u at corners $\{\mathtt{rank}_1(A,g), \mathtt{rank}_1(A,g)+1\}$ and $\{\mathtt{rank}_1(A,k), \mathtt{rank}_1(A,k)+1\}$. As for T^*, promoting (w,x) to primal corresponds to deleting the edge connecting the faces on either side of (w,x), so we delete the Euler tour edge in T^* corresponding to $\mathtt{rank}_0(A,g)$. This creates two subtrees in T^*, $T^*{}_F$ and $T^*{}_{F'}$, where one subtree contains F and the other contains F'. Deleting (v,u) corresponds to merging faces F and F' and thereby reconnecting T^*. To merge these faces in T^* we first add an edge to connect the two subtrees, $T^*{}_F$ and $T^*{}_{F'}$, at the corners $\{\mathtt{rank}_0(A,i), \mathtt{rank}_0(A,i)+1\}$ and $\{\mathtt{rank}_0(A,j), \mathtt{rank}_0(A,j)+1\}$ and temporarily store a reference to this newly added edge, ℓ, and its inverse, ℓ'. Then, we merge F and F' by performing $\mathtt{merge}(T^*, \mathtt{vertex}(T^*, \ell), \mathtt{vertex}(T^*, \ell'))$. By Lemma 2, merging vertices and deleting edges in a succinct Euler-Tour tree takes at most $O(\lg^{1+\epsilon} n)$ time.

Finally, we show how to update A. There are two cases. In the first case, inequality 1 holds. In this case, we update A to $A[1, g-1].1.A[k+1, j-1].A[i+1, k-1].1.A[g+1, i-1].A[j+1, 2m]$, where "." is the concatenation operator for bitvectors. This can be done using a constant number of $\mathtt{insert}$, $\mathtt{delete}$, $\mathtt{cut}$, and $\mathtt{link}$ operations over A in $O(\lg^{1+\epsilon} n)$ time. The correctness can be shown by analyzing how the Turán traversal works after edge deletion; details are omitted due to space constraints. With respect to the second case, inequality 2 holds, and we update A to $A[1, i-1].A[j+1, k-1].1.A[g+1, j-1].A[i+1, g-1].1.A[k+1, 2m]$. This bitvector is obtained by similar reasoning as the first case above.

Lemma 8. *An edge can be deleted from G in $O(\lg^{1+\epsilon} n)$ time, so long as G remains connected.*

References

1. Akram, M., Mohsan Dar, J., Farooq, A.: Planar graphs under Pythagorean fuzzy environment. Mathematics **6**(12), 278 (2018)
2. Aleardi, L.C., Devillers, O., Schaeffer, G.: Dynamic updates of succinct triangulations. Technical report (2005)
3. Barbay, J., Castelli Aleardi, L., He, M., Munro, J.I.: Succinct representation of labeled graphs. Algorithmica **62**, 224–257 (2012)
4. Blelloch, G.E., Farzan, A.: Succinct representations of separable graphs. In: Amir, A., Parida, L. (eds.) CPM 2010. LNCS, vol. 6129, pp. 138–150. Springer, Heidelberg (2010). https://doi.org/10.1007/978-3-642-13509-5_13
5. Chiang, Y.T., Lin, C.C., Lu, H.I.: Orderly spanning trees with applications. Soc. Ind. Appl. Math. J. Comput. **34**(4), 924–945 (2005)
6. Farzan, A., Munro, J.I.: Dynamic succinct ordered trees. In: Albers, S., Marchetti-Spaccamela, A., Matias, Y., Nikoletseas, S., Thomas, W. (eds.) ICALP 2009. LNCS, vol. 5555, pp. 439–450. Springer, Heidelberg (2009). https://doi.org/10.1007/978-3-642-02927-1_37
7. Ferres, L., Fuentes-Sepúlveda, J., Gagie, T., He, M., Navarro, G.: Fast and compact planar embeddings. Comput. Geom. **89**, 101630 (2020)
8. Fuentes-Sepúlveda, J., Navarro, G., Seco, D.: Navigating planar topologies in near-optimal space and time. Comput. Geom. **109**, 101922 (2023)
9. Gagie, T., Wild, S.: Succinct Euler-Tour trees. In: He, M., Sheehy, D. (eds.) Proceedings of the 33rd Canadian Conference on Computational Geometry, Dalhousie University, Halifax, Nova Scotia, Canada, 10–12 August 2021, pp. 368–376 (2021)
10. Holm, J., Rotenberg, E.: Dynamic planar embeddings of dynamic graphs. Theory Comput. Syst. **61**, 1054–1083 (2017)
11. Irribarra-Cortés, A., Fuentes-Sepúlveda, J., Seco, D., Asín, R.: Speeding up compact planar graphs by using shallower trees. In: 2022 Data Compression Conference, pp. 282–291. IEEE (2022)
12. Jacobson, G.: Space-efficient static trees and graphs. In: 30th Annual Symposium on Foundations of Computer Science, pp. 549–554. IEEE Computer Society (1989)
13. Kammer, F., Meintrup, J.: Succinct planar encoding with minor operations. arXiv Computing Research Repository abs/2301.10564 (2023). https://doi.org/10.48550/arXiv.2301.10564
14. Keeler, K., Westbrook, J.: Short encodings of planar graphs and maps. Discret. Appl. Math. **58**(3), 239–252 (1995)
15. Levin, O.: Discrete mathematics: an open introduction (2021)
16. Munro, J.I.: Tables. In: Chandru, V., Vinay, V. (eds.) FSTTCS 1996. LNCS, vol. 1180, pp. 37–42. Springer, Heidelberg (1996). https://doi.org/10.1007/3-540-62034-6_35
17. Navarro, G., Sadakane, K.: Fully functional static and dynamic succinct trees. Assoc. Comput. Mach. Trans. Algorithms **10**(3), 1–39 (2014)
18. von Staudt, K.G.C.: Geometrie de Lage. Bauer und Raspe, Nürnberg (1847)
19. Turán, G.: On the succinct representation of graphs. Discret. Appl. Math. **8**(3), 289–294 (1984)
20. Tutte, W.T.: A census of planar maps. Can. J. Math. **15**, 249–271 (1963)
21. Wilson, R.J.: Introduction to Graph Theory. Prentice Hall/Pearson, New York (2010)

A Simple Grammar-Based Index for Finding Approximately Longest Common Substrings

Travis Gagie[1,3]([✉]) [iD], Sana Kashgouli[1] [iD], and Gonzalo Navarro[2,3] [iD]

[1] Faculty of Computer Science, Dalhousie University, Halifax, Canada
travis.gagie@dal.ca
[2] Department of Computer Science, University of Chile, Santiago, Chile
[3] CeBiB—Center for Biotechnology and Bioengineering, Santiago, Chile

Abstract. We show how, given positive constants ϵ and δ, and an α-balanced straight-line program with g rules for a text $T[1..n]$, we can build an $O(g)$-space index that, given a pattern $P[1..m]$, in $O(m \log^\delta g)$ time finds w.h.p. a substring of P that occurs in T and whose length is at least a $(1 - \epsilon)$ fraction of the longest common substring of P and T. The correctness can be ensured within the same expected query time.

Keywords: Grammar-based indexing · Approximately longest common substrings · alpha-balanced grammars

1 Introduction

Recent years have witnessed a sustained effort for indexing highly repetitive text collections within compressed space and supporting exact pattern matching [10,11]. Exact pattern matching is however insufficient in some applications. In Bioinformatics, for example when storing repetitive collections formed by genomes of the same species, matching strings is rarely useful. Instead, one may be interested in finding long substrings of a string that appear in the sequence collections, to find for example conserved regions of a genome in a population.

The research on matching the longest possible substrings using these indices is scarce, however. A recent result [12] finds all the maximal exact matches (MEMs) of a pattern $P[1..m]$ in a text $T[1..n]$ that is indexed with a grammar. By building on an arbitrary (run-length) context-free grammar of size g, the index is of size $O(g)$ and finds all the MEMs in time $O(m^2 \log^\delta g)$, for any constant $\delta > 0$ (see also [6]). If the grammar is of a kind called locally consistent, the time improves to $O(m \log m (\log m + \log^\delta n))$. Other results (see [3,12]) require larger indices.

In this paper we consider the simpler problem of finding one longest common substring between P and T (i.e., a longest MEM). Further, we are satisfied with a common substring whose length is at least $1 - \epsilon$ times the longest one, for

Funded in part by NSERC grant RGPIN-07185-2020; NSF/BIO grant DBI-2029552; NIH/NHGRI grant R01HG011392; and Basal Funds FB0001, ANID, Chile.

F. M. Nardini et al. (Eds.): SPIRE 2023, LNCS 14240, pp. 246–252, 2023.
https://doi.org/10.1007/978-3-031-43980-3_19

some fixed $0 < \epsilon < 1$. We show that, on α-balanced grammars [4,14], this can be solved with high probability in time $O(m \log^\delta g)$ for any fixed constant $\delta > 0$. The correctness of the answer can be ensured in $O(m \log^\delta g)$ expected time.

2   Preliminaries

Our index uses grammar-based compression, which compresses a text $T[1..n]$ by building and storing a context-free grammar that generates only T [9]. We focus in particular on *straight-line programs (SLPs)*, where each rule is of the form $X \to Y Z$, where Y and Z are terminals or nonterminals (called symbols). If T is repetitive, then it can be represented with an SLP of g rules, with $g \ll n$. Grammar-based indices [5] aim to use space linear in the grammar size while offering indexed searches for patterns $P[1..m]$, that is, enumerating all the positions in T where P occurs. Following Charikar et al. [4], we write $\langle X \rangle$ and $[X]$ to denote the string symbol X expands to and the length of that expansion, respectively. Our work builds on α-balanced SLPs, defined next. There exist practical constructions of small α-balanced grammars from repetitive texts [14].

Definition 1 ([4]).   *For a constant $0 < \alpha \leq 1/2$, an SLP is said to be α-balanced if, for every rule $X \to Y Z$, it holds that*

$$\frac{\alpha}{1-\alpha} \leq \frac{[Y]}{[Z]} \leq \frac{1-\alpha}{\alpha} .$$

3   Data Structure

Our data structure is built from an α-balanced SLP G. For each nonterminal X in this SLP, the structure stores a set of prefixes and suffixes of $\langle X \rangle$, of exponentially increasing lengths. Those are called prefix and suffix blocks, respectively.

Definition 2. *Let X be a symbol in G and fix a constant $0 < \epsilon < 1$. Then, for each $0 \leq k \leq \log_{1/(1-\epsilon)}[X]$, we call $\langle X \rangle[1..\lceil 1/(1-\epsilon)^k \rceil]$ a prefix block and $\langle X \rangle[[X]-\lceil 1/(1-\epsilon)^k \rceil+1..[X]]$ a suffix block.*

Precisely, given ϵ, consider the following sets:

$$\mathcal{X} = \{\langle X \rangle, X \text{ is a symbol in } G\},$$
$$\mathcal{B}_{\text{pref}} = \{B, B \text{ is a prefix block of a symbol } X \text{ in } G\},$$
$$\mathcal{B}_{\text{suff}} = \{B, B \text{ is a suffix block of a symbol } X \text{ in } G\}.$$

For every prefix block $B \in \mathcal{B}_{\text{pref}}$, we compute B's Karp-Rabin [8] hash $h(B)$ and the lexicographic range $[s_B, e_B]$ of the strings in $\mathcal{X}$ that are prefixed by B. We store each pair $(h(B), [s_B, e_B])$ in a perfect hash table H_{pref}, with $h(B)$ as the key and $[s_B, e_B]$ as the value. Symmetrically, for each suffix block $B \in \mathcal{B}_{\text{suff}}$, we compute B's Karp-Rabin hash $h(B)$ and the co-lexicographic range $[s_B, e_B]$ of the strings in $\mathcal{X}$ that are suffixed by B, storing each pair $(h(B), [s_B, e_B])$

in a perfect hash table H_{suff} with $h(B)$ as the key and $[s_B, e_B]$ as the value. The Karp-Rabin hash function $h(B)$ is designed to have no collision between substrings of T, which can be built in $O(n \log n)$ expected time [1]. With low probability, however, there may be collisions between substrings of a pattern P and blocks of T.

We now show that $|\mathcal{B}_{\text{pref}}|$ and $|\mathcal{B}_{\text{suff}}|$ are $O(g)$, and therefore our hash tables are of size $O(g)$ as well.

Lemma 1. *If $X \to YZ$ is a rule in G, then only $O(1)$ prefix blocks $B \in \mathcal{B}_{\text{pref}}$ are prefixes of $\langle X \rangle$ but not of $\langle Y \rangle$, and only $O(1)$ suffix blocks $B \in \mathcal{B}_{\text{suff}}$ are suffixes of $\langle X \rangle$ but not of $\langle Z \rangle$.*

Proof. By Definition 1, we have

$$[X] = [Y] + [Z] \leq \left(1 + \frac{1-\alpha}{\alpha}\right) \cdot [Y] = \frac{[Y]}{\alpha},$$

so the number of prefix blocks that are prefixes of $\langle X \rangle$ but not $\langle Y \rangle$ is, by Definition 2,

$$\log_{\frac{1}{1-\epsilon}}[X] - \log_{\frac{1}{1-\epsilon}}[Y] + O(1) = \log_{\frac{1}{1-\epsilon}}\frac{[X]}{[Y]} + O(1) \leq \log_{\frac{1}{1-\epsilon}}\frac{1}{\alpha} + O(1) = O(1).$$

Symmetrically, because $[X] \leq [Z]/\alpha$, the number of suffix blocks that are suffixes of $\langle X \rangle$ but not of $\langle Z \rangle$ is $O(1)$. □

Corollary 1. *The number of prefix and suffix blocks is $|\mathcal{B}_{\text{pref}}| + |\mathcal{B}_{\text{suff}}| = O(g)$.*

Proof. By Lemma 1, each symbol X of G, of which there are g, contributes $O(1)$ prefix blocks to $\mathcal{B}_{\text{pref}}$ and $O(1)$ suffix blocks to $\mathcal{B}_{\text{suff}}$. □

The final component of our data structure is a discrete two-dimensional grid $\mathcal{G}$, with one row and one column per element of $\mathcal{X}$. Let

- $X \to YZ$ be a rule in G,
- $\langle Y \rangle$ have co-lexicographic position i in $\mathcal{X}$, and
- $\langle Z \rangle$ have lexicographic position j in $\mathcal{X}$,

then we set a point at position (i, j) in the grid. We label this point with the position where $\langle Y \rangle$ ends inside an occurrence of $\langle X \rangle$ in T (i.e., if we choose the occurrence $T[a..b] = \langle X \rangle$, then the label of the point is $a + [Y] - 1$). The grid has g points, thus it can be represented in $O(g)$ space and answer range emptiness queries in $O(\log^\delta g)$ time, for any constant $\delta > 0$ [2].

Our whole data structure then comprises H_{pref}, H_{suff}, and $\mathcal{G}$, which add up to $O(g)$ space. We note that the values $[s_B, e_B]$ stored in H_{pref} are the lexicographic ranges of grid columns corresponding to strings in $\mathcal{X}$ prefixed with B, and those stored in H_{suff} are the co-lexicographic ranges of grid rows corresponding to strings in $\mathcal{X}$ suffixed with B.

4   Queries

Our searches build on a key result used in all grammar-based indices [5].

Lemma 2. *Let string S, of length $|S| > 1$, appear in T. Then, there is an index $1 \leq p < |S|$ and a point (i, j) in $\mathcal{G}$ such that*

- *i is the co-lexicographic range of a string $\langle Y \rangle \in \mathcal{X}$ suffixed by $S[1..p]$ and*
- *j is the lexicographic range of a string $\langle Z \rangle \in \mathcal{X}$ prefixed by $S[p+1..|S|]$.*

Proof. Note that S appears as a substring of the expansion of the initial symbol and, possibly, of others. If we order the rules $X \to YZ$ so that Y and Z are listed before X, then the first time S appears as a substring of $\langle X \rangle$, it must appear as the concatenation of a nonempty suffix of $\langle Y \rangle$ and a nonempty prefix of $\langle Z \rangle$. The lemma then follows from the definition of $\mathcal{G}$. □

Now let L be the longest common substring of P and T and assume $|L| > 1$. Per Definition 2, let $k = \lfloor \log_{1/(1-\epsilon)} |L| \rfloor$. We note that

$$\left(\frac{1}{1-\epsilon} \right)^k > \left(\frac{1}{1-\epsilon} \right)^{\left(\log_{\frac{1}{1-\epsilon}} |L| \right) - 1} = (1 - \epsilon) \cdot |L|.$$

Thus, for our purposes, it suffices to find a substring of length $\ell = (1/(1-\epsilon))^k$ of L. By Lemma 2, there exists an index $1 \leq p < |L|$ such that $L_Y = L[1..p]$ suffixes some $\langle Y \rangle \in \mathcal{X}$, $L_Z = L[p+1..|L|]$ prefixes some $\langle Z \rangle \in \mathcal{X}$, and there is a rule $X \to YZ$ in G. Further, let $k_Y = \lfloor \log_{1/(1-\epsilon)} |L_Y| \rfloor$ and $k_Z = \lfloor \log_{1/(1-\epsilon)} |L_Z| \rfloor$. By the same argument above, it follows that

$$\left(\frac{1}{1-\epsilon} \right)^{k_Y} > (1 - \epsilon) \cdot |L_Y| \text{ and } \left(\frac{1}{1-\epsilon} \right)^{k_Z} > (1 - \epsilon) \cdot |L_Z|.$$

Therefore, it suffices to find a suffix of length $\ell_Y = \lceil (1/(1-\epsilon))^{k_Y} \rceil$ of $\langle Y \rangle$ and a prefix of length $\ell_Z = \lceil (1/(1-\epsilon))^{k_Z} \rceil$ of $\langle Z \rangle$ to form a substring of L of length $\ell_Y + \ell_Z > (1 - \epsilon) \cdot (|L_Y| + |L_Z|) = (1 - \epsilon) \cdot |L|$, because $L = L_Y \cdot L_Z$.

Per Definition 2, those suffixes $L'_Y = L_Y[|L_Y| - \ell_Y + 1..\ell_Y]$ are suffix blocks, and those prefixes $L'_Z = L_Z[1..\ell_Z]$ are prefix blocks, and therefore they are stored in our hash tables. Thus, if we search H_{suff} for L'_Y and retrieve the associated range $[s_Y, e_Y]$, and search H_{pref} for L'_Z and retrieve the associated range $[s_Z, e_Z]$, we will find a point in the (row,column) range $[s_Y, e_Y] \times [s_Z, e_Z]$ of $\mathcal{G}$.

The correctness of Algorithm 1 stems from this discussion. A position of the common substring found is obtained by noticing that, when we assign ℓ in line 12, the string occurs at $P[p - \ell_Y + 1..p + \ell_Z]$ and $T[t - \ell_Y + 1..t + \ell_Z]$, where t is the label of any point in the grid range.

Since we do not know $|L|$ beforehand, the algorithm tries all the possible values for k_Y and k_Z, which yields a time complexity dominated by $O(m \log^2 m)$ range emptiness queries, that is, $O(m \log^2 m \log^\delta n)$ [2]. We note that, since the hashes are of Karp-Rabin type, we can precompute in $O(m)$ time the hash of

Algorithm 1. The simple algorithm returning an approximation to the length of the longest common substring between T and $P[1..m]$.

1: $\ell \leftarrow 0$
2: **for** $p \leftarrow 1$ to m **do**
3:     **for** $k_Y \leftarrow 0$ to $\lfloor \log_{1/(1-\epsilon)} p \rfloor$ **do**
4:         $\ell_Y \leftarrow \lceil (1/(1-\epsilon))^{k_Y} \rceil$
5:         $[s_Y, e_Y] \leftarrow$ search H_{suff} for $P[p-\ell_Y+1..p]$
6:         **if** $[s_Y, e_Y]$ was found **then**
7:             **for** $k_Z \leftarrow 0$ to $\lfloor \log_{1/(1-\epsilon)} (m-p) \rfloor$ **do**
8:                 $\ell_Z \leftarrow \lceil (1/(1-\epsilon))^{k_Z} \rceil$
9:                 $[s_Z, e_Z] \leftarrow$ search H_{pref} for $P[p+1..p+\ell_Z]$
10:                 **if** $[s_Z, e_Z]$ was found **then**
11:                     **if** $\mathcal{G}$ has a point in $[s_Y, e_Y] \times [s_Z, e_Z]$ **then**
12:                         $\ell \leftarrow \max(\ell, \ell_Y + \ell_Z)$
13: **return** ℓ

---

every prefix, $h(P[1..p])$, and then we can compute in constant time the hash of every substring of P by operating with the modular inverses of the hashes [13]. If there is a collision we may find a false positive.

Note that Algorithm 1 will find only the empty string if $|L| = 1$, as we assumed $|L| > 1$. In case the algorithm returns zero, we must determine if $|L| = 1$ by checking if some symbol of P appears as a terminal in G; this is easily done with additional $O(m)$ time and $O(g)$ space.

5   Faster Queries

We can reduce the time complexity of Algorithm 1 by decreasing the number of combinations (k_Y, k_Z) we explore. The algorithm may try out $\Theta(\log^2 m)$ combinations per value of p, but several of those are redundant. For example, if the range $[s_Y, e_Y] \times [s_Z, e_Z]$ corresponding to the pair (k_Y, k_Z) is empty, then so is the range $[s'_Y, e'_Y] \times [s_Z, e_Z]$ corresponding to $(k_Y + 1, k_Z)$, as well as the range $[s_Y, e_Y] \times [s'_Z, e'_Z]$ corresponding to $(k_Y, k_Z + 1)$. It then suffices to explore *maximal* combinations (k_Y, k_Z). Further redundant work is done among values of p: we may be working on maximal combinations (k_Y, k_Z) that nevertheless yield shorter strings than one we had already obtained with a previous value of p.

To avoid redundant work, we will visit only the combinations (k_Y, k_Z) for which $\ell_Y + \ell_Z > \ell$; recall that ℓ is the maximum length $\ell_Y + \ell_Z$ obtained so far. Therefore, every time we find a nonempty range in $\mathcal{G}$, the value of ℓ increases. We say those combinations are *useful*. The other combinations, where either the searches in H_{pref} or in H_{suff} fail, or they succeed but the resulting range in $\mathcal{G}$ is empty, are *useless*. We will count useful and useless combinations separately.

Since there are only $O(\log^2 m)$ combinations (k_Y, k_Z), there exist $O(\log^2 m)$ different values $\ell_Y + \ell_Z$. Since the value of ℓ never decreases along the process, there are only $O(\log^2 m)$ situations in which a new value of $\ell_Y + \ell_Z$ can increase ℓ. This implies that the total number of useful combinations we visit is $O(\log^2 m)$.

To keep the number of useless combinations low, we will visit the space (k_Y, k_Z) in some suitable order. We first consider all the combinations where $k_Y \geq k_Z$, and then where $k_Z > k_Y$. We analyze the former case; the other is symmetric. We visit the values of k_Y in increasing order, and the values of k_Z in increasing order for each value of k_Y. Each new visited value k_Y is first combined with the smallest k_Z for which $\ell_Y + \ell_Z > \ell$. If this leads to a nonempty range in $\mathcal{G}$, then this is a useful combination, for which we have already accounted. The successive values of k_Z we try out from there are all useful, until we finally fail to find a nonempty range—and this then a useless combination— or until $k_Z > k_Y$. We do not consider further values $k_Z \leq k_Y$ in the first case because they will also fail to produce a nonempty range in $\mathcal{G}$.

Thus, each value of k_Y we visit leads to zero or more useful combinations possibly followed by a single useless one. We say that k_Y *succeeds* if it produces at least one useful combination; otherwise it *fails*. If k_Y succeeds, then the cost of its last useless combination, if any, can be charged to the useful ones it produced. Therefore we only need to count the number of values k_Y that fail. We will now show that a sequence of consecutive values of k_Y that fail has $O(1)$ combinations (all of them useless), and therefore their cost can also be charged to the preceding or following value of k_Y that succeeds. Only a sequence of all-failing values of k_Y cannot be accounted for in that way, but this can only be one sequence per value of p, adding up to $O(m)$ cost for the useless combinations.

The value of ℓ does not change across a sequence of failing values of k_Y. We never visit values $\ell_Y \leq \ell/2$: since $\ell_Z \leq \ell_Y$, they could not increase ℓ. A failing sequence of visited values k_Y then starts with some $\ell_Y > \ell/2$ and increments k_Y successively, combining it with nonincreasing values of k_Z. In this sequence, the first combination (k_Y, k_Z) we try for each k_Y, with the smallest k_Z that yields $\ell_Y + \ell_Z > \ell$, is useless, so we visit only that smallest value of k_Z per value of k_Y. We proceed increasing k_Y, always failing, until ℓ_Y exceeds ℓ, at which point the smallest value of k_Z that makes $\ell_Y + \ell_Z > \ell$ is 0. If such combination also fails, there is no point in continuing with larger values of ℓ_Y, because even combined with $k_Z = 0$ will not yield a useful combination. Since ℓ_Y is exponential in k_Y, there are only $O(1)$ values of k_Y that yield values $\ell/2 < \ell_Y \leq \ell$. Only $O(1)$ combinations are then tried along a sequence of failing values of k_Y.

Overall, we have $O(\log^2 m)$ steps charged to useful combinations and $O(m)$ to useless ones. Multiplied by the range emptiness time complexity, this yields $O(m \log^\delta g)$ total time. Note that we obtain a correct result only with high probability, because we check only that $h(L_Y)$ and $h(L_Z)$ match the hash values of the corresponding block prefixes and suffixes. To ensure correctness, we can store the nonterminal $X \to Y Z$ associated with the point connecting $\langle Y \rangle$ and $\langle Z \rangle$ in $\mathcal{G}$, so as to verify the correctness our answer in $O(m)$ time by extracting a suffix of $\langle Y \rangle$ and a prefix of $\langle Z \rangle$ in optimal time [7]. If our answer turns out to be incorrect (which happens with low probability) we can re-run the algorithm, this time verifying every potentially useful combination, in total time $O(m^2)$. We can thus ensure correct results by making our time $O(m \log^\delta n + m + n^{-c} m^2) = O(m \log^\delta n)$ in expectation (for any constant $c > 2$).

The construction time of our structure is dominated by the construction of the Karp-Rabin hash function with no collisions between blocks of T [13, Sec. 4].

Theorem 1. *Given positive constants ϵ and δ, and an α-balanced straight-line program with g rules for a text $T[1..n]$, we can build in $O(n \log n)$ expected time an $O(g)$-space index with which, given a pattern $P[1..m]$, in $O(m \log^\delta g)$ time we can find with high probability a substring of P that occurs in T and whose length is at least a $(1 - \epsilon)$ fraction of the longest common substring of P and T. The correctness can be guaranteed with time still $O(m \log^\delta g)$, yet in expectation.*

References

1. Bille, P., Gørtz, I.L., Sach, B., Vildhøj, H.W.: Time-space trade-offs for longest common extensions. J. Discrete Algorithms **25**, 42–50 (2014)
2. Chan, T.M., Larsen, K.G., Pǎtraşcu, M.: Orthogonal range searching on the RAM, revisited. In: Proceedings of the 27th ACM Symposium on Computational Geometry (SoCG), pp. 1–10 (2011)
3. Charalampopoulos, P., Kociumaka, T., Pissis, S.P., Radoszewski, J.: Faster algorithms for longest common substring. In: Proceedings of the 29th Annual European Symposium on Algorithms (ESA), pp. 30:1–30:17 (2021)
4. Charikar, M., et al.: The smallest grammar problem. IEEE Trans. Inf. Theory **51**(7), 2554–2576 (2005)
5. Claude, F., Navarro, G., Pacheco, A.: Grammar-compressed indexes with logarithmic search time. J. Comput. Syst. Sci. **118**, 53–74 (2021)
6. Gao, Y.: Computing matching statistics on repetitive texts. In: Proceedings of the 32nd Data Compression Conference (DCC), pp. 73–82 (2022)
7. Gasieniec, L., Kolpakov, R., Potapov, I., Sant, P.: Real-time traversal in grammar-based compressed files. In: Proceedings of the 15th Data Compression Conference (DCC), p. 458 (2005)
8. Karp, R.M., Rabin, M.O.: Efficient randomized pattern-matching algorithms. IBM J. Res. Dev. **2**, 249–260 (1987)
9. Kieffer, J.C., Yang, E.H.: Grammar-based codes: a new class of universal lossless source codes. IEEE Trans. Inf. Theory **46**(3), 737–754 (2000)
10. Navarro, G.: Indexing highly repetitive string collections, part I: repetitiveness measures. ACM Comput. Surv. **54**(2) (2021). Article 29
11. Navarro, G.: Indexing highly repetitive string collections, part II: compressed indexes. ACM Comput. Surv. **54**(2) (2021). Article 26
12. Navarro, G.: Computing MEMs on repetitive text collections. In: Proceedings of the 34th Annual Symposium on Combinatorial Pattern Matching (CPM), p. article 22 (2023)
13. Navarro, G., Prezza, N.: Universal compressed text indexing. Theor. Comput. Sci. **762**, 41–50 (2019)
14. Ohno, T., Goto, K., Takabatake, Y., I, T., Sakamoto, H.: LZ-ABT: a practical algorithm for α-balanced grammar compression. In: Iliopoulos, C., Leong, H.W., Sung, W.-K. (eds.) IWOCA 2018. LNCS, vol. 10979, pp. 323–335. Springer, Cham (2018). https://doi.org/10.1007/978-3-319-94667-2_27

On the Number of Factors in the LZ-End Factorization

Paweł Gawrychowski[1(✉)], Maria Kosche[2], and Florin Manea[2]

[1] Faculty of Mathematics and Computer Science, University of Wrocław, Wrocław, Poland
gawry@cs.uni.wroc.pl
[2] Computer Science Department and CIDAS, Göttingen University, Göttingen, Germany
{maria.kosche,florin.manea}@cs.uni-goettingen.de

Abstract. Kreft and Navarro [DCC 2010] introduced a restricted variant of the well-known Lempel-Ziv factorization, called the LZ-End factorization. Only recently Kempa and Saha [SODA2022] were able to obtain a good upper bound on the size of the LZ-End factorization in terms of the size of the LZ factorization. We extend their approach to improve the upper bound by a doubly-logarithmic factor.

Keywords: LZ factorization · LZ-End factorization · compressibility

1 Introduction

The Lempel-Ziv factorization is considered to provide a natural measure of compressibility of texts. It was introduced in [14] in order to define the Lempel-Ziv (LZ77) lossless compression algorithm. Currently, this is among the most commonly used such algorithms (as reflected in [13]), and many popular compression formats, such as gzip or png, are based on it (see also [1,3]). While, by now, we have algorithms for computing the LZ factorization of a given string that are efficient both in theory and practice [4,6,9], it is desirable to augment the stored compressed representation with a structure allowing random access to the underlying string. Denoting by z the size of the LZ factorization of a string of length n, we know how to support such queries in $O(\log n)$ time with a structure of size $O(z \log(n/z))$ by building a balanced grammar, and it is not known if the extra $\log(n/z)$ factor is necessary. Kreft and Navarro [11] proposed a variant of the LZ factorization, dubbed the LZ-End factorization, which allows decompressing arbitrary phrases in optimal time with a linear-size structure. While computable in linear time [8] and compressed space [7], for quite some time it was not known how the size of the new factorization, denoted z_e, relates to that of the LZ factorization, except that z_e/z_{no} has a lower bound approaching 2 [5,12], where z_{no} denotes the number of phrases in the so-called nonoverlapping

Maria Kosche's work was supported by the DFG project number 389613931. Florin Manea's work was supported by the DFG Heisenberg-project number 466789228.

LZ factorization. Then, Kempa and Saha [10] provided the first upper bound of $z_e = O(z \log^2(n/z))$. Later, computing an optimal LZ-End factorization was shown to be NP-hard [2].

Previous Work. The high-level idea of the proof of Kempa and Saha [10] is as follows. We call a phrase of the LZ-End factorization special when its length is at least half of that of the previous phrase. It is immediate that the total number of phrases is larger than the number of special phrases by at most a factor of $\log n$, hence it is enough to upper bound the number of special phrases. This is done by charging each special phrase of length ℓ to roughly ℓ distinct substrings of the same length 2^k, and arguing that each distinct substring will be charged at most twice. On the other hand, it is known that the number of distinct substrings of the same length m is at most mz. Altogether, this upper bounds the number of special phrases by $O(z \log n)$. Refining the upper bound by replacing n with n/z follows by standard arguments.

Our Result. We improve the upper bound of Kempa and Saha [10] by a factor of $\log\log(n/z)$. This is achieved by following their approach, however we are more careful about upper bounding the number of phrases that are not special. For an integer $b > 2$, we call a phrase b-shrinking when its length is smaller than that of the previous phrase by a factor of b. Firstly, we show that the number of b-shrinking phrases cannot exceed $O\left(\frac{z \log n}{\log b}\right)$. Secondly, we argue that the number of remaining phrases is $O(zb \log n)$ by again charging each such phrase of length ℓ to roughly ℓ distinct substrings of the same length 2^k, except that now each distinct substring is charged at most b times. Altogether, this allows us to upper bound the total number of phrases by $O\left(\frac{z \log n}{\log b} + zb \log n\right)$. By adjusting b we obtain that $z_e = O\left(\frac{z \log^2 n}{\log\log n}\right)$, and finally by a more careful analysis we reach $z_e = O\left(\frac{z \log^2(n/z)}{\log\log(n/z)}\right)$.

2 Preliminaries

We consider strings over an alphabet Σ. For two strings S and T, ST denotes their concatenation. For a string $S = s_1 s_2 \cdots s_\ell$, with $s_1, s_2, \ldots, s_\ell \in \Sigma$, $|s|$ denotes its length ℓ, $S[i]$ refers to the i-th character of S, and $S[i:j]$ refers to the substring $s_i s_{i+1} \cdots s_j$. By convention, $S(i:j) = S[i+1:j]$. When $i = 1$ (respectively, $j = |S|$) then $S[i:j]$ is called a prefix (respectively, a suffix) of S. We say that p is a period of $S[1:n]$ when $S[i] = S[i+p]$ for every $i \in [1, n-p]$.

For a string T of length n ending with a character $\$$ that does not appear anywhere else, we consider two factorizations of T.

The LZ factorization is a factorization $T = F_1 F_2 \ldots F_z$, where $F_1, \ldots, F_z$ are called the LZ phrases. The i-th phrase F_i is chosen as the longest prefix of $T[k:|T|]$, where $k = 1 + |F_1 \cdots F_{i-1}|$, that also occurs starting at a position smaller than k in T. If there is no such prefix then $F_i = T[k]$. This definition is slightly different than the one of Kreft and Navarro [11], who define the next phrase to always include an additional character. The bounds achieved by Kempa and Saha [10] and our improvement easily extend to the original definition.

The LZ-End factorization is given by $T = \alpha_1\alpha_2 \ldots \alpha_{z_e}$, where $\alpha_1, \ldots, \alpha_{z_e}$ are called the LZ-End phrases. In this case, the i-th phrase α_i is chosen as the longest prefix of $T[k : |T|]$, where $k = 1 + |\alpha_1 \cdots \alpha_{i-1}|$, that also occurs earlier in T ending at the end of a previous phrase. Formally, we want α_i to be a suffix of $T[1 : |\alpha_1 \cdots \alpha_j|]$ for some $j < i$. If there is no such prefix, then $\alpha_i = T[k]$.

All logarithms appearing in this paper are in base 2.

3   Our Result

In this section, we analyse the number of factors in the LZ-End factorization of a word T, of length n, and, in particular, their relation to the number of factors in the LZ factorization of T. In this setting, we want to show the following theorem.

Theorem 1. *Let $b > 2$ be a natural number. Then $z_e = O\left(\frac{z \log^2 n}{\log b} + zb \log n\right)$.*

Framework. We start by introducing several concepts and notations which are useful in our proof.

A substring $T[i : i+2\ell-1]$ of the string T is centered in $T[x : y]$ (respectively, in $T(x : y]$) if and only if $x \leq i + \ell \leq y$ (respectively, $x < i + \ell \leq y$).

Let $\alpha_1 = T(0 : e_1], \ldots, \alpha_{z_e} = T(e_{z_e-1} : e_{z_e}]$ be the phrases of the LZ-End factorization of T. That is, $T = \alpha_1\alpha_2 \cdots \alpha_{z_e} = T(0 : e_1]T(e_1 : e_2] \cdots T(e_{z_e-1}, e_{z_e}]$. For uniformity, let $e_0 = 0$. It is immediate that, according to the definition of the LZ-End factorization, we have $|\alpha_1| = |\alpha_{z_e}| = 1$; in particular, $\alpha_{z_e} = \$$ and $e_{z_e} = n$. Further, let $\ell_j = |\alpha_j| = e_j - e_{j-1}$, for $j \in [1 : z_e]$.

Let $S_m = \{w \mid w \text{ is a length-}m \text{ substring of } T\#^\infty\}$, where $\#$ is a letter not occurring in T. In [10] it is shown that $|S_m| \leq mz$. For simplicity of exposure, we use the following notation: for some $a \leq n$ and $\ell \in \mathbb{N}$, $T[a : a + \ell - 1]$ is defined as the prefix of length ℓ of $T\#^\infty$; that is, if $a + \ell - 1 > n$ then we simply define $T[a : a + \ell - 1]$ as the suffix $T[a : n]$ of T padded with $\#$ up to the desired length. Also, to avoid unnecessary case analyses, we call strings $T[a : a + \ell - 1]$ defined as above substrings of T although, in fact, they might be substrings of $T\#^\infty$, which only start in T but end after the last position of T in $T\#^\infty$.

Special Phrases. Following [10], we define a phrase α_j of the LZ-End factorization of T to be special if and only if $j \in \{1, z_e\}$, or $j \in [2 : z_e - 1]$ and $2\ell_j \geq \ell_{j-1}$. Let z_e' be the number of special phrases in the LZ-End factorization of T. In [10] it is shown that $z_e' = O(z \log n)$. Let $S \subseteq [1 : z_e]$ be the set such that $j \in S$ if and only if α_j is special.

b-Shrinking Phrases. For $j \geq 2$, the phrase α_j is called b-shrinking if $b\ell_j \leq \ell_{j-1}$. Let z_b be the number of b-shrinking phrases in the LZ-End factorization of T. Let $B \subseteq [1 : z_e]$ be the set such that $j \in B$ if and only if α_j is b-shrinking.

Lemma 1. $z_b = O\left(\frac{z \log^2 n}{\log b}\right)$.

Proof. We first note that for any phrase α_j with $j \geq 2$ we trivially have $\log \ell_j - \log \ell_{j-1} \leq \log n$. If α_j is b-shrinking, then $\log \ell_j + \log b \leq \log \ell_{j-1}$, so $\log b \leq \log \ell_{j-1} - \log \ell_j$ holds. Clearly, special phrases are not b-shrinking, so $S \subseteq [1 : z_e] \setminus B$. If α_j is not special then $2\ell_j < \ell_{j-1}$, so in particular $\log \ell_j - \log \ell_{j-1} \leq 0$.

We observe that $\log \ell_1 = \log \ell_{z_e} = 0$, so we have the following equality:

$$\sum_{j=1}^{z_e} (\log \ell_j - \log \ell_{j-1}) = \log \ell_{z_e} - \log \ell_1 = 0.$$

This implies:

$$\sum_{j \in B} (\log \ell_{j-1} - \log \ell_j) = \sum_{j \in [1:z_e] \setminus B} (\log \ell_j - \log \ell_{j-1}).$$

Further, by the definition of B we have $z_b \log b \leq \sum_{j \in B} (\log \ell_{j-1} - \log \ell_j)$, hence $z_b \log b \leq \sum_{j \in [1:z_e] \setminus B} (\log \ell_j - \log \ell_{j-1})$. Next, we observe that the following holds:

$$\sum_{j \in [1:z_e] \setminus B} (\log \ell_j - \log \ell_{j-1})$$
$$= \sum_{j \in S} (\log \ell_j - \log \ell_{j-1}) + \sum_{j \in [1:z_e] \setminus (S \cup B)} (\log \ell_j - \log \ell_{j-1})$$
$$\leq \sum_{j \in S} (\log \ell_j - \log \ell_{j-1}) \leq z_e' \log n,$$

using that for all $j \in [1 : z_e] \setminus (B \cup S)$ we have $\log \ell_j - \log \ell_{j-1} \leq 0$.

Altogether, $z_b \log b \leq z_e' \log n$. Thus, $z_b = O\left(\frac{z_e' \log n}{\log b}\right)$ and, as $z_e' = O(z \log n)$, the conclusion follows. ∎

Analysis of Phrases Which are Neither Special Nor b-Shrinking.

Let α_j be a phrase of the LZ-End factorization of T which is neither special nor b-shrinking. In this case, we have $2\ell_j < \ell_{j-1} < b\ell_j$. It follows that $2 < j < z_e$ holds; in particular, $j > 2$ because $\ell_1 = 1$, so $2\ell_2$ cannot be smaller than 1.

For α_j, let k_j be such that $2^{k_j} \leq 10\ell_{j-1} < 2^{k_j+1}$. We now consider all substrings $x = T[i - 2^{k_j-1} : i + 2^{k_j-1} - 1]$ with $i \in (e_{j-2} : e_{j-1}]$. That is $x = T[i - 2^{k_j-1} : i + 2^{k_j-1} - 1]$ is a substring of T of length 2^{k_j} centered in α_{j-1}. Let X_j be the multiset containing all such substrings.

Note that if $T[i - 2^{k_j-1} : i + 2^{k_j-1} - 1] \in X_j$, then $T[i : i + 2^{k_j-1} - 1]$ has length 2^{k_j-1} and $2^{k_j-1} \geq 2\ell_{j-1} + \ell_j$. Thus, $i - 2^{k_j-1} \leq e_{j-2} < e_{j-1} < e_j \leq i + 2^{k_j-1} - 1$, i.e., $T[i - 2^{k_j-1} : i + 2^{k_j-1} - 1]$ contains $\alpha_{j-1}\alpha_j$ as a substring.

We want to show the following two claims:

A: Each two strings of X_j are distinct.
B: Each substring contained in some set X_j is contained in at most $10b$ sets X_t.

Lemma 2 (Claim A). *Each two strings of X_j are distinct.*

Proof. Consider two substrings $T[x - 2^{k_j-1} : x + 2^{k_j-1} - 1]$ and $T[y - 2^{k_j-1} : y + 2^{k_j-1} - 1]$ from X_j, with $x < y$. Assume, for the sake of contradiction, that $T[x - 2^{k_j-1} : x + 2^{k_j-1} - 1] = T[y - 2^{k_j-1} : y + 2^{k_j-1} - 1]$. See Fig. 1.

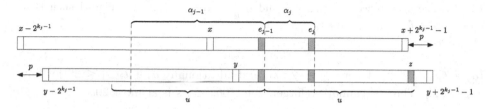

Fig. 1. The alignment of $T[x - 2^{k_j-1} : x + 2^{k_j-1} - 1]$ and $T[y - 2^{k_j-1} : y + 2^{k_j-1} - 1]$ in the proof of Lemma 2. Note that, in this figure, $|u|$ is divisible by p.

It is immediate that $p = y - x \le \ell_{j-1}$ is a period of $T[x - 2^{k_j-1} : y + 2^{k_j-1} - 1]$. Let z be the position of T such that $(z - e_{j-1}) \bmod p = 0$ and $z \in [x + 2^{k_j-1} : y + 2^{k_j-1} - 1]$. Then, $u = T(e_{j-1} : z]$ is a suffix of $T[x - 2^{k_j-1} : e_{j-1}]$ and has α_j as a proper prefix. This is a contradiction with the choice of α_j as a factor of the LZ-End factorization of T: the jth phrase of this factorization should have been u. Therefore, our assumption that $T[x - 2^{k_j-1} : x + 2^{k_j-1} - 1] = T[y - 2^{k_j-1} : y + 2^{k_j-1} - 1]$ is false. The conclusion of the lemma now follows. $\square$

Consequently, $|X_j| \ge \frac{2^{k_j}}{10}$, by the choice of k_j and $|X_j| = \ell_{j-1}$.

Lemma 3 (Claim B). *Assume that w is a word that appears in $X_{i_1}, \dots, X_{i_s}$, for some $s \ge 1$, with $i_1 < i_2 < \cdots < i_s$. Then, $s \le 20b$.*

Proof. Assume $|w| = 2^k$. Let $s' = \lfloor (s+1)/2 \rfloor$, and define $i'_j = i_{2j-1}$ for every $j \in [1, s']$. Recall that, for every $j \in [2, s']$, if $w \in X_{i'_j}$ then w has its occurrence $w = T[x_j : x_j + 2^k - 1]$ that fully contains the phrases $\alpha_{i'_j-1}\alpha_{i'_j}$. Let $a_j = e_{i'_j-2} - x_j + 1$. Then, $w[a_j : 2^k]$ starts with $\alpha_{i'_j-1}\alpha_{i'_j}$.

For any $j \in [2 : s' - 1]$, we want to show that $a_j + \ell_{i'_j-1} + \ell_{i'_j} \le a_{j+1} + \ell_{i'_{j+1}-1}$; that is, the starting position of $\alpha_{i'_{j+1}}$ from the occurrence of $\alpha_{i'_{j+1}-1}\alpha_{i'_{j+1}}$ in w is after the ending position of the occurrence of $\alpha_{i'_j-1}\alpha_{i'_j}$ in w.

Assume otherwise, so $a_j + \ell_{i'_j-1} + \ell_{i'_j} - 1 \ge a_{j+1} + \ell_{i'_{j+1}-1}$; see Fig. 2. Then, $w[a_{j+1} : a_j + \ell_{i'_j-1} + \ell_{i'_j} - 1]$ is a suffix of $\alpha_1 \cdots \alpha_{i'_j}$ and $|w[a_{j+1} : a_j + \ell_{i'_j-1} + \ell_{i'_j} - 1]| > \ell_{i'_{j+1}-1}$. We observe that $i'_j < i'_{j+1} - 1$ by the definition of the i'_js, so this this means that the phrase $\alpha_{i'_{j+1}-1}$ was not chosen correctly when the LZ-End factorization of T was constructed, a contradiction.

Then, $a_j + \ell_{i'_j-1} + \ell_{i'_j} \le a_{j+1} + \ell_{i'_{j+1}-1}$, in other words $(a_{j+1} + \ell_{i'_{j+1}-1}) - (a_j + \ell_{i'_j-1}) \ge \ell_{i'_j}$. By the choice of k, $\ell_{i'_j} \ge \frac{2^k}{10b}$, so in fact $(a_{j+1} + \ell_{i'_{j+1}-1}) -$

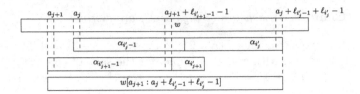

Fig. 2. The occurrences of $\alpha_{i'_j-1}\alpha_{i'_j}$ and $\alpha_{i'_{j+1}-1}\alpha_{i'_{j+1}}$ within w, aligned such that $a_j + \ell_{i'_j-1} + \ell_{i'_j} - 1 \geq a_{j+1} + \ell_{i'_{j+1}-1} - 1$.

$(a_j + \ell_{i'_j-1}) \geq \frac{2^k}{10b}$ for any $j \in [2 : s]$, and additionally $a_j + \ell_{i'_j-1} \leq 2^k$ for any $j \in [1 : s']$. Thus, s' cannot be larger than $10b$, and s is at most $20b$.   □

We are now ready to analyse the number of phrases which are neither special nor b-shrinking. We first note that, for each $j \in [1 : z_e]$, we have $k_j \leq \lceil \log n \rceil + 4$. We choose k such that $k \leq \lceil \log n \rceil + 4$, and let $P_k = \{j \in [1 : z_e] \mid k_j = k\}$. If $j \in P_k$ then, by the corollary of Lemma 2, we have $|X_j| \geq \frac{2^k}{10}$. Consequently, $\sum_{j \in P_k} |X_j| \geq \frac{|P_k|2^k}{10}$. However, $\sum_{j \in P_k} |X_j| \leq 10b|S_{2^k}|$, as each distinct string of S_{2^k} may appear in at most $20b$ sets X_j by Lemma 3. Thus, $\frac{|P_k|2^k}{10} \leq 20b|S_{2^k}|$. As observed in [10], $|S_{2^k}| \leq 2^k z$, so altogether $|P_k| = O(bz)$. Summing this up over all $k \leq \lceil \log n \rceil + 4$, we get that the number of phrases which are neither special nor b-shrinking is $O(zb\log n)$.

Proof (of Theorem 1). We separately analyse the number of special phrases, the number of b-shrinking phrases, and the number of phrases which are neither special nor b-shrinking. Altogether, we get that the total number of phrases in the LZ-End factorization of T is $O(z\log n) + O\left(\frac{z\log^2 n}{\log b}\right) + O(zb\log n)$.   □

Taking, for instance, $b = \sqrt{\log n}$ in the statement of Theorem 1, we immediately obtain the following result:

Theorem 2. $z_e = O\left(\frac{z\log^2 n}{\log\log n}\right)$

Refined Upper Bound. We conclude with a refined version of the upper bound from Theorem 2. Consider the proof of Lemma 1, where we have upper bounded $\sum_{j \in S}(\log \ell_j - \log \ell_{j-1})$ by $z'_e \log n$, and then z'_e by $O(z\log n)$. The former can be actually upper bounded by $O(z'_e \log(n/z'_e))$, because $|S| = z'_e$ and $\sum_{j \in S}\ell_j \leq n$, and by [10, Lemma 3.2], we have $z'_e = O(z\log(n/z))$. Altogether:

$$\sum_{j \in S}(\log \ell_j - \log \ell_{j-1}) = O(z'_e\log(n/z'_e)) = O(z\log(n/z)\log(n/z'_e))$$

$$= O(z\log(n/z)\log\frac{n}{z\log(n/z)}) = O(z\log^2(n/z)).$$

Hence, $z_b = O\left(\frac{z\log^2(n/z)}{\log b}\right)$. Next, when analysing the number of phrases that are neither special nor b-shrinking we summed up over all $k \leq \lceil\log n\rceil + 4$, but in fact it is enough to sum up over all $k \leq \lceil\log(n/z)\rceil + 4$, as the number of phrases of length exceeding n/z is at most z. Thus, the number of such phrases is $O(zb\log(n/z))$. The overall number of phrases is now $O(z\log(n/z)) + O\left(\frac{z\log^2(n/z)}{\log b}\right) + O(zb\log(n/z))$, which is $O\left(\frac{z\log^2(n/z)}{\log\log(n/z)}\right)$, for $b = \sqrt{\log(n/z)}$.

References

1. Alakuijala, J., et al.: Brotli: a general-purpose data compressor. ACM Trans. Inf. Syst. (TOIS) **37**(1), 1–30 (2018)
2. Bannai, H., Funakoshi, M., Kurita, K., Nakashima, Y., Seto, K., Uno, T.: Optimal LZ-End parsing is hard. In: CPM. LIPIcs, vol. 259, pp. 3:1–3:11. Schloss Dagstuhl - Leibniz-Zentrum für Informatik (2023)
3. Collet, Y., Kucherawy, M.: Zstandard compression and the application/zstd media type. Techical report (2018)
4. Goto, K., Bannai, H.: Simpler and faster Lempel Ziv factorization. In: DCC, pp. 133–142. IEEE (2013)
5. Ideue, T., Mieno, T., Funakoshi, M., Nakashima, Y., Inenaga, S., Takeda, M.: On the approximation ratio of LZ-End to LZ77. In: Lecroq, T., Touzet, H. (eds.) SPIRE 2021. LNCS, vol. 12944, pp. 114–126. Springer, Cham (2021). https://doi.org/10.1007/978-3-030-86692-1_10
6. Kärkkäinen, J., Kempa, D., Puglisi, S.J.: Linear time Lempel-Ziv factorization: simple, fast, small. In: Fischer, J., Sanders, P. (eds.) CPM 2013. LNCS, vol. 7922, pp. 189–200. Springer, Heidelberg (2013). https://doi.org/10.1007/978-3-642-38905-4_19
7. Kempa, D., Kosolobov, D.: LZ-end parsing in compressed space. In: DCC, pp. 350–359. IEEE (2017)
8. Kempa, D., Kosolobov, D.: LZ-end parsing in linear time. In: ESA. LIPIcs, vol. 87, pp. 53:1–53:14. Schloss Dagstuhl - Leibniz-Zentrum für Informatik (2017)
9. Kempa, D., Puglisi, S.J.: Lempel-Ziv factorization: simple, fast, practical. In: ALENEX, pp. 103–112. SIAM (2013)
10. Kempa, D., Saha, B.: An upper bound and linear-space queries on the LZ-End parsing. In: SODA, pp. 2847–2866. SIAM (2022)
11. Kreft, S., Navarro, G.: LZ77-like compression with fast random access. In: DCC, pp. 239–248. IEEE Computer Society (2010)
12. Kreft, S., Navarro, G.: On compressing and indexing repetitive sequences. Theor. Comput. Sci. **483**, 115–133 (2013)
13. Mahoney, M.: Large text compression benchmark (2011)
14. Ziv, J., Lempel, A.: A universal algorithm for sequential data compression. IEEE Trans. Inf. Theory **23**(3), 337–343 (1977)

Non-overlapping Indexing in BWT-Runs Bounded Space

Daniel Gibney[1] [iD], Paul Macnichol[2] [iD], and Sharma V. Thankachan[2($\boxtimes$)] [iD]

[1] Department of CS, University of Texas at Dallas, Dallas, TX, USA
daniel.gibney@utdallas.edu
[2] Department of CS, North Carolina State University, Raleigh, NC, USA
{pemacnic, svalliy}@ncsu.edu

Abstract. We revisit the non-overlapping indexing problem for an efficient repetition-aware solution. The problem is to index a text $T[1..n]$, such that whenever a pattern $P[1..p]$ comes as a query, we can report the largest set of non-overlapping occurrences of P in T. A previous index by Cohen and Porat [ISAAC 2009] takes linear space and optimal $O(p + \text{occ}_{no})$ query time, where occ_{no} denotes the output size. We present an index of size $O(r)$, where r denotes the number of runs in the Burrows Wheeler Transform (BWT) of T. The parameter r is significantly smaller than n for highly repetitive texts. The query time of our index is $O(p \log \log_w \sigma + \text{sort}(\text{occ}_{no}))$, where σ denotes the alphabet size, w denotes the machine word size in bits and $\text{sort}(x)$ denotes the time for sorting x integers within the range $[1, n]$.

1 Introduction and Related Work

Text indexing is a well-studied problem in computer science with many applications in information retrieval and bioinformatics. The basic version is defined as follows: Preprocess a given text $T[1..n]$ into a data structure (called index) such that whenever a pattern $P[1..p]$ comes as an input, we can efficiently support both *counting* queries and *reporting* queries. A reporting query asks to output $Occ(T, P) = \{i \mid T[i..i+p] = P\}$, the set of occurrences of P in T and a counting query asks for its size occ. We assume that the characters in T and P are from an alphabet $\Sigma = \{0, 1, 2, \ldots, \sigma - 1\}$ and $\sigma = n^{O(1)}$. Our model of computation is word RAM with a machine word of size $w = \Omega(\log n)$ bits.

By maintaining the classic suffix tree data structure over T, we can perform both counting and reporting in optimal times $O(p)$ and $O(p + occ)$, respectively [25]. Alternatively, we can use the suffix array of T for counting in time $O(p \log n)$ and reporting in time $O(p \log n + occ)$ [19]. The space complexity of both structures is $O(n)$ words, equivalently $O(n \log n)$ bits, which can be orders of magnitude more than the size of text, which is $n \lceil \log \sigma \rceil$ bits. Therefore, obtaining space-efficient encoding of these fundamental data structures has been an active line of research. Two important results on this topic from early 2000 are the Compressed Suffix Arrays and the FM index—encodings in succinct

© The Author(s), under exclusive license to Springer Nature Switzerland AG 2023
F. M. Nardini et al. (Eds.): SPIRE 2023, LNCS 14240, pp. 260–270, 2023.
https://doi.org/10.1007/978-3-031-43980-3_21

or entropy-compressed space [7,13]. Compressed suffix tree was also introduced later [24]. These initial results have witnessed various improvements over time; we refer to [20] for further reading. One of the recent breakthroughs in (compressed) text indexing is the *r-index* by Gagie, Navarro, and Prezza [8]. Its $O(r)$ space version can perform counting and reporting in times $O(p \log \log_w(\sigma + n/r))$ and $O((p + occ) \log \log_w(\sigma + n/r))$ respectively, where r denotes the number of runs in the text's Burrows-Wheeler Transform (BWT). The parameter r is a popular measure of compressibility that captures repetitiveness. It can be significantly smaller than n for highly repetitive texts. In a new result by Nishimoto and Tabei [22], the r-index's query times for counting and reporting were improved to $O(p \log \log_w \sigma)$ time and $O(p \log \log_w \sigma + occ)$, respectively. Another result by Gagie et al. [9] shows that the suffix tree can be encoded in $O(r \log(n/r))$ space and support most of its operations in time $O(\log(n/r))$, which includes random access to suffix array, inverse suffix array, longest common prefix array, etc.

We now formally define the main problem considered in this paper.

Problem 1 (Non-overlapping indexing). *Preprocess a given text $T[1..n]$ over an integer alphabet of size $\sigma = n^{O(1)}$ into a data structure (called index) such that whenever a pattern $P[1..p]$ comes as a query, we can report the largest set $Occ_{no}(T, P) \subseteq Occ(T, P)$ of occurrences of P in T, such that the difference between any two occurrences in $Occ_{no}(T, P)$ is at least p.*

Keller et al. [16] introduced this problem and presented an $O(n \log n)$ space solution with $O(p + occ_{no} \cdot \log \log n)$ query time, where $occ_{no} = |Occ_{no}(T, P)|$. In 2009, Cohen and Porat proposed an improved solution with space $O(n)$ and optimal $O(p + occ_{no})$ query time [4]. Later, Ganguly et al. [10,11] showed that all we need is a suffix tree (or any of its space-efficient variants) of T. The time complexity of their query algorithm is $O(search(P) + occ_{no} \cdot t_{SA} + sort(occ_{no}))$, where $search(P)$ denotes the time for computing the suffix range of P and t_{SA} denotes the time for accessing a given entry in the suffix array or inverse suffix array, and $sort(x)$ denotes the time for sorting a subset of $\{1, 2, \ldots, n\}$ of size x. Many space-time trade-offs are immediate from this general result, including a repetition-aware index of size $O(r \log(n/r))$ and query time $O(p + occ_{no} \cdot \log(n/r) + sort(occ_{no}))$ using the suffix tree of Gagie et al. [9]. The interesting question is, can we improve the space complexity to $O(r)$? Note that Ganguly et al.'s algorithm [11] needs random access to the suffix array and its inverse array, and whether suffix trees can be encoded in $O(r)$ space is still open. To that end, we present the following result.

Theorem 1. *For the non-overlapping indexing problem, there exists an $O(r)$ space index that can report $Occ_{no}(T, P)$ in time $O(p \log \log_w \sigma + sort(occ_{no}))$.*

Our result is based on the work by Hooshmand et al. [15], where the authors proposed modifying Ganguly et al.'s algorithm [11], which led to an efficient external memory solution; also see [14]. This modified algorithm avoids much of the random accesses but requires some additional structures, specifically the suffix array of the reverse of T, for its implementation. The critical insight we make in this paper is that the (less general) operations supported by the r-index of T suffice to efficiently implement the algorithm by Hooshmand et al. [15].

2   Preliminaries

For a string $S[1..m] \in \Sigma^m$, we denote its i-th character by $S[i]$, and a substring starting at position i and ending at position j by $S[i..j]$, which is an empty string if $i > j$. When $S[i..j]$ is a suffix of S (i.e., $j = m$), we denote it by $S[i..]$ and when $S[i..j]$ is a prefix of S (i.e., $i = 1$), we denote it by $S[..j]$. The reverse of S is denoted by $\overleftarrow{S}$. The concatenation of two strings (or characters) S_1 and S_2 is denoted by $S_1 S_2$.

2.1   Rank and Select

For any string $S[1..m] \in \Sigma^m$, $rank_S(i, c)$ denotes the number of occurrences of c in $S[1..i]$, where $i \in [1, m]$, $c \in \Sigma$. Also, $select_A(j, c)$ denotes the ith occurrence of c in S. A rank query of the form $rank_S(i, S[i])$ is called a partial query.

   If S is a binary string with t 1's, we can maintain a $t \log(m/t) + O(t)$-bit structure (known as indexible dictionary) and find $rank_S(i, 1)$ for any i with $S[i] = 1$ in $O(1)$ time [23]. It can also support select queries in $O(1)$ time.

2.2   Suffix Array

The suffix array of a text $T[1..n]$ is an array $SA[1, n]$, such that $SA[i]$ represents the starting position of the ith smallest suffix of T in lexicographic order. For convenience, we assume that the last character of T, denoted by \$, does not appear anywhere else in the text or in the pattern and is lexicographically smaller than all other symbols in Σ. The suffix range of a pattern $P[1..p]$, denoted by $[sp(P), ep(P)]$ is the maximal range, such that $Occ(T, P) = \{SA[i] \mid i \in [sp(P), ep(P)]\}$. The suffix range is empty if P does not appear in T. The suffix range, hence the number of occurrences, can be computed in $O(p \log n)$ time. The inverse suffix array ISA is also an array of length n, such that $ISA[SA[i]] = i$ for all $i \in [1, n]$; equivalently, $ISA[i]$ is the lexicographic rank of the suffix $T[i..]$.

2.3   Burrows–Wheeler Transform

The Burrows-Wheeler Transform (BWT) [3] of a text T is a (reversible) permutation of the symbols of T such that $BWT[i] = T[SA[i] - 1]$ if $SA[i] \neq 1$ and is $T[n]$ otherwise (recall that $T[n] = \$$ appears only once in T and is smaller than all other symbols in lexicographic order). The BWT can be encoded in $n \log \sigma$ bits or even in $O(r)$ words by applying run-length encoding, where $r \in [\sigma, n]$ denotes the number of runs (maximal unary substrings) in BWT. For example, the BWT of the text $mississippi\$$ is $ipssm\$pissii$ with 9 runs. The LF-mapping is a function defined as follows: $LF[i]$ is $ISA[SA[i] - 1]$ if $SA[i] \neq 1$ and is 1 otherwise. The LF-mapping can be computed using rank queries on BWT as follows: $LF[i] = Count[BWT[i]] + rank_{BWT}(i, BWT[i])$, where $Count[c] = |\{k \in [1, n] \mid T[k] < c\}|$ for any $c \in \Sigma$. We call $i \in [1, n]$ a run boundary, if $i \in \{1, n\}$ or $BWT[i] \neq BWT[i - 1]$ or $BWT[i] \neq BWT[i + 1]$.

2.4 The r-Index and Some Related Results

Using the r-**index** by Gagie et al. [8,9] and refinements by Bannai et al. [2], we can support the following operations:

1. Given a pattern $P[1..p]$, for each $j \in [1, p]$, we can compute the suffix range of $P[j..p]$. i.e., $[sp(P[j..]), ep(P[j..])]$, in total time $O(p \log \log_w (\sigma + n/r))$. In addition to this, we can get $SA[sp(P[j..])]$ and $SA[ep(P[j..])]$ for each $j \in [1, p]$ in the same time.
2. Given any i, we can compute $LF[i]$ in $O(\log \log_w (n/r))$ time.
3. Given any $(i, SA[i])$, we can compute $\phi^{-1}(SA[i]) = SA[i + 1]$ in $O(\log \log_w (n/r))$ time.

Nishimoto and Tabei [22] improved the time complexity of operation 1 to $O(p \log \log_w \sigma)$, and operations 2 and 3 to $O(1)$ time. As a result, given any $(i, i+h, SA[i])$, we can report $\{SA[k] \mid i \leq k \leq i+h\}$ in $O(h+1)$ time. Since the result for operation 1 is not explicitly stated in their paper, especially $SA[ep(\cdot)]$ part, we provide a short proof here.

Lemma 1 (Modified Toehold Lemma). *By maintaining some additional information with r-index in $O(r)$ space, we can support the following query: given a pattern $P[1..p]$, we can output $SA[sp(P[j..])]$ and $SA[ep(P[j..])]$ for all $j \in [1, p]$ in time $O(p \log \log_w \sigma)$.*

Proof. We store a bit vector $B[1..n]$ and a sampled suffix array SA'. The vector B is defined as follows: $B[LF[i]] = 1$ iff i is a run boundary. Therefore, number of 1's in B is $\Theta(r)$. By maintaining B in space $O(r \log(n/r))$ bits, i.e., $O(r)$ words, we can compute $rank_B(i, 1)$ for any i with $B[i] = 1$ in $O(1)$ time (via a partial rank query) [23]. The sampled suffix array SA' is defined as, $SA'[j] = SA[select_B(j, 1)]$ and its size is $O(r)$. Therefore, $SA[LF[i]] = SA'[rank_B(LF[i], 1)]$ for any run boundary i can be retrieved in $O(1)$ time. We also explicitly store $Count[c]$ for all $c \in \Sigma$.

We process a query $P[1..p]$ as follows. Inductively, assume that we have already computed $sp(P[k..]), ep(P[k..])$, $SA[sp(P[k..])]$ and $SA[ep(P[k..])]$ for all $k \in [j, p]$ for some $j \leq p$ (the base case where $k = p$ is easy). The r-index can give us $[sp(P[k - 1..]), ep(P[k - 1..])]$ in $O(\log \log_w \sigma)$ time. Let α be the first and β be the last occurrences of $P[k - 1]$ in the range $[sp(P[k..]), ep(P[k..])]$ in BWT. Note that since BWT is run-length encoded form; finding α and β is costly, however we have $LF[\alpha] = sp(P[k - 1..])$ and $LF[\beta] = ep(P[k - 1..])$. Also observe that finding $BWT[x]$ for an arbitrary x is costly. However, we can utilize the $O(1)$ time LF-mapping operation to determine if $\mathsf{BWT}[x]$ equals $P[k - 1]$, since $BWT[x] = P[k - 1]$ iff $Count(P[k - 1]) < LF[x] \leq Count(P[k - 1] + 1)$. We have the following cases:

- If $BWT[sp(P[k..])] = P[k-1]$, then $SA[sp(P[k-1..])] = SA[sp(P[k..])] - 1$. Else, α will be a run boundary and $SA[sp(P[k-1..])] = SA[LF[\alpha]] = SA'[rank_B(LF[\alpha],1)]$ can be obtained in constant time.
- If $BWT[ep(P[k..])] = P[k-1]$, then $SA[ep(P[k-1..])] = SA[ep(P[k..])] - 1$. Else, β will be a run boundary and $SA[ep(P[k-1..])] = SA[LF[\beta]] = SA'[rank_B(LF[\beta],1)]$ can be obtained in constant time.

This completes the proof. □

3 The Data Structures

Let $x_1, x_2, \ldots, x_{occ}$ denotes the occurrences of $P[1..p]$ in T in the ascending order. We say x_i and x_j, where $i < j$ are overlapping occurrences if $0 < x_j - x_i < p$ and non-overlapping occurrences otherwise. Define, $Overlap(x_i, x_j) = \max\{p - (x_j - x_i), 0\}$. The following simple algorithm can report the largest set of non-overlapping occurrences. First, find all occurrences of P and sort them to obtain $x_1, x_2, \ldots, x_{occ}$. Report the last occurrence x_{occ}. Then scan the remaining occurrences in the right-to-left order, and report an occurrence if it does not overlap with the last reported occurrence. Although this algorithm correctly reports Occ_{no}, its time complexity is equal to the time for reporting all occurrences of P plus sort(occ). For a better solution, we exploit the pattern's periodicity.

The period of $P[1..p]$ is its shortest prefix Q, such that we can write P as a concatenation of several copies of Q and a proper prefix R of Q. Note that R can be an empty string. For example, we can write $P = abcabcab$ as $Q^2 R$, where $Q = abc$ and $R = ab$. Also, define $\lambda = \lceil p/|Q| \rceil$. Also, we say P is *periodic* if $\lambda > 2$ and *aperiodic* otherwise. We can determine P's period in $O(p)$ time [5]. If P is aperiodic, then $occ = \Theta(occ_{no})$ and the result of Theorem 1 is immediate using r-index and the simple algorithm described before. The rest of this paper focuses only on the more involved periodic case.

If P is periodic and $Overlap(x_{i+1}, x_i) \geq |Q|$, then $x_{i+1} - x_i = |Q|$. Based on this, we have the following definition from [11].

Definition 1 (Cluster). *Let $1 \leq i \leq j \leq occ$ and P is periodic. We call a subset $\{x_i, x_{i+1}, \ldots, x_j\}$ of consecutive occurrences a cluster, iff*

1. $i = 1$ or $Overlap(x_{i-1}, x_i) < |Q|$,
2. $x_{k+1} - x_k = |Q|$ for all $k \in [i, j)$, and
3. $j = occ$ or $Overlap(x_j, x_{j+1}) < |Q|$.

Additionally, we call x_i (resp., x_j) the head (resp, tail) of the cluster.

We use π to denote the number of clusters. Let $h_1, h_2, \ldots, h_\pi$ denotes the clusters heads and $t_1, t_2, \ldots, t_\pi$ denotes clusters tails, where $h_1 \leq t_1 < h_2 \leq t_2 < \ldots, < h_\pi \leq t_\pi$. Define $C_i = \{h_i, h_i + |Q|, h_i + 2|Q|, \ldots, t_i\}$, which call the ith cluster. Note that two consecutive non-overlapping occurrences within the same cluster must be exactly $\lambda|Q|$ characters apart.

3.1   An $O(r \log(n/r))$ Space Solution

We obtain the following result in this section via a direct implementation of Ganguly et al.'s algorithm [11] using the $O(r \log(n/r))$ space suffix tree of Gagie et al. [9]. The algorithm is based on the following observations:

- The number of clusters $\pi = O(\mathsf{occ_{no}})$; follows from the fact that $\{h_1, h_3, h_5, \dots\}$ is a set of non-overlapping occurrences of size $\lceil \pi/2 \rceil$.
- The set $\{t_1, t_2, \dots, t_\pi\}$ of all cluster tails can be obtained using a suffix tree (or an equivalent data structure) efficiently as described below.
- Once we have sorted the list of all cluster tails, we can find Occ_{no} via $O(\mathsf{occ_{no}})$ number of ISA queries.

We now present the algorithm formally.

---

1. Find all cluster tails and sort them to obtain $t_1, t_2, \dots, t_\pi$ (also let $t_0 = 0$).
2. Initialize $x = \infty$ (we use this variable to keep track of the last reported occurrence).
3. For $i = \pi$ to 1, process C_i as follows:
   (a) If x and t_i are non-overlapping, then $x = t_i$; otherwise $x = t_i - |Q|$ (this new x is potentially the rightmost output from C_i).
   (b) While $x \in C_i$ (i.e., $ISA[x] \in [sp(P), ep(P)]$ and $t_{i-1} < x$) report x and $x \leftarrow x - |Q|\lambda$.

To find all cluster tails, observe that an occurrence of P is a cluster tail iff it is not an occurrence of QP. Therefore, $\{t_1, t_2, \dots, t_\pi\} = \{SA[k] \mid k \in [sp(P), ep(P)]$ and $k \notin [sp(QP), ep(QP)]\}$. Since P is a prefix of QP, we have $[sp(QP), ep(QP)] \subseteq [sp(P), ep(P)]$. Therefore,

$$\{t_1, t_2, \dots, t_\pi\} = \{SA[k] \mid sp(P) \le k < sp(QP), ep(QP) < k \le ep(P)\}.$$

The implementation is straightforward; step-1 takes $O(\pi)$ number of SA queries and the step-3 takes $O(\mathsf{occ_{no}})$ number of SA queries. This combined with the time initial pattern search and the sorting of all cluster tails, the query time can be bounded by $O(p + \mathsf{occ_{no}} \cdot \log(n/r) + \mathsf{sort}(\mathsf{occ_{no}}))$.

3.2   An $O(r + r^R)$ Space Solution

This result is based on a slight "modification" of Ganguly et al.'s algorithm [11], which was proposed by Hooshmand et al. [15] for efficiently solving the non-overlapping indexing problem in the external memory model by minimizing the number of SA/ISA queries. Some key observations on the previous algorithm are as follows:

- Step-1 (of finding all cluster tails) can be implemented using r-index (using ϕ^{-1} queries instead of SA queries).

– Once we have the sorted list of all cluster heads, we can avoid the ISA query
  in Step-3(b), because $x \in C_i$ iff $h_i \le x$.

Formally, we have the following algorithm with a slight modification.

---

1. Find all cluster tails and sort them to obtain $t_1, t_2, \ldots, t_\pi$.
2. Find all cluster heads and sort them to obtain $h_1, h_2, \ldots, h_\pi$.
3. Initialize $x = \infty$.
4. For $i = \pi$ to 1, process C_i as follows:
   (a) If x and t_i are non-overlapping, then $x = t_i$; otherwise $x = t_i - |Q|$
       (this new x is potentially the rightmost non-overlapping occur-
       rence from C_i).
   (b) While $x \in C_i$ (i.e., $h_i \le x$), report x and $x \leftarrow x - |Q|\lambda$.

We now present the implementation details. We execute step-1 using the r-
index of T as follows. Find the suffix range $[sp(P), ep(P)]$ of P and the suffix
range $[sp(QP), ep(QP)]$ of QP. We also obtain $SA[sp(P)]$ and $SA[ep(QP)]$ (refer
to Lemma 1). Then, all cluster tails can be obtained by applying ϕ^{-1} function
π times. The time complexity is $O(p \log \log_w \sigma + \pi)$ plus $\mathsf{sort}(\pi)$. For Step-2, we
use the following strategy by Hooshmand et al. [15]. An occurrence x of P is
a cluster head iff $(x - |Q|)$ is not an occurrence of QP. Alternatively, we can
say, $i \in [1, n]$ is a cluster head iff a substring of T ending at $(i + p - 1)$ matches
with P, but not QP. The position $(i + p - 1)$ in T corresponds to an occurrence
$(n - (i + p - 1) + 1)$ of $\overleftarrow{P}$, but $\overleftarrow{QP}$ in $\overleftarrow{T}$. This means, cluster heads are equivalent
to cluster tails in the reverse text, and we can retrieve them using the strategy
used before, but on the suffix tree (or r-index) of the reverse text. Therefore,
the time complexity is also $O(p \log \log_w \sigma + \pi)$ plus $\mathsf{sort}(\pi)$. Step-4 takes $(\mathsf{occ}_{\mathsf{no}})$
time and the overall time is $O(p \log \log_w \sigma + \mathsf{sort}(\mathsf{occ}_{\mathsf{no}}))$.

Since we maintain two r-indexes, the space complexity is $O(r + r^R)$, where
r^R is the number of runs in the BWT of $\overleftarrow{T}$. Note that r^R can be more than r
(see [12]) although a recent result shows that $r^R = O(r \log^2 n)$ [17]. Therefore,
the space complexity (in terms of r and n) is $O(r \log^2 n)$.

3.3 Our Final $O(r)$ Space Solution

In this section, we prove that by maintaining an $O(r)$ space structure and the
r-index of T, we can find all cluster heads in time $O(p \log \log_w \sigma + \pi)$. Therefore,
for implementing Step-2 of the previous algorithm in Sect. 3.2, the r-index for
the text's reverse is no longer required; hence Theorem 1 is immediate.

Recall that a position x is a cluster head iff x is an occurrence of P and
$x - |Q|$ is not an occurrence of QP. This means, x is a cluster head, iff there
exists a proper (possibly empty) suffix $Q[j..]$ of Q (i.e., $j \in [2, |Q| + 1]$), such
that $y = (x - (|Q| - j + 1))$ is an occurrence of $Q[j..]P$ and $T[y - 1] \ne Q[j - 1]$.
We have the following observation by substituting $SA[i] = y$.

Observation 1. *For some $j \in [2, |Q| + 1]$, $SA[i]$ is an occurrence of $Q[j..]P$ and
$BWT[i] \ne Q[j - 1]$ iff $SA[i] + (|Q| - j + 1)$ is a cluster head.*

The set of cluster heads is given by the union of $\Pi_2, \Pi_3, \ldots, \Pi_{|Q|+1}$, where

$$\Pi_j = \{SA[i] + (|Q| - j + 1) \mid i \in [sp(Q[j..]P), ep(Q[j..]P)] \text{ and } BWT[i] \neq Q[j-1]\}.$$

Lemma 2 presents our structure for finding Π_j for any j in optimal $O(1 + |\Pi_j|)$ time, given $sp(Q[j..]P), ep(Q[j..]P)$ and $SA[sp(Q[j..]P)]$. Finding these input parameters for all values of $j \in [2, |Q| + 1]$ using r-index takes $O(p \log \log_w \sigma)$ time. Thus, the overall time for finding all cluster heads is $O(p \log \log_w \sigma + |Q| + \sum_j |\Pi_j|) = O(p \log \log_w \sigma + \pi)$ as desired.

Lemma 2. *By maintaining an $O(r)$ space structure with r-index, we can support the following query: given a range $[sp, ep]$, $SA[sp]$ and a character $c \in \Sigma$, we can output the elements in $X = \{SA[i] \mid i \in [sp, ep] \text{ and } BWT[i] \neq c\}$ in optimal $O(1 + |X|)$ time.*

Proof. We maintain a sorted list $L[1, r]$ of the start of all run boundaries (i.e., i's, where $i = 1$ or $BWT[i - 1] \neq BWT[i]$). We also maintain a sampled suffix array $SA'[1, r]$, where $SA'[i] = SA[L[i]]$. We now present the query algorithm.

If $ep - sp = LF[ep] - LF[sp]$, we conclude that all characters in $BWT[sp, ep]$ are the same. Then, if $BWT[sp] \neq c$, we report $SA[sp]$ and all the remaining entries in $SA[sp, ep]$ using Φ^{-1} function, else, we report none of them. On the other hand, if $ep - sp \neq LF[ep] - LF[sp]$, there exists two values f and h, such that $L[f - 1] \leq sp < L[f] \leq L[f + h] \leq ep < L[f + h + 1]$. We find f via binary search in time $O(\log r)$ and then find h in $O(h)$ time. Then, perform the steps below.

1. If $BWT[sp] \neq c$, then report $SA[sp]$, compute the remaining entries in $SA[sp, L[f])$ using Φ^{-1} function, and report them.
2. For all $g \in [f, f + h)$, if $BWT[L[g]] \neq c$, then report $SA[L[g]] = SA'[g]$, compute the remaining entries in $SA[L[g], L[g + 1])$ using Φ^{-1} function, and report them.
3. If $BWT[L[f + h]] \neq c$, then report $SA[L[f + h]] = SA'[f + h]$, compute the remaining entries in $SA[L[g], ep]$ using Φ^{-1} function, and report them.

The time complexity is $O(\log r + h + |X|)$. Also note that for any g, $BWT[L[g]] \neq BWT[L[g + 1]]$. Therefore, $|X| \geq (h - 1)/2$.

Finally, to remove the term $\log r$, we maintain some additional structures: (i) the optimal one-dimensional range reporting structure by Alstrup et al. [1] over L in $O(r)$ space and (ii) a bit vector $B[1..n]$, such that $B[j] = 1$ iff $j = L[i]$ for some $i \in [1, r]$. We maintain B in space $O(r \log(n/r))$ bits, i.e., $O(r)$ words, so that partial rank queries ($rank_B(j, 1)$ when $B[j] = 1$) can be computed in $O(1)$ time [23]. Now, for computing f and h, we use the following procedure: report all $L[i]$'s within $(sp, ep]$ in time $O(h)$. The smallest among them is $L[f]$ and the largest among them is $L[f + h]$. Then compute $f = rank_B(L[f], 1)$ and $f + h = rank_B(L[f + h], 1)$ using two partial rank queries. The overall time complexity is optimal as desired. □

4   Open Problems

We conclude with some follow-up questions for future research.

1. Can we design an efficient index for *counting* the largest number of non-overlapping occurrences of P in T? i.e., an index that can quickly output occ_{no}. No nontrivial result is known for this problem; therefore, it is interesting to know whether there exists an $O(n \cdot poly \log(n))$ space index with query time $O(p \cdot poly \log(n))$.
2. Can we design new space-time trade-offs for the non-overlapping indexing problem, where space is in terms of other measures of repetitiveness, like the number of Lempel-Ziv factors [26] or δ-measure [18] (a.k.a. substring complexity)?
3. Can we design repetition-aware indexes for the *range non-overlapping indexing* problem, which is a generalization of the non-overlapping indexing problem? Here the input consists of a pattern P and a range $[\alpha, \beta]$, and the task is to output the largest set of non-overlapping occurrences within the range $[\alpha, \beta]$. Several solutions exist to this problem [4,6,16], including an $O(n \log^\epsilon n)$ space index with optimal query time [11] and a linear-space index with near-optimal query time [21], where $\epsilon > 0$ denotes an arbitrarily small constant. An orthogonal range query data structure is a part of these indexes, which makes it challenging to encode them in repetition-aware space.

Acknowledgements. This research is supported in part by the U.S. National Science Foundation (NSF) award CCF-2315822.

References

1. Alstrup, S., Brodal, G.S., Rauhe, T.: Optimal static range reporting in one dimension. In: Proceedings on 33rd Annual ACM Symposium on Theory of Computing, 6–8 July 2001, Heraklion, Crete, Greece, pp. 476–482 (2001). http://doi.acm.org/10.1145/380752.380842, https://doi.org/10.1145/380752.380842
2. Bannai, H., Gagie, T., Tomohiro, I.: Refining the r-index. Theor. Comput. Sci. **812**, 96–108 (2020). https://doi.org/10.1016/j.tcs.2019.08.005
3. Burrows, M., Wheeler, D.J.: A block-sorting lossless data compression algorithm. SRC Research Report, 124 (1994)
4. Cohen, H., Porat, E.: Range non-overlapping indexing. In: Proceedings of the Algorithms and Computation, 20th International Symposium, ISAAC 2009, Honolulu, Hawaii, USA, 16–18 December 2009, pp. 1044–1053 (2009). http://dx.doi.org/10.1007/978-3-642-10631-6_105, https://doi.org/10.1007/978-3-642-10631-6_105
5. Crochemore, M.: String-matching on ordered alphabets. Theoret. Comput. Sci. **92**(1), 33–47 (1992)
6. Crochemore, M., Iliopoulos, C.S., Kubica, M., Rahman, M.S., Walen, T.: Improved algorithms for the range next value problem and applications. In: Proceedings of the STACS 2008, 25th Annual Symposium on Theoretical Aspects

of Computer Science, Bordeaux, France, 21–23 February 2008, pp. 205–216 (2008). http://dx.doi.org/10.4230/LIPIcs.STACS.2008.1359, https://doi.org/10.4230/LIPIcs.STACS.2008.1359

7. Ferragina, P., Manzini, G.: Indexing compressed text. J. ACM **52**(4), 552–581 (2005). http://doi.acm.org/10.1145/1082036.1082039, https://doi.org/10.1145/1082036.1082039

8. Gagie, T., Navarro, G., Prezza, N.: Optimal-time text indexing in BWT-runs bounded space. In: Czumaj, A. (ed.) Proceedings of the Twenty-Ninth Annual ACM-SIAM Symposium on Discrete Algorithms, SODA 2018, New Orleans, LA, USA, 7–10 January 2018, pp. 1459–1477. SIAM (2018). https://doi.org/10.1137/1.9781611975031.96

9. Gagie, T., Navarro, G., Prezza, N.: Fully functional suffix trees and optimal text searching in BWT-runs bounded space. J. ACM **67**(1), 2:1–2:54 (2020). https://doi.org/10.1145/3375890

10. Ganguly, A., Shah, R., Thankachan, S.V.: Succinct non-overlapping indexing. In: Cicalese, F., Porat, E., Vaccaro, U. (eds.) CPM 2015. LNCS, vol. 9133, pp. 185–195. Springer, Cham (2015). https://doi.org/10.1007/978-3-319-19929-0_16

11. Ganguly, A., Shah, R., Thankachan, S.V.: Succinct non-overlapping indexing. Algorithmica **82**(1), 107–117 (2020). https://doi.org/10.1007/s00453-019-00605-5

12. Giuliani, S., Inenaga, S., Lipták, Z., Prezza, N., Sciortino, M., Toffanello, A.: Novel results on the number of runs of the burrows-wheeler-transform. In: Bureš, T., et al. (eds.) SOFSEM 2021. LNCS, vol. 12607, pp. 249–262. Springer, Cham (2021). https://doi.org/10.1007/978-3-030-67731-2_18

13. Grossi, R., Vitter, J.S.: Compressed suffix arrays and suffix trees with applications to text indexing and string matching. SIAM J. Comput. **35**(2), 378–407 (2005). https://doi.org/10.1137/S0097539702402354

14. Hooshmand, S., Abedin, P., Külekci, M.O., Thankachan, S.V.: Non-overlapping indexing - cache obliviously. In: Navarro, G., Sankoff, D., Zhu, B. (eds.) Annual Symposium on Combinatorial Pattern Matching, CPM 2018, 2–4 July 2018 - Qingdao, China. LIPIcs, vol. 105, pp. 8:1–8:9. Schloss Dagstuhl - Leibniz-Zentrum für Informatik (2018). https://doi.org/10.4230/LIPIcs.CPM.2018.8

15. Hooshmand, S., Abedin, P., Külekci, M.O., Thankachan, S.V.: I/O-efficient data structures for non-overlapping indexing. Theor. Comput. Sci. **857**, 1–7 (2021). https://doi.org/10.1016/j.tcs.2020.12.006

16. Keller, O., Kopelowitz, T., Lewenstein, M.: Range non-overlapping indexing and successive list indexing. In: Dehne, F., Sack, J.-R., Zeh, N. (eds.) WADS 2007. LNCS, vol. 4619, pp. 625–636. Springer, Heidelberg (2007). https://doi.org/10.1007/978-3-540-73951-7_54

17. Kempa, D., Kociumaka, T.: Resolution of the burrows-wheeler transform conjecture. In: Irani, S. (ed.) 61st IEEE Annual Symposium on Foundations of Computer Science, FOCS 2020, Durham, NC, USA, 16–19 November 2020, pp. 1002–1013. IEEE (2020). https://doi.org/10.1109/FOCS46700.2020.00097

18. Kociumaka, T., Navarro, G., Prezza, N.: Toward a definitive compressibility measure for repetitive sequences. IEEE Trans. Inf. Theory **69**(4), 2074–2092 (2023). https://doi.org/10.1109/TIT.2022.3224382

19. Manber, U., Myers, E.W.: Suffix arrays: a new method for on-line string searches. SIAM J. Comput. **22**(5), 935–948 (1993). https://doi.org/10.1137/0222058

20. Navarro, G., Mäkinen, V.: Compressed full-text indexes. ACM Comput. Surv. **39**(1), 2 (2007). https://doi.org/10.1145/1216370.1216372

21. Nekrich, Y., Navarro, G.: Sorted range reporting. In: Fomin, F.V., Kaski, P. (eds.) SWAT 2012. LNCS, vol. 7357, pp. 271–282. Springer, Heidelberg (2012). https://doi.org/10.1007/978-3-642-31155-0_24

22. Nishimoto, T., Tabei, Y.: Optimal-time queries on BWT-runs compressed indexes. In: Bansal, N., Merelli, E., Worrell, J. (eds.) 48th International Colloquium on Automata, Languages, and Programming, ICALP 2021, 12–16 July 2021, Glasgow, Scotland (Virtual Conference). LIPIcs, vol. 198, pp. 101:1–101:15. Schloss Dagstuhl - Leibniz-Zentrum für Informatik (2021). https://doi.org/10.4230/LIPIcs.ICALP.2021.101

23. Raman, R., Raman, V., Satti, S.R.: Succinct indexable dictionaries with applications to encoding k-ary trees, prefix sums and multisets. ACM Trans. Algorithms **3**(4), 43 (2007). https://doi.org/10.1145/1290672.1290680

24. Sadakane, K.: Compressed suffix trees with full functionality. Theory Comput. Syst. **41**(4), 589–607 (2007). https://doi.org/10.1007/s00224-006-1198-x

25. Weiner, P.: Linear pattern matching algorithms. In: 14th Annual Symposium on Switching and Automata Theory, Iowa City, Iowa, USA, 15–17 October 1973, pp. 1–11 (1973). http://dx.doi.org/10.1109/SWAT.1973.13, https://doi.org/10.1109/SWAT.1973.13

26. Ziv, J., Lempel, A.: A universal algorithm for sequential data compression. IEEE Trans. Inf. Theory **23**(3), 337–343 (1977)

Efficient Parameterized Pattern Matching in Sublinear Space

Haruki Ideguchi$^{(\boxtimes)}$, Diptarama Hendrian, Ryo Yoshinaka,
and Ayumi Shinohara

Graduate School of Information Sciences, Tohoku University, Sendai, Japan
haruki.ideguchi.q3@dc.tohoku.ac.jp,
{diptarama,ryoshinaka,ayumis}@tohoku.ac.jp

Abstract. The parameterized matching problem is a variant of string
matching, which is to search for all *parameterized* occurrences of a pattern P in a text T. In considering matching algorithms, the combinatorial
natures of strings, especially *periodicity*, play an important role. In this
paper, we analyze the properties of periods of parameterized strings and
propose a generalization of Galil and Seiferas's exact matching algorithm
(1980) into parameterized matching, which runs in $O(\pi|T| + |P|)$ time
and $O(\log|P| + |\Pi|)$ space in addition to the input space, where Π is
the parameter alphabet and π is the number of parameter characters
appearing in P plus one.

Keywords: Parameterized matching · String matching · Sublinear
space · Combinatorics on words

1 Introduction

String matching is a problem to search for all occurrences of a pattern P in a text
T. Since it is one of the most important computer applications, many efficient
algorithms for the problem have been proposed. Let us denote the length of T
and P by n and m, respectively. While a naive algorithm takes $O(nm)$ time to
solve the problem, Knuth, Morris, and Pratt [13] gave an algorithm which runs
in only $O(n+m)$ time by constructing auxiliary arrays called *border arrays*. After
that, various algorithms to solve the problem in linear time have been proposed,
which use auxiliary data structures, such as suffix trees [19], suffix arrays [15],
LCP arrays [15]. All of those algorithms outperform the naive algorithm in terms
of time complexity. They require additional space to store their auxiliary data,
whose sizes are typically $\Theta(n)$ or $\Theta(m)$. On the other hand, studies for reducing
such extra space were conducted. Firstly, Galil and Seiferas reduced extra space
usage to $O(\log m)$ [11], and later several time-space-optimal, $O(n+m)$ time and
$O(1)$ extra-space algorithms were devised [5,6,12].

In this paper, we consider a variant of string matching: *parameterized matching*. It is a pattern matching paradigm in which two strings are considered a
match if we can map some characters (*parameter characters*) in one string to

F. M. Nardini et al. (Eds.): SPIRE 2023, LNCS 14240, pp. 271–283, 2023.
https://doi.org/10.1007/978-3-031-43980-3_22

characters in another string. This paradigm was first introduced by Baker [4] for use in software maintenance by the ability to detect 'identical' computer programs renaming their variables. For solving the parameterized matching problem, a number of linear-time algorithms have been proposed that extend algorithms for exact matching [2,7,8,10,14,17,18]. See also [16] for a survey. However, we know of no previous attempt to reduce extra space usage to sublinear for time-efficient parameterized matching algorithms, although one can solve the problem in constant extra space if the time efficiency does not matter.

The main contribution of this paper is to give a sublinear-extra-space algorithm for the parameterized matching problem by extending Galil and Seiferas's exact matching algorithm [11]. It runs in $O(|\Pi_P|n + m)$ time and $O(\log m + |\Pi|)$ space in addition to the input space, where Π is the set of parameter characters and Π_P is the non-empty[1] set of parameter characters appearing in P.

In order to provide the basis for our algorithm, we also investigate the properties of periodicity of parameterized strings in this paper. It is widely known that periods of strings are useful for exact matching algorithms [5,6,11–13], which is also the case for parameterized matching [2]. We extend previous work on parameterized periods by Apostolico and Giancarlo [3] and derive several properties for our algorithm. In particular, we focus on 'sufficiently short' periods of parameterized strings having properties useful for matching algorithms. Those results contain a parameterized version of Fine and Wilf's periodicity lemma [9].

Remark 1. The time and space complexities of our algorithm stated above are based on a computing model in which functions $\Pi \to \mathbb{N}$ can be stored as arrays. If not, one can use AVL trees [1] instead of arrays to store such functions. Then, our algorithm runs in $O((|\Pi_P|n + m) \log |\Pi_P|)$ time and $O(\log m + |\Pi_P|)$ extra space.

2   Preliminaries

Let $\mathbb{N}$ and $\mathbb{N}^+$ be the set of natural numbers including and excluding 0, respectively. For $x, y \in \mathbb{N}$, we denote by $x \mid y$ that y is a multiple of x.

For $n \in \mathbb{N}$ and a function f whose domain and codomain are the same, we denote by f^n the composite of the function n times.

2.1   Parameterized Matching Problem

In parameterized matching, we consider two disjoint alphabets: the *constant alphabet* Σ and the *parameter alphabet* Π. A string over $\Sigma \cup \Pi$ is called a *parameterized string* or a *p-string*. Consider a p-string $w \in (\Sigma \cup \Pi)^*$. We denote the length of w by $|w|$. For $0 \le i < |w|$, let us denote i-th letter of w by $w[i]$, where the index i is 0-based. For $0 \le i \le j \le |w|$, we denote the substring $w[i]w[i+1] \cdots w[j-1]$ by $w[i : j]$. (Note that $w[i : j]$ does not contain $w[j]$.)

---

[1] We can assume $\Pi_P \neq \emptyset$ without loss of generality. See Remark 2.

We denote the set of permutations of Π by S_Π. Throughout this paper, for a permutation $f \in S_\Pi$ and a constant character $c \in \Sigma$, let $f(c) = c$. Then, the map f is naturally expanded as a bijection over p-strings: $(\Sigma \cup \Pi)^* \to (\Sigma \cup \Pi)^*$.

Definition 1 (Baker [4]). *Two p-strings x and y are called a* parameterized-match *or a p-match if and only if there exists a permutation $f \in S_\Pi$ such that $f(x) = y$. Denote this relation by $x \equiv y$.*

Example 1. Let $\Sigma = \{a, b, c\}$ and $\Pi = \{A, B, C\}$. We have ABaCBCa $\equiv$ BCaACAa with a permutation f such that $f(A) = B$, $f(B) = C$, and $f(C) = A$.

Clearly, the relation $\equiv$ is an equivalence relation over $(\Sigma \cup \Pi)^*$. Note that if $x \equiv y$, we have $|x| = |y|$ and $x[i : j] \equiv y[i : j]$ for any $0 \le i \le j \le |x|$. By this relation, the problem we consider in this paper, the *parameterized matching problem*, is defined as follows.

Problem 1 ([4]). *Given two p-strings T (text) and P (pattern), find all $0 \le i \le |T| - |P|$ such that $T[i : i + |P|] \equiv P$.*

Remark 2. For Problem 1, we can assume that P contains at least one parameter character without loss of generality. If $P \in \Sigma^*$, choose any $c \in \Sigma$ appearing in P and let constant and parameter alphabets be $\Sigma \cup \Pi \setminus \{c\}$ and $\{c\}$, respectively. Our algorithm presented in Sect. 4 is based on this assumption.

2.2 Periodicity of Parameterized Strings

Periodicity is one of the most fundamental concepts in combinatorics of strings and a wealth of applications. In exact matching, the Knuth-Morris-Pratt algorithm and various algorithms based on it rely on the properties of periods [5,6,11–13]. It is also the case for parameterized matching [2], where periods of parameterized strings are defined as follows:

Definition 2 (Apostolico and Giancarlo [3]). *Consider $w \in (\Sigma \cup \Pi)^*$ and $p \in \mathbb{N}^+$ with $p \le |w|$. Then, p is called a* period *of w if and only if $w[0 : |w| - p] \equiv w[p : |w|]$.*

If p is a period of w, there exists $f \in S_\Pi$ satisfying $f(w[0 : |w| - p]) = w[p : |w|]$ by definition. We denote this relation by $p \parallel_f w$ or simply by $p \parallel w$ when f is not specified.

In general, a p-string w can have multiple periods. We denote the shortest period of w as *period(w)*. It is clear that a period p of a p-string w is also a period of any substring w' of w such that $|w'| \ge p$.

Example 2. Let $\Sigma = \{a, b, c\}$ and $\Pi = \{A, B, C\}$. For $w := $ ABaCBCaaACAa, we have $4 \parallel_f w$ as ABaCBCa $\equiv$ BCaACAa with $f(A) = B$, $f(B) = C$, and $f(C) = A$.

Instead of Definition 2, one can use the following equivalent definition for periods, which is a more intuitive representation of the repetitive structure of strings:

Lemma 1 ([3]). *Consider $w \in (\Sigma \cup \Pi)^*$, $p \in \mathbb{N}^+$, and $f \in S_\Pi$. Then, $p \parallel_f w$ holds if and only if w can be written as*

$$w = f^0(v) \cdot f^1(v) \cdot f^2(v) \cdots f^{\lfloor \rho \rfloor - 1}(v) \cdot f^{\lfloor \rho \rfloor}(v'),$$

where $\rho = \frac{|w|}{p}$, $v = w[0:p]$ and v' is a prefix of v (allowing the case v' is empty).

The following lemma has important applications for various matching algorithms. Particularly, it is used to shift the pattern string safely in the Knuth-Morris-Pratt algorithm and variants [2,13].

Lemma 2. *Consider $x, y \in (\Sigma \cup \Pi)^*$ with $x \equiv y$. For any $0 < \delta < period(y)$, we have $x[\delta : |x|] \not\equiv y[0 : |y| - \delta]$.*

Proof. We give a proof by contraposition. Suppose $x[\delta : |x|] \equiv y[0 : |y| - \delta]$. Then we have $y[0 : |y| - \delta] \equiv x[\delta : |x|] \equiv y[\delta : |y|]$, which means $\delta \parallel y$. Hence, $\delta \geq period(y)$ holds. □

One of the main interest regarding string periodicity is what holds when a string w has two different periods p and q. For ordinary strings, Fine and Wilf's periodicity lemma [9] gives an answer: $\gcd(p, q)$ is also a period when $|w| \geq p + q - \gcd(p, q)$, where $\gcd(p, q)$ is the greatest common divisor of p and q. Apostolico and Giancarlo showed a similar property for parameterized strings.

Lemma 3 ([3]). *For $w \in (\Sigma \cup \Pi)^*$, $p, q \in \mathbb{N}^+$, and $f, g \in S_\Pi$, assume that $p \parallel_f w$ and $q \parallel_g w$. If $|w| \geq p + q$ and $fg = gf$, we have $\gcd(p, q) \parallel w$.*

It is known that the length $|w| = p + q - \gcd(p, q)$ is not sufficient for this lemma unlike in the case of ordinary strings [3].

3   Properties of Parameterized Periods

In this section, we show some properties of periods of parameterized strings. They play an important role in our algorithm presented in Sect. 4.

3.1   Alternative Periodicity Lemma

The requirements of Lemma 3 are slightly different from Fine and Wilf's lemma for ordinary strings. Particularly, the commutativity of f and g is essential (Lemma 5 in [3]). In this section, we show a new periodicity lemma for parameterized strings which does not assume the commutativity.

Firstly, we focus on parameter characters contained in a given p-string and its substrings. For $w \in (\Sigma \cup \Pi)^*$, we denote by Π_w the set of parameter characters appearing in w.

Example 3. Let $\Sigma = \{a, b, c\}$ and $\Pi = \{A, B, C\}$. For $w := ABabAca$, we have $\Pi_w = \{A, B\}$.

Lemma 4. *Consider $w \in (\Sigma \cup \Pi)^*$ and any of its substrings w' and w''. Then, the following hold:*

- *If $|w'| \geq period(w) \cdot (|\Pi_w| - 1)$, we have $|\Pi_{w'}| \geq |\Pi_w| - 1$.*
- *If $|w''| \geq period(w) \cdot |\Pi_w|$, we have $\Pi_{w''} = \Pi_w$.*

Proof. The case $\Pi_w = \emptyset$ is trivial. Suppose $\Pi_w \neq \emptyset$. Let $p := period(w)$ and f be a permutation of Π such that $p \parallel_f w$. It suffices to show the lemma for the cases $|w'| = p \cdot (|\Pi_w| - 1)$ and $|w''| = p \cdot |\Pi_w|$. By Lemma 1, w' and w'' can be written as $w' = v' \cdot f(v') \cdots f^{|\Pi_w|-2}(v')$ and $w'' = v'' \cdot f(v'') \cdots f^{|\Pi_w|-1}(v'')$, where v' and v'' are the prefixes of w' and w'' of length p, respectively. Now, we consider the cyclic decomposition of f.

Suppose the characters in Π_w make one cyclic permutation in f. Let a be any parameter character contained in v'. Note that $a, f(a), \cdots, f^{|\Pi_w|-2}(a)$ are all different characters and all appear in w'. Therefore, we have $|\Pi_{w'}| \geq |\Pi_w| - 1$. The analogous argument shows $|\Pi_{w''}| = |\Pi_w|$.

Suppose the characters in Π_w make two or more cyclic permutations in f. Then, those cyclic permutations are all of length $|\Pi_w| - 1$ or less. For $0 \leq i < |w|$, there exists an integer k such that $w[i + kp], w[i + (k + 1)p], \cdots, w[i + (k + |\Pi_w| - 2)p]$ are all contained in w'. Then, those characters can be represented as $f^k(w[i]), f^{k+1}(w[i]), \cdots, f^{k+|\Pi_w|-2}(w[i])$, and by the assumption about f, at least one of them is equal to $w[i]$. Therefore, we have $w[i] \in \Pi_{w'}$. Since i is arbitrary, we end up with $\Pi_w \subseteq \Pi_{w'}$, as required. $\qquad\square$

Now, we show a variant of Lemma 3. It does not require any assumption on the permutations, in exchange of a stricter requirement for the length of strings.

Lemma 5. *Suppose $w \in (\Sigma \cup \Pi)^*$ with $\Pi_w \neq \emptyset$ has periods p and q. If $|w| \geq p + q + \min(p, q) \cdot (|\Pi_w| - 1)$, we have $\gcd(p, q) \parallel w$.*

Proof. Let f and g be permutations of Π such that $p \parallel_f w$ and $q \parallel_g w$. Without loss of generality, we suppose $f(a) = a$ and $g(a) = a$ for any $a \in \Pi \setminus \Pi_w$. By Lemma 3, it suffices to show that $fg = gf$. Let $w' := w[0 : |w| - p - q]$. Then, notice that $fg(w') = f(w[q : |w| - p]) = w[p + q : |w|] = g(w[p : |w| - q]) = gf(w')$, which claims $fg(a) = gf(a)$ for any $a \in \Pi_{w'}$. Moreover, given $|w'| = |w| - p - q \geq \min(p, q) \cdot (|\Pi_w| - 1) \geq period(w) \cdot (|\Pi_w| - 1)$, we have $|\Pi_{w'}| \geq |\Pi_w| - 1$ by Lemma 4. Hence, the permutations fg and gf behave the same for at least $|\Pi| - 1$ parameter characters. This implies $fg = gf$. $\qquad\square$

Corollary 1. *Suppose $w \in (\Sigma \cup \Pi)^*$ with $\Pi_w \neq \emptyset$ has a period q. If $q \leq \frac{|w|}{|\Pi_w|+1}$, then $period(w) \mid q$.*

Proof. Let $p := period(w)$. By $p \leq q \leq \frac{|w|}{|\Pi_w|+1}$, we have $p \cdot |\Pi_w| + q \leq q \cdot (|\Pi_w| + 1) \leq \frac{|w|}{|\Pi_w|+1}(|\Pi_w|+1) = |w|$. Hence, we can use Lemma 5 to obtain $\gcd(p, q) \parallel w$. Then, since p is the smallest period of w, we have $\gcd(p, q) \geq p$, which means $\gcd(p, q) = p$ i.e. $p \mid q$, as required. $\qquad\square$

Table 1. Let $\Pi = \{\mathtt{A},\mathtt{B}\}$. A p-string $w := \mathtt{ABABBABAABABBABAABBA}$ has prefix periods 1 and 4. Circled numbers in the table below are prefix periods of w with $w[0:i+1]$ as witnesses. For instance, 4 is a prefix period of w with $w[0:18]$ as a witness because $period(w[0:18]) = 4$ and $4 \leq \frac{|w[0:18]|}{k}$. (Note that $k = |\Pi_w| + 2 = 4$.)

i	0	1	2	3	4	5	6	7	8	9	10	11	12	13	14	15	16	17	18	19
$w[i]$	A	B	A	B	B	A	B	A	A	B	A	B	B	A	B	A	A	B	B	A
$period(w[0:i+1])$	1	1	1	①	4	4	4	4	4	4	4	4	4	4	4	④	④	④	18	18

$reach_w(1) = 4$ $\qquad\qquad\qquad\qquad\qquad\qquad\qquad reach_w(4) = 18$

3.2 Prefix Periods

Galil and Seiferas's exact matching algorithm [11] can be regarded as an extension of the Knuth-Morris-Pratt algorithm [13]. The main idea of their algorithm is to deal with only periods of pattern prefixes which are 'short enough.' They pointed out that periods shorter than $\frac{1}{k}$ times the length of the string have useful properties for saving space usage in exact string matching for an arbitrarily fixed $k \geq 3$. We show, in this section, that similar properties hold for parameterized strings as well when k is set to be $|\Pi_w| + 2$. Most of those properties come from Lemma 5 we proved in the previous section.

Lemma 6. *Suppose $w \in (\Sigma \cup \Pi)^*$ has a period p. If $p \leq \frac{|w|}{|\Pi_w|+1}$, there exists only one character $a \in \Sigma \cup \Pi$ such that $p \parallel wa$.*

Proof. Consider the prefix $w' := w[0:|w|-p]$. By $p \leq \frac{|w|}{|\Pi_w|+1}$, we have $|w|-p \geq p|\Pi_w| \geq period(w)|\Pi_w|$. By Lemma 4, $\Pi_{w'} = \Pi_w$. Therefore, $w[|w|-p]$ already appears in w' as $w[i] = w[|w|-p]$ for some $i < |w|-p$. Hence, for any f such that $p \parallel_f w$, it holds that $p \parallel_f wa$ if and only if $a = w[i+p]$. □

Corollary 2. *Suppose $w \in (\Sigma \cup \Pi)^*$ has a period p. For any $\ell \in \mathbb{N}^+$ such that $\ell p \leq \frac{|w|}{|\Pi_w|+1}$, we have $p \parallel wa \iff \ell p \parallel wa$ for any $a \in \Sigma \cup \Pi$.*

Proof. By Lemma 6, the characters a_1 and a_2 such that $p \parallel wa_1$ and $\ell p \parallel wa_2$ are unique respectively. Then, since $p \parallel wa_1 \implies \ell p \parallel wa_1$ (shown immediately by Lemma 1), we get $a_1 = a_2$, as required. □

Now, we introduce the key concept for our algorithm: *prefix periods*. This is a natural extension of the one introduced in [12] for parameterized strings. Hereafter in this section, we consider a fixed p-string $w \in (\Sigma \cup \Pi)^*$ with $\Pi_w \neq \emptyset$ and let $k := |\Pi_w| + 2$.

Definition 3. *A positive integer $p \in \mathbb{N}^+$ is called a* prefix period *of w if and only if there exists a prefix w' of w such that $period(w') = p$ and $p \leq \frac{|w'|}{k}$.*

We give an example of prefix periods in Table 1. For a fixed p, only prefixes w' of w satisfying $|w'| \geq kp$ can be a witness for p being a prefix period. We show in the following lemmas that it suffices to consider only one prefix $w' = w[0:kp]$ for checking whether p is a prefix period.

Lemma 7. *For any $a \in \Sigma \cup \Pi$, if $period(wa) \neq period(w)$, we have $period(wa)$ $> \frac{|w|}{|\Pi_w|+1}$.*

Proof. We show the lemma by contraposition. Suppose $period(wa) \leq \frac{|w|}{|\Pi_w|+1}$. Since $period(wa)$ is also a period of w, we can use Corollary 1 to obtain $period(w) \mid period(wa)$. Therefore, we get $period(w) \parallel wa$ by Corollary 2, which implies $period(w) \geq period(wa)$. On the other hand, we have $period(w) \leq period(wa)$ by definition. Thus $period(w) = period(wa)$ holds. □

Lemma 8. *Consider any $0 < p \leq \frac{|w|}{k}$. Then, p is a prefix period of w if and only if $period(w') = p$ where $w' := w[0 : kp]$.*

Proof. ($\Longleftarrow$) Immediate by the definition of prefix periods.
($\Longrightarrow$) Let v be a prefix of w that witnesses p being a prefix period, i.e., $|v| \geq kp$ and $period(v) = p$. If $|v| = kp$, we are done. Suppose $|v| > kp$ and let $u := v[0 : |v| - 1]$. Then, $period(v) = p < \frac{|v|}{k} \leq \frac{|v|}{|\Pi_u|+2} < \frac{|u|}{|\Pi_u|+1}$. By Lemma 7, we have $period(u) = period(v) = p$. By repeatedly applying this discussion, we can shorten the witness up to length kp. □

Next, we introduce an auxiliary function $reach_w$.

Definition 4. *For any $0 < p \leq |w|$, let*

$$reach_w(p) := \max\{r \in \mathbb{N} : r \leq |w| \text{ and } p \parallel w[0 : r]\}.$$

Note that $p \parallel w[0 : r] \iff reach_w(p) \geq r$ holds by definition. Using $reach_w$, we get an equivalent definition of prefix periods as follows, which is directly used in our searching algorithm.

Lemma 9. *Consider any $0 < p \leq \frac{|w|}{k}$. Then, p is a prefix period of w if and only if all the following hold:*

(1) $reach_w(p) \geq kp$,
(2) $reach_w(q) < reach_w(p)$ for any $0 < q < p$.

Proof. ($\Longrightarrow$) (1) is by definition. We show (2). By Lemma 8, $period(w[0 : kp]) = p$. Thus, $q < p$ is not a period of $w[0 : kp]$, i.e., $reach_w(q) < kp \leq reach_w(p)$ by (1).
($\Longleftarrow$) Let $w' := w[0 : reach_w(p)]$. (2) implies $period(w') = p$ since any q satisfying $0 < q < p$ is not a period of w'. Additionally, we have $p \leq \frac{|w'|}{k}$ by (1). Thus p is a prefix period of w with w' as a witness. □

Galil and Seiferas [11] in Corollary 1 pointed out that the number of prefix periods of a word w is $O(\log |w|)$. We show in the following lemma that it is the case for parameterized strings. It contributes directly to reducing the space complexity of our algorithm.

Lemma 10. *Suppose w has prefix periods p and q. If $p < q$, then $2p \leq q$.*

Proof. We prove the lemma by contradiction. Suppose $p < q < 2p$. By definition, $p \parallel w[0 : kp]$ and $q \parallel w[0 : kq]$ hold. Let $w' := w[0 : kp]$. By Lemma 8, p is the shortest period of w'. Since both p and q are periods of w' and $p \cdot |\Pi_{w'}| + q < p \cdot |\Pi_w| + 2p = kp = |w'|$, we get $\gcd(p, q) \parallel w'$ by Lemma 5. Hence, we have $\gcd(p, q) \geq period(w') = p$, which claims $\gcd(p, q) = p$ i.e. $p \mid q$. However, this contradicts to the assumption $p < q < 2p$. $\square$

Corollary 3. *The number of prefix periods of $w \in (\Sigma \cup \Pi)^*$ is at most $\log_2 |w|$.*

4   Proposed Algorithm

In this section, we propose a sublinear-extra-space algorithm for the parameterized matching problem. Throughout this section, let T and P be p-strings whose lengths are n and m respectively, and let $k := |\Pi_P| + 2$. Besides, we suppose $\Pi_P \neq \emptyset$. Our algorithm is an extension of Galil and Seiferas's exact string matching algorithm [11] and runs in $O(|\Pi_P|n + m)$ time and $O(\log m + |\Pi|)$ extra space. When $|\Pi| = |\Pi_P| = 1$, our algorithm behaves exactly as theirs.

Firstly, we introduce a method for testing whether two p-strings match. While it is common to use the *prev-encoding* [4] for this purpose, it is not suitable for our goal since it requires additional space proportional to the input size. Thus we use an alternative method as follows, which requires only $O(|\Pi|)$ extra space.

Lemma 11. *Consider a prefix x of P and $y \in (\Sigma \cup \Pi)^*$ with $x \equiv y$ and any $a, b \in \Sigma \cup \Pi$. We have $xa \equiv yb$ if and only if one of the following holds:*

1. *$a \in \Sigma$ and $a = b$,*
2. *$a \in \Pi$ and $first_P(a) \geq |x|$ and $b \in \Pi$ and $count_y(b) = 0$,*
3. *$a \in \Pi$ and $first_P(a) < |x|$ and $y[first_P(a)] = b$,*

where $first_P : \Pi \to \mathbb{N}$ and $count_y : \Pi \to \mathbb{N}$ are defined as follows:

$$first_P(c) = \begin{cases} \min\{i \in \mathbb{N} : i < |P| \text{ and } P[i] = c\} & \text{if } c \in \Pi_P , \\ |P| & \text{if } c \in \Pi \setminus \Pi_P , \end{cases}$$

$$count_y(c) = |\{i \in \mathbb{N} : i < |y| \text{ and } y[i] = c\}|$$

Proof. By definition, we have $xa \equiv yb$ if and only if $b = f(a)$, where f satisfies $y = f(x)$. If a is a constant character or appears in x, the value $f(a)$ is determined (Cases 1 and 3). Otherwise, b must be a parameter character not appearing in y (Case 2). $\square$

Let $\text{MATCH}(x, y, a, b, first_P, count_y)$ be the function which returns whether $xa \equiv yb$ under the condition $x \equiv y$ using Lemma 11. Clearly, one can compute it in constant time if $first_P$ and $count_y$ are given as arrays. Note that $first_P$ can be computed in $O(m)$ time and $O(|\Pi|)$ space.

Algorithm 1: PREFIX_PERIODS
___

Input: $P \in (\Sigma \cup \Pi)^*$

Output: a list of all prefix periods of P and their reaches

1 **begin**
2      $k \leftarrow |\Pi_P| + 2$
3      $first \leftarrow first_P$
4      $PP \leftarrow$ empty list   // PP is a list of pairs (val, reach)
5      $idx \leftarrow -1$
6      $(p, r) \leftarrow (1, 1)$
7   **foreach** $a \in \Pi$ **do** $count[a] \leftarrow 0$
8      $max\_reach \leftarrow 0$
9   **while** $kp \leq |P|$ **do**
10   **while** MATCH$(P[0:r-p], P[p:r], P[r-p], P[r], first, count)$ **do**
11   Increment $count[P[r]]$
12             $r \leftarrow r + 1$
13   **if** $idx + 1 < |PP|$ *and* $PP[idx + 1].val \leq \frac{r-p}{k}$ **then** Increment idx

14   **if** $r \geq kp$ *and* $r > max\_reach$ **then**
15   Push (p, r) into PP
16         $max\_reach \leftarrow \max\{max\_reach, r\}$

17   **if** $0 \leq idx < |PP|$ *and* $PP[idx].reach \geq r - p > 0$ **then**
18   **for** $p \leq i < p + PP[idx].val$ **do** Decrement $count[P[i]]$
19             $p \leftarrow p + PP[idx].val$
20   **else**
21   **for** $p \leq i < r$ **do** Decrement $count[P[i]]$
22             $p \leftarrow p + \lfloor \frac{r-p}{k} \rfloor + 1$
23             $r \leftarrow p$
24   **until** $PP[idx].val \leq \frac{r-p}{k}$ *or* $idx = -1$ **do** Decrement idx

25   **return** PP

4.1 Pattern Preprocessing

In this section, we show the preprocessing for the pattern P for our matching algorithm. The output of the preprocessing is the list of pairs of a prefix period of P (in ascending order) and its reach, just like Galil and Seiferas [11] introduced for exact string matching. The list plays a similar role to the *border array* in the parameterized Knuth-Morris-Pratt algorithm [2]. While border array uses $\Theta(m)$ space to memorize the shortest periods of all prefixes of P, the prefix period list requires only $O(\log m)$ space by Corollary 3.

We present the preprocess in Algorithm 1. The algorithm finds prefix periods and their reaches in order from the smallest to the largest and put them into the list PP. By $PP[idx].val$ and $PP[idx].reach$, we denote the idx-th prefix period and its reach in PP, respectively. Starting with $p = 1$, it monotonically increases p and checks whether an integer p is a prefix period based on Lemma 9. Throughout the algorithm run, we maintain the invariant

$$p \parallel P[0:r], \text{ i.e., } P[0:r-p] \equiv P[p:r] \qquad (\spadesuit)$$

We calculate $reach_P(p)$ by increasing r as long as $P[0:r-p] \equiv P[p:r]$ holds (Lines 10–13). To let the function MATCH decide $P[0:r-p] \equiv P[p:r]$, we use two auxiliary arrays $first$ and $count$ that satisfy $first[a] = first_P(a)$ and $count[a] = count_{P[p:r]}(a)$, defined in Lemma 11. Moreover, we maintain the variable max_reach to be the largest reach calculated so far. By Lemma 9, the condition of Line 14 is satisfied if and only if p is a prefix period. One can construct the list PP by incrementing p one by one, but it takes too much time. Instead, we use a more efficient way explained later to make the algorithm run in linear time.

The following lemmas justify the behavior of our algorithm.

Lemma 12. *Throughout Algorithm 1, the value of the variable idx is always the upper bound that satisfies $PP[idx].val \leq \frac{r-p}{k}$. If there exists no such index, we have $idx = -1$.*

Proof. The variable idx is updated in conjunction with p and r to preserve the condition. See Lines 13 and 24. $\square$

Lemma 13. *Let $\spadesuit$ hold at Line 17 in Algorithm 1. If $period(P[0:r-p]) \leq \frac{r-p}{k}$, we have $PP[idx].val = period(P[0:r-p])$.*

Proof. Let $w' := P[0:r-p]$, $p' := period(w')$, $p'' := PP[idx].val$, and $w'' = P[0:kp'']$. By the assumption, p' is a prefix period of P. Additionally, we have $p' \leq p$ since $p \parallel w'$. Thus p' is in the list PP, and thus we have $p' \leq p''$ by Lemma 12. On the other hand, we have $period(w'') = p''$ by Lemma 8. Since $|w''| = kp'' \leq r - p = |w'|$, we have $period(w'') \leq period(w')$, i.e. $p'' \leq p'$. Hence we get $p' = p''$. $\square$

Lemma 14. *Let $\spadesuit$ hold at Line 17 in Algorithm 1. We have $PP[idx].reach \geq r - p \iff period(P[0:r-p]) \leq \frac{r-p}{k}$.*

Proof. Let $w' := P[0:r-p]$ and $p' := PP[idx].val$.
($\implies$) We have $p' \parallel w'$ by the assumption. Then $period(w') \leq p' \leq \frac{r-p}{k}$ holds by Lemma 12.
($\impliedby$) By Lemma 13, we have $p' = period(w')$. Then $PP[idx].reach = reach_P(p') = reach_P(period(w')) \geq |w'| = r - p$. $\square$

Now, we show that the invariant $\spadesuit$ always holds.

Lemma 15. *Throughout Algorithm 1, we have $P[0:r-p] \equiv P[p:r]$.*

Proof. One must see the condition preserved at the lines in which p or r is updated. The update at Lines 22–23 is trivial. Line 12 preserves the condition, ensured by the condition of Line 10. For Line 19, let $q := PP[idx].val$. Since $q =$

$period(P[0:r-p])$ by Lemma 13, we have $P[0:r-(p+q)] \equiv P[q:r-p] \equiv$
$P[p+q:r]$. Note that Lemma 13 requires ♠ only at Line 17, so the argument
does not circulate.   □

The following lemma plays a key role to avoid incrementing p one by one.

Lemma 16. *Consider $P \in (\Sigma \cup \Pi)^*$, $p \in \mathbb{N}^+$ and let $r := reach_P(p)$. Then, no
prefix period q of P exists such that $p < q < p + period(P[0:r-p])$.*

Proof. We use Lemma 2 for $x := P[p:r]$, $y := P[0:r-p]$, $\delta := q-p$ to obtain
$P[q:r] \not\equiv P[0:r-q]$, which means $q \nmid P[0:r]$. Thus we have $reach_P(q) < r =$
$reach_P(p)$, which implies that q is not a prefix period of P by Lemma 9.   □

We now present the way to compute the list of prefix periods efficiently, in
which we skip calculating $reach_P(p)$ if we are sure that p is not a prefix period.
For realizing an efficient shift, we maintain a variable idx so that it points at the
largest index of PP such that $PP[idx].val \leq \frac{r-p}{k}$ (Lemma 12). The shift amount
is determined in the following manner. If $PP[idx].reach \geq r - p > 0$ at Line 17,
Lemmas 14 and 13 imply $PP[idx].val = period(P[0:r-p])$. Hence, Lemma 16
justifies the shift amount $PP[idx].val$ of p at Line 19. On the other hand, if
$PP[idx].reach < r - p$, by Lemma 14, we have $period(P[0:r-p]) > \frac{r-p}{k}$. This
justifies the shift $\lfloor \frac{r-p}{k} \rfloor + 1$ of p at Line 22 again by Lemma 16. If $r - p = 0$,
then p is incremented by just one.

Now, we show that the algorithm runs in $O(m)$ time. Firstly, notice that the
while loops at Line 9 and 10 are repeated only $O(m)$ times in total, since the
quantity $kp + r$ keeps increasing and $kp + r \leq k \cdot \frac{m}{k} + m = O(m)$. Hence, the
fact we must show is that decrementing $count$ and idx at Line 18, 21, and 24
takes $O(m)$ time in total. As their values are always greater than or equal to
their initial values, the number of decrements does not exceed the number of
increments, which is $O(m)$ since they are in Line 11–13.

Theorem 1. *All prefix periods of P and their reaches can be calculated in $O(m)$
time and $O(\log m + |\Pi|)$ extra space.*

4.2 Searching for Parameterized Matches

Our matching algorithm is shown in Algorithm 2. As it is the case for the Galil-
Seiferas algorithm, it resembles the preprocess. Now, the invariants in Algorithm 2 are obtained by replacing p, r, and $P[p:r]$ in Lemma 12–15 with i, j,
and $T[i:j]$, respectively. Particularly, by the invariant that $P[0:j-i] \equiv T[i:j]$,
one can find matching positions i when $j = i + |P|$ (Line 13). The shift amounts
are also justified by using Lemma 2 for $x := T[i:j]$ and $y := P[0:j-i]$, whose
conclusion $T[i+\delta:j] \not\equiv P[0:j-i-\delta]$ implies $T[i+\delta:i+\delta+|P|] \not\equiv P$ for
any δ smaller than the shift by the algorithm. We can show that the searching
phase (Line 8–22) runs in $O(|\Pi_P|n)$ time in the same way as for the preprocess
with the increasing quantity $ki + j$.

Theorem 2. *The parameterized matching problem can be solved in $O(|\Pi_P|n + m)$ time and $O(\log m + |\Pi|)$ extra space.*

Algorithm 2: SEARCH

Input: $T, P \in (\Sigma \cup \Pi)^*$
Output: all $0 \le i \le |T| - |P|$ such that $T[i : i + |P|] \equiv P$

1 **begin**
2 $\quad k \leftarrow |\Pi_P| + 2$
3 $\quad first \leftarrow first_P$
4 $\quad PP \leftarrow \text{PREFIX\_PERIODS}(P)$
5 $\quad idx \leftarrow -1$
6 $\quad (i, j) \leftarrow (0, 0)$
7 $\quad$ **foreach** $a \in \Pi$ **do** $count[a] \leftarrow 0$
8 $\quad$ **while** $i < |T| - |P|$ **do**
9 $\quad\quad$ **while** $\text{MATCH}(P[0 : j - i], T[i : j], P[j - i], T[j], first, count)$ **do**
10 $\quad\quad\quad$ Increment $count[T[j]]$
11 $\quad\quad\quad$ $j \leftarrow j + 1$
12 $\quad\quad\quad$ **if** $idx + 1 < |PP|$ *and* $PP[idx + 1].val \le \frac{j-i}{k}$ **then** Increment idx

13 $\quad\quad$ **if** $j - i = |P|$ **then**
14 $\quad\quad\quad$ output i

15 $\quad\quad$ **if** $0 \le idx < |PP|$ *and* $PP[idx].reach \ge j - i > 0$ **then**
16 $\quad\quad\quad$ **for** $i \le u < i + PP[idx].val$ **do** Decrement $count[T[u]]$
17 $\quad\quad\quad$ $i \leftarrow i + PP[idx].val$
18 $\quad\quad$ **else**
19 $\quad\quad\quad$ **for** $i \le u < j$ **do** Decrement $count[T[u]]$
20 $\quad\quad\quad$ $i \leftarrow i + \lfloor \frac{j-i}{k} \rfloor + 1$
21 $\quad\quad\quad$ $j \leftarrow i$
22 $\quad\quad$ **until** $PP[idx].val \le \frac{j-i}{k}$ *or* $idx = -1$ **do** Decrement idx

---

5   Conclusion and Future Work

We studied the periodicity of parameterized strings and extended the Galil-Seiferas algorithm [11] for parameterized matching. The proposed algorithm requires only sublinear extra space. The properties of periods of parameterized strings we presented in this paper may be used to design more space-efficient algorithms for parameterized matching, as Galil and Seiferas [12] used prefix periods to design a constant-extra-space algorithm for exact matching.

Acknowledgements. The authors deeply appreciate the anonymous reviewers helpful comments. This work was supported by JSPS KAKENHI Grant Numbers JP19K20208 (DH), JP18K11150 (RY), JP20H05703 (RY), JP23K11325 (RY), and JP21K11745 (AS).

References

1. AdelsonVelskii, M., Landis, E.M.: An algorithm for the organization of information. Joint Publications Research Service Washington DC, Technical report (1963)
2. Amir, A., Farach, M., Muthukrishnan, S.: Alphabet dependence in parameterized matching. Inf. Process. Lett. **49**(3), 111–115 (1994)
3. Apostolico, A., Giancarlo, R.: Periodicity and repetitions in parameterized strings. Discret. Appl. Math. **156**(9), 1389–1398 (2008)
4. Baker, B.S.: Parameterized pattern matching: algorithms and applications. J. Comput. Syst. Sci. **52**(1), 28–42 (1996)
5. Crochemore, M.: String-matching on ordered alphabets. Theor. Comput. Sci. **92**(1), 33–47 (1992)
6. Crochemore, M., Perrin, D.: Two-way string-matching. J. ACM **38**(3), 650–674 (1991)
7. Deguchi, S., Higashijima, F., Bannai, H., Inenaga, S., Takeda, M.: Parameterized suffix arrays for binary strings. In: Proceedings of the Prague Stringology Conference 2008, pp. 84–94 (2008)
8. Diptarama, Katsura, T., Otomo, Y., Narisawa, K., Shinohara, A.: Position heaps for parameterized strings. In: Proceedings of the 28th Annual Symposium on Combinatorial Pattern Matching (CPM 2017), pp. 8:1–8:13 (2017)
9. Fine, N.J., Wilf, H.S.: Uniqueness theorems for periodic functions. Proc. Am. Math. Soc. **16**(1), 109–114 (1965)
10. Fujisato, N., Nakashima, Y., Inenaga, S., Bannai, H., Takeda, M.: Right-to-left online construction of parameterized position heaps. In: Proceedings of the Prague Stringology Conference 2018 (PSC 2018), pp. 91–102 (2018)
11. Galil, Z., Seiferas, J.: Saving space in fast string-matching. SIAM J. Comput. **9**(2), 417–438 (1980)
12. Galil, Z., Seiferas, J.: Time-space-optimal string matching. J. Comput. Syst. Sci. **26**(3), 280–294 (1983)
13. Knuth, D.E., Morris, J.H., Jr., Pratt, V.R.: Fast pattern matching in strings. SIAM J. Comput. **6**(2), 323–350 (1977)
14. Kosaraju, S.R.: Faster algorithms for the construction of parameterized suffix trees. In: Proceedings of the 36th Annual Symposium on Foundations of Computer Science, pp. 631–638 (1995)
15. Manber, U., Myers, G.: Suffix arrays: a new method for on-line string searches. SIAM J. Comput. **22**(5), 935–948 (1993)
16. Mendivelso, J., Thankachan, S.V., Pinzón, Y.: A brief history of parameterized matching problems. Discret. Appl. Math. **274**, 103–115 (2020)
17. Nakashima, K., et al.: Parameterized DAWGs: efficient constructions and bidirectional pattern searches. Theor. Comput. Sci. **933**, 21–42 (2022)
18. Nakashima, K., Hendrian, D., Yoshinaka, R., Shinohara, A.: An extension of linear-size suffix tries for parameterized strings. In: SOFSEM 2020 Student Research Forum, pp. 97–108 (2020)
19. Weiner, P.: Linear pattern matching algorithms. In: Proceedings of the 14th Annual Symposium on Switching and Automata Theory, pp. 1–11 (1973)

Largest Repetition Factorization of Fibonacci Words

Kaisei Kishi[1]($\boxtimes$), Yuto Nakashima[2] (ID), and Shunsuke Inenaga[2] (ID)

[1] Department of Information Science and Technology, Kyushu University,
Fukuoka, Japan
kishi.kaisei.216@s.kyushu-u.ac.jp
[2] Department of Informatics, Kyushu University, Fukuoka, Japan
{nakashima.yuto.003,inenaga.shunsuke.380}@m.kyushu-u.ac.jp

Abstract. A factorization of a string w is said to be a *repetition factorization* of w if every factor in the factorization is a repetition (i.e., the factor has a period shorter than or equal to the half of its length). Inoue et al. [TOCS 2022] showed how to compute the largest/smallest repetition factorization of a given string w of length n in $O(n \log n)$ time and $O(n)$ space, by reducing the problems to the longest/shortest path problems on the repetition graph built on w. Inoue et al. also considered repetition factorizations on Fibonacci words, and posed a conjecture on the size S_{F_k} of the largest repetition factorization of the k-th Fibonacci word F_k. In this work, we provide a complete proof for this problem, by showing that S_{F_k} is given by the recurrence $S_{F_k} = S_{F_{k-1}} + S_{F_{k-2}} + 1$ for every $k \geq 15$.

1 Introduction

Various factorizations (or parsings) of strings play important roles in stringology and are well-studied. A sequence of m-strings $f_1, \ldots, f_m$ is said to be a factorization of a string w if $w = f_1 \cdots f_m$ holds. We call each f_i a factor of the factorization and m the size of the factorization. One of the most significant applications of factorizations is data compression. For instance, each of the factorizations in the Lempel-Ziv family [9–12], lexparse [8] produces a compact representation of a string whose size depends on the size of the factorization (see also a nice survey [7]). Also, many variants of string factorizations by combinatorial properties or structures are considered, such as the Lyndon factorization [1], palindromic factorizations, etc.

In this paper, we deal with a factorization such that each factor is a repetitive structure called a repetition. Factorizations by repetitive structures were studied by Dumitran et al. [2]. They considered the two types of factorizations called square factorizations and repetition factorizations, which are factorized into squares and repetitions, respectively. For square factorizations, they presented an $O(n \log n)$-time algorithm that computes a square factorization of a given string of length n. After that, a linear-time algorithm on the word RAM model of machine word size $\Omega(\log n)$ were presented by Matsuoka et al. [6]. For

F. M. Nardini et al. (Eds.): SPIRE 2023, LNCS 14240, pp. 284–296, 2023.
https://doi.org/10.1007/978-3-031-43980-3_23

repetition factorizations, Dumitran et al. [2] claimed that a repetition factorization of a given string of length n can be computed in $O(n)$ time.

Inoue et al. [4] extended the repetition factorization problem. The new problem aims to find the largest/smallest repetition factorizations with the maximum/minimum number of factors, respectively. They proposed an algorithm that computes an arbitrary such factorization in $O(n \log n)$ time and $O(n)$ space using a reduction from the largest/smallest factorization problems to the longest/shortest path problems on a graph representing repetitive structures in the input text. They also showed that the size of the graph is in $\Theta(n \log n)$ when the input string is the Fibonacci word. This bound introduced a problem about the size of the largest/smallest repetition factorizations of the Fibonacci word. They proved that the size of the smallest repetition factorization of the k-th Fibonacci word is 2 for any $k \geq 8$, and also conjectured that the size S_{F_k} of the largest repetition factorization of the k-th Fibonacci word F_k can be represented as $S_{F_k} = S_{F_{k-1}} + S_{F_{k-2}} + 1$ for any sufficiently large k.

In this paper, we proved that the conjecture is true. More formally, $S_{F_k} = S_{F_{k-1}} + S_{F_{k-2}} + 1$ holds for $k \geq 15$. The main ideas for our proof can be explained as follows. First, we give a parsing (representing a rough repetition factorization) by specific substrings of the Fibonacci word. Because every specific substring has a repetition factorization, we consider the sum of the size of the largest repetition factorization of every phrase of the parsing as a candidate of the size of the largest repetition factorization of the Fibonacci word. Next, we show that the length of factors of the largest repetition factorization is at most 18. Then, by using the property, we prove that the size of other repetition factorizations cannot exceed the candidate.

The rest of this paper is organized as follows. First, we give notation and definitions on strings in Sect. 2. In Sect. 3, we present the candidates of the largest repetition factorizations. Finally, in Sect. 4, we describe the maximality of the size of the candidate repetition factorization.

2   Preliminaries

Strings. Let Σ be a binary alphabet. An element of Σ^* is called a *string*. The length of a string w is denoted by $|w|$. The empty string ε is a string of length 0, namely, $|\varepsilon| = 0$. Let Σ^+ be the set of non-empty strings, i.e., $\Sigma^+ = \Sigma^* - \{\varepsilon\}$. For a string $w = xyz$, x, y and z are called a *prefix*, *substring*, and *suffix* of w, respectively. The i-th character of a string w is denoted by $w[i]$, where $1 \leq i \leq |w|$. For a string w and two integers $1 \leq i \leq j \leq |w|$, let $w[i..j]$ denote the substring of w that begins at position i and ends at position j. For convenience, let $w[i..j] = \varepsilon$ when $i > j$. $Pre(w, k)$ and $Suf(w, k)$ denote the prefix and suffix of length k of w, respectively. Namely, $Pre(w, k) = w[1..k]$ and $Suf(w, k) = w[|w| - k + 1..|w|]$. For a string w and an integer $k \geq 2$, let $w^1 = w$ and $w^k = ww^{k-1}$.

Repetitions and Factorizations. An integer $p \geq 1$ is said to be a *period* of a string w if $w[i] = w[i + p]$ for all $1 \leq i \leq |w| - p$. If p is a period of a string

w with $p < |w|$, then $w[1..|w| - p] = w[p + 1..|w|]$ is said to be a *border* of w. A non-empty string s is said to be a *repetition*, if $s = x^k x'$ for some string x, some integer $k \geq 2$, and some proper prefix x' of x. A sequence $r_1, \ldots, r_m$ of non-empty strings is said to be a *repetition factorization* of a string w, if $w = r_1 \cdots r_m$ and each r_i ($1 \leq i \leq m$) is a repetition. Each r_i in a repetition factorization $r_1, \ldots, r_m$ of a string w is called a *factor* of the factorization. The *size* of the repetition factorization is the number m of factors in the factorization. We refer to a repetition factorization that has the maximum number of factors as a *largest repetition factorization*. In this paper, we will represent a factorization (parsing) as a concatenation of substrings for simplicity. For example, consider the string $w = $ abaabaababababababa. There are ten largest repetition factorizations of w as follows.

> abaaba|abab|abababa   abaaba|ababa|bababa   abaaba|ababab|ababa
> abaaba|abababa|baba   abaabaa|baba|bababa   abaabaa|babab|ababa
> abaabaa|bababa|baba   abaabaab|abab|ababa   abaabaab|ababa|baba
> abaabaaba|baba|baba

We remark that there exist strings that have no repetition factorization. However, the k-th Fibonacci word has a repetition factorization for every $k \geq 8$ [4].

Fibonacci Words. The k-th (finite) Fibonacci word $F_k(a, b)$ over an alphabet $\{a, b\}$ is defined as follows: $F_1(a, b) = b$, $F_2(a, b) = a$, and $F_k(a, b) = F_{k-1}(a, b) \cdot F_{k-2}(a, b)$ for any $k > 2$ (cf. [5]). Let f_k be the length of k-th Fibonacci word. In this paper, we sometimes use strings u, v as characters in the notation of Fibonacci words. For example, $F_6(u, v) = uvuuvuvu$ for any strings u, v. We also deal with the infinite Fibonacci word $\mathcal{F}(a, b) = \lim_{k \to \infty} F_k(a, b)$ over an alphabet $\{a, b\}$. We will drop the alphabet whenever it is clear from context.

Regular Expression. Let A be the set of special symbols $\{ (,), \varepsilon, \emptyset, \cdot, +, {}^* \}$. Notice that $\Sigma \cap A = \emptyset$. A regular expression over Σ is a string over $\Sigma \cup A$ that is recursively defined as follows.

- a ($\in \Sigma$), ε, and $\emptyset$ are regular expressions.
- For any regular expressions x, y, $(x \cdot y)$ is a regular expression.
- For any regular expressions x, y, $(x + y)$ is a regular expression.
- For any regular expression x, x^* is a regular expression.

For any sets W, Z of strings, let $W \cdot Z = \{wz \mid w \in W, z \in Z\}$. A regular expression represents a (regular) language which will be explained in the following. For any regular expression x, let $\|x\|$ denotes a language represented by x.

- For any $a \in \Sigma$, $\|a\| = \{a\}$.
- $\|\varepsilon\| = \{\varepsilon\}$.
- $\|\emptyset\| = \emptyset$.
- For any regular expressions x, y, $\|(x \cdot y)\| = \|x\| \cdot \|y\|$.
- For any regular expressions x, y, $\|(x + y)\| = \|x\| \cup \|y\|$.
- For any regular expression x, $\|x^*\| = \|x\|^*$.

We will sometimes drop $(,)$ if it is clear. We say that a string w matches a regular expression x if $w \in \|x\|$.

3 Candidate of Largest Factorizations

In this paper, we show the size of the largest repetition factorizations of Fibonacci words. Our goal is stated in the following theorem.

Theorem 1. *Let S_w be the size of the largest repetition factorization of a string w. Then $S_{F_k} = S_{F_{k-1}} + S_{F_{k-2}} + 1$ holds for every $k \geq 15$.*

In this section, we explain a repetition factorization that has the largest number of factors. Notice that the proof of its maximality will be given in Sect. 4.

First, we introduce a base-factor parsing of finite Fibonacci words. Roughly speaking, we consider specific substrings of Fibonacci words and represent Fibonacci words as a concatenation of the substrings. First, we present a set of specific substrings which are called *base-factors* in our parsing.

Definition 1 (Base-factors). *Let BFs be the set of base-factors. The seven elements of BFs are given in the following.*

$$
\begin{aligned}
&\mathsf{oddpre} = F_8 && \mathsf{oddsuf} = baabaab \\
&\mathsf{evenpre} = F_{12} \cdot F_8 && \mathsf{evensuf} = baabaab \cdot F_{12} \\
&\mathsf{F}_{11}^- = F_{11}[f_8 + 1 .. f_{11} - 7] && \mathsf{F}_{13}^- = F_{13}[f_8 + 1 .. f_{13} - 7] \\
&\mathsf{core} = baabaab \cdot F_{12} \cdot F_8
\end{aligned}
$$

By the definition, we can see the following relations:

$$\mathsf{core} = \mathsf{evensuf} \cdot \mathsf{oddpre} = \mathsf{oddsuf} \cdot \mathsf{evenpre},$$

$$F_{11} = \mathsf{oddpre} \cdot \mathsf{F}_{11}^- \cdot \mathsf{oddsuf},$$

$$F_{13} = \mathsf{oddpre} \cdot \mathsf{F}_{13}^- \cdot \mathsf{oddsuf},$$

$$\mathsf{oddpre} \cdot \mathsf{F}_{13}^- = \mathsf{evenpre} \cdot \mathsf{F}_{11}^-.$$

Next, we consider a parsing of F_k with *BFs*. We give a parsing by using a regular expression such that the largest repetition factorization which we will propose is a refinement of this parsing:

$$Reg_\Sigma = (\mathsf{oddpre} + \mathsf{evenpre}) \cdot (\mathsf{F}_{11}^- + \mathsf{F}_{13}^-) \cdot (\mathsf{core} \cdot (\mathsf{F}_{11}^- + \mathsf{F}_{13}^-))^* \cdot (\mathsf{oddsuf} + \mathsf{evensuf}).$$

Due to a formal description of the parsing, we use meta-strings such that each meta-character corresponds to a base-factor. Let $\Pi = \{b_{\mathsf{oddpre}}, \ldots, b_{\mathsf{F}_{13}^-}\}$ be an alphabet such that $\Sigma \cap \Pi = \emptyset$. We also consider a function $g : BFs \to \Pi$ as follows.

$$
\begin{aligned}
&g(\mathsf{oddpre}) = b_{\mathsf{oddpre}} && g(\mathsf{oddsuf}) = b_{\mathsf{oddsuf}} \\
&g(\mathsf{evenpre}) = b_{\mathsf{evenpre}} && g(\mathsf{evensuf}) = b_{\mathsf{evensuf}} \\
&g(\mathsf{F}_{11}^-) = b_{\mathsf{F}_{11}^-} && g(\mathsf{F}_{13}^-) = b_{\mathsf{F}_{13}^-} \\
&g(\mathsf{core}) = b_{\mathsf{core}}
\end{aligned}
$$

We extend the regular expression for Π as follows:

$$Reg_\Pi = (\mathsf{b_{oddpre}} + \mathsf{b_{evenpre}}) \cdot (\mathsf{b_{F_{11}^-}} + \mathsf{b_{F_{13}^-}}) \cdot (\mathsf{b_{core}} \cdot (\mathsf{b_{F_{11}^-}} + \mathsf{b_{F_{13}^-}}))^* \cdot (\mathsf{b_{oddsuf}} + \mathsf{b_{evensuf}}).$$

Then we can give a rough parsing of a largest repetition factorization (Definition 2) because every base-factor has a repetition factorization (cf. Fact 1), and we also show that F_k can be written as a concatenation of base-factors (Lemma 1).

Definition 2 (Base-factor parsing (BFparse)). *Let w be a string over Σ. A factorization $w_1, \ldots, w_k$ of w is a base-factor parsing of w if there exists a string $x \in \|Reg_\Pi\|$ that has $g(w_1) \cdots g(w_k)$ as a substring.*

Lemma 1. *For every $k \geq 15$, $F_k \in \|Reg_\Sigma\|$.*

Proof. We show that F_k has a base-factor parsing by induction on k. In other words, we show the following parsing $\mathsf{BFparse}(F_k)$ of F_k:

$$\mathsf{BFparse}(F_k) = \begin{cases} \mathsf{oddpre} \cdot \mathsf{F_{13}^-} \cdot X_k \cdot \mathsf{core} \cdot \mathsf{F_{13}^-} \cdot \mathsf{oddsuf} & \text{for odd } k \quad (1) \\ \mathsf{evenpre} \cdot \mathsf{F_{11}^-} \cdot X_k \cdot \mathsf{core} \cdot \mathsf{F_{11}^-} \cdot \mathsf{evensuf} & \text{for even } k \quad (2) \end{cases}$$

where X_k is a base-factor parsing such that $g(X_k) \in \|(\mathsf{b_{core}} \cdot (\mathsf{b_{F_{11}^-}} + \mathsf{b_{F_{13}^-}}))^*\|$.
For $k = 15$,

$$F_{15} = F_{13} \cdot F_{12} \cdot F_{13} = \mathsf{oddpre} \cdot \mathsf{F_{13}^-} \cdot \mathsf{oddsuf} \cdot F_{12} \cdot \mathsf{oddpre} \cdot \mathsf{F_{13}^-} \cdot \mathsf{oddsuf}$$

$$= \mathsf{oddpre} \cdot \mathsf{F_{13}^-} \cdot \mathsf{core} \cdot \mathsf{F_{13}^-} \cdot \mathsf{oddsuf}.$$

Thus Eq. (1) holds for $k = 15$ since $X_{15} = \varepsilon$.
For $k = 16$,

$$F_{16} = F_{15} \cdot F_{14} = \mathsf{oddpre} \cdot \mathsf{F_{13}^-} \cdot \mathsf{core} \cdot \mathsf{F_{13}^-} \cdot \mathsf{oddsuf} \cdot F_{14}$$

$$= \mathsf{evenpre} \cdot \mathsf{F_{11}^-} \cdot \mathsf{core} \cdot \mathsf{F_{13}^-} \cdot \mathsf{oddsuf} \cdot F_{12} \cdot F_{11} \cdot F_{12}$$

$$= \mathsf{evenpre} \cdot \mathsf{F_{11}^-} \cdot \mathsf{core} \cdot \mathsf{F_{13}^-} \cdot \mathsf{oddsuf} \cdot F_{12} \cdot \mathsf{oddpre} \cdot \mathsf{F_{11}^-} \cdot \mathsf{oddsuf} \cdot F_{12}$$

$$= \mathsf{evenpre} \cdot \mathsf{F_{11}^-} \cdot \mathsf{core} \cdot \mathsf{F_{13}^-} \cdot \mathsf{core} \cdot \mathsf{F_{11}^-} \cdot \mathsf{evensuf}.$$

Thus Eq. (2) holds for $k = 16$ since $X_{16} = \mathsf{core} \cdot \mathsf{F_{13}^-}$.
Suppose that there exists a base-factor parsing that is represented by Eqs. (1) and (2) for every $k < c$ for some integer $c \geq 17$. Let c be an odd. By the induction hypothesis, there are base-factor parsings as follows:

$$\mathsf{BFparse}(F_{c-1}) = \mathsf{evenpre} \cdot \mathsf{F_{11}^-} \cdot X_{c-1} \cdot \mathsf{core} \cdot \mathsf{F_{11}^-} \cdot \mathsf{evensuf},$$

$$\mathsf{BFparse}(F_{c-2}) = \mathsf{oddpre} \cdot \mathsf{F_{13}^-} \cdot X_{c-2} \cdot \mathsf{core} \cdot \mathsf{F_{13}^-} \cdot \mathsf{oddsuf}.$$

Then,

$$F_c = \mathsf{evenpre} \cdot \mathsf{F_{11}^-} \cdot X_{c-1} \cdot \mathsf{core} \cdot \mathsf{F_{11}^-} \cdot \mathsf{evensuf}$$

$$\cdot \mathsf{oddpre} \cdot \mathsf{F_{13}^-} \cdot X_{c-2} \cdot \mathsf{core} \cdot \mathsf{F_{13}^-} \cdot \mathsf{oddsuf}$$

$$= \mathsf{oddpre} \cdot \mathsf{F_{13}^-} \cdot X_{c-1} \cdot \mathsf{core} \cdot \mathsf{F_{11}^-} \cdot \mathsf{core}$$

$$\cdot \mathsf{F_{13}^-} \cdot X_{c-2} \cdot \mathsf{core} \cdot \mathsf{F_{13}^-} \cdot \mathsf{oddsuf}.$$

Since $X_{c-1} \cdot \text{core} \cdot \text{F}_{11}^- \cdot \text{core} \cdot \text{F}_{13}^- \cdot X_{c-2}$ is a base-factor parsing that corresponds to $(\text{b}_{\text{core}} \cdot (\text{b}_{\text{F}_{11}^-} + \text{b}_{\text{F}_{13}^-}))^*$, let $X_c = X_{c-1} \cdot \text{core} \cdot \text{F}_{11}^- \cdot \text{core} \cdot \text{F}_{13}^- \cdot X_{c-2}$. Then

$$\text{BFparse}(F_c) = \text{oddpre} \cdot \text{F}_{13}^- \cdot X_c \cdot \text{core} \cdot \text{F}_{13}^- \cdot \text{oddsuf}.$$

Let c be an even. We can prove the statement for this case in a similar way, where $X_c = X_{c-1} \cdot \text{core} \cdot \text{F}_{13}^- \cdot \text{core} \cdot \text{F}_{11}^- \cdot X_{c-2}$. Therefore, the lemma holds for any $k \geq 15$. ☐

We refer to the BFparse described in the lemma as *canonical BFparse* of the Fibonacci word. Let us denote the canonical parsing of F_k by Fib_k. Formally, the equations are given as follows.

$$\text{Fib}_k = \begin{cases} \text{oddpre} \cdot \text{F}_{13}^- \cdot X_k \cdot \text{core} \cdot \text{F}_{13}^- \cdot \text{oddsuf} & \text{for odd } k \\ \text{evenpre} \cdot \text{F}_{11}^- \cdot X_k \cdot \text{core} \cdot \text{F}_{11}^- \cdot \text{evensuf} & \text{for even } k \end{cases}$$

$$X_k = \begin{cases} X_k = X_{k-1} \cdot \text{core} \cdot \text{F}_{11}^- \cdot \text{core} \cdot \text{F}_{13}^- \cdot X_{k-2} & \text{for odd } k \\ X_k = X_{k-1} \cdot \text{core} \cdot \text{F}_{13}^- \cdot \text{core} \cdot \text{F}_{11}^- \cdot X_{k-2} & \text{for even } k \end{cases}$$

In the rest of this section, we give several properties and notations on *BFs* and base-factor parsings.

Fact 1. *Every base-factor has a repetition factorization. The sizes of largest repetition factorizations of base-factors are* $S_{\text{oddpre}} = 4$, $S_{\text{oddsuf}} = 1$, $S_{\text{evenpre}} = 32$, $S_{\text{evensuf}} = 29$, $S_{\text{F}_{11}^-} = 12$, $S_{\text{F}_{13}^-} = 40$, *and* $S_{\text{core}} = 34$. *Note that* $S_{\text{core}} = S_{\text{evensuf}} + S_{\text{oddpre}} + 1 = S_{\text{oddsuf}} + S_{\text{evenpre}} + 1$ *(this structure gives an additional factor).*

Example 1. The largest repetition factorizations of oddpre, evenpre, evensuf, and F_{11}^- are as follows.

oddpre	abaaba|baabaa|baba|ababa
evenpre	abaaba|baabaa|baba|ababa|abaaba|baabaab|abaaba|baabaa|baba| abab|aabaab|abaaba|abab|aa|baba|abaaba|baabaa|baba|abab| aabaab|abaaba|baabaa|baba|abaaba|baababaaba|abab|aa|baba| abaaba|baabaa|baba|ababa
evensuf	baabaab|abaaba|baabaa|baba|ababa|abaaba|baabaab|abaaba| baabaa|baba|abab|aabaab|abaaba|abab|aa|baba|abaaba|baabaa| baba|abab|aabaab|abaaba|baabaa|baba|abaaba|baababaaba|abab| aa|baba
F_{11}^-	abaaba|baabaab|abaaba|baabaa|baba|abab|aabaab| abaaba|abab|aa|baba|abaaba

Fact 2. *A string* babaabaababaababa *(of length 17) is a suffix of* oddpre, evenpre, core, *and a string* aababaababaabaaba *(of length 17) is a suffix of* F_{11}^-, F_{13}^-.

We can check the above facts by a computer search. In the rest of this paper, for a base-factor x, $\mathcal{L}(x)$ denote an arbitrary fixed largest repetition factorization of x (i.e., Fact 1). If a string x has a base-factor parsing $x_1, \ldots, x_i$, then x has a repetition factorization $\mathcal{L}(x_1), \ldots, \mathcal{L}(x_i)$. We call this factorization the repetition factorization of a base-factor parsing $x_1, \ldots, x_i$. Especially, we call the repetition factorization of Fib_k the candidate repetition factorization of F_k. Then we consider the size of the specific repetition factorization as follows.

Definition 3. *Let X be a base-factor parsing of a string $x \in \Sigma^*$. We define B_X as the size of the repetition factorization of X.*

In the next section, we prove that the candidate factorization of F_k is a largest repetition factorization. Since $B_{\mathrm{Fib}_k} \leq S_{F_k}$ is clearly holds, our task is to show $B_{\mathrm{Fib}_k} \geq S_{F_k}$. We conclude this section with the following property about the candidate repetition factorization of F_k.

Lemma 2. *For every $k \geq 17$, $B_{\mathrm{Fib}_k} = B_{\mathrm{Fib}_{k-1}} + B_{\mathrm{Fib}_{k-2}} + 1$.*

Proof. Due to the discussions in the proof of Lemma 1, we replaced $\mathsf{evenpre} \cdot F_{11}^-$ with $\mathsf{oddpre} \cdot F_{13}^-$ and $\mathsf{evensuf} \cdot \mathsf{oddpre}$ with core for odd k. In the former case, the number of factors cannot be changed since $S_{\mathsf{evenpre}} + S_{F_{11}^-} = S_{\mathsf{oddpre}} + S_{F_{13}^-} = 44$. In the latter case, the number of factors increases by one since $S_{\mathsf{evensuf}} + S_{\mathsf{oddpre}} + 1 = S_{\mathsf{core}} = 34$. Thus $B_{\mathrm{Fib}_k} = B_{\mathrm{Fib}_{k-1}} + B_{\mathrm{Fib}_{k-2}} + 1$ holds for any odd k. We can prove for even k in a similar argument. □

4   Maximality of Candidate Repetition Factorizations

In this section, we prove that the candidate repetition factorization of F_k which was given in the previous section (Definition 3) is a largest repetition factorization of F_k. Namely, we show the following lemma.

Lemma 3. *For every $k \geq 17$, $B_{\mathrm{Fib}_k} = S_{F_k}$.*

Clearly, by Lemmas 2 and 3, $S_{F_k} = S_{F_{k-1}} + S_{F_{k-2}} + 1$ holds for every $k \geq 17$. By a computer search, we know that $S_{13} = 45, S_{14} = 73, S_{13} = 119, S_{13} = 193$ (cf. [4]). This implies that $S_{F_k} = S_{F_{k-1}} + S_{F_{k-2}} + 1$ also holds for $k = 15, 16$. Then we can obtain our main result Theorem 1. This section is organized as follows. In Subsect. 4.1, we show an upper bound of the length of factors of the largest repetition factorizations of Fibonacci words. Finally, in Subsect. 4.2, we present our main result by using careful analysis for the candidate repetition factorization.

4.1   Upper Bound of the Length of Factors

We give the upper bound of the length of factors of the largest repetition factorizations of Fibonacci words (Lemma 6). To prove the lemma, we use Lemma 5.

Lemma 4 (Lemma 2.2 of [3]). *For any integers ℓ and k satisfying $0 \leq \ell < k$,*

$$F_k(a,b) = F_{k-\ell}(F_{\ell+2}(a,b), F_{\ell+1}(a,b)).$$

Moreover, $\mathcal{F}(a,b) = \mathcal{F}(F_{k+1}(a,b), F_k(a,b))$ for any $k \geq 1$.

Lemma 5. *Let $k \geq 3$, and d_k be an integer satisfying $f_k < d_k \leq f_{k+1}$. Then every substring of length d_k of $\mathcal{F}$ occurs in $\mathcal{F}[1..f_{k+2} + d_k - 1]$.*

Proof. By Lemma 4, $\mathcal{F}$ can be represented as

$$\mathcal{F} = F_{k+2} \cdot F_{k+1} \cdot F_{k+2} \cdot F_{k+2} \cdot F_{k+1} \cdot F_{k+2} \cdot F_{k+1} \cdot F_{k+2} \cdots$$

for every k. Let x be a substring of length d_k of $\mathcal{F}$.

1. Suppose that x is a substring of F_{k+2}. Then x has an occurrence in $\mathcal{F}[1..f_{k+2} + d_k - 1]$ since F_{k+2} is a prefix of $\mathcal{F}$.
2. Suppose that x is a substring of F_{k+1}. Then x has an occurrence in $\mathcal{F}[1..f_{k+2} + d_k - 1]$ since F_{k+1} is also a prefix of $\mathcal{F}$.
3. Suppose that x has an occurrence in $F_{k+2} \cdot F_{k+1}$ that contains the boundary of F_{k+2} and F_{k+1}. Then x has an occurrence in $\mathcal{F}[1..f_{k+2} + d_k - 1]$ since $d_k \leq f_{k+1}$ and $F_{k+2} \cdot F_{k+1}$ is also a prefix of $\mathcal{F}$.
4. Suppose that x has an occurrence in $F_{k+1} \cdot F_{k+2}$ that contains the boundary of F_{k+1} and F_{k+2}. Then x is a substring of $F_{k+1} \cdot F_{k+1}$ since $F_{k+1} \cdot F_{k+2} = F_{k+1} \cdot F_{k+1} \cdot F_k$. Moreover, $\mathcal{F}$ has $F_{k+1} \cdot F_{k+1}$ as a prefix because

$$\mathcal{F} = F_{k+2} \cdot F_{k+1} \cdots = F_{k+1} \cdot F_k \cdot F_{k+1} \cdots$$
$$= F_{k+1} \cdot F_k \cdot (F_{k-1} \cdot F_{k-2} \cdot F_{k-1}) \cdots$$
$$= F_{k+1} \cdot F_{k+1} \cdot F_{k-2} \cdot F_{k-1} \cdots .$$

Thus x has an occurrence in $\mathcal{F}[1..f_{k+2} + d_k - 1]$.

Therefore the lemma holds. □

Lemma 6. *For any largest repetition factorization of Fibonacci words, the length of every factor is at most 18.*

Proof. Let us consider a substring of length d of $\mathcal{F}$ where $19 \leq d \leq 37$. By Lemma 5, every substring of length d of $\mathcal{F}$ is a substring of $\mathcal{F}[1..125]$ since $f_9 < 37 \leq f_{10}$. By an exhaustive enumeration of length-d substrings of $\mathcal{F}[1..125]$, we can see that the size of the largest repetition factorization of every length-d substring of $\mathcal{F}$ is at least 2. This implies that the size of the largest repetition factorization of substring of length more than 37 of $\mathcal{F}$ is at least 4 (since each substring can be represented as a concatenation of substrings of length in $[19, 37]$). Thus every substring of length more than 18 of $\mathcal{F}$ has a repetition factorization of size at least 2. Assume on the contrary that a largest repetition factorization of F_k has a factor of length more than 18. Then the factor can be factorized into at least 2 repetitions, a contradiction. Therefore the lemma holds. □

4.2   Analysis of Candidate Factorizations

To prove Lemma 3, we consider other parsings which can be obtained by shifting boundaries of candidate parsings. Furthermore, we show that the size of the largest repetition factorizations of any of such parsings cannot exceed the candidate.

First, by Lemma 6, a simple but a significant property for our proof is given as the following lemma.

Lemma 7. *For any substring xy of F_k such that $S_x, S_y \geq 1$, there is an end-position of a phrase of any largest repetition factorization of F_k in an interval $[|x| - 17, |x|]$ of xy.*

If we know $S_{Pre(x,|x|-i)}$ for all $i \in [0,17]$ and $S_{Suf(x,i) \cdot y}$ for all $i \in [0,17]$, then we can obtain S_{xy} as follows: $S_{xy} = \max_{0 \leq i \leq 17}(S_{Pre(x,|x|-i)} + S_{Suf(x,i) \cdot y})$. In our proof, we use this idea for every boundary of base-factor parsings. Now we will define such strings which are obtained by shifting boundaries in the following way. We consider *shifted base-factors* in a base-factor parsing.

Definition 4 (Shifted base-factors (SBF)). *Let $X = t_1, \ldots, t_m$ be a base-factor parsing of a string x in $\|Reg_\Sigma\|$ and $Y = t_p \ldots t_q$ be a base-factor parsing of y which is a substring of x $(1 \leq p \leq q \leq m)$. For any integers i, j satisfying $1 \leq i, j \leq 18$, the shifted base-factor $SBF_{X,Y}(i,j)$ is defined as follows:*

$$SBF_{X,Y}(i,j) = Suf(t_{p-1}, i - 1) \cdot Pre(y, |y| - j + 1).$$

Notice that $Suf(t_{p-1}, i - 1) = \varepsilon$ if $p = 1$ and $i = 1$, $SBF_{X,Y}(i,j)$ is undefined if $p = 1$ and $i \geq 2$, or $p = m$ and $j \geq 2$.

Lemma 8. *Let $X = t_1, \ldots, t_m$ be a base-factor parsing of a string x in $\|Reg_\Sigma\|$ and $Y = t_p \ldots t_q$ be a base-factor parsing of y which is a substring of x $(1 \leq p \leq q \leq m)$. If $SBF_{X,Y}(i,j)$ is defined, then the following equations hold.*

1. *If $t_p \in \{\mathsf{F}_{11}^-, \mathsf{F}_{13}^-\}$, then*
   $SBF_{X,Y}(i,j) = Suf(aababaababaabaaba, i - 1) \cdot Pre(y, |y| - j + 1)$.
2. *If $t_p \in \{\mathsf{oddpre}, \mathsf{evenpre}, \mathsf{core}\}$, then*
   $SBF_{X,Y}(i,j) = Suf(babaabaababaababa, i - 1) \cdot Pre(y, |y| - j + 1)$.
3. *If $t_p \in \{\mathsf{oddsuf}, \mathsf{evensuf}\}$, then*
   $SBF_{X,Y}(i,1) = Suf(babaabaababaababa, i - 1) \cdot t_p$.

Proof. By the definition of parsings, $\mathsf{F}_{11}^-, \mathsf{F}_{13}^-$ appear as even numbered base-factors in X, and the other factors appear as odd-numbered base-factors in X. If $t_p \in \{\mathsf{F}_{11}^-, \mathsf{F}_{13}^-\}$, then $t_{p-1} \in \{\mathsf{oddpre}, \mathsf{evenpre}, \mathsf{core}\}$. Thus $Suf(t_{p-1}, i - 1) = Suf(aababaababaabaaba, i - 1)$ by Fact 2. If $t_p = \mathsf{core}$, then $t_{p-1} \in \{\mathsf{F}_{11}^-, \mathsf{F}_{13}^-\}$. On the other hand, $t_p \in \{\mathsf{oddpre}, \mathsf{evenpre}\}$ implies that $p = 1$ and $i = 1$. Thus $Suf(t_{p-1}, i - 1) = Suf(babaabaababaababa, i - 1)$ by Fact 2. Otherwise, $t_{p-1} \in \{\mathsf{F}_{11}^-, \mathsf{F}_{13}^-\}$, $p = q = m$, and $j = 1$. These conditions implies that $SBF_{X,Y}(i,1) = Suf(babaabaababaababa, i - 1) \cdot t_p$. Therefore the lemma holds. $\square$

Table 1. MBF$_{core}$

i \ j	1	2	3	4	5	6	7	8	0	10	11	12	13	14	15	16	17	18
1	0	-1	-2	-3	-1	-2	-2	-3	-3	-3	-3	-4	-3	-4	-4	-4	-4	-5
2	0	0	-1	-2	-1	-1	-2	-2	-2	-2	-3	-3	-2	-3	-3	-3	-4	-4
3	0	0	-1	-2	-1	-1	-2	-2	-2	-2	-3	-3	-2	-3	-3	-3	-4	-4
4	0	0	-1	-2	-1	-1	-2	-2	-2	-2	-3	-3	-2	-3	-3	-3	-4	-4
5	1	1	0	-1	0	0	-1	-1	-1	-1	-2	-2	-1	-2	-2	-2	-3	-3
6	0	0	-1	-2	-1	-1	-2	-2	-2	-2	-3	-3	-2	-3	-3	-3	-4	-4
7	1	0	-1	-2	0	-1	-1	-2	-2	-2	-2	-3	-2	-3	-3	-3	-3	-4
8	1	1	0	-1	0	0	-1	-1	-1	-1	-2	-2	-1	-2	-2	-2	-3	-3
9	1	1	0	-1	0	0	-1	-1	-1	-1	-2	-2	-1	-2	-2	-2	-3	-3
10	1	1	0	-1	0	0	-1	-1	-1	-1	-2	-2	-1	-2	-2	-2	-3	-3
11	2	1	0	-1	1	0	0	-1	-1	-1	-1	-2	-1	-2	-2	-2	-2	-3
12	2	2	1	0	1	1	0	0	0	0	-1	-1	0	-1	-1	-1	-2	-2
13	3	2	1	0	2	1	1	0	0	0	0	-1	0	-1	-1	-1	-1	-2
14	1	1	0	-1	0	0	-1	-1	-1	-1	-2	-2	-1	-2	-2	-2	-3	-3
15	2	2	1	0	1	1	0	0	0	0	-1	-1	0	-1	-1	-1	-2	-2
16	3	3	2	1	2	2	1	1	1	1	0	0	1	0	0	0	-1	-1
17	4	3	2	1	3	2	2	1	1	1	1	0	1	0	0	0	0	-1
18	4	4	3	2	3	3	2	2	2	2	1	1	2	1	1	1	0	0

This lemma implies that $SBF_{X,Y}(i,j)$ depends on the first phrase of Y. Hence, we will use $SBF_Y(i,j)$ by dropping a subscript X.

Definition 5 (Matrix for base-factors (MBF)). *Let t be a base-factor. For any integers i,j satisfying $1 \leq i,j \leq 18$, we define a matrix MBF$_t$ as follows:*

$$\text{MBF}_t(i,j) = S_{SBF_t(i,j)} - S_t.$$

For convenience, $S_{SBF_t(i,j)} = -\infty$ if $SBF_t(i,j)$ is undefined or $SBF_t(i,j)$ has no repetition factorization.

The matrix for core is given in Table 1 (matrices for other factors are omitted due to the lack of space).

Intuitively, MBF$_t(i,j)$ represents the difference of the size of repetition factorizations between a base factor t and its shifted factor. Since the size and the number of *MBFs* are constant, we can check all values in the matrices by an exhaustive search. Then we can obtain the following fact.

Fact 3 (MBF property). *For any integers i,j satisfying $1 \leq i,j \leq 18$,*

- MBF$_{oddpre}(i,j) \leq$ MBF$_{evenpre}(i,j)$,
- MBF$_{oddsuf}(i,j) \leq$ MBF$_{evensuf}(i,j)$,
- MBF$_{F_{11}^-}(i,j) \leq$ MBF$_{F_{13}^-}(i,j)$.

Table 2. $\mathrm{MCBF}_{\mathrm{core \cdot F_{13}^-}} = \mathrm{MCBF}_{(\mathrm{core \cdot F_{13}^-}) \cdot (\mathrm{core \cdot F_{13}^-})}(i, j)$

i	j																	
	1	2	3	4	5	6	7	8	9	10	11	12	13	14	15	16	17	18
1	0	−1	−1	−2	−2	−1	−1	−2	−2	−3	−2	−3	−3	−3	−3	−4	−4	−5
2	0	−1	−1	−2	−2	−1	−1	−2	−2	−3	−2	−3	−3	−3	−3	−4	−4	−5
3	0	−1	−1	−2	−2	−1	−1	−2	−2	−3	−2	−3	−3	−3	−3	−4	−4	−5
4	0	−1	−1	−2	−2	−1	−1	−2	−2	−3	−2	−3	−3	−3	−3	−4	−4	−5
5	1	0	0	−1	−1	0	0	−1	−1	−2	−1	−2	−2	−2	−2	−3	−3	−4
6	0	−1	−1	−2	−2	−1	−1	−2	−2	−3	−2	−3	−3	−3	−3	−4	−4	−5
7	1	0	0	−1	−1	0	0	−1	−1	−2	−1	−2	−2	−2	−2	−3	−3	−4
8	1	0	0	−1	−1	0	0	−1	−1	−2	−1	−2	−2	−2	−2	−3	−3	−4
9	1	0	0	−1	−1	0	0	−1	−1	−2	−1	−2	−2	−2	−2	−3	−3	−4
10	1	0	0	−1	−1	0	0	−1	−1	−2	−1	−2	−2	−2	−2	−3	−3	−4
11	2	1	1	0	0	1	1	0	0	−1	0	−1	−1	−1	−1	−2	−2	−3
12	2	1	1	0	0	1	1	0	0	−1	0	−1	−1	−1	−1	−2	−2	−3
13	3	2	2	1	1	2	2	1	1	0	1	0	0	0	0	−1	−1	−2
14	1	0	0	−1	−1	0	0	−1	−1	−2	−1	−2	−2	−2	−2	−3	−3	−4
15	2	1	1	0	0	1	1	0	0	−1	0	−1	−1	−1	−1	−2	−2	−3
16	3	2	2	1	1	2	2	1	1	0	1	0	0	0	0	−1	−1	−2
17	4	3	3	2	2	3	3	2	2	1	2	1	1	1	1	0	0	−1
18	4	3	3	2	2	3	3	2	2	1	2	1	1	1	1	0	0	−1

A matrix MBF is defined for base-factors. Next we extend the notion of matrix for base-factor parsings (i.e., concatenation of base-factors).

Definition 6 (Matrix for concatenated base-factors (MCBF)). *Let Z be a base-factor parsing of a string. For any integers i, j satisfying $1 \le i, j \le 18$, we define $\mathrm{MCBF}_Z(i, j)$ as follows. If $Z \in BFs$, then $\mathrm{MCBF}_Z(i, j) = \mathrm{MBF}_Z(i, j)$. If $Z = X \cdot Y$ where X and Y are base-factor parsings, then*

$$\mathrm{MCBF}_{X \cdot Y}(i, j) = \max\{\mathrm{MCBF}_X(i, \ell) + \mathrm{MCBF}_Y(\ell, j) \mid 1 \le \ell \le 18\}.$$

Intuitively, when $Z = \mathsf{Fib}_k$, $\mathrm{MCBF}_{X \cdot Y}(1, 1)$ represents the difference size between the largest repetition factorization and the candidate factorization of F_k. This claim will be shown as Corollary 1. Then we show properties of MCBF as Lemmas 9 and 10. The first one can be obtained by an exhaustive search, and the second one can be obtained by the definition of MCBF.

Lemma 9 (MCBF property 1). *For any integers i, j satisfying $1 \le i, j \le 18$, the following equation holds:*

$$\mathrm{MCBF}_{\mathrm{core \cdot F_{13}^-}}(i, j) = \mathrm{MCBF}_{(\mathrm{core \cdot F_{13}^-}) \cdot (\mathrm{core \cdot F_{13}^-})}(i, j).$$

The matrix $\mathrm{MCBF}_{\mathrm{core \cdot F_{13}^-}}(i, j)$ is given in Table 2. This property indicates why we consider the matrices as the difference of values instead of the maximum values.

Lemma 10 (MCBF property 2). *Let $X, X', Y, Y', X \cdot Y, X' \cdot Y$, and $X \cdot Y'$ be base-factor parsings.*

1. *If $\mathrm{MCBF}_X(i,j) \leq \mathrm{MCBF}_{X'}(i,j)$ for all i,j satisfying $1 \leq i,j \leq 18$, then $\mathrm{MCBF}_{X \cdot Y}(m,n) \leq \mathrm{MCBF}_{X' \cdot Y}(m,n)$ for all m,n satisfying $1 \leq m,n \leq 18$.*
2. *If $\mathrm{MCBF}_Y(i,j) \leq \mathrm{MCBF}_{Y'}(i,j)$ for all i,j satisfying $1 \leq i,j \leq 18$, then $\mathrm{MCBF}_{X \cdot Y}(m,n) \leq \mathrm{MCBF}_{X \cdot Y'}(m,n)$ for all m,n satisfying $1 \leq m,n \leq 18$.*

Finally, in the rest of this section, we discuss the matrices for the Fibonacci word.

Lemma 11. *For every integer $k \geq 15$, $\mathrm{MCBF}_{\mathrm{Fib}_k}(1,1) \leq 0$.*

Proof. Let $\mathrm{Fib}_k = t_1 \cdots t_m$. Then

$$\mathrm{MCBF}_{\mathrm{Fib}_k}(1,1) = \mathrm{MCBF}_{t_1 \cdots t_m}(1,1)$$
$$\leq \mathrm{MCBF}_{\mathrm{evenpre} \cdot F_{13}^- \cdot (\mathrm{core} \cdot F_{13}^-)^\alpha \cdot \mathrm{evensuf}}(1,1)$$
$$= \mathrm{MCBF}_{\mathrm{evenpre} \cdot F_{13}^- \cdot \mathrm{core} \cdot F_{13}^- \cdot \mathrm{evensuf}}(1,1)$$
$$= 0,$$

where α is the number of occurrences of core in the parsing Fib_k. $\qquad\square$

Lemma 12. *Let $\mathrm{Fib}_k = t_1, \ldots, t_m$ and $Z = t_p, \ldots, t_q$ where p,q satisfies $1 \leq p \leq q \leq m$. Then*
$$\mathrm{MCBF}_Z(i,j) = S_{SBF_Z(i,j)} - B_Z$$
holds for all integers i,j satisfying $1 \leq i,j \leq 18$.

Proof. Let $d = q - p + 1$ be the number of phrases of a parsing Z. Namely, we write $Z = t_p, \ldots, t_q = s_1, \ldots, s_d$. We prove this lemma by induction on d. For $d = 1$,
$$\mathrm{MCBF}_Z(i,j) = \mathrm{MBF}_Z(i,j) = S_{SBF_Z(i,j)} - B_Z$$
by Definitions 5 and 6. Suppose that the statement holds for every $d \leq c$ for some integer $c \geq 1$. We consider the case when $d = c + 1$. Let $X = s_1, \ldots, s_\alpha$ and $Y = s_{\alpha+1}, \ldots, s_{c+1}$ for some integer α that satisfying $1 \leq \alpha \leq c$. By the definition of B, $B_Z = B_X + B_Y$ holds. By Lemma 7,
$$S_{SBF_Z(i,j)} = \max_{1 \leq \ell \leq 18} (S_{SBF_X(i,q)} + S_{SBF_Y(q,j)})$$
holds. Then

$$\mathrm{MCBF}_Z(i,j) = \max_{1 \leq \ell \leq 18} (\mathrm{MCBF}_X(i,q) + \mathrm{MCBF}_Y(q,j))$$
$$= \max_{1 \leq \ell \leq 18} ((S_{SBF_X(i,q)} - B_X) + (S_{SBF_Y(q,j)} - B_Y))$$
$$= \max_{1 \leq \ell \leq 18} ((S_{SBF_X(i,q)} + S_{SBF_Y(q,j)}) - (B_X + B_Y))$$
$$= \max_{1 \leq \ell \leq 18} (S_{SBF_X(i,q)} + S_{SBF_Y(q,j)}) - B_Z$$
$$= S_{SBF_Z(i,j)} - B_Z.$$

Thus $\mathrm{MCBF}_Z(i,j) = S_{SBF_Z(i,j)} - B_Z$ holds for any d, and the lemma holds. $\square$

Corollary 1. *For every $k \geq 17$, $\mathrm{MCBF}_{\mathsf{Fib}_k}(1,1) = S_{F_k} - B_{\mathsf{Fib}_k}$.*

Finally, we can obtain our main result with the proof of Lemma 3.

Proof of Lemma 3. Lemma 11 and Corollary 1 imply that $S_{F_k} - B_{\mathsf{Fib}_k} \leq 0$. Thus $B_{\mathsf{Fib}_k} \geq S_{F_k}$. On the other hand, it is clear from the definition that $B_{\mathsf{Fib}_k} \leq S_{F_k}$. Therefore $B_{\mathsf{Fib}_k} = S_{F_k}$ holds for any $k \geq 17$. □

Acknowledgments. We gratefully acknowledge the comments of anonymous reviewers for improving our paper. This work was supported by JSPS KAKENHI Grant Numbers JP21K17705, JP23H04386 (YN), JP22H03551 (SI).

References

1. Chen, K.T., Fox, R.H., Lyndon, R.C.: Free differential calculus. IV. The quotient groups of the lower central series. Ann. Math. **68**(1), 81–95 (1958)
2. Dumitran, M., Manea, F., Nowotka, D.: On prefix/suffix-square free words. In: Iliopoulos, C., Puglisi, S., Yilmaz, E. (eds.) SPIRE 2015. LNCS, vol. 9309, pp. 54–66. Springer, Cham (2015). https://doi.org/10.1007/978-3-319-23826-5_6
3. Iliopoulos, C.S., Moore, D., Smyth, W.: A characterization of the squares in a Fibonacci string. Theor. Comput. Sci. **172**(1), 281–291 (1997). https://doi.org/10.1016/S0304-3975(96)00141-7. https://www.sciencedirect.com/science/article/pii/S0304397596001417
4. Inoue, H., Matsuoka, Y., Nakashima, Y., Inenaga, S., Bannai, H., Takeda, M.: Factorizing strings into repetitions. Theory Comput. Syst. **66**(2), 484–501 (2022). https://doi.org/10.1007/s00224-022-10070-3
5. Lothaire, M.: Combinatorics on Words. Addison-Wesley, Boston (1983)
6. Matsuoka, Y., Inenaga, S., Bannai, H., Takeda, M., Manea, F.: Factorizing a string into squares in linear time. In: Proceedings of the CPM 2016, pp. 27:1–27:12 (2016)
7. Navarro, G.: Indexing highly repetitive string collections, part I: repetitiveness measures. ACM Comput. Surv. **54**(2), 29:1–29:31 (2022). https://doi.org/10.1145/3434399
8. Navarro, G., Ochoa, C., Prezza, N.: On the approximation ratio of ordered parsings. IEEE Trans. Inf. Theory **67**(2), 1008–1026 (2021). https://doi.org/10.1109/TIT.2020.3042746
9. Storer, J., Szymanski, T.: Data compression via textual substitution. J. ACM **29**(4), 928–951 (1982)
10. Welch, T.A.: A technique for high performance data compression. IEEE Comput. **17**, 8–19 (1984)
11. Ziv, J., Lempel, A.: A universal algorithm for sequential data compression. IEEE Trans. Inf. Theory **IT-23**(3), 337–349 (1977)
12. Ziv, J., Lempel, A.: Compression of individual sequences via variable-length coding. IEEE Trans. Inf. Theory **24**(5), 530–536 (1978)

String Covers of a Tree Revisited

Łukasz Kondraciuk[(✉)] [ID]

University of Warsaw, Warsaw, Poland
lk385775@students.mimuw.edu.pl

Abstract. We consider covering labeled trees with a collection of paths with the same string label, called a (string) cover of a tree. This problem was originated by Radoszewski et al. (SPIRE 2021), who show how to compute all covers of a directed rooted labeled tree in $O(n \log n / \log \log n)$ time and all covers of an undirected labeled tree in $O(n^2)$ time and space, or $O(n^2 \log n)$ time and $O(n)$-space. (Here n denotes the number of nodes of a given tree). We improve those results by proposing a linear time algorithm for reporting all covers of a directed tree, and showing an $O(n^2)$ time and $O(n)$-space algorithm for computing undirected tree covers. Both algorithms assume that labeling characters come from an integer alphabet.

1 Introduction

String C is a cover of string S if every character of S belongs to at least one substring of S equal to C. Cover-related problems have been studied since at least 1990. Apostolico and Ehrenfeucht introduced this feature of a string in [2]. Apostolico, Farach, and Iliopoulos in [3] discovered an algorithm for checking if string S contains any covers, other than S, in $O(|S|)$ time. They called S *superprimitive* if it doesn't contain any cover other than itself. Dany Breslauer in [4] extended this algorithm to work online - it tests if each prefix of the input string is superprimitive as soon as the prefix is given. Please note that *cover* and *quasiperiod* are equivalent terms. However, it is not clear how to define periodicity when we switch from words to trees, thus later in this paper we will only use the term *cover*.

Moore and Smyth [15] were the first to discover a way to report all covers of a string S in $O(|S|)$ time – our algorithm and the previous algorithm [16] use their result as a starting point for the directed cover problem. Czajka and Radoszewski in [8] evaluated the practical performance of algorithms computing covers of strings. For a very recent survey on other variants of covers, see [14].

Let us consider a rooted tree T, consisting of n nodes. Each of its edges is labeled by a single character $\in \Sigma$. A simple directed path (let us denote it as p) is a non-repeating sequence of nodes. Each pair of consecutive nodes is connected by an edge. The first node of this sequence is a start point of p - let us denote it as s. The last node is an endpoint of p - let us denote it as e. We will define $s \rightarrow e$ as a sequence of nodes on a single path from s to e, equivalent

© The Author(s), under exclusive license to Springer Nature Switzerland AG 2023
F. M. Nardini et al. (Eds.): SPIRE 2023, LNCS 14240, pp. 297–309, 2023.
https://doi.org/10.1007/978-3-031-43980-3_24

to p. Any simple directed path of T is uniquely identified by a two-element tuple (aka ordered pair) (startpoint, endpoint). The label of a directed simple path is a string constructed by concatenating characters on edges connecting consecutive nodes. String C is a cover of a tree T, if there exists a set of simple paths M, each of them having a label equal to C, so that each edge of T belongs to at least one path from M. We will consider two variants of this problem, which were first proposed and studied by Radoszewski et al. in [16]. Similar problems, involving labeled trees and computing their runs, powers, and palindromes, were extensively studied in [5, 7, 9, 10, 12, 13, 19], and most recently in [11].

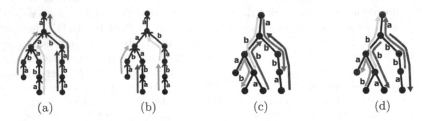

(a)   (b)   (c)   (d)

Fig. 1. a) and b): *aba* is a directed cover of this tree, *ab* is not: 3 edges cannot be covered by this string. c) and d): *abb*, *bba* = (*abb*)^R, and *ba* = (*ab*)^R are undirected covers this tree (only *aab* and *bba* are visualized).

For the directed cover problem the considered tree is rooted at some node, and we add a restriction that all paths from M should be only going up. It means that for any $p \in M$ its endpoint is an ancestor of its startpoint. The algorithm presented in [16] runs in time $O(n \log n / \log \log n)$ time and $O(n)$-space, and requires the labeling character alphabet to be integer. Our improvement involves preprocessing the input tree by compacting branchless paths, which later bounds the total number of iterations of an inner loop of an algorithm, instead of utilizing a data structure for the dynamic marked ancestor problem. This gives us $O(n)$ time and space algorithm. Unfortunately, it still requires the labeling alphabet to be integer.

For the undirected cover problem, there are no more restrictions. Algorithms described in [16] were $O(n^2)$ time $O(n^2)$ space (divides paths of a tree between cover-candidates all at once), and $O(n^2 \log n)$ time $O(n)$-space (this one verifies only one cover-candidate at the time). Our algorithm will combine both ideas by, for one candidate at a time, finding a compressed set of paths that this candidate covers. This way we can achieve $O(n^2)$ time $O(n)$-space complexity.

Figure 1 shows examples of a directed and an undirected cover. Each cover will be reported by our algorithms as an ordered pair of nodes. (v, w) will be representing a string - label on a path $v \to w$. This way we can report all covers (we will later prove that in both variants there are at most $O(n)$ of them) in $o(n^2)$ time, even though their straightforward representation can be as large as $\Theta(n^2)$. This is the case for instance when a given tree forms a simple path, and all of its edges are labeled by the same letter.

We provide reference Python implementations of the described algorithms,[1] as well as the extended edition of this publication containing proofs and listings omitted due to page limit.[2]

2   Preliminaries

A string S is a sequence of characters $S[1], S[2], ..., S[|S|] \in \Sigma$. A substring of S is any string of the form $S[i..j] = S[i], S[i+1], ..., S[j]$. If $i = 1$ ($j = |S|$), it is called a prefix (a suffix, respectively), $S^R = S[|S|], S[|S| - 1], ..., S[1]$.

Let us consider any string S. A cover C is a substring of S, which occurs at some positions of S, and each letter of S is covered by at least one occurrence of C. Formally, string C is a cover of S, if there exists a set of positions $M \subseteq \{1, ..., |S| - |C| + 1\}$, such that for every $i \in M$, $S[i..(i + |C| - 1)] = C$ (M represents a – not necessarily proper – subset of occurrences of C), and for every $1 \leq i \leq |S|$, there exists $j \in M$, such that $i - |C| + 1 \leq j \leq i$ (every letter of S should be covered by an occurrence).

When we consider a rooted tree and algorithms processing it, it is helpful to define a few properties of a tree and its nodes. $\text{path}(u \to v)$ is a simple path connecting nodes u and v. Sometimes to simplify notation we will omit $\text{path}(*)$ and denote it as $u \to v$. $|p|$ denotes the number of nodes in a path p. $\text{parent}(v)$ is the first node on the path from v to the root ($\text{parent}(\text{root}) = \text{null}$). $\text{children}(v) = \{u \mid \text{parent}(v) = u\}$. $\text{dist}(u, v)$ is the number of edges on a simple path connecting u and v. Please note that $\text{dist}(v, u) = \text{dist}(u, v) = |u \to v| - 1$, $\text{depth}(v) = \text{dist}(v, \text{root})$, $\text{subtree}(v) = \{u \mid v \in \text{path}(u \to \text{root})\}$, $\text{height}(v) = \max_{l \in \text{subtree}(v)} \text{dist}(v, l)$, $\text{label}(u \to v)$ is a label of $\text{path}(u \to v)$ constructed as a concatenation of characters labeling its consecutive edges, $\text{label}_d(u \to v) = \text{label}(u \to v)[1..d]$.

Let us define $\text{childrenHeights}(v) = \sum_{w \in \text{children}(v)} \text{height}(w)$, $\text{maxChildHeight}(v) = \max_{w \in \text{children}(v)} \text{height}(w)$ (0 if $\text{children}(v) = \emptyset$), and $\text{superHeight}(v) = \text{childrenHeights}(v) - \text{maxChildHeight}(v)$.

Lemma 1. *For a rooted tree T with n nodes, we have*

$$\sum_v \text{superHeight}(v) = \sum_v (\text{childrenHeights}(v) - \text{maxChildHeight}(v)) \leq n$$

Proof. The following proof is based on Second Heights lemma proof from [16]. For a node v we define $\text{MaxPath}(v)$ as the longest path from v to a leaf in $\text{subtree}(v)$. ($|\text{MaxPath}(v)| = \text{height}(v)$). Initially, we choose (one of possibly many) $\text{MaxPath}(\text{root})$, then we remove this path (both nodes and edges) and choose the longest paths for roots of resulting subtrees. We continue in this way and obtain a decomposition of the tree into node-disjoint longest paths.

---

[1] https://students.mimuw.edu.pl/~lk385775/string_tree_covers_ref_impl.zip.
[2] https://students.mimuw.edu.pl/~lk385775/string_tree_covers_extended.pdf.

Let FirstChild(v) denote a child of v which belongs to the same path in the decomposition and OtherChildren(v) = $\{w \in \text{children}(v) : w \neq \text{FirstChild}(v)\}$. We have $\sum_v \text{superHeight}(v) = \sum_{v;w \in \text{OtherChildren}(v)} |\text{MaxPath}(w)| \leq n$ since all selected longest paths are node-disjoint. This sum is a sum of the lengths of all removed paths, excluding the one removed in the first step of the algorithm. $\square$

Let us define secondHeight(v) as the height of a second highest child of v (or 0 if $|\text{children}(v)| < 2$). We have childrenHeights(v) − maxChildHeight(v) $\geq$ secondHeight(v), so

Lemma 2 *(Second height lemma, also used in* [16]*). For a rooted tree T with n nodes, the following inequality holds:*

$$\sum_v \text{secondHeight}(v) \leq n.$$

Pref table is a data structure that is used to store and retrieve information about prefixes of a given string. It is defined as

$$\text{Pref}_S[i] = \max\{d \geq 0 : S[i..(i+d-1)] = S[1..d]\}$$

This data structure can be generalized to rooted trees with character-labeled edges. For a string S, rooted tree T and node $v \in T$, we denote

$$\text{TreePref}_S[v] = \max\{d \geq 0 : \text{label}_d(v \to \text{root}) = S[1..d]\}$$

Lemma 3 *(*[16] *using* [17]*). TreePref$_S$ can be computed in $O(n)$ time for a rooted tree T with n nodes over an integer alphabet.*

3   Directed Tree Cover

Let us consider a directed variant of the string tree cover. For a given rooted tree T, we direct each edge towards the root. For this problem, we will assume that all edge labels are characters over an integer alphabet. Let us fix some leaf node l. An edge between l and parent(l) can only be covered by a path starting in l and going upwards.

Observation 1 [16]. *The cover must be a prefix of* label($l \to$ root).

Let $L = \text{label}(l \to \text{root})$. Let us define up($v, k$) = parent(up($v, k − 1$)), up($v, 0$) = v. For each $1 \leq d \leq |L|$, we will check if $L[1..d]$ covers all edges of the tree. Let us calculate TreePref$_L$. Set $\{v : \text{TreePref}_L[v] \geq d\}$ contains nodes (let us denote any of those nodes as v), for which label($v \to$ root) matches L on the first d positions. If label$_d(v \to \text{root}) = L[1..d]$ holds, then then the first d edges on path $v \to$ root are covered by path $v \to$ up(v, d).

For each fixed d, we will consider $L[1..d]$. Let us mark all w's, for which TreePref$_L[w] \geq d$. Edge $w \to$ parent(w) is covered (by a path having label = $L[1..d]$), if and only if there exists a marked node $u \in$ subtree(w), having dist(w, u) < d.

We will maintain a data structure to store marked nodes. It will be able to:

- initialize itself with any set of marked nodes (all leaves need to be in this set),
- unmark a marked node,
- query for the largest distance between any node and its closest marked descendant.

3.1 Gaps

All leaves have to be marked at all times - it is the only way to cover leaf edges. For a set M of marked nodes we define $\mathrm{gap}(v) = \min_{u \in M \cap \mathrm{subtree}(v)} \mathrm{dist}(v, u)$ and $\mathrm{MaxGap} = \max_{v \in nodes \setminus \{\mathrm{root}\}} \mathrm{gap}(v)$.

In a data structure, nodes will keep (either directly or indirectly) their current gap value. Let v be any marked node, which is not a leaf. Let $M' = M \setminus \{v\}$ be the set of marked nodes after unmarking v. Let us define $\mathrm{gap}'(w) = \min_{u \in M' \cap \mathrm{subtree}(w)} \mathrm{dist}(w, u)$ – it is a gap function after unmarking v.

Observation 2. *If* $\mathrm{gap}(w) \neq \mathrm{gap}'(w)$, *then* $w \in \mathrm{path}(v \to \mathrm{root})$.

Proof. A gap could change only for those nodes w, for which $v \in \mathrm{subtree}(w)$. □

Lemma 4. *Let* $u, w \in \mathrm{path}(v \to \mathrm{root})$, *and* $\mathrm{depth}(u) > \mathrm{depth}(w)$. *If* $\mathrm{gap}'(u) = \mathrm{gap}(u)$, *then* $\mathrm{gap}'(w) = \mathrm{gap}(w)$.

Proof. If $\mathrm{dist}(w, v) \neq \mathrm{gap}(w)$ then there exists $x \in \mathrm{subtree}(w) \cap M$ ($x \neq v$), for which $\mathrm{dist}(w, x) = \mathrm{gap}(w)$. Since $x \in M' = M \setminus \{v\}$, then $\mathrm{gap}'(w) = \mathrm{dist}(w, x) = \mathrm{gap}(w)$.

Otherwise, if $\mathrm{dist}(w, v) = \mathrm{gap}(w)$, then $\mathrm{dist}(u, v) = \mathrm{gap}(u)$. Since $\mathrm{gap}'(u) = \mathrm{gap}(u)$, then there exists $x \in \mathrm{subtree}(u) \cap M'$, for which $\mathrm{dist}(u, x) = \mathrm{dist}(u, v) = \mathrm{gap}(u)$. $\mathrm{dist}(w, x) = \mathrm{dist}(w, u) + \mathrm{dist}(u, x) = \mathrm{dist}(w, u) + \mathrm{dist}(u, v) = \mathrm{dist}(w, v)$. And since $x \in \mathrm{subtree}(u) \cap M' \subseteq \mathrm{subtree}(w) \cap M'$, then $\mathrm{gap}'(w) = \mathrm{dist}(w, x) = \mathrm{dist}(w, v) = \mathrm{gap}(w)$. □

Corollary 1. *After unmarking* v, *we only need to update* gap *for some prefix of nodes on* $\mathrm{path}(v \to \mathrm{root})$.

3.2 Binarisation and Path Compaction

For the data structure representation, we apply two transformations on tree T:

1. **binarisation** - we will insert artificially created nodes, so that every node, except for the root, has at most 2 children. Let us denote the resulting tree as T'. The number of inserted nodes is $\sum_{v \neq \mathrm{root}} \min(0, \mathrm{degree}(v) - 3) \leq n$, so $|T'| \leq 2n = O(n)$.
2. **path compaction** - we will replace each non-branching path of T' by a single entity, called **compacted path**. Here by non-branching path, we refer to a group of nodes, forming a simple path, each of them having only one child.

The result of those transformations will be called a pseudotree.

The defined gap function will be invariant to those transformations. We will calculate and update its values as if all nodes were located in the original input tree T. We will not take into account gap values calculated for nodes added during binarisation.

The resulting pseudotree will consist of four types of nodes: root, leaves, binary nodes, and implicit nodes. Each non-branching vertical path will be replaced by a compacted path. The remaining edges will be replaced by trivial (that is not having any implicit nodes inside) compacted paths.

Each binary node holds two compacted paths going down and one going up. The root holds some number (possibly one) of compacted paths going down. Each leaf holds only one compacted path going up. Implicit nodes are connected together and contained inside compacted paths. Compacted paths will hold both: nodes they are connected to, and a collection of implicit nodes inside of it.

3.3 Updates

The key observation here is that we do not need to explicitly keep and update the gap sizes of implicit nodes.

Listing 1: Auxilary functions needed for gap values updates.

```
global maxGap = 1
function calcGapBinaryNode (v)
    if v.isMarked then
        return 0
    else
        return min(v.pathLeft.topGap, v.pathRight.topGap)+!v.isFake;
function walkAndUpdate (v)
    while (not isRoot(v)) and v.gap < calcGapBinaryNode(v) do
        v.gap = calcGapBinaryNode(v)
        path = v.pathUp
        if path.lowestMarked ≠ null then
            gap = v.gap + dist(path.bottom, path.lowestMarked) + 1
            maxGap = max(maxGap, gap)
            break
        else
            path.topGap = v.gap + dist(path.top, path.bottom) + 1
            maxGap = max(maxGap, path.topGap)
            v = path.nodeUp
```

---

unmark(v) is an update entry point of the data structure. It will be described below, please refer to the extended edition of this paper for its pseudocode. On Listing 1 we attach the pseudocode of walkAndUpdate function, which is called by unmark to propagate new gap values upwards.

Let us denote any implicit node v, and its compacted path as p. When we unmark v:

- If v is the only marked node on p, then the new gap value for $p.top$ is $p.length+$ $p.nodeDown.gap$ and this value needs to be passed to `walkAndUpdate` to propagate it upwards.
- Otherwise, if v is the highest marked node on p, then we can calculate the new longest gap as a distance between $p.top$ and the new highest marked node on p. Then it needs to be propagated upwards using `walkAndUpdate`.
- Otherwise, if v is the lowest marked node on p, then the longest new gap is equal to $p.nodeDown.gap$ plus the distance between $p.bottom$ and child of the new lowest marked node on p. This value does not get propagated upwards, since there exist marked nodes on p, other than v.
- Otherwise, v is located between two other marked implicit nodes. We can calculate the longest new gap using the distance between them.

Similarly, whenever `walkAndUpdate` tries to traverse a compacted path p (to proceed from a binary node on its lower end to the one on its upper end), it checks if there exists any marked node on this path. It can use the gap calculated for the binary node connected to the lower end of that path, in order to calculate the new gap size for implicit nodes of p. It needs this value to update `maxGap`.

Listing 2 contains the final high-level algorithm for computing all directed covers of a given rooted tree.

Listing 2: Computing all directed covers of a given rooted tree.

Fix any leaf l. Let us denote $L = \text{label}(l \to \text{root})$

Calculate TreePref$_L$ - this can be done in $O(n)$ as we are working with integer alphabet

Initialize gap data structure - apply binarization and path compression

Set all real nodes (that is those which were not created during binarisation) as marked

for $k := 0$ to $\min_{v \in leaves}$ TreePref$_L[v] - 1$ **do**
    for v $\in$ nodes and TreePref$_L[\text{v}] = k$ **do**
        unmark(v)
    if maxGap $<$ k **then**
        report that $L[1..k]$ is a cover

Lemma 5. *The total amortized time cost of maintaining the gap data structure is $O(n)$.*

Proof. unmark itself does only $O(1)$ amount of work. `walkAndUpdate` runs in time proportional to the number of touched binary nodes. By *touched nodes*, we refer to those nodes, for which node.gap has changed. Since node.gap can only increase, the total amount of time used by `walkAndUpdate` is limited by $\sum_{v \in \text{binary nodes}} \text{FinalGap}(v)$. (By FinalGap($v$) we refer to v.gap after termination of the algorithm).

All leaves are always marked, so $\text{FinalGap}(v) \leq \min_{l \in \text{leaves} \cap \text{subtree}(v)} \text{dist}(v, l)$ (here we refer to leaves, subtree, and dist in regards to T'). Each binary node

v has two children, so FinalGap$(v) \leq$ secondHeight(v). Second heights lemma (Lemma 2) applied to T' implies that $\sum_v$ secondHeight$(v) \leq 2n$. This indicates that the total number of binary node gap updates done by `walkAndUpdate` is bounded by $2n$. This also proves the complexity of the complete directed cover computing algorithm. □

Theorem 1. *All covers of a directed rooted tree labeled with characters over an integer alphabet can be computed in $O(n)$ time and $O(n)$-space.*

The only step of the algorithm, which requires characters labeling edges to be over an integer alphabet, is the subroutine computing TreePref$_L$ 3 from [17]. Thus

Corollary 2. *Covers of a directed tree labeled with characters over a general alphabet can be computed in $O(n)$ time and $O(n)$-space if TreePref$_L$, for $L =$ label of some path from a leaf to the root, is given.*

Comparison with the algorithm from [16]. Instead of maintaining gap data structure over a transformed tree – they maintain the so-called chain decomposition of an input tree, where each marked node is an end of some chain, and the chain from any unmarked node leads to the closest marked node in its subtree – our maxGap is equal to the length of the longest chain in their data structure. Chain description is stored in the top node of each chain. None of the other chain nodes store any information about the chain they currently belong to. A data structure from [1] is used to query for a top node of each chain. The time complexity of such a query is $O(\log n/\log \log n)$. This impacts the overall time complexity of their algorithm, which is $O(n \log n/\log \log n)$.

4   Undirected Tree Cover

The cover of an undirected tree is quite a different problem than that of a directed tree. String S is a cover of an undirected tree T, if we can pick a set of simple paths M, each of them having a label equal to S, such that for each edge from the tree, there should exist at least one path from M, going through that edge.

Let us root the input tree in any node.

Observation 3. *For any leaf l, edge $l \rightarrow$ parent(l) can only be covered by a path starting or finishing in l.*

We will fix some leaf l. Let us denote set of paths that can cover edge $l \rightarrow$ parent(l) as $P = \{l \rightarrow v \mid v \in T \setminus \{l\}\} \cup \{v \rightarrow l \mid v \in T \setminus \{l\}\}$. We have $|P| = 2n - 2$.

The set of candidates for a cover is naturally induced by P. Let us denote it by $C = \{$label$(p) \mid p \in P \}$. Thus $|C| \leq 2n - 2 = O(n)$.

Corollary 3 ([16]). *The set of candidates has $O(n)$ elements.*

4.1   Match Tables

Let us fix a candidate S. For each node v we will consider those paths, for which v is the highest point (v is the highest point of p if it has the smallest depth among all its nodes). We will try to match a prefix of S, coming from a subtree of some $u \in$ children(v), with a prefix of S^R coming from another subtree. A prefix of S concatenated with the reverse of a prefix of S^R of proper length makes S. Found matches-paths will be saved on the side.

We will be going from the bottom to the top of the tree (the top being root, and the bottom being leaves). For each node v we will compute two *match tables*: dynamic arrays of linked lists: A and B. They have the following properties:

- $|A| = |B| = \text{height}(v)$
- $A[i]$ and $B[i]$ contain nodes from subtree(v).
- For each $u \in A[i]$ (and for each $u \in B[i]$), dist(u, v) = i.
- If there exists node u having label($u \to v$) = $S[1..i]$, then $A[i]$ is not empty and contains a node y such that label($y \to v$) = $S[1..i]$. Otherwise, it might be empty or might contain some nodes with label different than $S[1..i]$.
- If there exists node u having label($u \to v$) $- S^R[1..i]$, then $B[i]$ is not empty and contains a node y such that label($y \to v$) = $S^R[1..i]$. Otherwise, it might be empty or might contain some nodes having label different than $S^R[1..i]$.
- For all nodes $u \in A[i]$, label($u \to v$) is the same.
- For all nodes $u \in B[i]$, label($u \to v$) is the same.

Each table is held by a data structure, which allows amortized $O(1)$-time insertions to the front and $O(1)$-time random access. Such data structure can be implemented similarly to std::vector from C++ STL [18] or Dynamic Table from [6]: by keeping a pointer to a chunk of allocating memory, and lazily moving its content to a chunk twice as big when space runs out. The only difference is that we will push to the front, not to the back, and use the current size to calculate the offset for $O(1)$-time random access. We will call it FrontVector.

If v is a leaf, then match tables for v are trivial: A.size() = B.size() = 1, and $A[0] = B[0] = \{v\}$. (By V.size() we denote the number of elements of a table V. For the match table represented by FrontVector it is the difference between its capacity and offset.) Otherwise, we need to calculate A and B for v. At first we "claim" match tables calculated for the highest child. That is from the child, whose subtree is the highest among all v's children, in case of ambiguity, whichever child can be picked. Match tables calculated for a node will be used only by its parent, so we can immediately claim its ownership.

Then we need to push $\{v\}$ to the front of A and B, iterate over the remaining children, and join its match tables. We will be also looking for matches during that process.

At the beginning of the algorithm, we precalculate TreePref_S and TreePref_{S^R}, which per Lemma 3 can be done in $O(n)$ [17].

This lemma requires the labeling alphabet to be integer. If this is not the case, then we can convert it to an equivalent integer alphabet in $O(n^2)$ time and $O(n)$-space. Equivalent in the sense, that equivalence relation on edges, based on equality of their labeling characters, will look exactly the same. Since we represent a cover as an ordered pair of endpoints of a path having its label as an actual cover, the output of our algorithms will not change after this transformation. Now we are able for a given v and $u \in \text{subtree}(v)$ check if $\text{label}(u \to v)$ is a prefix of S (or S^R). This condition is equivalent to checking if $\text{TreePref}_S[u] \geq \text{dist}(u,v)$ ($\text{TreePref}_{S^R}[u] \geq \text{dist}(u,v)$).

Listing 3: auxiliary procedures used to find matches in match tables and mark found matches.

```
function matchAndMark (v, A, B, i, j)
    // i + j == len(W)
    clearA(v, A, i)
    clearB(v, B, j)
    if !A[i].empty() and !B[j].empty() then
        for u ∈ (A[i] ∪ B[j]) do
            markVerticalPath(u, v) // markVerticalPath works in O(1) and
            will be described later
        clearButOne(A, i)
        clearButOne(B, j)
function findMatches (v, A, B, A', B', height)
    lowerBound = max(len(S) - A.size() + 1, 0)
    upperBound = min(height, len(S)) + 1
    for i := lowerBound to upperBound do
        matchAndMark(v, A, B', len(S) - i, i)
        matchAndMark(v, A', B, i, len(S) - i)
```

Let us denote current tables as A and B, and match tables calculated for some other child as A' and B'. Let $h = A'.\text{size}() = B'.\text{size}() = $ height of the subtree rooted in that child. For $1 \leq i \leq h$, we consider matches between $A'[i]$ and $B[|S|-i]$. To do that we first need to check if $A'[i]$ contains valid candidates, that is nodes u, for which $\text{label}(u \to v) = S[1..i]$. Please recall that every $u \in A'[i]$ has the same $\text{label}(u \to v)$, so it is sufficient to check that condition for any element of $A'[i]$ - we will use the first one. In our pseudocode, we will use `clearA` (and `clearB` for match table B) procedures to refer to this step.

Similar procedure must be performed for $B[|S|-i]$ - it should contain nodes u having $\text{label}(u \to v) = S^R[1..(|S|-i)]$.

After that step, for each $u_1 \in A'[i]$ and $u_2 \in B[1..(|S|-i)]$ we have $\text{label}(u_1 \to u_2) = S$, so we can mark both $u_1 \to v$ and $u_2 \to v$ as covered. If both $A'[i]$ and $B[1..(|S|-i)]$ are not empty, then we will delete all elements of $A'[i]$ except for one and $B[1..(|S|-i)]$ except for one. This a crucial step to the complexity of the algorithm. Here we will only present simplified intuition why we can do this, proof of correctness will be discussed later.

Imagine that we would not delete any elements of $A'[i]$. When seeking matches for some v' - ancestor of v, in terms of $A'[i]$ we will care about two things:

- $A'[i]$ (which will become $A[i + \mathrm{dist}(v, v')]$) is not empty - then we can find match it with some $B[j]$ coming from some other child of v' and, as a consequence mark all paths $v' \to u$ for $u \in B[j]$.
- if we indeed find a match, all paths $v' \to u$ for $u \in A'[i]$ will become marked.

Please note that since each path $v \to u$ is already marked, if we mark $v' \to v$ then all paths $v' \to u$ will become marked, so marking any path $v' \to u$ marks all of them.

Both of those objectives will still be satisfied if we delete all but one element of $A'[i]$. Similar argument can be conducted for $B[i]$, $A[i]$ and $B'[i]$. In pseudocodes, we will use `clearButOne` subroutine to refer to this step.

Then we follow the same procedure to find matches between $B'[i]$ and $A[|S| - i]$.

After we have considered all valid matches, we can merge A' into A, and B' into B. For $1 \leq i \leq h$, we have to check if $A[i]$ contains nodes u, having label$(u \to v) = S[1..i]$. To do that, it is enough to check that condition for $u =$ first element of $A[i]$ (all nodes in $A[i]$ have the same path to the root). If not, we can clear $A[i]$. We apply the exact same procedure to $A'[i]$. After that, both $A'[i]$ and $A[i]$ contain only valid candidates, so we can move all nodes from $A'[i]$ into $A[i]$ (in $O(1)$-time).

To merge B' into B we execute the same algorithm, but instead of comparing labels of paths to root with S, we compare them with S^R.

4.2   Complexity and Correctness

Let us denote `calcAB(root)` as an entry point for the entire procedure. Please refer to the extended edition of this publication for its formal pseudocode. It calls recursively itself, resulting in a call to `calcAB(v)` once for every v of the tree ($O(n)$ calls). It also iterates over all children ($O(|\mathrm{edges}|) = O(n)$ in total). A single call consumes amortized constant time for inserting $\{v\}$ to the front of match tables - implemented as a `push_front` on an instance of `FrontVector`. So $O(|\mathrm{edges}|) + O(|\mathrm{nodes}|) = O(n)$ of total time consumed.

For each other child (let us denote its height as h) it calls: `findMatches` once, and `clearA`, `clearB` h times - to clear each entry of match tables propagated from that child. `findMatches` calls `matchAndMark` h times. If we disregard time spent on `clearA`, `clearB`, `clearButOne` and `matchAndMark`, then total time spent in `findMatches` and `calcAB` is bounded by $O(\sum_v \mathrm{superHeight}(v)) + O(n) = O(n)$ (by Sum of heights, Lemma 1).

`clearA`, `clearB`, and `clearButOne` run in time proportional to the number of deleted nodes. Each node is inserted only twice - each node v is inserted only in `calcAB(v)`. When a node is deleted, it is deleted permanently. Match tables are never copied - only moved and merged. It implies that the total number of deletions is bounded by the total number of insertions $= 2n$, so the time cost of clearing functions is bounded by $O(n)$. Please refer to the extended edition of this publication for pseudocodes of those functions.

`matchAndMark` calls: `clearA`, `clearB`, and `clearButOne`. If we disregard that, it runs in time proportional to the number of deleted nodes by auxiliary subroutines: loop for $u \in (A[i] \cup B[j])$ iterates over $|A[i]| + |B[j]|$ elements. Calls to `clearButOne(A, i)` delete $|A[i]| - 1$ elements, calls to `clearButOne(B, j)` delete $|B[j]| - 1$ elements. Thus the total cost of `matchAndMark` is also bounded by $O(n)$.

Corollary 4. *We can compute match tables for all nodes of a tree using $O(n)$ time.*

Finally, we verify if the found set of marked paths indeed covers the whole tree. This procedure comes from [16], we repeat it here for completeness. We will store a counter for each node. All marked paths are vertical - one of their ends is a descendant of the other. For each marked path we will add 1 to the counter of the lower, and add -1 to the counter of the higher end - this is what `markVerticalPath` from pseudocode serves for. Let us fix any node v. The sum of counters in the subtree rooted in v is equal to the number of marked paths going through the edge from v to its parent. With a single DFS from the root we can calculate sums for every subtree. All edges of a tree are covered if and only if $\sum_{u \in \text{subtree}(v)} \text{counter}(u)$ is positive for every node v except for the root.

Now all that is left, is the proof of correctness.

Lemma 6. `matchAndMark` *marks only simple paths p having* $\text{label}(p) = S$.

Lemma 7. *If* `matchAndMark` *did not delete nodes using* `clearButOne` *subroutine, then we would mark every path p in the tree having* $\text{label}(p) = S$.

The next lemma establishes that deleting some nodes in `matchAndMark` and leaving only one representative, even if labels of their paths to v are still valid prefixes of S/S^R, is indeed a correct action.

Lemma 8. *If* $\text{label}(a \rightarrow v) = \text{label}(b \rightarrow v)$ *and both $a \rightarrow v$ and $b \rightarrow v$ are marked, then we can skip propagating either one upwards.*

Please refer to the extended edition of this publication for proofs of Lemmas 6, 7, and 8. Those three lemmas combined together prove that `matchAndMark` will mark all, and only edges covered by S. Thus they are completing a proof of the main theorem of this section.

Theorem 2. *We can compute all undirected covers in $O(n^2)$ time and $O(n)$-space.*

The general idea of the presented algorithm is similar to the centroid decomposition algorithm from [16]. We rely on the same set of candidates, and in the same way, we check if the set of marked paths covers all edges of the tree. However, calculating match tables A and B to mark all covered edges is a completely new idea.

Achieved $O(n^2)$ time $O(n)$-space complexity improves results from [16] ($O(n^2)$ time and space, or $O(n^2 \log n)$ time $O(n)$-space), but is still superlinear. Further work will be focused on research on $o(n^2)$ time algorithm or finding a conditional lower bound.

References

1. Alstrup, S., Husfeldt, T., Rauhe, T.: Marked ancestor problems. In: 39th Annual Symposium on Foundations of Computer Science, FOCS 1998, 8–11 November 1998, Palo Alto, California, USA, pp. 534–544. IEEE Computer Society (1998)
2. Apostolico, A., Ehrenfeucht, A.: Efficient detection of quasiperiodicities in strings. Theor. Comput. Sci. **119**(2), 247–265 (1993)
3. Apostolico, A., Farach, M., Iliopoulos, C.S.: Optimal superprimitivity testing for strings. Inf. Process. Lett. **39**(1), 17–20 (1991)
4. Breslauer, D.: An on-line string superprimitivity test. Inf. Process. Lett. **44**(6), 345–347 (1992)
5. Brlek, S., Lafrenière, N., Provençal, X.: Palindromic complexity of trees. In: Potapov, I. (ed.) DLT 2015. LNCS, vol. 9168, pp. 155–166. Springer, Cham (2015). https://doi.org/10.1007/978-3-319-21500-6_12
6. Cormen, T.H., Leiserson, C.E., Rivest, R.L., Stein, C.: Introduction to Algorithms, 2nd edn. The MIT Press, Cambridge (2001)
7. Crochemore, M., et al.: The maximum number of squares in a tree. In: Kärkkäinen, J., Stoye, J. (eds.) CPM 2012. LNCS, vol. 7354, pp. 27–40. Springer, Heidelberg (2012). https://doi.org/10.1007/978-3-642-31265-6_3
8. Czajka, P., Radoszewski, J.: Experimental evaluation of algorithms for computing quasiperiods. Theor. Comput. Sci. **854**, 17–29 (2021)
9. Funakoshi, M., Nakashima, Y., Inenaga, S., Bannai, H., Takeda, M.: Computing maximal palindromes and distinct palindromes in a trie. In: Holub, J., Žďárek, J. (eds.) Prague Stringology Conference 2019, Prague, Czech Republic, 26–28 August 2019, pp. 3–15. Czech Technical University in Prague, Faculty of Information Technology, Department of Theoretical Computer Science (2019)
10. Gawrychowski, P., Kociumaka, T., Rytter, W., Waleń, T.: Tight bound for the number of distinct palindromes in a tree. In: Iliopoulos, C., Puglisi, S., Yilmaz, E. (eds.) SPIRE 2015. LNCS, vol. 9309, pp. 270–276. Springer, Cham (2015). https://doi.org/10.1007/978-3-319-23826-5_26
11. Gawrychowski, P., Kociumaka, T., Rytter, W., Waleń, T.: Tight bound for the number of distinct palindromes in a tree. Electron. J. Comb. **30**, 04 (2023)
12. Kociumaka, T., Pachocki, J., Radoszewski, J., Rytter, W., Waleń, T.: Efficient counting of square substrings in a tree. Theor. Comput. Sci. **544**, 60–73 (2014)
13. Kociumaka, T., Radoszewski, J., Rytter, W., Waleń, T.: String powers in trees. Algorithmica **79**(3), 814–834 (2017)
14. Mhaskar, N., Smyth, W.F.: String covering: a survey. CoRR, abs/2211.11856 (2022)
15. Moore, D.W.G., Smyth, W.F.: A correction to "an optimal algorithm to compute all the covers of a string". Inf. Process. Lett. **54**(2), 101–103 (1995)
16. Radoszewski, J., Rytter, W., Straszyński, J., Waleń, T., Zuba, W.: String covers of a tree. In: Lecroq, T., Touzet, H. (eds.) SPIRE 2021. LNCS, vol. 12944, pp. 68–82. Springer, Cham (2021). https://doi.org/10.1007/978-3-030-86692-1_7
17. Shibuya, T.: Constructing the suffix tree of a tree with a large alphabet. In: ISAAC 1999. LNCS, vol. 1741, pp. 225–236. Springer, Heidelberg (1999). https://doi.org/10.1007/3-540-46632-0_24
18. Stroustrup, B.: The C++ Programming Language - Special Edition, 3rd edn. Addison-Wesley (2007)
19. Sugahara, R., Nakashima, Y., Inenaga, S., Bannai, H., Takeda, M.: Efficiently computing runs on a trie. Theor. Comput. Sci. **887**, 143–151 (2021)

Compacting Massive Public Transport Data

Benjamín Letelier[2] , Nieves R. Brisaboa[1] , Pablo Gutiérrez-Asorey[1] ,
José R. Paramá[1] , and Tirso V. Rodeiro[1(✉)]

[1] Universidade da Coruña, CITIC, Campus Elviña, 15071 A Coruña, Spain
`tirso.varela.rodeiro@udc.es`
[2] Instituto de Informática, Universidad Austral de Chile, Valdivia, Chile

Abstract. In this work, we present a compact method for storing and indexing users' trips across transport networks. This research is part of a larger project focused on providing transportation managers with the tools to analyze the need for improvements in public transportation networks. Specifically, we focus on addressing the problem of grouping the massive amount of data from the records of traveller cards as coherent trips that describe the trajectory of users from one origin stop to a destination using the transport network, and the efficient storage and querying of those trips. We propose two alternative methods capable of achieving a space reduction between 60 to 80% with respect to storing the raw trip data. In addition, our proposed methods are auto-indexed, allowing fast querying of the trip data to answer relevant questions for public transport administrators, such as how many trips have been made from an origin to a destination or how many trips made a transfer in a certain station.

Keywords: Compression · Public Transport · Trip analysis

1 Introduction

The widespread use of public transport cards has opened the door to a wide range of possibilities for analysing people's movements through cities. Each time a passenger boards any means of public transport, his card number, time, and transport mean get registered. This enables not only the analysis of individuals (where, when, etc.) but also aggregated analysis to study the movement patterns in cities (rush hours, flows between zones, etc.). However, the storage and

This work was partially supported by the CITIC research center funded by Xunta de Galicia, FEDER Galicia 2014-2020 80%, SXU 20% [CSI: ED431G 2019/01]; MCIN/ AEI/10.13039/501100011033 ([EXTRA-Compact: PID2020-114635RB-I00]; "NextGenerationEU"/PRTR [SIGTRANS: PDC2021-120917-C21], [PLAGEMIS: TED2021-129245B-C21]; EU/ERDF A way of making Europe [OASSIS-UDC: PID2021-122554OB-C3]); by GAIN/Xunta de Galicia [GRC: ED431C 2021/53]; by UE FEDER [CO3: IN852D 2021/3]; by Xunta de Galicia [ED481A/2021-183], and by the Fondecyt grant #11221029 of Universidad Austral de Chile.

F. M. Nardini et al. (Eds.): SPIRE 2023, LNCS 14240, pp. 310–322, 2023.
https://doi.org/10.1007/978-3-031-43980-3_25

management of these data are not trivial due to their spatio-temporal nature and their size.

Usually, static information about transport networks and their topology is publicly shared to be the basis of geographic information systems used for public transport webs and applications. There are several queries of interest in this context that could be solved using directly the network information (e.g. *What route should be followed to travel from point A to point B using the transport network?*). This is not the target of this work, our goal is the representation of user behaviour, i.e. movement patterns, preferred routes, etc.

Our current line of research involves the creation of a tool for efficiently storing and analyzing the vast amount of data related to the use of public transport networks. While this research faces three main challenges pertaining to the grouping of traveller cards transaction records into trips, the efficient storage of trips on a novel Compact Data Structure (CDS) [9], and the design and implementation of user interfaces for querying the stored data based on Geographical Information System (GIS) technologies, this work will focus primarily on introducing our compact method for the storage of the trips.

2 Background

Compact Data Structures have gained a lot of notoriety as an efficient strategy for Big Data scenarios, and in particular, we have already done some work at our research group regarding trajectories and mobility data on transport networks [6]. Our proposal follows those footsteps, using a variety of well-known data structures as the building blocks for our approach.

First, it should be noted that, given that CDS operate at bit level, bitvectors (sequences of bits) are one fundamental element in most of these data structures. Bitvectors can be processed in a highly efficient manner due to two well-known basic operations: *rank* and *select* [8]. For this work, we used *rank* and *select* implementations included in [7], both with a temporal cost of $O(1)$, where the *rank* operation uses $0.25n$ additional bits (with n being the size of the bitvector), and *select* at most $0.2n$ bits of additional space.

We also used an indexing structure known as K^n tree. The K^2 structure was first conceived as a compact web graph representation [5]. However, it turns out to be a more efficient version of a region quadtree [10], and thus, it can be used as a spatial index [2–4].

It represents a binary matrix of size $n \times n$ using a conceptual K^2-ary tree for a given K. The root node represents the entire matrix. Then, the matrix is divided into K^2 submatrices of size $n/K \times n/K$. The order for those submatrices is established as left-to-right and top-to-bottom, adding a child node to the root of the tree for each submatrix. Each node is labelled with a bit. If a submatrix contains at least a 1-bit, the corresponding node stores a 1-bit, otherwise, the node stores a 0-bit. For the submatrices with at least one 1-bit, the procedure continues recursively until reaching a submatrix with only 0-bits or the individual cells of the original matrix. The K^2-*tree* divides that bitmap in two: L, which is formed by the bits corresponding to the last level of the tree, and T, which contains the rest.

The K^2-*tree* can efficiently answer several queries, like retrieving the value of a single cell, row/column queries, or window queries. Queries are solved by navigating T and L with *rank* and *select* operations.

Note that the K^2 structure can be generalized to any n. In this work, we used a variant for indexing three-dimensional binary matrices called the $K^3 - tree$ [2].

One last concept that is necessary in order to fully understand the following sections is the *prefix sums*. Given an array $S[1..n]$ of non-negative numbers, the prefix sum of S is defined as a new array $S_{PS}[1..n+1]$ that stores the sum of prefixes of the input sequence in the form of $S_{PS}[1] = 0$ and $S_{PS}[j] = \sum_{i=1}^{j-1} S[i]$, with $2 \leq j \leq n+1$. This representation allows for the recovery of the sum in any range on S by using $\sum_{i'=i}^{j} S[i'] = (S_{PS}[j+1] - S_{PS}[i])$, with $1 \leq i \leq j \leq n$, in $O(1)$. Note that, in the case of $i = j$, the value returned is $S[i]$. One apparent problem with this representation is that S_{PS} requires much more space than S. Therefore, this work used the sampling technique described by Navarro [9] to space optimize prefix sums.

3   Previous Concepts on Transport Networks

This section addresses the basic vocabulary related to transport networks needed to understand the following sections and shows some of the difficulties associated with constructing *trips* from traveller card records.

The fundamental element in any public transport network is the *stop*. A stop is where a traveller can board (or alight) a particular means of transport following a given route (consecutive sequence of stops). Using this basic element, public networks build an entire topology combining stops (routes, lines, schedules, etc.).

We define a *trip* as the act of a user travelling across a transport network from one stop (origin) to a different stop (destination). A trip always consists of, at least, one boarding at an origin and an alighting at a destination. However, a trip may not be a direct trip, that is, it could have intermediate stops where the user alights a vehicle and boards a different one to continue towards its final destination. Each individual pair of boarding-alighting will be denoted as *trip stage* from now on.

The prime difficulty in grouping the travellers' card records into trips resides in the fact that, in many cities around the world, the traveller cards are normally only validated when boarding a means of transport. Thus, there is hardly any alighting information. To tackle this problem, we designed a deduction algorithm for the alighting stops following the general strategy presented in previous works [1]. This strategy exploits the expected space-time continuity of consecutive stages of the same trip, as well as the movement symmetry of travellers (each day's movements start where they left off the previous day) to estimate alighting stops based on the boarding data and the topology of the transport network.

As a final addendum to our definition of trips, note that in practice, we are only interested in storing the alighting of the last trip stage, i.e. the final destination of the whole trip. Applying the same continuity principle that allows us to estimate alightings by using boarding data, we can consider the intermediate alighting stops implicit on a consecutive sequence of boardings. Therefore, we can represent a complete trip as a sequence of stops $T = B_1, T_1, ..., T_n, A_1$, with the first stop always corresponding to the boarding at the origin stop, the last stop with the alighting at the destination, and any intermediate stop being a boarding in a *transfer stop*.

4   Our Proposal

We need to store all the ways users travelled between each possible origin-destination pair of a transport network in a compact and indexed way.

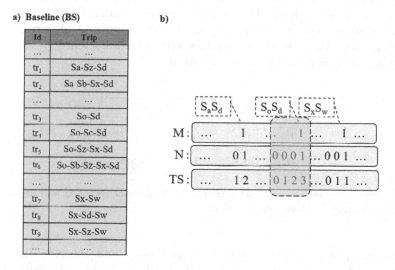

Fig. 1. A readable trip matrix describing each trip and the number of passengers that made it on the left; and its proposed compact representation on the right.

Let $S = [S_1, S_2, ..., S_\sigma]$ be the ordered set of stops in a public transport network, and let $BS[1..v]$ be the ordered list of different possible trips that travellers actually follow across the considered network. Each of these trips is an ordered list of stops, we denote each of them as *type of trip*. In Fig. 1.a, we can see a list of types of trips. For example, the type of trip tr_1 starts at S_a, includes a transfer at stop S_z, and ends at stop S_d. It is possible to compactly store BS by using three vectors:

1. The bitvector $M[1..\sigma^2]$ specifies for which *origin-destination* pairs there are types of trips in BS. Bitvector M simulates a conceptual origin-destination

matrix of size $\sigma \times \sigma$, where the cell at coordinates (i, j) stores a 1-bit if there is at least one trip between S_i and S_j in BS. M is simply the result of storing the contents of that matrix row by row in a vector. If at least one type of trip starts at S_i and ends at S_j, then $M[((i-1)*\sigma)+j]$ stores a 1-bit and a 0-bit otherwise.

2. The bitvector $N[1..v]$ indicates for each pair of origin-destination stops how many different types of trips exist in BS connecting them. More precisely, for each 1-bit in M, N stores in unary the number of different trips between two particular stops. For example, the N bitvector in Fig. 1.b shows that there are four ways (0001) to get to S_d from S_o.

3. The integer vector $TS[1..v]$ contains the number of transfer stops for every type of trip represented in N. In the example of Fig. 1.b, the four ways to get to S_d from S_o require 0, 1, 2, and 3 transfers.

Any extra information related to a trip can be represented as an array aligned with N or TS. This can be any type of information that may be of interest to transport managers, such as the example we have chosen for this paper: total number of passengers Q that made that trip last month. Additionally, in our implementation, we store Q as its prefix sum representation Q_{PS} in order to allow aggregated and individual queries on it.

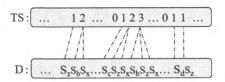

Fig. 2. Plain Transfer Representation (**PTR**). An additional vector D and its alignment with TS in order to recover transfer stops.

This simple representation is enough for storing and querying direct trips. However, it does not allow recovering information related to transfer stops of a type of trip. The straightforward approach to achieve this, henceforth *Plain Transfer Representation* (**PTR**), is by just adding a new vector D with all those intermediate stop identifiers (see Fig. 2). For the implementation of this method, we replace TS with its prefix sum representation TS_{PS}.

5   Improving Our Proposal

The previous approach PTR requires sequential searches on D in case of querying data about transfer stops. To improve this, our second proposal, *Tree Transfer Representation* (**TTR**), indexes the transfer stops identifiers of the existing trips using $K^3 - trees$, by considering a transfer at stop S_t of a trip that begins in S_o and ends in S_d as a 3D point $\langle S_o, S_d, S_t \rangle$ over a 3D-grid. TTR uses two

$K^3 - trees$, one for trips with exactly one transfer stop and the other one for every trip with two or more transfer stops, as illustrated in Fig. 3. Note that direct trips can already be retrieved by using only M, N and TS (see Fig. 1).

From now on, we will refer to the $K^3 - tree$ indexing the trips with only 1 transfer stop as $T1 - tree$ (Fig. 3.a), and the $K^3 - tree$ indexing trips with 2 or more transfer stops as $T2^+ - tree$ (Fig. 3.c).

$T1$-tree: Representing a Single Transfer Stop. Any type of trip with a single transfer stop that starts at S_o, ends at S_d, and passes through S_t can be represented as the 3-tuple $\langle S_o, S_d, S_t \rangle$ in a 3D-grid, which can be represented and indexed with a $K^3 - tree$. Observe that each 1-bit in the last level of the tree (L_1) corresponds to one triplet.

For example, in Fig. 3.b, the first (from left to right) shadowed 1-bit in L_1 indicates that there is a type of trip that starts at S_x, ends at S_w, and has only one transfer stop at S_d.

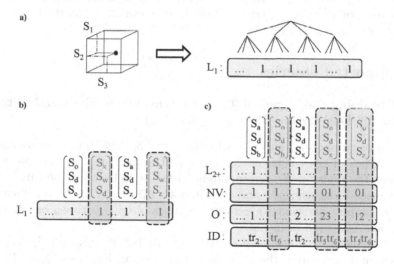

Fig. 3. Tree Transfer Representation (**TTR**). Additional $K^3 - trees$ ($T1$ and $T2+$) to efficiently index transfer stops.

$T2^+$-tree: Representing Two or more Transfer Stops. This tree is built following the same idea of $T1-tree$, but considering the type of trips with two or more transfer stops. However, observe that now, for a given triplet $\langle S_o, S_d, S_t \rangle$, we may have several types of trips in BS that start at S_o, end at S_d, and make a transfer in S_t.

This particularity requires three additional data structures:

1. A bitvector NV for storing the number of types of trips (in unary) that start at S_o, end at S_d, and make a transfer in S_t.

2. For each of those types of trips, the array O indicates the position of the transfer stop at S_t within the stops of the corresponding type of trip. Observe in Fig. 3.c that these values are aligned with the values on NV.
3. For each position of NV, the array ID stores the identifier of the corresponding type of trip, which is the position of the trip within BS.

Figure 3.c. shows an example, we only display the last level of the $K^3 - tree$ (L_{2+}). Each 1-bit in L_{2+} corresponds to the triplet of stops displayed above it (shown only for ease of understanding). Focusing on the triplet $\langle S_o, S_d, S_x \rangle$, its 1-bit in L_{2+} means that there is at least one type of trip with origin in S_o, a transfer in S_x, and destination S_d. However, there can be more than one type of trip having that property, and each of them may have other transfers at different stops. In our example, there are two types of trips connecting S_o and S_d through S_x, signalled with the bits 01 in NV. The array O indicates that in the first type of trip, S_x is the second transfer stop of the type of trip whereas, in the second type of trip, S_x is the third transfer stop. The array ID stores the position in BS of the corresponding type of trip, observe that in our example, the two types of trips are tr_5 and tr_6 (see Fig. 1).

6 Supported Queries

6.1 Obtaining the Types of Trips to Travel from the Origin Stop S_o to the Destination Stop S_d (getTrips)

Given an origin stop S_o and a destination stop S_d, this query returns all the existing types of trips between S_o and S_d. The first step in this query is to determine the range in N corresponding to the trips between the target stops. This can be solved using M and N, for both PTR and TTR. As an illustrative example, suppose we want to recover the types of trips between origin S_o and destination S_d, as shown in Fig. 1.

First, we need to determine if at least one trip between S_o and S_d does exist. This is true if and only if the value in the in the position $pos = ((o - 1)\sigma) + d$ of M is equal to 1, where o and d are the indexes of the origin and destination stops respectively, and σ the total number of stops in the network. Then, the number of existing 1-bits in $M[1..pos]$ is computed by using $q = rank_1(M, pos)$. After that, the positions of the q-th and the $(q+1)$-th 1-bits in N are calculated by using $\langle i, j \rangle = \langle select_1(N, q) + 1, select_1(N, q + 1) \rangle$, returning the range $\langle i, j \rangle$, the number of positions of this range is the number of different types of trips. As it can be seen highlighted in Fig. 1.b., there are four different types of trips between the pair (S_o, S_d).

The next step is to recover the transfer stops of those trips. This process varies depending on the version:

PTR: Conceptually, we retrieve the range $[i..j]$ of TS to obtain the number of transfer stops on each trip, and then search D for the corresponding stop identifiers. Given that PTR (unlike in the figure that uses a plain representation) uses the prefix sum representation TS_{PS}, for each position k within $[i..j]$, we recover the range $\langle l, r \rangle = \langle TS_{PS}[k], TS_{PS}[k+1] \rangle$, which corresponds to the range in D that stores the transfer stops of the type of trip represented by position k. Then, for each k, the type of trip $\langle S_o, D[l..r], S_d \rangle$ is included in the response. Note that a direct trip occurs whenever $l = r$, in which case the trip $\langle S_o, S_d \rangle$ is included in the response.

TTR: First, if there is a 0 within $TS[i..j]$ (direct trip), $\langle S_o, S_d \rangle$ is included in the response. Second, if there is a 1 within $TS[i..j]$ (meaning trips with only 1 transfer), then the region $\langle S_o, S_d, S_1 \rangle \times \langle S_o, S_d, S_\sigma \rangle$ is recovered from the $T1 -$ *tree*. We add all the points within this region to the response. In the example of Fig. 3.b, we can see that there are two types of trips between S_x and S_w, one uses S_d as a transfer stop while the other uses S_z.

Finally, if there is a 2 or a larger number within $TS[i..j]$ (trips with 2 or more transfer stops), then the same region is recovered from the $T2^+ - tree$ (Fig. 3.c), obtaining the p points $\langle S_o, S_d, S_t \rangle$ that exist in the region and their $rank_1$ value over L_{2+}. For each point p and its $rank_1$ value l_r, a $select_1(NV, l_r) + 1$ and a $select_1(NV, l_r + 1)$ would be needed to obtain the range $O[o_l..o_r]$ (highlighted in Fig. 3.c) assuming NV was stored as a bitvector, but in practice, given that we store NV as its prefix sum representation NV_{PS}, only one subtraction $NV_{PS}[l_r + 1] - NV_{PS}[l_r]$ is needed to obtain the same range.

Observe in Fig. 3.c that there are three 1-bits in L_{2+} corresponding to types of trips with origin S_o and destination S_d. The first one (from left to right) indicates that S_b is the first transfer stop of the type of trip tr_6, the second 1-bit, indicates that the third transfer stop of tr_6 is S_x and that the second stop of tr_5 is also S_x. Finally, the third 1-bit, signals that the second stop of tr_6 is S_z and that the first stop of tr_5 is also S_z.

6.2   Obtaining the Total Number of People Who Made a Trip Starting at S_o and Ending at S_d (*getPeople*)

The algorithm for this query is the same for both PTR and TTR. The first step is again to find the range in N $\langle i, j \rangle$ that corresponds with the trips between the target stops. Using the structures depicted in Fig. 1.b, the results would be the sum of all the values within that range in the vector Q (for the example of origin S_o and destination S_d, see the highlighted range in Fig. 1.b), given that Q uses a prefix sum representation, the result is computed simply as $Q_{PS}[j+1] - Q_{PS}[i]$.

6.3   Obtaining All the Origin and Destination Stops that Uses S_t as Transfer Stop (*getOriginDestinations*)

PTR: First, the range containing all the trips starting at S_t must be discarded, as we are only interested in trips where S_t was boarded as a transfer stop.

The range $\langle i, j \rangle$ in M, representing the trips starting at S_t, can be calculated as $i = ((t-1)\sigma) + 1$ and $j = 1 + i + \sigma$. Then by using $rank$ and $select$ operations on M and N, in a similar way to the first step of the *getTrips* query, we obtain the corresponding range in N.

Then, we search sequentially D for S_t appearances and, for each of those positions, its corresponding positions in N (excluding those falling in the range where S_t is the starting stop). For each position k of N obtained in the previous step, we compute $r_1 = rank_1(N, k) + 1$ and then, $pos = select_1(M, r_1) - 1$. This recovers the position in M that corresponds to the k-th trip in N. Finally, the origin-destination pair is calculated by using $\langle S_o, S_d \rangle = \langle \lfloor pos/\sigma \rfloor + 1, (pos\%\sigma) + 1 \rangle$. If the obtained destination is different from S_t, the pair is added to the result with a count of 1 unless it was already present, in that case, then a 1 is added to its count.

TTR: First, the region $\langle S_1, S_1, S_t \rangle \times \langle S_\sigma, S_\sigma, S_t \rangle$ is searched on each tree. In the case of the $T1 - tree$, just the returned p points $\langle S_o, S_d, S_t \rangle$ are needed. All the pairs $\langle S_o, S_d \rangle$ are added to the result with a count of 1.

In the case of $T2^+ - tree$, for each position p with a 1-bit in L_{2+} returned by the search within $\langle S_1, S_1, S_t \rangle \times \langle S_\sigma, S_\sigma, S_t \rangle$, we compute l as the $rank_1(L_{2+}, p)$. Then, for each l, the value $NV_{PS}[l + 1] - NV_{PS}[l]$ is added to the count corresponding to the origin-destination designed by the considered position p of L_{2+}.

7   Experimental Evaluation

The experiments were performed on a server with an Intel(R) Xeon(R) E5-2470 @ 2.30 GHz CPU and a main memory of 4×16 GB DDR3-1067 MHz MHz. All algorithms were programmed using C++11 language, with several data structures of the SDSL Library [7], and compiled using g++ version 8.3.0 with -O3 -DNDEBUG -march=native options. All given runtimes are real (wallclock) times.

7.1   Input Data

As explained, the computation of real trips from only the boarding information is a complex task, which is being carried out in parallel in the same project. Since those trips are not ready yet, we opted to test our developed methods using trips simulated over the A Coruña (Spain) bus transport network instead, containing 24 different bus lines and a total number of stops $\sigma = 1,101$.

Partial representations of the transport network were used to create four datasets of different sizes (detailed in Table 1). Each dataset contains the trip of 150 million users that were simulated through a random-walk algorithm over the corresponding transport network.

Table 1. Description of the datasets.

Dataset	# bus lines	# stops (σ)	# trips (v)
coruna-25	6	259	7,898,453
coruna-50	12	531	15,591,174
coruna-75	18	845	21,412,796
coruna-100	24	1,101	22,898,895

7.2   Used Methods

This work includes the `baseline` method proposed in Fig. 1.a, containing all the registered trips ordered by origin stop, then by destination stop and then by trip length. This order allows answering *getTrips* and *getPeople* by using binary searches, while *getOriginDestinations* is solved using a sequential search.

In addition to our proposed methods, `PTR` and `TTR`, we also include space-efficient versions for each one, called `PTR-C` and `TTR-C`. In `PTR-C`, the array D was stored using the minimum amount of bits needed, that is, $\log_2 \sigma$ bits to represent each element, while the prefix sums TS_{PS} and Q_{PS} were stored using the sampling technique. The same technique was used in the arrays TS, O, ID and the prefix sums NV_{PS} and Q_{PS} of `TTR-C`. In all methods, the data type used in each array is the smallest integer data type of `C++` that can represent the biggest value stored in the array.

All the tested methods are constructed once for each dataset and stored on disk. To run an experiment over any method, first, the data structure is loaded into main memory and then the corresponding experiment is executed.

7.3   Results

Table 2 shows the resulting sizes of every approach tested for the four experimental datasets, in MBs, as well as the compression ratios compared to the baseline.

Table 2. Space usage on each dataset (MB).

	baseline	PTR	PTR-C	TTR	TTR-C
coruna-25	159.90	70.00 (43.78%)	**28.01** (17.52%)	63.98 (40.01%)	35.51 (22.21%)
coruna-50	313.20	134.77 (43.03%)	**56.77** (18.12%)	126.61 (40.42%)	68.52 (21.88%)
coruna-75	426.75	181.40 (42.51%)	**75.16** (17.61%)	177.82 (41.67%)	98.44 (23.07%)
coruna-100	453.34	191.18 (42.17%)	**83.74** (18.47%)	189.90 (41.89%)	109.54 (24.16%)

Notice that all of our proposals occupy less space than the `baseline` on each dataset. The non-compressed versions use between 40.01% and 43.78% the size of the `baseline`. The compressed version `PTR-C` uses between 17.52% and 18.47% the space of the baseline, while `TTR-C` needs between 22.21% and 24.16%.

After building the data structures, our first experiment consisted in testing the speed of the query *getTrips*. Each method had to reconstruct all the kinds of trips that started on a random origin S_i and ended on a random S_j. This was repeated 1,000,000 times on each method and the result is the average time to reconstruct one kind of trip that exists between an origin and a destination. Each method searched exactly the same values on each query. Table 3 shows the result of this experiment in $\mu s/trip$.

Table 3. Time of *getTrips* to recover one trip ($\mu s/trip$).

	baseline	PTR	PTR-C	TTR	TTR-C
coruna-25	0.077	**0.067**	0.313	0.322	0.332
coruna-50	0.115	**0.076**	0.318	0.777	0.777
coruna-75	0.18	**0.078**	0.317	1.804	1.761
coruna-100	0.252	**0.074**	0.317	2.612	2.602

The obtained results of this first experiment prove that PTR and PTR-C recover single trips in $O(1)$. The time difference between both versions can only be explained by the access time to the arrays since both use exactly the same algorithm. The baseline method is always slower than PTR and always faster than every other method. TTR and TTR-C do not perform well on this query, since both versions need to access both $T1$ and $T2+$ trees in the worst case (if there are trips with one or more transfer stops to be recovered). Note that the access time of the arrays that affected the results in PTR and PTR-C does not affect TTR and TTR-C, achieving similar times.

The next experiment consisted in testing the speed of *getPeople* query. Each method had to recover the total number of people that started their trip on a random origin S_i and ended on a random S_j. This was repeated 1,000,000 times on each method, using exactly the same values in each query. The displayed value is the average time to answer one query. Table 4 shows the result of the experiment in $\mu s/query$.

Table 4. Time of *getPeople* query to recover one query ($\mu s/query$).

	baseline	PTR	PTR-C	TTR	TTR-C
coruna-25	2.688	0.735	0.734	**0.719**	0.727
coruna-50	3.497	**0.619**	0.628	0.766	0.770
coruna-75	3.975	0.773	0.770	0.682	**0.639**
coruna-100	4.22	**0.469**	0.478	0.729	0.472

Theoretically, our proposals should answer the *getPeople* query in $O(1)$, since the range in N can be obtained in $O(1)$ and, given that Q is stored as a prefix sum, we can also return any range of data in $O(1)$. In the case of the `baseline`, it answers the query in $O(k + \log_2(v))$, with k being the number of trips that start in S_o and end in S_d and $\log_2(v)$ the additional time to return the range using the binary searches. The results of this particular experiment show that PTR, PTR-C, TTR and TTR-C can answer the *getPeople* query at least 3.6× faster than the `baseline`.

The final experiment consisted in testing the speed of the methods in the *getOriginDestinations* query. Each method had to recover a list containing all the origin-destination pairs that used the stop S_t as a transfer stop. This was repeated 1,000 times on each method, using exactly the same values in each query. The results are shown in Table 5, containing the average time to answer one query in *ms/query*.

Table 5. Time of *getOriginDestinations* query to recover one query (*ms/query*).

	baseline	PTR	PTR-C	TTR	TTR-C
coruna-25	53.079	31.147	1927.08	6.898	**6.777**
coruna-50	101.962	58.446	3749.51	14.649	**14.449**
coruna-75	141.806	85.286	5027.86	22.070	**22.040**
coruna-100	152.269	95.884	5252.03	22.409	**22.329**

As it can be seen in Table 5, tree-based methods always achieve better performance as the intermediate stops are indexed while the baseline and PTR need extra steps to reach them. Thus, TTR and TTR-C can return all the existing points that contain the searched transfer stop in logarithmic time, while the `baseline`, PTR and PTR-C need to do a sequential search over all the stored trips. Note that PTR is always faster than the `baseline`, while PTR-C access times lead to much worse results.

8   Conclusions and Future Work

This work introduced two compact data structures that represent user trips over a public transport network, PTR and TTR. We fed these data structures with the simulated data of 150 million users travelling across the entire A Coruña bus transport network. The results showed that our main approaches use less than 44% of the space required by the `baseline` to store these trips, while the space-efficient versions we implemented, PTR-C and TTR-C, uses less than 18.47% and 24.16% of the space used by the `baseline` respectively.

The proposed representations are able to efficiently answer three queries of interest: retrieve trips from an origin stop S_i to a destination stop S_j (*getTrips*), get the total number of people who travelled between any pair origin-destination (*getPeople*), and calculate a list containing all the origin-destination pairs who used a certain stop S_t as transfer stop (*getOriginDestination*). In terms of results, it is advisable to use our compressed tree version TTR-C if transfer stops retrieval is needed and the plain version PTR in all other cases.

As future work, we plan to create a hybrid approach between PTR and TTR-C that could allow faster performance on all the queries, while still using less space compared to our baseline. We also intend to expand our query capabilities using feedback from transport administrators and test our methods using real trip data from the city of Madrid.

References

1. Alsger, A., Assemi, B., Mesbah, M., Ferreira, L.: Validating and improving public transport origin-destination estimation algorithm using smart card fare data. Transp. Res. Part C Emerg. Technol. **68**, 490–506 (2016)
2. de Bernardo, G., Álvarez-García, S., Brisaboa, N.R., Navarro, G., Pedreira, O.: Compact querieable representations of raster data. In: Kurland, O., Lewenstein, M., Porat, E. (eds.) SPIRE 2013. LNCS, vol. 8214, pp. 96–108. Springer, Cham (2013). https://doi.org/10.1007/978-3-319-02432-5_14
3. Brisaboa, N.R., Bernardo, G.D., Gutiérrez, G., Luaces, M.R., Paramá, J.R.: Efficiently querying vector and raster data. Comput. J. **60**(9), 1395–1413 (2017)
4. Brisaboa, N.R., Gómez-Brandón, A., Navarro, G., Paramá, J.R.: GraCT: a grammar-based compressed index for trajectory data. Inf. Sci. **483**, 106–135 (2019)
5. Brisaboa, N.R., Ladra, S., Navarro, G.: Compact representation of web graphs with extended functionality. Inf. Syst. **39**, 152–174 (2014)
6. Brisaboa, N., Fariña, A., Galaktionov, D., V Rodeiro, T., Rodriguez, A.: Improved structures to solve aggregated queries for trips over public transportation networks. Inf. Sci. **584** (2021)
7. Gog, S., Beller, T., Moffat, A., Petri, M.: From theory to practice: plug and play with succinct data structures. In: Gudmundsson, J., Katajainen, J. (eds.) SEA 2014. LNCS, vol. 8504, pp. 326–337. Springer, Cham (2014). https://doi.org/10.1007/978-3-319-07959-2_28
8. Jacobson, G.: Space-efficient static trees and graphs. In: 30th Annual Symposium on Foundations of Computer Science, pp. 549–554. IEEE Computer Society (1989)
9. Navarro, G.: Compact Data Structures: A Practical Approach. Cambridge University Press, USA (2016)
10. Samet, H.: Foundations of Multimensional and Metric Data Structures. Morgan Kaufmann, San Francisco (2006)

Constant Time and Space Updates for the Sigma-Tau Problem

Zsuzsanna Lipták[1] , Francesco Masillo[1](✉) , Gonzalo Navarro[2] ,
and Aaron Williams[3]

[1] Department of Computer Science, University of Verona, Verona, Italy
{zsuzsanna.liptak,francesco.masillo}@univr.it
[2] CeBiB and Department of Computer Science, University of Chile, Santiago, Chile
gnavarro@dcc.uchile.cl
[3] Computer Science Department, Williams College, Williamstown, MA, USA
aaron.williams@williams.edu

Abstract. Sawada and Williams in [SODA 2018] and [ACM Trans. Alg. 2020] gave algorithms for constructing Hamiltonian paths and cycles in the Sigma-Tau graph, thereby solving a problem of Nijenhuis and Wilf that had been open for over 40 years. The Sigma-Tau graph is the directed graph whose vertex set consists of all permutations of n, and there is a directed edge from π to π' if π' can be obtained from π either by a cyclic left-shift (sigma) or by exchanging the first two entries (tau). We improve the existing algorithms from $\mathcal{O}(n)$ time per permutation to $\mathcal{O}(1)$ time per permutation. Moreover, our algorithms require only $\mathcal{O}(1)$ extra space. The result is the first combinatorial generation algorithm for n-permutations that is optimal in both time and space, and lists the objects in a Gray code order using only two types of changes. The simple C code ($\sim$50 lines) can be found at https://github.com/fmasillo/sigma-tau.

Keywords: permutations · sigma-tau problem · dynamic data structures · combinatorial generation · combinatorial Gray codes

1 Introduction

The problem of efficiently generating all permutations of $[n] = \{1, 2, \ldots, n\}$ (in one-line notation) is one of the oldest in combinatorial generation. When surveying permutation generation algorithms in 1977, Sedgewick [37] remarked that "It was actually one of the first nontrivial nonnumeric problems to be attacked by computer". Updated surveys on generating combinatorial objects, including permutations, have been written by Savage [33], and more recently by Mütze [27].

Permutations are of fundamental importance in all areas of computer science. In string algorithms, they form the basis of compressed data structures such as compressed suffix arrays [19], compressed suffix trees [15,16,26], and BWT-based data structures, such as the FM-index [13], the RLFM-index [25], the

G. Navarro—Funded in part by Basal Funds FB0001, ANID, Chile.

F. M. Nardini et al. (Eds.): SPIRE 2023, LNCS 14240, pp. 323–330, 2023.
https://doi.org/10.1007/978-3-031-43980-3_26

r-index [17], or the extended r-index [5]. Permutations are also of central interest in computational biology, where they have been used extensively to model genome rearrangements [1–3,7,12,14,20,21].

In this paper, we provide iterative permutation generation algorithms that update the current permutation in worst-case $\mathcal{O}(1)$ time (i.e., *loopless*) using $\mathcal{O}(1)$ (additional) space. (We use the transdichotomous RAM model, where a word has $\Theta(\log n)$ bits. So $\mathcal{O}(1)$ space is $\mathcal{O}(\log n)$ total bits; the current permutation's memory is not counted [39].) They create combinatorial Gray codes, where consecutive permutations differ by one of two operations (one type of swap or rotation). To the best of our knowledge, no existing permutation generation algorithm has this set of features, see [27,33,37].

Loopless algorithms for permutations rarely use $\mathcal{O}(1)$ space as it cannot support $n!$ different internal states: $\log(n!) = \Theta(n \log n)$. Thus, an $\mathcal{O}(1)$ space algorithm cannot count to $n!$ or compute natural sequences of length $n!$ like the factorial ruler sequence $\mathcal{L}_n$ (OEIS A055881 [28]). This discounts frameworks by Ganapathi and Chowdhurysee [18], which generalize 19 previous algorithms using $\mathcal{L}_n$ or a similar sequence $\mathcal{R}_n$; also see Knuth's framework [24]. (As a specific example, Zaks[1] uses two additional arrays.) Thus, an $\mathcal{O}(1)$ space algorithm must at times *read* from the current permutation. This is true of cool-lex order's simple successor rule [30], which can be generated by a loopless $\mathcal{O}(1)$ space algorithm for multiset permutations [39], but it uses $n - 1$ different changes. Shorthand universal cycles [23,31,36] give simple Gray codes with two change types, but no existing loopless implementation uses $\mathcal{O}(1)$ space.

Our algorithms generate (σ, τ)-Gray codes by Sawada and Williams [34,35]. Here τ swaps the first two values, and σ rotates the full permutation one position to the left. Hamilton paths are given for all n [34,35] and Hamilton cycles for odd n [35] in the underlying directed Cayley graph $\mathcal{G}_n$. Figure 1 shows $\mathcal{G}_4$ and a Hamilton path; Hamilton cycles do not exist for even n [29,38]. Both papers give successor rules and worst-case $\mathcal{O}(n)$ time array-based C programs. Egan created length $n! + (n - 1)! + (n - 2)! + (n - 3)! + n - 3$ *superpermutations* [9,11] using (σ, τ)-Gray codes from an earlier manuscript [41]. Prior work had found Hamilton cycles in the *undirected shuffle exchange network* (i.e., $\mathcal{G}_n$ plus σ^{-1} edges) [4,8].

In his pioneering work on loopless algorithms, Ehrlich [10] differentiates between (a) changing the current object into its successor, and (b) deciding which change to apply in (a). Note that both types of computation must be completed in worst-case $\mathcal{O}(1)$ time to obtain a loopless algorithm. Our loopless σ-τ algorithms use circular data structures to address (a) since the σ operation requires $\Theta(n)$ time in a conventional array. To address (b), we must carefully introduce additional variables that can be updated in worst-case $\mathcal{O}(1)$ time. The output of one of our algorithms for $n = 4$ (see Sect. 4) is visualized in Fig. 2.

Our contribution is summarized in Theorem 1 (subsuming Lemmas 3, 6, and 7). Full C code can be found at https://github.com/fmasillo/sigma-tau.

Theorem 1. *There is a data structure implementing the Hamilton path successor rule of* [34], *as well as the Hamilton path and Hamilton cycle successor rules of* [35], *in worst-case $\mathcal{O}(1)$ time per permutation, using $\mathcal{O}(1)$ additional space.*

---

[1] His *pancake flip order* dates to the 1700 s [22] (see [6]) and is loopless in a BLL [40].

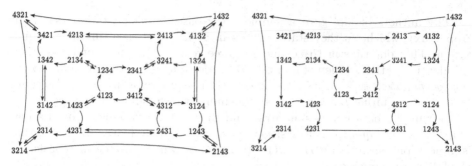

(a) The Sigma-Tau graph $\mathcal{G}_4$ [34].   (b) Hamilton path HP from 3421 to 2314.

Fig. 1. Our loopless algorithms traverse Hamilton paths and cycles in the Sigma-Tau graph $\mathcal{G}_n$ in worst-case $\mathcal{O}(1)$ time per vertex. The path in (b) follows [35].

σ	τ	σ	τ	σ	σ	σ	τ	σ	σ	σ	τ	σ	σ	σ	τ	σ	τ	σ	τ	σ	σ	σ
3	4	2	4	1	4	3	2	1	2	4	3	1	3	2	3	4	1	2	1	3	1	4	2
4	2	4	1	4	3	2	1	2	4	3	1	3	2	3	4	1	2	1	3	1	4	2	3
2	1	1	3	3	2	1	4	4	3	1	2	2	4	4	1	2	3	3	4	4	2	3	1
1	3	3	2	2	1	4	3	3	1	2	4	4	1	1	2	3	4	4	2	2	3	1	4

Fig. 2. An alternate order of permutations HP' from [34]. Each τ transition swaps the first (i.e., topmost) pair of elements. Each σ left-rotates all elements one position, with the leftmost visualized as wrapping around from top to bottom.

2 Constant Time Successor Rule for Hamilton Paths

Sawada and Williams in [34] provided a successor rule for Hamilton paths, which they later modified slightly in [35] to harmonize with the Hamilton cycle successor rule in the same paper. We first look at the latter rule, and will discuss the original rule [34] in Sect. 4.

In the new version [35], the following successor rule was given for constructing a Hamilton path in $\mathcal{G}_n$, for any $n > 1$:

Hamilton path successor rule for $\mathcal{G}_n$ ([35]) Let $\pi = \pi(1)\pi(2)\cdots\pi(n)$ be a permutation and let r be the symbol to the right of n when π is considered cyclically and skipping over $\pi(2)$. Define the successor rule HP on $\mathcal{G}_n$ as follows:

$$HP(\pi) = \begin{cases} \tau(\pi) & \text{if } (r, \pi(2)) \in \{(1,2), (2,3), \ldots, (n-2, n-1), (n-1, 2)\} \\ & \text{and } \pi \neq n(n-1)(n-2)\cdots 1; \\ \sigma(\pi) & \text{otherwise.} \end{cases}$$

Let $p = \pi^{-1}(n)$, then the definition of r in the successor rule above is: $r = \pi(3)$ if $p = 1$, $r = \pi(1)$ if $p = n$, and $r = \pi(p+1)$ otherwise.

The authors of [35] gave a simple array-based implementation (see their Appendix), which, as they state, results in $\mathcal{O}(n)$ worst-case time per permutation. The code runs in $\Omega(n)$ time for three reasons: (1) the sigma-operation, (2) identifying the position of n in π, and (3) deciding if π is the *special (decreasing) permutation* $\pi_{sp} = n(n-1)(n-2)\cdots 321$. The programs in [34,35] also count to $n!$, thus requiring $\Omega(n \log n)$ bits of memory, which is not $O(1)$ space.

Our implementation uses an array and three integer variables. Given a permutation π, an *up-step*[2] is a position where π, taken circularly, increases, that is, a position i such that $\pi(i) < \pi(1 + (i \bmod n))$. Our data structure consists of the following components:

1. an array $C[1, n]$ containing a rotation of π,
2. a pointer b to the position of $\pi(1)$, $C[b] = \pi(1)$,
3. a pointer p to the position of n, $C[p] = n$, and
4. a counter u, giving the number of up-steps of π.

Example 1. Let $n = 7$ and $\pi = 5624137$. Then the following are two possible implementations: $C_1 = [4, 1, 3, 7, 5, 6, 2], b_1 = 5, p_1 = 4, u_1 = 4$, or $C_2 = [5, 6, 2, 4, 1, 3, 7], b_2 = 1, p_2 = 7, u_2 = 4$. Note that u remains invariant.

Note that any value $\pi(i)$ can be accessed in constant time, since $\pi(i) = C[1 + (b + i - 2 \bmod n)]$. In particular, permutation π can be listed as $C[b], C[b+1], \ldots, C[n], C[1], \ldots, C[b-1]$, and can thus be returned in $\mathcal{O}(n)$ time, if required.

We can check the conditions whether to apply τ or σ in constant time:

Lemma 1. *Let π be a permutation and C, b, p, u as defined. We can test in $\mathcal{O}(1)$ time if (1) $(r, \pi(2)) \in \{(1, 2), (2, 3), \ldots, (n-2, n-1), (n-1, 2)\}$, and (2) $\pi = \pi_{sp}$.*

Proof. 1. Recall that $\pi(i) = C[1 + (b + i - 2 \bmod n)]$ is computed in constant time. In particular $\pi(2) = C[1 + (b \bmod n)]$. On the other hand, we compute r in constant time as $C[1 + (p \bmod n)]$ if $p \neq b$ and $C[1 + (b + 1 \bmod n)]$ otherwise. So we test whether $r < n - 1$ and $\pi(2) = r + 1$, or $r = n - 1$ and $\pi(2) = 2$.

2. It is easy to see that $\pi = \pi_{sp}$ if and only if $u = 1$ and $p = b$.   □

We next show how to implement σ and τ using our data structure:

Lemma 2. *Both operations σ and τ can be executed in constant time using the data structure C, b, p, u.*

Proof. A σ-operation is implemented in constant time by simply incrementing b circularly, $b = 1 + (b \bmod n)$. A τ-operation, which exchanges $\pi(1)$ with $\pi(2)$, is implemented, again in constant time, as follows:

1. decrement u once if $\pi(n) < \pi(1)$, once if $\pi(1) < \pi(2)$, and once if $\pi(2) < \pi(3)$;
2. increment u once if $\pi(n) < \pi(2)$, once if $\pi(2) < \pi(1)$, and once if $\pi(1) < \pi(3)$;
3. set $p = 1 + (b \bmod n)$ if $p = b$, or set $p = b$ if $p = 1 + (b \bmod n)$; and
4. exchange $C[b]$ with $C[1 + (b \bmod n)]$.   □

---

[2] Note that this definition differs from *ascent*, which is not taken circularly.

The total space occupied by our data structure is the permutation itself (array C), and in addition the three integer variables, each taking $\Theta(\log n)$ bits. Our algorithm needs $\mathcal{O}(n)$ time to write the initial permutation $\pi_{op} \cdot \tau$ to C and $\mathcal{O}(1)$ time to initialize the variables b, p, u. (Alternatively, we can view the initial permutation as the input, in which case we refer to the input array as C.) Thus:

Lemma 3. *Using the data structure consisting of array $C[1, n]$ and the variables b, p, u, which are initialized in $\mathcal{O}(n)$ time, we can construct a Hamilton path in $\mathcal{G}_n$, starting from the permutation $\pi_{sp} \cdot \tau = (n-1)n(n-2)\cdots 321$ and implementing the HP successor rule of [35], in $\mathcal{O}(1)$ worst-case time per permutation, using $\mathcal{O}(1)$ extra words.*

3 Constant Time Successor Rule for Hamilton Cycles

In [35], a successor rule for Hamiltonian cycles for odd n is provided. The authors define the *special set* R_n, included in the conditions for applying τ rather than σ. For a permutation π, we define $\pi_{\backslash 2}$ the $(n-1)$-length string obtained from π by removing the element in position 2 (which is also an $(n-1)$-permutation in case $\pi(2) = n$). Then the special set is defined as $R_n = \{\pi \mid \pi(2) = n$ and $\pi_{\backslash 2}$ is a rotation of $id_{n-1}\}$. E.g., $R_5 = \{15234, 25341, 35412, 45123\}$.

Hamilton cycle successor rule for $\mathcal{G}_n$, where n is odd ([35]) Let $\pi = \pi(1)\pi(2)\cdots\pi(n)$ be a permutation and let r be the symbol to the right of n when π is considered cyclically and skipping over $\pi(2)$. Define:

$$HC(\pi) = \begin{cases} \tau(\pi) & \text{if } (r, \pi(2)) \in \{(1,2),(2,3),\ldots,(n-2,n-1),(n-1,2)\} \\ & \text{or } \pi \in R_n \\ \sigma(\pi) & \text{otherwise.} \end{cases}$$

Again, the array based implementation given in [35] results in $\mathcal{O}(n)$ time per permutation in the worst case. In order to achieve constant time, we slightly modify our data structure, replacing counter u by counter u', the number of up-steps of $\pi_{\backslash 2}$, that is, we count up-steps skipping over position 2.

Lemma 4. *It can be checked in constant time whether $\pi \in R_n$.*

Proof. It is clear that an $(n-1)$-permutation is a rotation of the identity $123\cdots(n-2)(n-1)$ if and only if the number of its up-steps is $n-2$. The fact that n is inserted in position 2 is equivalent to $p = 1 + (b \bmod n)$. If $\pi(2) = n$ then $\pi_{\backslash 2}$ is an $(n-1)$-permutation, and therefore, $\pi_{\backslash 2}$ is a rotation of $123\cdots(n-2)(n-1)$ if and only if $u' = n-2$. Both checks can be done in constant time. □

Lemma 5. *Both operations σ and τ can be executed in constant time, using the modified data structure C, b, p, u'.*

Proof. For the σ-operation, before setting $b = 1 + (b \bmod n)$ we need to update u' in constant time as follows:

1. decrement u' once if $\pi(1) < \pi(3)$, and once if $\pi(3) < \pi(4)$;
2. increment u' once if $\pi(1) < \pi(2)$, and once if $\pi(2) < \pi(4)$.

For the τ-operation we do as follows, also in constant time:

1. decrement u' once if $\pi(n) < \pi(1)$, and once if $\pi(1) < \pi(3)$;
2. increment u' once if $\pi(n) < \pi(2)$, and once if $\pi(2) < \pi(3)$;
3. set $p = 1 + (b \bmod n)$ if $p = b$, or set $p = b$ if $p = 1 + (b \bmod n)$;
4. exchange $C[b]$ with $C[1 + (b \bmod n)]$.   □

Similarly to the Hamilton path data structure, we use additional $\mathcal{O}(1)$ words. Note that the Hamilton cycle can be started at any permutation. From this discussion and Lemmas 1, 4, and 5, we have:

Lemma 6. *Using the data structure consisting of array $C[1, n]$ and the variables b, p, u', which are initialized in $\mathcal{O}(n)$ time, we can construct a Hamilton cycle in $\mathcal{G}_n$, starting from the identity permutation and implementing the HC successor rule of [35], in $\mathcal{O}(1)$ worst-case time per permutation, using $\mathcal{O}(1)$ extra words.*

4   Simpler Rule for Hamilton Paths and Termination

The original Hamilton path successor rule HP' given in [34] differs in only one detail from the one in [35], namely that in the condition for τ, $(n - 1, 2)$ is replaced by $(n - 1, 1)$. The resulting Hamilton paths in $\mathcal{G}_4$ is visualized in Fig. 2.

This change can be easily accommodated using our data structure, by a simple change in the condition for applying τ. Alternatively, insights from [32] on this Hamilton path can be used for a further simplification: The *syntactic sequence* of a Hamilton path in $\mathcal{G}_n$ is a string over the alphabet $\{\tau, \sigma\}$ which specifies the sequence of operations applied. Rytter and Zuba [32] showed that for the Hamilton path resulting from successor rule HP', this sequence is highly compressible.

Lemma 7. *Using the data structure consisting of array $C[1, n]$ and variables b, p, u, which are initialized in $\mathcal{O}(n)$ time, we can construct a Hamilton path in $\mathcal{G}_n$, starting from the permutation $\pi_{sp} \cdot \tau = (n - 1)n(n - 2) \cdots 321$ and implementing the HP' successor rule of [34], in $\mathcal{O}(1)$ worst-case time per permutation, using $\mathcal{O}(1)$ extra words.*

Termination. To terminate our algorithms, we cannot resort to a counter maintaining the number of permutations, as is done in [34,35], since this would exceed the $\mathcal{O}(1)$ space restriction. Instead, we apply termination conditions identifying the final permutation. For example, as we start the HC algorithm at the identity $id = 123 \cdots n$, the final permutation is $n123 \cdots (n - 1)$. This is the unique permutation with $\pi(1) = n$, $\pi(2) = 1$, and $u' = n - 2$ up-steps (skipping over $\pi(2)$). Similar tests terminate HP and HP': starting from $\pi_{sp} \cdot \tau$, we have to detect when the last permutation $(n - 2)(n - 1)(n - 3)(n - 4) \cdots 21n$ occurs. This can be done again in constant time and space by checking whether $u = 2, \pi(1) = n - 2, \pi(2) = n - 1, \pi(n - 1) = 1$, and $\pi(n) = n$: the only permutation with those extreme values fixed and with no further up-steps is the one containing the descending sequence $(n - 3) \cdots 2$ in between.

References

1. Bader, D.A., Moret, B.M.E., Yan, M.: A linear-time algorithm for computing inversion distance between signed permutations with an experimental study. J. Comput. Biol. **8**(5), 483–491 (2001)
2. Bafna, V., Pevzner, P.A.: Genome rearrangements and sorting by reversals. In: Proceedings of the 34th Annual Symposium on Foundations of Computer Science (FOCS 1993), pp. 148–157. IEEE Computer Society (1993)
3. Bafna, V., Pevzner, P.A.: Sorting by transpositions. SIAM J. Discret. Math. **11**(2), 224–240 (1998)
4. Bass, D.W., Sudborough, I.H.: On the shuffle-exchange permutation network. In: Proceedings of the 1997 International Symposium on Parallel Architectures, Algorithms and Networks (I-SPAN 1997), pp. 165–171. IEEE (1997)
5. Boucher, C., Cenzato, D., Lipták, Z., Rossi, M., Sciortino, M.: r-indexing the eBWT. In: Lecroq, T., Touzet, H. (eds.) SPIRE 2021. LNCS, vol. 12944, pp. 3–12. Springer, Cham (2021). https://doi.org/10.1007/978-3-030-86692-1_1
6. Cameron, B., Sawada, J., Therese, W., Williams, A.: Hamiltonicity of k-sided pancake networks with fixed-spin: efficient generation, ranking, and optimality. Algorithmica **85**(3), 717–744 (2023)
7. Cerbai, G., Ferrari, L.S.: Permutation patterns in genome rearrangement problems: The reversal model. Discret. Appl. Math. **279**, 34–48 (2020)
8. Compton, R.C., Gill Williamson, S.: Doubly adjacent gray codes for the symmetric group. Linear Multilinear Algebra **35**(3–4), 237–293 (1993)
9. Egan, G.: Superpermutations (2018). http://www.gregegan.net/SCIENCE/Superpermutations/Superpermutations.html
10. Ehrlich, G.: Loopless algorithms for generating permutations, combinations, and other combinatorial configurations. J. ACM **20**(3), 500–513 (1973)
11. Engen, M., Vatter, V.: Containing all permutations. Am. Math. Mon. **128**(1), 4–24 (2020)
12. Feng, J., Zhu, D.: Faster algorithms for sorting by transpositions and sorting by block interchanges. ACM Trans. Algorithms **3**(3), 25 (2007)
13. Ferragina, P., Manzini, G.: Indexing compressed text. J. ACM **52**, 552–581 (2005)
14. Fertin, G., Labarre, A., Rusu, I., Tannier, E., Vialette, S.: Combinatorics of Genome Rearrangements. Computational Molecular Biology, MIT Press, Cambridge (2009)
15. Fischer, J., Mäkinen, V., Navarro, G.: Faster entropy-bounded compressed suffix trees. Theoret. Comput. Sci. **410**(51), 5354–5364 (2009)
16. Gagie, T., Navarro, G., Prezza, N.: Fully-functional suffix trees and optimal text searching in BWT-runs bounded space. J. ACM **67**(1), article 2 (2020)
17. Gagie, T., Navarro, G., Prezza, N.: Optimal-time text indexing in BWT-runs bounded space. In: Proceedings of the 29th Annual ACM-SIAM Symposium on Discrete Algorithms (SODA 2018), pp. 1459–1477 (2018)
18. Ganapathi, P., Chowdhury, R.: A unified framework to discover permutation generation algorithms. Comput. J. **66**(3), 603–614 (2023)
19. Grossi, R., Vitter, J.S.: Compressed suffix arrays and suffix trees with applications to text indexing and string matching. SIAM J. Comput. **35**(2), 378–407 (2005)
20. Hannenhalli, S., Pevzner, P.A.: Transforming cabbage into turnip: polynomial algorithm for sorting signed permutations by reversals. In: Proceedings of the 27th Annual ACM Symposium on Theory of Computing (STOC 1995), pp. 178–189. ACM (1995)

21. Hartman, T., Shamir, R.: A simpler and faster 1.5-approximation algorithm for sorting by transpositions. Inf. Comput. **204**(2), 275–290 (2006)
22. Hindenburg, C.F.: Sammlung combinatorisch-analytischer Abhandlungen, vol. 1. ben Gerhard Fleischer dem Jungern (1796)
23. Holroyd, A.E., Ruskey, F., Williams, A.: Shorthand universal cycles for permutations. Algorithmica **64**, 215–245 (2012)
24. Knuth, D.E.: The Art of Computer Programming, Volume 4, Fascicle 2: Generating All Tuples and Permutations (Art of Computer Programming). Addison-Wesley Professional (2005)
25. Mäkinen, V., Navarro, G.: Succinct suffix arrays based on run-length encoding. Nord. J. Comput. **12**(1), 40–66 (2005)
26. Munro, J.I., Raman, R., Raman, V., Rao, S.S.: Succinct representations of permutations and functions. Theoret. Comput. Sci. **438**, 74–88 (2012)
27. Mütze, T.: Combinatorial gray codes–an updated survey. Electron. J. Combin. **30**(3-DS26) (2023)
28. OEIS Foundation Inc.: Sequence A055881 in the On-line Encyclopedia of Integer Sequences. https://oeis.org/A055881. Accessed 2 June 2023
29. Rankin, R.A.: A campanological problem in group theory. In: Mathematical Proceedings of the Cambridge Philosophical Society, vol. 44, pp. 17–25. Cambridge University Press (1948)
30. Ruskey, F., Williams, A.: The coolest way to generate combinations. Discret. Math. **309**(17), 5305–5320 (2009)
31. Ruskey, F., Williams, A.: An explicit universal cycle for the $(n-1)$-permutations of an n-set. ACM Trans. Algorithms (TALG) **6**(3), 1–12 (2010)
32. Rytter, W., Zuba, W.: Syntactic view of sigma-tau generation of permutations. Theor. Comput. Sci. **882**, 49–62 (2021)
33. Savage, C.D.: A survey of combinatorial Gray codes. SIAM Rev. **39**(4), 605–629 (1997)
34. Sawada, J., Williams, A.: A Hamilton path for the Sigma-Tau problem. In: Proceedings of the 29th Annual ACM-SIAM Symposium on Discrete Algorithms (SODA 2018), pp. 568–575. SIAM (2018)
35. Sawada, J., Williams, A.: Solving the Sigma-Tau problem. ACM Trans. Algorithms **16**(1), 11:1–11:17 (2020)
36. Sawada, J., Williams, A.: Constructing the first (and coolest) fixed-content universal cycle. Algorithmica **85**, 1–32 (2022)
37. Sedgewick, R.: Permutation generation methods. ACM Comput. Surv. (CSUR) **9**(2), 137–164 (1977)
38. Swan, R.G.: A simple proof of Rankin's campanological theorem. Am. Math. Mon. **106**(2), 159–161 (1999)
39. Williams, A.: Loopless generation of multiset permutations using a constant number of variables by prefix shifts. In: Proceedings of the 20th Annual ACM-SIAM Symposium on Discrete Algorithms (SODA 2009), pp. 987–996. SIAM (2009)
40. Williams, A.: O(1)-time unsorting by prefix-reversals in a boustrophedon linked list. In: Boldi, P., Gargano, L. (eds.) FUN 2010. LNCS, vol. 6099, pp. 368–379. Springer, Heidelberg (2010). https://doi.org/10.1007/978-3-642-13122-6_35
41. Williams, A.: Hamiltonicity of the Cayley digraph on the symmetric group generated by $\sigma = (1\ 2 \ldots n)$ and $\tau = (1\ 2)$. CoRR abs/1307.2549 (2013)

Linear-Time Computation of Generalized Minimal Absent Words for Multiple Strings

Kouta Okabe[1], Takuya Mieno[2] ⓘ, Yuto Nakashima[3] ⓘ,
Shunsuke Inenaga[3(✉)] ⓘ, and Hideo Bannai[4] ⓘ

[1] Department of Information Science and Technology, Kyushu University,
Fukuoka, Japan
[2] Department of Computer and Network Engineering,
University of Electro-Communications, Tokyo, Japan
tmieno@uec.ac.jp
[3] Department of Informatics, Kyushu University, Fukuoka, Japan
{nakashima.yuto.003,inenaga.shunsuke.380}@m.kyushu-u.ac.jp
[4] M&D Data Science Center, Tokyo Medical and Dental University, Tokyo, Japan
hdbn.dsc@tmd.ac.jp

Abstract. A string w is called a *minimal absent word (MAW)* for a string S if w does not occur as a substring in S and all proper substrings of w occur in S. MAWs are well-studied combinatorial string objects that have potential applications in areas including bioinformatics, musicology, and data compression. In this paper, we generalize the notion of MAWs to a set $\mathcal{S} = \{S_1, \ldots, S_k\}$ of multiple strings. We first describe our solution to the case of $k = 2$ strings, and show how to compute the set M of MAWs in optimal $O(n+|\mathsf{M}|)$ time and with $O(n)$ working space, where n denotes the total length of the strings in $\mathcal{S}$. We then move on to the general case of $k > 2$ strings, and show how to compute the set M of MAWs in $O(n\lceil k/\log n\rceil + |\mathsf{M}|)$ time and with $O(n(k + \log n))$ bits of working space, in the word RAM model with machine word size $\omega = \log n$. The latter algorithm runs in optimal $O(n + |\mathsf{M}|)$ time for $k = O(\log n)$.

1 Introduction

A non-empty string w is said to be an *absent word* (a.k.a. *a forbidden word*) for a string S if w is *not* a substring of S. An absent word w for S is said to be a *minimal absent word (MAW)* for S if all proper substrings of w occur in S. For instance, for string $S = $ bbacccbaa over an alphabet $\Sigma = \{$a, b, c, d$\}$, the set MAW(S) of all MAWs for S is $\{$aaa, bbb, cccc, d, ab, ca, bc, aac, acb, cbb, accb, cbac, bbaa$\}$. MAWs are combinatorial string objects, and their interesting mathematical properties have extensively been studied in the literature (see [1,7,16,17,19,23] and references therein). MAWs also enjoy several applications including phylogeny [11], data compression [3,15,18], musical information retrieval [14], and bioinformatics [2,12,22,24].

F. M. Nardini et al. (Eds.): SPIRE 2023, LNCS 14240, pp. 331–344, 2023.
https://doi.org/10.1007/978-3-031-43980-3_27

It is known that the number $|\mathsf{MAW}(S)|$ of MAWs for a string S of length n over an alphabet of size σ is $O(\sigma n)$ and that this bound is tight [17]. Crochremore et al. [17] gave an algorithm that computes $\mathsf{MAW}(S)$ in $O(\sigma n)$ time with $O(n)$ working space. Fujishige et al. [20] showed an improved algorithm for computing $\mathsf{MAW}(S)$ in optimal $O(n + |\mathsf{MAW}(S)|)$ time with $O(n)$ working space, for an input string S of length n over an integer alphabet of polynomial size in n. Both of the two aforementioned algorithms utilize an $O(n)$-size string data structure called the *(directed acyclic word graph) DAWG* [9], which recognizes the set of substrings of S, and can be built in $O(n \log \sigma)$ time for general ordered alphabets [9], and in $O(n)$ time for integer alphabets of polynomial size in n [20]. There also exist other efficient algorithms for computing MAWs with other string data structures such as suffix arrays and Burrows-Wheeler transforms [4,8].

The aim of this paper is to extend the notion of MAWs to a set $\mathcal{S} = \{S_1, \ldots, S_k\}$ of multiple k strings. We are aware of a few related attempts in earlier work: Chairungsee and Crochemore [11] introduced a string similarity measure based on the symmetric difference $\mathsf{MAW}(S_1) \triangle \mathsf{MAW}(S_2)$ of the sets of MAWs for two strings S_1 and S_2 to compare. They introduced a length threshold $\ell \geq 1$, and described an approach for computing $(\mathsf{MAW}(S_1) \triangle \mathsf{MAW}(S_2)) \cap \Sigma^\ell$ with the two following steps: First, the tries of size $O(n\ell)$ each representing the substrings of S_1 and S_2 of length up to ℓ are built, where $n = |S_1| + |S_2|$. Then, two tries each representing $\mathsf{MAW}(S_1) \cap \Sigma^\ell$ and $\mathsf{MAW}(S_2) \cap \Sigma^\ell$ are built, which require $O(n\sigma)$ space. Finally, the length-bounded symmetric difference $(\mathsf{MAW}(S_1) \triangle \mathsf{MAW}(S_2)) \cap \Sigma^\ell$ is computed from $\mathsf{MAW}(S_1) \cap \Sigma^\ell$ and $\mathsf{MAW}(S_2) \cap \Sigma^\ell$, but the authors did not explicitly describe how this computation is done in their method. Overall, their algorithm requires $\Omega(n(\ell + \sigma))$ time and space [11][1]. Charalampapaulose et al. [12] tackled the same problem of computing the symmetric difference $\mathsf{MAW}(S_1) \triangle \mathsf{MAW}(S_1)$ (without length threshold ℓ), and proposed a solution that requires $O(\sigma n)$ time and space. Their method firstly computes $\mathsf{MAW}(S_1)$ and $\mathsf{MAW}(S_2)$ separately, and then removes the elements that are in $\mathsf{MAW}(S_1) \cap \mathsf{MAW}(S_2)$. Charalampopoulos, Crochemore, and Pissis [13] presented how to count the number $|\mathsf{MAW}(S_1) \triangle \mathsf{MAW}(S_2)|$ of elements in the symmetric difference $\mathsf{MAW}(S_1) \triangle \mathsf{MAW}(S_2)$ in $O(n)$ time in the case of integer alphabets of polynomial size in n, by avoiding to list the elements explicitly.

Let $\mathcal{S} = \{S_1, \ldots, S_k\}$ be the input set of k strings, and $\mathbf{B} \in \{0,1\}^k$ be a given bit vector of length k. Our problem is to list (generalized) MAWs w for $\mathcal{S}$ and $\mathbf{B}$ such that $w \in \mathsf{MAW}(S_i)$ for every $\mathbf{B}[i] = 1$, and $w \notin \mathsf{MAW}(S_i)$ for every $\mathbf{B}[i] = 0$. For $k = 2$, the aforementioned problem of computing $\mathsf{MAW}(S_1) \triangle \mathsf{MAW}(S_2)$ is equivalent to solving our problem for $\mathbf{B} = 01$ and $\mathbf{B} = 10$. In Sect. 4 and Sect. 5, we deal with the case with $k = 2$, and present an algorithm running in $O(n + |\mathsf{M_B}|)$ time with $O(n)$ working space, where $\mathsf{M_B}$ denotes the set of (generalized) MAWs to output for a given bit vector $\mathbf{B}$ (Theorem 2). This immediately gives us an algorithm for listing the elements of the symmetric

---

[1] The claimed time bound for computing the trie is $O(n\sigma)$ (Theorem 1 of [11]). It seems that the authors regarded the length threshold ℓ as a constant.

difference $\mathsf{MAW}(S_1) \triangle \mathsf{MAW}(S_2)$ in optimal $O(n + |\mathsf{MAW}(S_1) \triangle \mathsf{MAW}(S_2)|)$ time (Corollary 1). In Sect. 6, we deal with the general case of $k > 2$, and extend our solution for $k = 2$ to the general case. Let n be the total length of the input k strings in $\mathcal{S}$. Our solution for general $k > 2$ works in $O(n\lceil k/\log n \rceil + |\mathbf{M_B}|)$ time with $O(n(k + \log n))$ *bits* of working space on the word RAM model with machine word size $\omega = \log n$. Thus, for $k = O(\log n)$, our algorithm runs in optimal $O(n + |\mathbf{M_B}|)$ time. All the bounds claimed in this paper are valid for linearly sortable alphabets, including integer alphabets of polynomial size in n.

As in the previous work [17,20,21], our key data structure is the DAWG for the input set $\mathcal{S}$ of strings. The best-known algorithm for constructing the DAWG for a set of strings of total length n takes $O(n \log \sigma)$ time [10], thus it can require $O(n \log n)$ time for large alphabets. We describe how the DAWG for a given set $\mathcal{S}$ of strings over an integer alphabet of polynomial size in n can be obtained in optimal $O(n)$ time (Theorem 1), which may be of independent interest.

2 Preliminaries

Strings. Let Σ be an ordered alphabet. An element of Σ is called a character. For characters $a, b \in \Sigma$, we write $a \prec b$ (or equivalently $b \succ a$) if a is lexicographically smaller than b. An element of Σ^* is called a string. The length of a string S is denoted by $|S|$. The empty string ε is the string of length 0. If $S = xyz$, then x, y, and z are called a *prefix*, *substring*, and *suffix* of S, respectively. They are called a *proper prefix*, *proper substring*, and *proper suffix* of S if $x \neq S$, $y \neq S$, and $z \neq S$, respectively. Let $\mathsf{Substr}(S)$ denote the set of substrings of string S. For any $1 \leq i \leq |S|$, the i-th character of S is denoted by $S[i]$. For any $1 \leq i \leq j \leq |S|$, $S[i..j]$ denotes the substring of S starting at i and ending at j. For convenience, let $S[i..j] = \varepsilon$ for $0 \leq j < i \leq |S| + 1$. We say that a string w *occurs* in a string S iff w is a substring of S. Note that by definition the empty string ε is a substring of any string S and hence ε always occurs in S.

For a set $\mathcal{S}$ of strings, let $\|\mathcal{S}\|$ denote the total length of the strings in $\mathcal{S}$, that is, $\|\mathcal{S}\| = \sum_{S \in \mathcal{S}} |S|$. Let $\mathsf{Substr}(\mathcal{S})$ denote the set of substrings of the strings in $\mathcal{S}$, that is, $\mathsf{Substr}(\mathcal{S}) = \left(\bigcup_{S \in \mathcal{S}} \{S[i..j] \mid 1 \leq i \leq j \leq |S|\} \right) \cup \{\varepsilon\}$.

Minimal Absent Words (MAWs). A string w is called an *absent word* for a string S if w does not occur in S. Let $\mathsf{AW}(S) = \Sigma^* \setminus \mathsf{Substr}(S)$ denote the set of absent words for a string S. An absent word $w \in \mathsf{AW}(S)$ for string S is called a *minimal absent word* or *MAW* for S if any proper substring of w occurs in S. We denote by $\mathsf{MAW}(S)$ the set of all MAWs for S. Let $\mathsf{nonMAW}(S) = \mathsf{AW}(S) \setminus \mathsf{MAW}(S)$ be the set of absent words for S which are not MAWs. Note that, for strings w and S, it holds that $w \notin \mathsf{MAW}(S)$ iff $w \in \mathsf{Substr}(S) \cup \mathsf{nonMAW}(S)$.

We extend the aforementioned notion of MAWs to a set $\mathcal{S} = \{S_1, \ldots, S_k\}$ of k strings for $k \geq 1$, as follows: Let $\mathbf{B}$ be a bit-vector of length k, and let $\mathcal{S}_{\mathbf{B}}$ be a subset of $\mathcal{S}$ such that $\mathcal{S}_{\mathbf{B}} = \{S_i \mid \mathbf{B}[i] = 1\}$. Let $\overline{\mathcal{S}_{\mathbf{B}}} = \{S_i \mid \mathbf{B}[i] = 0\} = \mathcal{S} \setminus \mathcal{S}_{\mathbf{B}}$. A string w is said to be a MAW for $\mathcal{S}_{\mathbf{B}}$ if (1) $w \in \bigcap_{S_i \in \mathcal{S}_{\mathbf{B}}} \mathsf{MAW}(S_i)$ and (2) $w \notin \bigcup_{S_i \in \overline{\mathcal{S}_{\mathbf{B}}}} \mathsf{MAW}(S_i)$. Condition (1) implies that w is a MAW for any string in $\mathcal{S}_{\mathbf{B}}$.

Condition (2) implies that w is *not* a MAW for any string in $\overline{\mathcal{S}_\mathbf{B}}$, which is equivalent to say that $w \in \bigcap_{S_i \in \overline{\mathcal{S}_\mathbf{B}}}(\mathsf{Substr}(S_i) \cup \mathrm{nonMAW}(S_i))$. Let $\mathrm{MAW}(\mathcal{S}_\mathbf{B})$ be the set of all MAWs for $\mathcal{S}_\mathbf{B}$. Here is some example: For string set $\mathcal{S} = \{\mathsf{abaab}, \mathsf{aacbba}\}$ over the alphabet $\Sigma = \{\mathsf{a}, \mathsf{b}, \mathsf{c}, \mathsf{d}\}$, $\mathrm{MAW}(\mathcal{S}_{10}) = \{\mathsf{aaba}, \mathsf{bab}, \mathsf{bb}, \mathsf{c}\}$, $\mathrm{MAW}(\mathcal{S}_{01}) = \{\mathsf{ab}, \mathsf{baa}, \mathsf{bac}, \mathsf{bbb}, \mathsf{bc}, \mathsf{ca}, \mathsf{cba}, \mathsf{cc}\}$, and $\mathrm{MAW}(\mathcal{S}_{11}) = \{\mathsf{aaa}, \mathsf{d}\}$.

The problem we consider in this paper is the following:

Problem 1 (MAWs for multiple input strings). *Given a set $\mathcal{S} = \{S_1, \ldots, S_k\}$ of k strings over an alphabet Σ and a bit vector $\mathbf{B}$ of length k, compute $\mathrm{MAW}(\mathcal{S}_\mathbf{B})$.*

3   The DAWG Data Structure

We use the *directed acyclic word graph (DAWG)* [9] data structure for a set $\mathcal{S} = \{S_1, \ldots, S_k\}$ of k strings, which is a DFA of size $O(\|\mathcal{S}\|)$ that recognizes all suffixes of the strings in $\mathcal{S}$.

To give a formal definition of $\mathsf{DAWG}(\mathcal{S})$, let $\mathsf{End\_Pos}_\mathcal{S}(w)$ denote the set of ending positions of all occurrences of a string w in the strings of $\mathcal{S}$, that is,

$$\mathsf{End\_Pos}_\mathcal{S}(w) = \{(i, j) \mid S_i[j - |w| + 1..j] = w, 1 \leq i \leq k, 1 \leq j \leq |S_i|\}.$$

We consider an equivalence relation $\equiv_\mathcal{S}$ of strings over Σ w.r.t. $\mathcal{S}$ such that, for any two strings w and u, $w \equiv_\mathcal{S} u$ iff $\mathsf{End\_Pos}_\mathcal{S}(w) = \mathsf{End\_Pos}_\mathcal{S}(u)$. For any string $x \in \Sigma^*$, let $[x]_\mathcal{S}$ denote the equivalence class for x w.r.t. $\equiv_\mathcal{S}$. All the non-substrings $x \notin \mathsf{Substr}(\mathcal{S})$ form a unique equivalence class, called the *degenerate* class.

Definition 1. *The DAWG of a set $\mathcal{S}$ of strings, denoted $\mathsf{DAWG}(\mathcal{S})$, is an edge-labeled DAG (V, E) such that*

$$V = \{[x]_\mathcal{S} \mid x \in \mathsf{Substr}(\mathcal{S})\},$$
$$E = \{([x]_\mathcal{S}, b, [xb]_\mathcal{S}) \mid x, xb \in \mathsf{Substr}(\mathcal{S}), b \in \Sigma\}.$$

We also define the set L of suffix links *of $\mathsf{DAWG}(\mathcal{S})$ by*

$$L = \{([ax]_\mathcal{S}, a, [x]_\mathcal{S}) \mid x, ax \in \mathsf{Substr}(\mathcal{S}), a \in \Sigma, [ax]_\mathcal{S} \neq [x]_\mathcal{S}\}.$$

Namely, two substrings x and y in $\mathsf{Substr}(\mathcal{S})$ are represented by the same node of $\mathsf{DAWG}(\mathcal{S})$ iff the ending positions of x and y in the strings of $\mathcal{S}$ are equal. Note that $\mathsf{DAWG}(\mathcal{S})$ does not contain the node for the degenerate class nor its in-coming edges. This is important for $\mathsf{DAWG}(\mathcal{S})$ to have a total linear number of edges [9], and for our linear-time algorithm for listing all the MAWs for a given query.

For convenience, assume that each string S_i in $\mathcal{S} = \{S_1, \ldots, S_k\}$ terminates with a unique end-marker $\#_i$ which does not occur elsewhere, where $\#_i \neq \#_j$ for $i \neq j$. Then $\mathsf{DAWG}(\mathcal{S})$ has exactly k sink nodes, each of which recognizes all the non-empty suffixes of S_i. For each $1 \leq i \leq k$, the sink that recognizes the suffixes of S_i is labeled by i.

The DAWG for a single string T is the DAWG for a singleton $\{T\}$ and is denoted by DAWG(T).

The state-of-the-art algorithm that builds DAWG($\mathcal{S}$) is Blumer et al 's online algorithm [9] which runs in $O(n \log \sigma)$ time with $O(n)$ space, where $n = \|\mathcal{S}\|$ is the total length of the strings in $\mathcal{S}$ and σ is the alphabet size. Below we describe a faster construction of DAWG($\mathcal{S}$) in the case of integer alphabets:

Theorem 1 (Linear-time DAWG construction for a set of strings). *For a given set $\mathcal{S} = \{S_1, \ldots, S_k\}$ of k strings of total length n over an integer alphabet Σ of polynomial size in n, one can build the edge-sorted DAWG($\mathcal{S}$) in $O(n)$ time and space.*

Proof. We first create a concatenated string $T = S_1 \cdots S_k$ of total length n from the strings in $\mathcal{S}$. We build DAWG(T) for the single string T in $O(n)$ time and space, using the algorithm of Fujishige et al. [20,21], where the out-going edges of every node are lexicographically sorted. Our goal is to convert $G_T = $ DAWG(T) to $G_{\mathcal{S}} = $ DAWG($\mathcal{S}$). For a set P of integer pairs and a pair (a, b) of integers, let $P \oplus (a, b) = \{(p + a, q + b) \mid (p, q) \in P\}$. Our key observation is that, for any substrings $w \in $ Substr($\mathcal{S}$) that *do not* contain separators $\#_i$ except for their last positions, it holds that

$$\text{End\_Pos}_{\mathcal{S}}(w)$$

$$= \text{End\_Pos}_{S_1}(w) \cup \left(\bigcup_{2 \le i \le k} \text{End\_Pos}_{S_i}(w) \oplus (i - 1, |S_1 \cdots S_{i-1}|) \right). \quad (1)$$

Equation (1) implies that the substrings w of $T = S_1 \cdots S_k$ which are also substrings of $\mathcal{S}$ are represented by essentially the same nodes in G_T and in $G_{\mathcal{S}}$, meaning that there is an injection from the nodes of $G_{\mathcal{S}}$ to the nodes of G_T.

What is left is how to remove the redundant nodes in G_T which represent the substrings y of T containing a separator $\#_i$ inside, which are thus not substrings of $\mathcal{S}$. Let us call the longest path of G_T that represents T as the *spine*. Since each $\#_i$ occurs exactly once in T, any substrings of T that contain $\#_i$ are represented by the spine of G_T. Thus, we can obtain $G_{\mathcal{S}}$ by removing the redundant nodes from the spine of G_T, but we ensure that for every i the suffixes of S_i ending with $\#_i$ are still represented in the graph. This can be achieved as follows: We process $i = k, \ldots, 2$ in decreasing order. We first split the spine into two parts each spelling out $S_1 \cdots S_{k-1}$ and S_k. We remove the nodes in the S_k part which are not reachable from the source of the modified graph, together with their out-going edges and suffix links. This gives us DAWG($\{S_1 \cdots S_{k-1}, S_k\}$). After processing $i = k$, we continue the same process for $i = k - 1$ with the remaining spine that spells out $S_1 \cdots S_{k-1}$. After processing $i = 2$, we obtain $G_{\mathcal{S}} = $ DAWG($\mathcal{S}$). See Fig. 1 for an example of our construction. It is trivial that all the redundant nodes can be removed in $O(n)$ time. $\square$

We remark that the order of concatenating the strings in $\mathcal{S}$ does not affect the correctness nor the complexity of our algorithm.

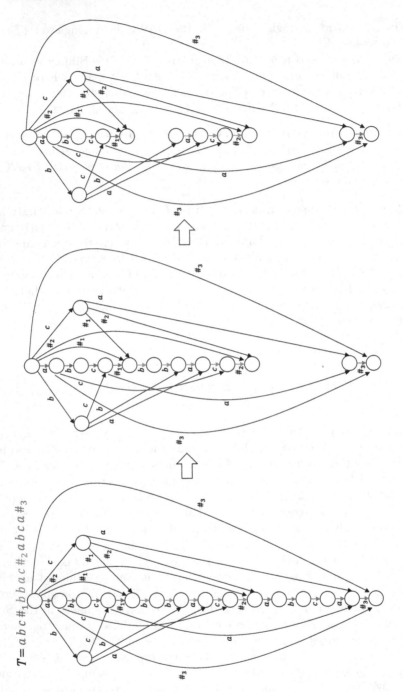

Fig. 1. Illustration for our linear-time construction of $\mathrm{DAWG}(\mathcal{S})$ for a set $\mathcal{S} = \{abc\#_1, bbac\#_2, abca\#_3\}$ of strings. We first build $\mathrm{DAWG}(T)$ for the concatenated string $T = abc\#_1 bbac\#_2 abca\#_3$. Then, we remove the redundant nodes in the spine of the DAWG for $i = 3$ and then for $i = 2$. This gives us $\mathrm{DAWG}(\mathcal{S})$.

4   Algorithm Overview for $k = 2$

In what follows, we consider the case where our input set S consists of two strings S_1 and S_2 which respectively terminate with special characters $\#_1$ and $\#_2$. We show how, given a bit vector $\mathbf{B} \in \{00, 01, 10, 11\}$ of length 2, we can compute $\mathsf{MAW}(S_{\mathbf{B}})$ in $O(n + |\mathsf{MAW}(S_{\mathbf{B}})|)$ time and $O(n)$ working space, where $n = \|S\|$.

We first build the edge-sorted $\mathsf{DAWG}(S)$ for a given $S = \{S_1, S_2\}$ in $O(n)$ time and space with Theorem 1. We label each node v of $\mathsf{DAWG}(S)$ by $\#_i$ iff v represents a substring of S_i ($1 \leq i \leq 2$). Let $\mathsf{label}(v) \in \{\#_1, \#_2, \#_1\#_2\}$ denote the label of node v. The labels of all nodes can be precomputed in $O(n)$ time.

Our algorithm is based on Fujishige et al.'s algorithm [20, 21] for computing all the MAWs in the case of a single input string. As such, for each node x of $\mathsf{DAWG}(S)$ we focus on the *shortest* string represented by x and denote it by au, where $a \in \Sigma$ and $u \in \Sigma^*$. We use the suffix link of the node x and its target node y whose *longest* member is u (namely, the first letter a of au is removed by following the suffix link from x to y). For ease of explanation, we identify the node x with the string au, and the node y with the string u.

Fujishige et al.'s algorithm compares the out-going edges of au and those of u one by one in the sorted order. Suppose au has an out-going edge labeled b. If u *does not* have an out-going edge labeled b, then their algorithm outputs aub as a MAW for the input string. Otherwise, it outputs nothing, and the cost is charged to the out-going edge of au labeled b. Each MAW aub in the output is encoded by a tuple (a, i, j) such that $w[i..j] = ub$, thus taking $O(1)$ space. This is how Fujishige et al.'s algorithm works in $O(n + |\mathsf{MAW}(S)|)$ time and with $O(n)$ working space for a single string S.

However, in our case of multiple strings, depending on the label of nodes au, aub and ub, and depending on the value of the given bit vector $\mathbf{B}$, there may exist some edge comparisons that cannot be charged either to the output MAWs or to the out-going edges of node au. It is also possible that even if there is a node representing aub in $\mathsf{DAWG}(S)$, still aub is a MAW for some string(s) in S. To overcome these difficulties, we introduce *skip links* that permit us to avoid unwanted edge character comparisons.

5   Skip Links for $k = 2$

We use the same conventions for the nodes au, aub and u on $\mathsf{DAWG}(S)$ as in the previous section, and also consider the node ub. We have three possible cases for the label of node au, where $\mathsf{label}(au) = \#_1\#_2$, $\mathsf{label}(au) = \#_1$, or $\mathsf{label}(au) = \#_2$. In each of the three cases, there are some sub-cases for the labels of node aub and node ub. By inspection, we obtain all the possible cases that need to be considered, as shown in Fig. 2.

When $\mathbf{B} = 00$, then since $\mathsf{MAW}(S_{00}) = \Sigma^* \setminus (\mathsf{MAW}(S_1) \cup \mathsf{MAW}(S_2))$, there are no MAWs to output. In what follows, we describe our solutions to the cases with $\mathbf{B} \in \{10, 11\}$. We remark that the case with $\mathbf{B} = 01$ is symmetric to the case with $\mathbf{B} = 10$.

5.1   When B = 10

There are four cases in which we output aub as a MAW for $\mathsf{MAW}(\mathcal{S}_{10})$ (see the table on the left of Fig. 2):

		au		
		$\#_1\#_2$	$\#_1$	$\#_2$
aub	ub	**B**		
$\#_1\#_2$	$\#_1\#_2$	00	-	-
$\#_1$	$\#_1$	00	00	-
	$\#_1\#_2$	01	00	-
$\#_2$	$\#_2$	00	-	00
	$\#_1\#_2$	10	-	00
absent	$\#_1$	10	10	00
	$\#_2$	01	00	01
	$\#_1\#_2$	11	10	01

Fig. 2. Left: All possible cases of the labels of the nodes au, aub, and ub, and their corresponding bit vectors **B**. "absent" refers to the case where there is no out-going edge labeled b from node au. The cells with "-" refer to impossible combinations of node labels. Middle: Illustration for $\mathsf{DAWG}(\mathcal{S})$ which shows the case where au is labeled $\#_1\#_2$, aub is labeled $\#_1$, and ub is labeled $\#_1\#_2$. In this case aub is a MAW in $\mathsf{MAW}(\mathcal{S}_\mathbf{B})$ with **B** = 01 (see the left table). Right: The regions corresponding to the bit vectors $\mathbf{B} \in \{00, 01, 10, 11\}$.

(1) $\mathsf{label}(au) = \#_1\#_2$, $\mathsf{label}(aub) = \#_2$, and $\mathsf{label}(ub) = \#_1\#_2$;
(2) $\mathsf{label}(au) = \#_1\#_2$, $aub \in \mathsf{AW}(\mathcal{S})$, and $\mathsf{label}(ub) = \#_1$;
(3) $\mathsf{label}(au) = \#_1$, $aub \in \mathsf{AW}(\mathcal{S})$, and $\mathsf{label}(ub) = \#_1$;
(4) $\mathsf{label}(au) = \#_1$, $aub \in \mathsf{AW}(\mathcal{S})$, and $\mathsf{label}(ub) = \#_1\#_2$.

When $\mathsf{label}(au) = \#_1\#_2$. We create skip links that simultaneously manage Cases (1) and (2), both having $\mathsf{label}(au) = \#_1\#_2$. We create a selected list $\mathsf{schar}(u)$ of out-going edge labels of node u such that $\mathsf{schar}(u) = \{b \mid \mathsf{label}(ub) = \#_1\}$, where the elements are lexicographically sorted. Let $\mathsf{char}(au)$ be the sorted list of all out-going edge labels of node au. For any list L of characters and any character $c \in \Sigma$, let $\mathsf{succ}(c, L)$ denote the lexicographical successor of c in L. Our algorithm for $\mathbf{B} = 10$ and $\mathsf{label}(au) = \#_1\#_2$ is described in Algorithm 1.

When $\mathsf{label}(au) = \#_1$. We create skip links that simultaneously manage Cases (3) and (4), both having $\mathsf{label}(au) = \#_1$. We create another selected list $\mathsf{schar}'(u)$ of out-going edge labels of node u such that $\mathsf{schar}'(u) = \{b \mid \mathsf{label}(ub) \in \{\#_1, \#_1\#_2\}\}$, where the elements are lexicographically sorted. We use the same $\mathsf{char}(au)$ in the previous case. Our algorithm for $\mathbf{B} = 10$ and $\mathsf{label}(au) = \#_1$ is described in Algorithm 2.

Lemma 1 (Linear-time MAW computation for B = 10). *Given* $\mathbf{B} = 10$, *one can compute* $\mathsf{MAW}(\mathcal{S}_{10})$ *in* $O(n + |\mathsf{MAW}(\mathcal{S}_{10})|)$ *time and* $O(n)$ *working space for integer alphabets of polynomial size in* $n = \|\mathcal{S}\|$.

Algorithm 1: Algorithm for $\mathbf{B} = 10$ and $\mathsf{label}(au) = \#_1\#_2$

Input: A node au of $\mathsf{DAWG}(\mathcal{S})$ such that $\mathsf{label}(au) = \#_1\#_2$, $\mathbf{B} = 10$.
Output: A subset M of MAWs aub with $b \subset \Sigma$.

1   $M \leftarrow \emptyset$;
2   $U \leftarrow \mathsf{char}(au) \cup \{\$_U\}$;   /* $\$_U$ is lex. largest in U */
3   $L \leftarrow \mathsf{schar}(u) \cup \{\$_L\}$;   /* $\$_L$ is lex. largest in L and $\$_L \prec \$_u$ */
4   $\hat{b} \leftarrow U[1]; b \leftarrow L[1]$;   /* start with lex. smallest characters */
5   **while** $b \neq \$_L$ **do**
6   **if** $\hat{b} = b$ **then**
7   **if** $\mathsf{label}(aub) = \#_2$ **and** $\mathsf{label}(ub) = \#_1\#_2$ **then**
8            $\lfloor \; M \leftarrow M \cup \{aub\}$;   /* output aub */
9         $\hat{b} \leftarrow \mathsf{succ}(\hat{b}, U)$;   /* move to the next character in U */
10        $b \leftarrow \mathsf{succ}(b, L)$;   /* move to the next character in L */
11   **else if** $\hat{b} \succ b$ **then**
12         $M \leftarrow M \cup \{aub\}$;   /* output aub */
13         $b \leftarrow \mathsf{succ}(b, L)$;   /* move to the next character in L */
14 **return** M;

Proof. We run Algorithm 1 and Algorithm 2 for every node au of $\mathsf{DAWG}(\mathcal{S})$.

In the preprocessing phase, we build the edge-sorted $\mathsf{DAWG}(\mathcal{S})$ in $O(\|\mathcal{S}\|)$ time and space by Theorem 1. Since the out-going edges of every node are sorted, we can easily compute the sorted lists $\mathsf{char}(au)$, $\mathsf{schar}(u)$, $\mathsf{schar}'(u)$, and $\mathsf{schar}''(u)$ for all nodes in $O(n)$ total time.

Let us consider the complexity of the scanning phase of Algorithm 1. Each edge-label comparison that falls into "$\hat{b} = b$" in line 6 of Algorithm 1 is associated either to the reported MAW aub if $\mathsf{label}(aub) = \#_2$ and $\mathsf{label}(ub) = \#_1\#_2$ (in line 7 and line 8), or to the out-going edge of node au labeled b otherwise. Each edge-label comparison that falls into "$\hat{b} \succ b$" in line 11 is associated to the reported MAW aub in line 12. This ensures the desired time complexity for Algorithm 1. The complexity for Algorithm 2 is similar to show.

The correctness of Algorithm 1 and Algorithm 2 is immediate from the tables in Fig. 2 and 3. $\square$

5.2   When $\mathbf{B} = 11$

There is a single case in which we output aub as a MAW for $\mathsf{MAW}(\mathcal{S}_{11})$ (see Fig. 2): $\mathsf{label}(au) = \#_1\#_2$, $aub \in \mathsf{AW}(\mathcal{S})$, and $\mathsf{label}(ub) = \#_1\#_2$.

Unwanted comparisons can occur here if $aub \in \mathsf{AW}(\mathcal{S})$, and $\mathsf{label}(ub) = \#_1$ or $\mathsf{label}(ub) = \#_2$. To avoid such comparisons, we consider another carefully selected list $\mathsf{schar}''(u)$ of out-going edge labels of node u such that $\mathsf{schar}''(u) = \{b \mid \mathsf{label}(ub) = \#_1\#_2\}$, where the elements are lexicographically sorted. We can use the same $\mathsf{char}(au)$ in the previous subsection.

We can modify Algorithm 2 for $\mathbf{B} = 01$ with $\mathsf{label}(au) = \#_1$ so that the modified algorithm computes MAWs for $\mathbf{B} = 11$, only by using $\mathsf{schar}''(u)$ in place of $\mathsf{schar}'(u)$. This leads us to the following lemma:

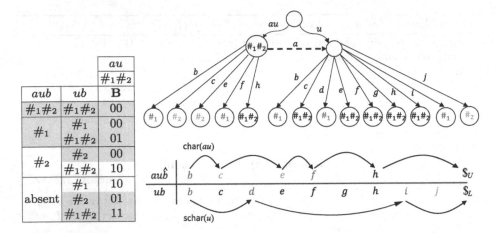

Fig. 3. Illustration for our algorithm for $\mathbf{B} = 10$ and label$(au) = \#_1\#_2$. The white cells in the table show the cases where we output elements of MAW$(\mathcal{S}_{10})$. We compare the labels of the selected out-going edges of node au and u which are connected by the skip links, in sorted order. In this diagram, aud and aui are output in line 12 and auc and aue are output in line 12 of Algorithm 1 as elements of MAW$(\mathcal{S}_{10})$.

Lemma 2 (Linear-time MAW computation for B = 11). *Given* $\mathbf{B} = 11$, *one can compute* MAW$(\mathcal{S}_{11})$ *in* $O(n + |\text{MAW}(\mathcal{S}_{11})|)$ *time and* $O(n)$ *working space for integer alphabets of polynomial size in* $n = \|\mathcal{S}\|$.

5.3 Our Main Result for $k = 2$

Finally we obtain the main result for a case of two strings with $k = 2$.

Theorem 2 (Linear-time MAW computation for a set of two strings). *Given a set* $\mathcal{S} = \{S_1, S_2\}$ *of two strings of total length* n *and a bit vector* $\mathbf{B} \in \{01, 10, 11\}$, *one can compute* MAW$(\mathcal{S}_{\mathbf{B}})$ *in* $O(n + |\text{MAW}(\mathcal{S}_{\mathbf{B}})|)$ *time and* $O(n)$ *working space for integer alphabets of polynomial size in* n.

The following corollary is immediate from Theorem 2.

Corollary 1. *Given a set* $\mathcal{S} = \{S_1, S_2\}$ *of two strings of total length* n, *one can compute* MAW$(S_1) \cap$ MAW(S_2), MAW$(S_1) \cup$ MAW(S_2), *and* MAW$(S_1) \triangle$ MAW(S_2) *in* $O(n + |\text{MAW}(S_1) \cap \text{MAW}(S_2)|)$ *time,* $O(n + |\text{MAW}(S_1) \cup \text{MAW}(S_2)|)$ *time, and* $O(n + |\text{MAW}(S_1) \triangle \text{MAW}(S_2)|)$ *time, respectively, using* $O(n)$ *working space, for integer alphabets of polynomial size in* n.

6 Algorithm for Arbitrary $k > 2$

In this section, we present our algorithm for computing MAW$(\mathcal{S}_{\mathbf{B}})$ in case where $\mathcal{S} = \{S_1, \ldots, S_k\}$ contains $k > 2$ strings.

Algorithm 2: Algorithm for $\mathbf{B} = 10$ and $\mathsf{label}(au) = \#_1$

Input: A node au of $\mathsf{DAWG}(\mathcal{S})$ such that $\mathsf{label}(au) = \#_1$, $\mathbf{B} = 10$.
Output: A subset M of MAWs aub with $b \in \Sigma$.

1 $M \leftarrow \emptyset$;
2 $U \leftarrow \mathsf{char}(au) \cup \{\$_U\}$;   /* $\$_U$ is lex. largest in U */
3 $L \leftarrow \mathsf{schar}'(u) \cup \{\$_L\}$;   /* $\$_L$ is lex. largest in L and $\$_L \prec \$_u$ */
4 $\hat{b} \leftarrow U[1]; b \leftarrow L[1]$;   /* start with lex. smallest characters */
5 while $b \neq \$_L$ do
6   if $\hat{b} = b$ then
7         $\hat{b} \leftarrow \mathsf{succ}(\hat{b}, U)$;   /* move to the next character in U */
8         $b \leftarrow \mathsf{succ}(b, L)$;   /* move to the next character in L */
9   else if $\hat{b} \succ b$ then
10        $M \leftarrow M \cup \{aub\}$;   /* output aub */
11        $b \leftarrow \mathsf{succ}(b, L)$;   /* move to the next character in L */
12 return M;

Let $\mathbf{B} \in \{0, 1\}^k \setminus \{0^k\}$ be an input bit vector of length $k > 2$. We redefine the labels of the nodes of $\mathsf{DAWG}(\mathcal{S})$ such that $\mathsf{label}(v)[i] = 1$ iff v is a substring of S_i for $1 \leq i \leq k$. Namely, $\mathsf{label}(v)$ is now also a bit vector of length k.

Let $aub \in \Sigma^*$ ($a, b \in \Sigma$ and $u \in \Sigma^*$) be a candidate of an element of $\mathsf{MAW}(\mathcal{S}_\mathbf{B})$ as in the previous sections, where the suffix link of node au points to node u and node u has an out-going edge labeled b. Then, it follows from the definition of $\mathsf{MAW}(\mathcal{S}_\mathbf{B})$ that $aub \in \mathsf{MAW}(\mathcal{S}_\mathbf{B})$ iff

(A) $\mathsf{label}(aub)[i] = 0$, $\mathsf{label}(au)[i] = 1$, and $\mathsf{label}(ub)[i] = 1$ (i.e. $aub \in \mathsf{MAW}(S_i)$), or

(A') au has no out-going edge labeled b, $\mathsf{label}(au)[i] = 1$, and $\mathsf{label}(ub)[i] = 1$ (i.e. $aub \in \mathsf{MAW}(S_i)$)

for all $1 \leq i \leq k$ with $\mathbf{B}[i] = 1$, and

(B) $\mathsf{label}(aub)[i] = 1$ (i.e. $aub \in \mathsf{Substr}(S_i)$), or
(C) $\mathsf{label}(aub)[i] = 0$, and $\mathsf{label}(au)[i] = 0$ or $\mathsf{label}(ub)[i] = 0$ (i.e. $aub \in \mathsf{nonMAW}(S_i)$), or

(C') au has no out-going edge labeled b, and $\mathsf{label}(au)[i] = 0$ or $\mathsf{label}(ub)[i] = 0$ (i.e. $aub \in \mathsf{nonMAW}(S_i)$)

for all $1 \leq i \leq k$ with $\mathbf{B}[i] = 0$.

For each node au in $\mathsf{DAWG}(\mathcal{S})$ whose suffix link points to node u, we create a united single skip link $\mathsf{schar}(ub)$ for the children ub of node u such that $b \in \mathsf{schar}(ub)$ iff $\mathsf{label}(ub)[i] = 1$ for every i with $\mathbf{B}[i] = 1$.

After the above preprocessing is finished, we proceed to the scanning phase of our algorithm. For each node au, we scan the skip links $\mathsf{char}(aub)$ and $\mathsf{schar}(ub)$ in parallel, analogously to the case with $k = 2$. Let $\hat{b} \in \mathsf{char}(aub)$ and $b \in \mathsf{schar}(ub)$. Our algorithm compares these characters in sorted order while keeping the invariant $\hat{b} \succeq b$ as in the case with $k = 2$.

When the comparison falls into the case "$\hat{b} = b$", then we output aub as an element of MAW($\mathcal{S}_\mathbf{B}$) if Case (A) is satisfied and if Case (B) or Case (C) is satisfied. When the comparison falls into the case "$\hat{b} \succ b$", then we output aub as an element of MAW($\mathcal{S}_\mathbf{B}$) if Cases (A') and (C') are both satisfied.

This already gives us an $O(nk)$-time algorithm for computing MAW($\mathcal{S}_\mathbf{B}$) using $O(n(k+\log n))$ bits of working space, or alternatively $O(n\lceil k/\log n\rceil)$ words of working space in the word RAM model with machine word size $\omega = \log n$.

We can speed up checking Cases (A), (B), (C) for each node au by using bit masks of size $\omega = \log n$ each stored at nodes aub, au, and ub, from $O(k)$ time to $O(\lceil k/\log n\rceil)$ time. For Cases (A') and (C'), it suffices for us to use only the bit masks stored at nodes au and ub, since node aub does not exist in these cases and we detect this as a result of "$\hat{b} \succ b$" comparison.

Theorem 3 (Efficient MAW computation for a set of k strings). *Given a set $\mathcal{S} = \{S_1, \ldots, S_k\}$ of k strings of total length n and a bit vector $\mathbf{B} \in \{0,1\}^k \setminus \{0^k\}$, one can compute MAW($\mathcal{S}_\mathbf{B}$) in $O(n\lceil k/\log n\rceil + |\text{MAW}(\mathcal{S}_\mathbf{B})|)$ time and $O(n(k + \log n))$ bits of working space (or alternatively $O(n\lceil k/\log n\rceil)$ words of working space), for integer alphabets of polynomial size in n.*

7   Discussions

Béal et al. [6] considered a different version of MAWs MAW'($\mathcal{S}$) for a set $\mathcal{S}$ of k strings, where a string $w = aub$ is a MAW for $\mathcal{S} = \{S_1, \ldots, S_k\}$ if $aub \notin$ Substr($\mathcal{S}$), $au \in$ Substr(S_i) and $ub \in$ Substr(S_j) for some $1 \le i, j \le k$. They gave an $O(\sigma n)$-time and space solution for computing MAW'($\mathcal{S}$). This version of MAWs can be computed in optimal $O(n + |\text{MAW}'(\mathcal{S})|)$ time, independently of k, by running our algorithm without skip links. Ayad et al. [3] considered the problem of computing the same version of MAWs of length up to $\ell > 1$.

Independently to our work, the recent work by Béal and Crochemore [5] considered the following problem: Let $\mathcal{T}$ and $\mathcal{R}$ be sets of strings, where $\mathcal{T}$ is called a target and $\mathcal{R}$ is called a reference. A $\mathcal{T}$-specific string with respect to $\mathcal{R}$ is a string u such that $u \in$ Substr($\mathcal{T}$), $u \notin$ Substr($\mathcal{R}$), $v \in$ Substr($\mathcal{R}$) for any proper substring v of u. By definition, a string u is a $\mathcal{T}$-specific string with respect to $\mathcal{R}$ if and only if $u \in$ MAW($\mathcal{R}$) $\cap$ Substr($\mathcal{T}$). Béal and Crochemore [5] showed an algorithm for finding all $\mathcal{T}$-specific strings w.r.t. $\mathcal{R}$ in $O(n\sigma)$-time and $O(n)$ space, where n is the total length of the strings in $\mathcal{T}$ and $\mathcal{R}$, assuming that the edges of the DAWG are represented by transition matrices (Proposition 2, [5]). Their algorithm also uses the DAWG built on $\mathcal{T}$ and $\mathcal{R}$ and marks its nodes in an appropriate way (Proposition 1, [5]). This marking technique is very similar to our skip links from Sect. 5 for the case of $k = 2$, and thus our algorithm can be extended to solve this problem in $O(n)$ time and space for integer alphabets.

Acknowledgments. This work was supported by JSPS KAKENHI Grant Numbers JP23H04381 (TM), JP21K17705, JP23H04386 (YN), JP22H03551 (SI), JP20H04141 (HB).

References

1. Akagi, T., et al.: Combinatorics of minimal absent words for a sliding window Theor. Comput. Sci. **927**, 109–119 (2022). https://doi.org/10.1016/j.tcs.2022.06.002
2. Almirantis, Y., et al.: On avoided words, absent words, and their application to biological sequence analysis. Algorithms Mol. Biol. **12**(1), 5 (2017)
3. Ayad, L.A.K., Badkobeh, G., Fici, G., Héliou, A., Pissis, S.P.: Constructing anti-dictionaries of long texts in output-sensitive space. Theory Comput. Syst. **65**(5), 777–797 (2021)
4. Barton, C., Heliou, A., Mouchard, L., Pissis, S.P.: Linear-time computation of minimal absent words using suffix array. BMC Bioinform. **15**(1), 388 (2014)
5. Béal, M., Crochemore, M.: Fast detection of specific fragments against a set of sequences. In: Drewes, F., Volkov, M. (eds.) Developments in Language Theory. DLT 2023. LNCS, vol. 13911, pp. 51–60. Springer, Cham (2023). https://doi.org/10.1007/978-3-031-33264-7_5
6. Béal, M., Crochemore, M., Mignosi, F., Restivo, A., Sciortino, M.: Computing forbidden words of regular languages. Fundam. Inform. **56**(1–2), 121–135 (2003)
7. Béal, M.-P., Mignosi, F., Restivo, A.: Minimal forbidden words and symbolic dynamics. In: Puech, C., Reischuk, R. (eds.) STACS 1996. LNCS, vol. 1046, pp. 555–566. Springer, Heidelberg (1996). https://doi.org/10.1007/3-540-60922-9_45
8. Belazzougui, D., Cunial, F., Kärkkäinen, J., Mäkinen, V.: Versatile Succinct Representations of the Bidirectional Burrows-Wheeler Transform. In: Bodlaender, H.L., Italiano, G.F. (eds.) ESA 2013. LNCS, vol. 8125, pp. 133–144. Springer, Heidelberg (2013). https://doi.org/10.1007/978-3-642-40450-4_12
9. Blumer, A., Blumer, J., Haussler, D., Ehrenfeucht, A., Chen, M.T., Seiferas, J.I.: The smallest automaton recognizing the subwords of a text. Theor. Comput. Sci. **40**, 31–55 (1985)
10. Blumer, A., Blumer, J., Haussler, D., McConnell, R., Ehrenfeucht, A.: Complete inverted files for efficient text retrieval and analysis. J. ACM **34**(3), 578–595 (1987). https://doi.org/10.1145/28869.28873
11. Chairungsee, S., Crochemore, M.: Using minimal absent words to build phylogeny. Theor. Comput. Sci. **450**, 109–116 (2012)
12. Charalampopoulos, P., Crochemore, M., Fici, G., Mercaş, R., Pissis, S.P.: Alignment-free sequence comparison using absent words. Inf. Comput. **262**, 57–68 (2018)
13. Charalampopoulos, P., Crochemore, M., Pissis, S.P.: On extended special factors of a word. In: Gagie, T., Moffat, A., Navarro, G., Cuadros-Vargas, E. (eds.) SPIRE 2018. LNCS, vol. 11147, pp. 131–138. Springer, Cham (2018). https://doi.org/10.1007/978-3-030-00479-8_11
14. Crawford, T., Badkobeh, G., Lewis, D.: Searching page-images of early music scanned with OMR: a scalable solution using minimal absent words. In: ISMIR 2018, pp. 233–239 (2018)
15. Crochemore, M., Mignosi, F., Restivo, A., Salemi, S.: Data compression using antidictionaries. Proc. IEEE **88**(11), 1756–1768 (2000)
16. Crochemore, M., Héliou, A., Kucherov, G., Mouchard, L., Pissis, S.P., Ramusat, Y.: Absent words in a sliding window with applications. Inf. Comput. **270**, 104461 (2020)
17. Crochemore, M., Mignosi, F., Restivo, A.: Automata and forbidden words. Inf. Process. Lett. **67**(3), 111–117 (1998)

18. Crochemore, M., Navarro, G.: Improved antidictionary based compression. In: 12th International Conference of the Chilean Computer Science Society, 2002. Proceedings, pp. 7–13. IEEE (2002)
19. Fici, G.: Minimal forbidden words and applications. Ph.D. thesis, Università di Palermo and Université Paris-Est Marne-la-Vallée (2006)
20. Fujishige, Y., Tsujimaru, Y., Inenaga, S., Bannai, H., Takeda, M.: Computing DAWGs and minimal absent words in linear time for integer alphabets. In: MFCS 2016, vol. 58, pp. 38:1–38:14 (2016)
21. Fujishige, Y., Tsujimaru, Y., Inenaga, S., Bannai, H., Takeda, M.: Linear-time computation of DAWGs, symmetric indexing structures, and MAWs for integer alphabets. Theor. Comput. Sci. (2023, to appear)
22. Koulouras, G., Frith, M.C.: Significant non-existence of sequences in genomes and proteomes. Nucleic Acids Res. 49(6), 3139–3155 (2021)
23. Mieno, T., et al.: Minimal unique substrings and minimal absent words in a sliding window. In: Chatzigeorgiou, A., et al. (eds.) SOFSEM 2020. LNCS, vol. 12011, pp. 148–160. Springer, Cham (2020). https://doi.org/10.1007/978-3-030-38919-2_13
24. Pratas, D., Silva, J.M.: Persistent minimal sequences of SARS-CoV-2. Bioinformatics 36(21), 5129–5132 (2020)

Frequency-Constrained Substring Complexity

Solon P. Pissis[1,2]([✉]), Michael Shekelyan[3], Chang Liu[4], and Grigorios Loukides[5]

[1] CWI, Amsterdam, The Netherlands
solon.pissis@cwi.nl
[2] Vrije Universiteit, Amsterdam, The Netherlands
[3] Queen Mary University of London, London, UK
m.shekelyan@qmul.ac.uk
[4] Zhejiang University, Medical Center, Zhejiang, China
0623541@zju.edu.cn
[5] King's College London, London, UK
grigorios.loukides@kcl.ac.uk

Abstract. We introduce the notion of frequency-constrained substring complexity. For any finite string, it counts the distinct substrings of the string per length *and* frequency class. For a string x of length n and a partition of $[n]$ in τ intervals, $\mathcal{I} = I_1, \ldots, I_\tau$, the *frequency-constrained substring complexity* of x is the function $f_{x,\mathcal{I}}(i,j)$ that maps i, j to the number of distinct substrings of length i of x occurring at least α_j and at most β_j times in x, where $I_j = [\alpha_j, \beta_j]$. We extend this notion as follows. For a string x, a dictionary $\mathcal{D}$ of d strings (documents), and a partition of $[d]$ in τ intervals $I_1, \ldots, I_\tau$, we define a 2D array $S = S[1 .. |x|, 1 .. \tau]$ as follows: $S[i,j]$ is the number of distinct substrings of length i of x occurring in at least α_j and at most β_j documents, where $I_j = [\alpha_j, \beta_j]$. Array S can thus be seen as the distribution of the substring complexity of x into τ document frequency classes. We show that after a *linear-time* preprocessing of $\mathcal{D}$, for any x and any partition of $[d]$ in τ intervals given online, array S can be computed in near-optimal $\mathcal{O}(|x|\tau \log \log d)$ time.

Keywords: Substring complexity · Suffix tree · Predecessor search

1 Introduction

The *substring complexity* or *subword complexity* of an infinite string x is the function that maps i to the number of distinct substrings (subwords) of length i in x. Substring complexity is one of the main topics in combinatorics on words [22]. The ultimate goal is to find explicit formulas for (or estimates of) the number of distinct fragments of length i occurring in a given infinite string [9,17]. Substring complexity in finite strings plays also a crucial role in data compression [21]; it underlies a promising compressibility measure for repetitive sequences [13,14].

We introduce the notion of frequency-constrained substring complexity of finite strings. For any finite string, it counts the distinct substrings of the string

F. M. Nardini et al. (Eds.): SPIRE 2023, LNCS 14240, pp. 345–352, 2023.
https://doi.org/10.1007/978-3-031-43980-3_28

per length *and* frequency class. For a string x of length n and a partition of $[n]^1$ in τ intervals $\mathcal{I} = I_1, \ldots, I_\tau$, the *frequency-constrained substring complexity* of x is the function $f_{x,\mathcal{I}}(i,j)$ that maps i,j to the number of distinct substrings of length i of x occurring at least α_j and at most β_j times in x, where $I_j = [\alpha_j, \beta_j]$. We extend this notion as follows. For a string x, a dictionary $\mathcal{D}$ of d strings (documents) and a partition of $[d]$ in τ intervals $\mathcal{I} = I_1, \ldots, I_\tau$, the function $f_{x,\mathcal{D},\mathcal{I}}(i,j)$ maps i,j to the number of distinct substrings of length i of x occurring in at least α_j and at most β_j documents in $\mathcal{D}$, where $I_j = [\alpha_j, \beta_j]$. In fact, computing $f_{x,\mathcal{D},\mathcal{I}}$ efficiently is the main problem we consider in this paper.

The frequency-constrained substring complexity of x is very descriptive as it provides subtle information about the substrings of x. It can thus help us tune string processing algorithms by setting bounds on the substrings length or on frequency; for example, when $\tau = 2$, the substrings of x are classified into frequent and infrequent [19]. We can also tune the output size of a document retrieval algorithm [20], the term's length used by a `tf-idf` algorithm [15], or the seed length used by seed-and-extend sequence alignment algorithms [5,16].

Example 1. Let $\mathcal{D} = \{$a, ananan, baba, ban, banna, nana$\}$. For $x = $ banana and $I_1 = [1,2], I_2 = [3,4], I_3 = [5,6]$, we have $f_{x,\mathcal{D},\mathcal{I}}(2,2) = 3$: ba occurs in $3 \in I_2$ documents; an occurs in $4 \in I_2$ documents; and na occurs in $3 \in I_2$ documents.

Our Contribution.   Let S be a 2D array such that $S[i,j] = f_{x,\mathcal{D},\mathcal{I}}(i,j)$. We show that after a *linear-time* preprocessing of $\mathcal{D}$, for any x and any partition $\mathcal{I}$ of $[d]$ in τ intervals given online, array S can be computed in near-optimal $\mathcal{O}(|x|\tau \log \log d)$ time. Since array S is of size $|x| \times \tau$, our data structure is nearly-optimal with respect to the preprocessing and query times. The main ingredients of our data structure are suffix trees [4,7,23] and predecessor search [6,18].

2   The Data Structure

Let us denote by $\mathcal{D} = \{y_1, \ldots, y_d\}$ the input dictionary consisting of $d = |\mathcal{D}|$ strings (documents). We assume that all strings in $\mathcal{D}$ are over an integer alphabet Σ of size $\sigma \leq ||\mathcal{D}||^{\mathcal{O}(1)}$, where $||\mathcal{D}||$ is the total length of all the strings in $\mathcal{D}$.

Let us denote by $y = y_1\$_1 \ldots y_d\$_d$ the concatenation of the d documents in $\mathcal{D}$ in some arbitrary but fixed order; the $\$_i$ letters, $i \in [1,d]$, are unique letters not from Σ. We construct the suffix tree $\mathsf{ST}(y)$ of y (with suffix links) in linear time [7]. We implement $\mathcal{O}(1)$-time transitions in the suffix tree in linear time using perfect hashing [10]. For any string w, we define its *document frequency* in $\mathcal{D}$ as the number of distinct documents in $\mathcal{D}$ in which w has at least one occurrence. We decorate each node u of $\mathsf{ST}(y)$ with the document frequency of the string spelled from the root of $\mathsf{ST}(y)$ to u. This is done in linear time [12].

Upon a query string x, we construct the suffix tree $\mathsf{ST}(x)$ of x in $\mathcal{O}(|x|)$ time [7]: if any letter of x is not in $\mathcal{D}$, which is checked using $\mathsf{ST}(y)$, we replace it with a unique letter not in Σ, and hash the letters of x into the range $[0, |x|]$ [10].

---

[1] By the notation $[u]$ we denote $\{1, 2, \ldots, u\}$.

We first show how to compute, for each node u of $\mathsf{ST}(x)$, the document frequency of the string spelled from the root of $\mathsf{ST}(x)$ to u in $\mathcal{O}(|x|)$ total time.

We perform a DFS on $\mathsf{ST}(x)$. Every leaf in a standard suffix tree is labeled with the starting position of the suffix it represents. While traversing $\mathsf{ST}(x)$, we propagate upwards the labels of the leaf nodes maintaining only the *smallest* label (starting position) in every node. For any $\mathsf{ST}(\cdot)$, we denote the smallest label i for node u by $\mathsf{start}(u) = i$. Consider now a node u of $\mathsf{ST}(x)$ which stores label $\mathsf{start}(u)$. Then the path from the root to u spells the string $x[\mathsf{start}(u) \mathrel{..} \mathsf{start}(u) + d(u) - 1]$, where $d(u)$ is the string depth of node u. At the end of the DFS, we group the nodes per label i, for all $i \in [1, |x|]$, using radix sort. Specifically, two nodes u, v of $\mathsf{ST}(x)$ are in group G_i if and only if $\mathsf{start}(u) = \mathsf{start}(v) = i$. By construction (i.e., by choosing the smallest label) one node represents a prefix of the other node. The whole process takes $\mathcal{O}(|x|)$ time.

We run the *matching statistics* algorithm [3,11] using x and $\mathsf{ST}(y)$: for each starting position i in x, we compute the longest match of length $\ell_i \geq 0$ in any document in $\mathcal{D}$. In particular, this algorithm gives us a locus on $\mathsf{ST}(y)$, which represents the longest match $x[i \mathrel{..} i + \ell_i - 1]$, for all $i \in [1, |x|]$. More formally, a *locus* in a suffix tree is a pair (v, ℓ_i) where $d(\mathsf{parent}(v)) < \ell_i \leq d(v)$, for some node v of the suffix tree and some string depth ℓ_i. Provided that $\mathsf{ST}(y)$ is already constructed, computing the matching statistics takes $\mathcal{O}(|x|)$ time [11].

Let this locus on $\mathsf{ST}(y)$ be (v, ℓ_i) and let it represent $y[\mathsf{start}(v) \mathrel{..} \mathsf{start}(v) + \ell_i - 1]$. In particular, substring $x[i \mathrel{..} i + \ell_i - 1] = y[\mathsf{start}(v) \mathrel{..} \mathsf{start}(v) + \ell_i - 1]$ is precisely this longest match. We consider G_i: the group of nodes from $\mathsf{ST}(x)$ having label i. Say we are processing such a node $u \in G_i$. We have two cases:

- If $\ell_i < d(u)$ the frequency assigned to node u is 0. This is correct because $x[\mathsf{start}(u) \mathrel{..} \mathsf{start}(u) + d(u) - 1]$ does not occur in any document in $\mathcal{D}$ otherwise a longer than the longest match would be output by the matching statistics.
- If $\ell_i \geq d(u)$ then we ask a weighted ancestor query [8] to locate the substring $y[\mathsf{start}(v) \mathrel{..} \mathsf{start}(v) + d(u) - 1]$ of y, and it gives us a locus $(w, d(u))$ in constant time after a linear-time preprocessing of y [2]. More formally, the *weighted ancestor* problem on suffix trees is defined as follows: given $\mathsf{ST}(y)$, we are asked to preprocess it so that can find the locus of any substring $y[p \mathrel{..} q]$ of y on $\mathsf{ST}(y)$. We read the frequency stored at node w, and this is precisely the frequency we assign to node u. This is correct, because $x[\mathsf{start}(u) \mathrel{..} \mathsf{start}(u) + d(u) - 1]$ is a prefix of $x[\mathsf{start}(u) \mathrel{..} \mathsf{start}(u) + \ell_i - 1]$, by $\ell_i \geq d(u)$, and because $x[\mathsf{start}(u) \mathrel{..} \mathsf{start}(u) + d(u) - 1] = y[\mathsf{start}(v) \mathrel{..} \mathsf{start}(v) + d(u) - 1]$.

Since the matching statistics algorithm finds a locus (v, ℓ_i) for every starting position i of x, we can assign the correct document frequency to every node of $\mathsf{ST}(x)$ in $\mathcal{O}(|x|)$ total time. We obtain the following result, which we refine next.

Lemma 1. *The document frequency for all nodes of $\mathsf{ST}(x)$ can be computed in the optimal $\mathcal{O}(|x|)$ time after a linear-time preprocessing of dictionary $\mathcal{D}$.*

Let us now describe in detail how we can efficiently compute array S. The first step is to construct $\mathsf{ST}(x)$ and compute for all of its nodes the document

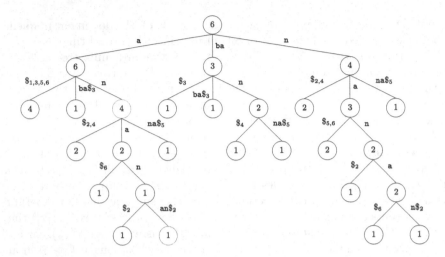

Fig. 1. $\mathsf{ST}(y)$ with nodes weighted by *document frequency* for $\mathcal{D} = \{\mathtt{a,ananan,baba,ban,banna,nana}\}$ from Example 1. A successor weighted ancestor query of the blue node with argument $\alpha_j = 3$, takes us to the ancestor of the blue node with the smallest frequency at least 3. This is the red node. Any such query can be answered in $\mathcal{O}(\log\log d)$ time after a linear-time preprocessing of $\mathsf{ST}(y)$ [1,8].

frequency using Lemma 1. This takes $\mathcal{O}(|x|)$ time after a linear-time preprocessing of dictionary $\mathcal{D}$. Up to this point, we have correctly identified the document frequency for every substring of x that is spelled from the root of $\mathsf{ST}(x)$ ending exactly at some node of $\mathsf{ST}(x)$. However, we have no access to the document frequency of the substrings of x that *end in the middle of an edge* of $\mathsf{ST}(x)$.

We thus need to have an efficient way to subdivide the edges of $\mathsf{ST}(x)$ accordingly. The crucial observation is that we have only τ frequency intervals $\mathcal{I} = I_1, \ldots, I_\tau$, and thus it suffices to split every edge of $\mathsf{ST}(x)$ in at most τ sub-edges. To achieve this, we also *preprocess* $\mathsf{ST}(y)$ for successor weighted ancestor queries with respect to document frequency as node weights. This is possible because of the *max-heap property*: any node on $\mathsf{ST}(y)$ has equal or smaller weight than any of its ancestors. Recall that for any node u in $\mathsf{ST}(x)$ we can find the corresponding locus $(w, d(u))$ in $\mathsf{ST}(y)$ (see Lemma 1) in $\mathcal{O}(|x|)$ total time. In the second step, we enhance $\mathsf{ST}(x)$ with at most τ new nodes per edge using τ weighted ancestor queries on $\mathsf{ST}(y)$. In particular, we ask one weighted ancestor query α_j per interval $I_j = [\alpha_j, \beta_j]$ (see Fig. 1). Each new node stores a document frequency and it takes $\mathcal{O}(\log\log d)$ time to find its locus on $\mathsf{ST}(y)$ using a weighted ancestor query, after a linear-time preprocessing of $\mathsf{ST}(y)$ [1,8].

The third step is to traverse the enhanced $\mathsf{ST}(x)$ and construct a collection of labeled length intervals $[i, j]_f$, one for each node u of $\mathsf{ST}(x)$, defined as follows: $i = d(\mathsf{parent}(u)) + 1$, $j = d(u)$, and f is the document frequency stored in u. We do this in $\mathcal{O}(\tau|x|)$ total time because we have $\mathcal{O}(\tau|x|)$ nodes in $\mathsf{ST}(x)$. We ignore labeled length intervals with $f = 0$ (no occurrence) or $j = 0$ (empty string).

The fourth step is to sort these intervals and view each interval, say of node u, as a line starting at point (i, u) and ending at point (j, u) on the $[0, |x|] \times [0, 2\tau|x|]$

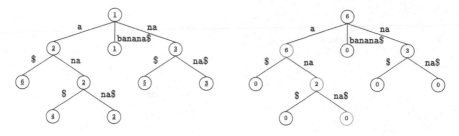

Fig. 2. Step 1 from Example 2. We assume that x ends with a unique letter $\$ \notin \Sigma$.

plane. The y-axis represents the distinct lines (we have no more than $2\tau|x|$ intervals because we have no more than $2\tau|x|$ nodes), and the x-axis represents the lengths of substrings x (the maximum length is $|x|$). We can do this in $\mathcal{O}(\tau|x|)$ time using radix sort because for any interval $[i,j]_f$, $i,j \in [|x|]$.

In the last step, for each length in $[|x|]$, we count how many lines it stabs after classifying the lines in frequency intervals. For the latter, we employ predecessor/successor search after $\mathcal{O}(d)$-time and space preprocessing [6]: we insert the endpoints of every interval in $\mathcal{O}(\tau \log \log d)$ total time as we have 2τ endpoints in total. A line with frequency f belongs to the frequency interval $I_j = [\alpha_j, \beta_j]$ if and only if the predecessor of f is α_j and its successor is β_j. The search takes $\mathcal{O}(\log \log d)$ time per line [6]. The endpoints are then deleted from the structure in $\mathcal{O}(\tau \log \log d)$ total time [6]. We sweep through the lines from left to right and maintain counters on the sum of currently "active" lines per frequency interval. We do this in $\mathcal{O}(\tau)$ time per length. We have arrived at the following result.

Theorem 1 (Main Result). *After a linear-time preprocessing of a dictionary $\mathcal{D}$ of d strings, for any query string x and any partition $\mathcal{I}$ of $[d]$ in τ intervals, array S, such that $S[i,j] = f_{x,\mathcal{D},\mathcal{I}}(i,j)$, can be computed in $\mathcal{O}(|x|\tau \log \log d)$ time.*

Example 2. Consider the dictionary $\mathcal{D}$, the query string $x = x[1 \mathinner{\ldotp\ldotp} |x|] =$ banana, and the partition $\mathcal{I}$ from Example 1. We show in Fig. 2 (on the left) the suffix tree $\mathsf{ST}(x)$ and the smallest label $\mathsf{start}(u)$ (underlined) for every node u after the DFS. We have three groups of nodes: $G_1 = \{-\text{banana}\}$, $G_2 = \{-\text{a}, -\text{ana}, -\text{anana}\}$ and $G_3 = \{-\text{na}, -\text{nana}\}$. (Here we use the notation $-$ before a string to denote a node.) From the matching statistics algorithm, we know the longest match for each position i of x: $\mathcal{L} = [3, 5, 4, 3, 2, 1]$; e.g., $\ell_1 = \mathcal{L}[1] = 3$ tells us that the longest match of $x[1 \mathinner{\ldotp\ldotp} 6] =$ banana in $\mathcal{D}$ is $x[1 \mathinner{\ldotp\ldotp} 3] =$ ban. For the only node in G_1 we have frequency 0 since $3 < |\text{banana}|$. For G_2, we first find the document frequency for the deepest node $-$anana and we reach the node $-$ananan in $\mathsf{ST}(y)$ (see Fig. 1). Then from this node ($-$ananan) we ask for depths 3 and 1 (using weighted ancestor queries), and get to nodes $-$ana and $-$a in $\mathsf{ST}(y)$, which give the corresponding document frequencies in $\mathsf{ST}(x)$. Similarly we process G_3 and get the $\mathsf{ST}(x)$ in Fig. 2 (on the right) with document frequencies (Lemma 1).

We next show in Fig. 3 how we enhance $\mathsf{ST}(x)$ (on the left) with at most τ nodes per edge (on the right) using weighted ancestor queries on $\mathsf{ST}(y)$. Let us

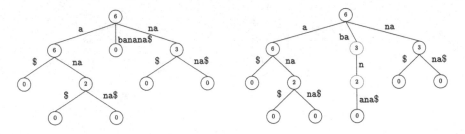

Fig. 3. In Step 2 from Example 2 the edge -banana is subdivided to -ba-n-ana.

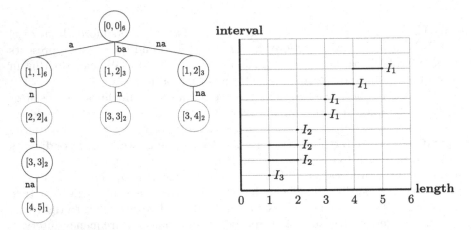

Fig. 4. Steps 3–4 from Example 2. On the left: the added nodes are in red; the nodes with $f = 0$ are pruned. On the right: the length intervals labeled by frequency interval.

consider the edge -banana, for which we need to add new nodes. First, we have $\mathcal{L}[1] = 3$. We add a node at depth 3 and separate -banana to -ban-ana. From $\mathsf{ST}(y)$, we know the frequency of node -ban is 2. Then we ask whether node -ban in $\mathsf{ST}(y)$ has an ancestor node with a frequency at least 3 (I_2) or at least 5 (I_3). Indeed we find that node -ba in $\mathsf{ST}(y)$ has frequency 3, and so we add a node in $\mathsf{ST}(x)$ subdividing -ban to -ba-n. No ancestor of -ban in $\mathsf{ST}(y)$ has a frequency of at least 5, so we do not need to add any more nodes in $\mathsf{ST}(x)$.

After the end of the second step, we construct a labeled length interval for each node of the enhanced $\mathsf{ST}(x)$, and so we get the tree in Fig. 4 (on the left). In the fourth step, we sort these length intervals and view them as lines (on the right). In the last step, after classifying the lines in frequency intervals, we sweep through them from left to right, and compute array S.

S	[1,2]	[3,4]	[5,6]
1	0	2	1
2	0	3	0
3	3	0	0
4	2	0	0
5	1	0	0
6	0	0	0

References

1. Amir, A., Landau, G.M., Lewenstein, M., Sokol, D.: Dynamic text and static pattern matching. ACM Trans. Algorithms **3**(2), 19 (2007). https://doi.org/10.1145/1240233.1240242
2. Belazzougui, D., Kosolobov, D., Puglisi, S.J., Raman, R.: Weighted ancestors in suffix trees revisited. In: Gawrychowski, P., Starikovskaya, T. (eds.) 32nd Annual Symposium on Combinatorial Pattern Matching, CPM 2021, 5–7 July 2021, Wrocław, Poland. LIPIcs, vol. 191, pp. 8:1–8:15. Schloss Dagstuhl - Leibniz-Zentrum für Informatik (2021). https://doi.org/10.4230/LIPIcs.CPM.2021.8
3. Chang, W.I., Lawler, E.L.: Sublinear approximate string matching and biological applications. Algorithmica **12**(4/5), 327–344 (1994). https://doi.org/10.1007/BF01185431
4. Crochemore, M., Hancart, C., Lecroq, T.: Algorithms on Strings. Cambridge University Press, Cambridge (2007)
5. Delcher, A.L., Kasif, S., Fleischmann, R.D., Peterson, J., White, O., Salzberg, S.L.: Alignment of whole genomes. Nucleic Acids Res. **27**(11), 2369–2376 (1999). https://doi.org/10.1093/nar/27.11.2369
6. van Emde Boas, P.: Preserving order in a forest in less than logarithmic time and linear space. Inf. Process. Lett. **6**(3), 80–82 (1977). https://doi.org/10.1016/0020-0190(77)90031-X
7. Farach, M.: Optimal suffix tree construction with large alphabets. In: 38th Annual Symposium on Foundations of Computer Science, FOCS '97, Miami Beach, Florida, USA, 19–22 October 1997, pp. 137–143. IEEE Computer Society (1997). https://doi.org/10.1109/SFCS.1997.646102
8. Farach, M., Muthukrishnan, S.: Perfect hashing for strings: formalization and algorithms. In: Hirschberg, D., Myers, G. (eds.) CPM 1996. LNCS, vol. 1075, pp. 130–140. Springer, Heidelberg (1996). https://doi.org/10.1007/3-540-61258-0_11
9. Ferenczi, S.: Complexity of sequences and dynamical systems. Discret. Math. **206**(1–3), 145–154 (1999). https://doi.org/10.1016/S0012-365X(98)00400-2
10. Fredman, M.L., Komlós, J., Szemerédi, E.: Storing a sparse table with 0(1) worst case access time. J. ACM **31**(3), 538–544 (1984). https://doi.org/10.1145/828.1884
11. Gusfield, D.: Algorithms on Strings, Trees, and Sequences - Computer Science and Computational Biology. Cambridge University Press, Cambridge (1997). https://doi.org/10.1017/cbo9780511574931
12. Chi, L., Hui, K.: Color set size problem with applications to string matching. In: Apostolico, A., Crochemore, M., Galil, Z., Manber, U. (eds.) CPM 1992. LNCS, vol. 644, pp. 230–243. Springer, Heidelberg (1992). https://doi.org/10.1007/3-540-56024-6_19
13. Kociumaka, T., Navarro, G., Prezza, N.: Toward a definitive compressibility measure for repetitive sequences. IEEE Trans. Inf. Theory **69**(4), 2074–2092 (2023). https://doi.org/10.1109/TIT.2022.3224382
14. Kutsukake, K., Matsumoto, T., Nakashima, Y., Inenaga, S., Bannai, H., Takeda, M.: On repetitiveness measures of thue-morse words. In: Boucher, C., Thankachan, S.V. (eds.) SPIRE 2020. LNCS, vol. 12303, pp. 213–220. Springer, Cham (2020). https://doi.org/10.1007/978-3-030-59212-7_15
15. Leskovec, J., Rajaraman, A., Ullman, J.D.: Mining of Massive Datasets, 2nd ed. Cambridge University Press, Cambridge (2014). https://www.mmds.org/
16. Loukides, G., Pissis, S.P.: Bidirectional string anchors: a new string sampling mechanism. In: Mutzel, P., Pagh, R., Herman, G. (eds.) 29th Annual European Symposium on Algorithms, ESA 2021, 6–8 September 2021, Lisbon, Portugal (Virtual

Conference). LIPIcs, vol. 204, pp. 64:1–64:21. Schloss Dagstuhl - Leibniz-Zentrum für Informatik (2021). https://doi.org/10.4230/LIPIcs.ESA.2021.64

17. Mignosi, F.: Infinite words with linear subword complexity. Theor. Comput. Sci. **65**(2), 221–242 (1989). https://doi.org/10.1016/0304-3975(89)90046-7

18. Navarro, G., Rojas-Ledesma, J.: Predecessor search. ACM Comput. Surv. **53**(5), 105:1–105:35 (2021). https://doi.org/10.1145/3409371

19. Pissis, S.P.: MoTeX-II: structured MoTif eXtraction from large-scale datasets. BMC Bioinform. **15**, 235 (2014). https://doi.org/10.1186/1471-2105-15-235

20. Puglisi, S.J., Zhukova, B.: Document retrieval hacks. In: Coudert, D., Natale, E. (eds.) 19th International Symposium on Experimental Algorithms, SEA 2021, 7–9 June 2021, Nice, France. LIPIcs, vol. 190, pp. 12:1–12:12. Schloss Dagstuhl - Leibniz-Zentrum für Informatik (2021). https://doi.org/10.4230/LIPIcs.SEA.2021.12

21. Raskhodnikova, S., Ron, D., Rubinfeld, R., Smith, A.D.: Sublinear algorithms for approximating string compressibility. Algorithmica **65**(3), 685–709 (2013). https://doi.org/10.1007/s00453-012-9618-6

22. Shallit, J.O., Shur, A.M.: Subword complexity and power avoidance. Theor. Comput. Sci. **792**, 96–116 (2019). https://doi.org/10.1016/j.tcs.2018.09.010

23. Weiner, P.: Linear pattern matching algorithms. In: 14th Annual Symposium on Switching and Automata Theory, Iowa City, Iowa, USA, 15–17 October 1973, pp. 1–11. IEEE Computer Society (1973). https://doi.org/10.1109/SWAT.1973.13

Chaining of Maximal Exact Matches in Graphs

Nicola Rizzo[ID], Manuel Cáceres[ID], and Veli Mäkinen[(✉)][ID]

Department of Computer Science, University of Helsinki,
P. O. Box. 68, Pietari Kalmin katu 5, 00014 Helsinki, Finland
{nicola.rizzo,manuel.caceresreyes,veli.makinen}@helsinki.fi

Abstract. We show how to chain *maximal exact matches* (MEMs) between a query string Q and a labeled directed acyclic graph (DAG) $G = (V, E)$ to solve the *longest common subsequence* (LCS) problem between Q and G. We obtain our result via a new symmetric formulation of chaining in DAGs that we solve in $O(m + n + k^2|V| + |E| + kN \log N)$ time, where $m = |Q|$, n is the total length of node labels, k is the minimum number of paths covering the nodes of G and N is the number of MEMs between Q and node labels, which we show encode full MEMs.

Keywords: sequence to graph alignment · longest common subsequence · sparse dynamic programming

1 Introduction

Due to recent developments in *pangenomics* [9] there is a high interest to extend the notion of string alignments to graphs. A common pangenome representation is a node-labeled directed acyclic graph (DAG), whose paths represent plausible individual genomes from a species. Unfortunately, even finding an exact occurrence of a query string as a subpath in a graph is a conditionally hard problem [12,13]: only quadratic time dynamic programming solutions are known and faster algorithms would contradict the Strong Exponential Time Hypothesis (SETH). Due to this theoretical barrier, parameterized solutions have been developed [5,10,11,21], and/or the task has been separated into finding short exact occurrences (anchors) and then *chaining* them into longer matches [8,15,16,18]. Although the chaining algorithms provide exact solutions to their internal chaining formulations and their solutions can be interpreted as alignments of queries to a graph with edit operations, so far they have not been shown to provide exact solutions to the corresponding alignment formulation.

In this paper, we integrate a symmetric formulation from string chaining [17,22] to graph chaining [18] yielding the first chaining-based parameterized exact alignment algorithm between a query string and a graph. Namely, we obtain an $O(m + n + k^2|V| + |E| + kN \log N)$ time algorithm for computing the length of a *longest common subsequence* (LCS) between a query string Q and a

F. M. Nardini et al. (Eds.): SPIRE 2023, LNCS 14240, pp. 353–366, 2023.
https://doi.org/10.1007/978-3-031-43980-3_29

path of G, where $m = |Q|$, n is the total length of node labels, k is the width (minimum number of paths covering the nodes) of G, and N is the number of *maximal exact matches* (MEMs) between Q and the node labels (node MEMs). While N can be quadratic (thus, the result is not breaking the known conditional LCS lower bounds [1,4]), there are also inputs where N grows slower. Moreover, when MEMs are limited by a length threshold, N can be made smaller; we show that in this setting the algorithm solves a variant of the LCS problem.

The paper is structured as follows. The preliminaries in Sect. 2 and the basic concepts in Sect. 3 follow the notions developed in our recent work [20], where we introduce the definition of a MEM between a string and a graph, and study the non-trivial problem of finding graph MEMs with a length threshold; for the purposes of this paper, we observe that node MEMs are sufficient. In Sect. 4.1, we revise the solution for an asymmetric chaining formulation in DAGs [18] for the case of node MEMs. Then, in Sect. 4.2, we tailor the string to string symmetric chaining algorithm [17,22] to use MEM anchors. In Sect. 4.3, we show how to integrate these two approaches to obtain our main result. Finally, in Sect. 5 we discuss the length threshold setting and cyclic graphs.

2   Preliminaries

Strings. We work with strings coming from a finite alphabet $\Sigma = [1..\sigma]$ and assume that σ is at most the length of the strings we work with. For two integers x and y we use $[x..y]$ to denote the integer interval $\{x, x+1, \ldots, y\}$ or the empty set $\emptyset$ when $x > y$. A *string* T is an element of Σ^n for a non-negative integer n, that is, a sequence of n symbols from Σ, where $n = |T|$ is the *length* of the string. We denote ε to the only string of length zero. We also denote $\Sigma^+ = \Sigma^* \setminus \{\varepsilon\}$. For two strings T_1 and T_2 we denote their *concatenation* as $T_1 \cdot T_2$, or just $T_1 T_2$. For a set of integers I and a string T, we use $T[I]$ to denote the *subsequence* of T made of the concatenation of the characters indicated by I in increasing order. If I is an integer interval $[x..y]$, then $T[x..y]$ is a *substring*: if $x = y$ then we also use $T[x]$, if $y < x$ then $T[x..y] = \varepsilon$, if $x \leq y = n$ we call it a *suffix* (*proper suffix* when $x > 1$) and if $1 = x \leq y$ we call it a *prefix* (*proper prefix* when $y < n$). A length-κ' substring $Q[x..x + \kappa' - 1]$ *occurs* in T if $Q[x..x + \kappa' - 1] = T[i..i + \kappa' - 1]$; in this case, we say that (x, i, κ') is an *(exact) match* between Q and T, and *maximal* (a MEM) if the match cannot be extended to the left (*left-maximality*), that is, $x_1 = 1$ or $x_2 = 1$ or $Q[x_1 - 1] \neq T[x_2 - 1]$ nor it can be extended to the right (*right-maximality*) $x_1 + \ell = |Q|$ or $x_2 + \ell = |T|$ or $Q[x_1 + \ell] \neq T[x_2 + \ell]$.

Labeled Graphs. We work with labeled directed acyclic graphs (DAGs) $G = (V, E, \ell)$, where V is the vertex set, E the edge set, and $\ell : V \to \Sigma^+$ a *labeling* function on the vertices. A length-k *path* P from v_1 to v_k is a sequence of nodes $v_1, \ldots, v_k$ such that $(v_1, v_2), (v_2, v_3), \ldots, (v_{k-1}, v_k) \in E$, in this case we say that v_1 *reaches* v_k. We extend the labeled function to paths by concatenating the corresponding node labels, that is, $\ell(P) := \ell(v_1) \cdots \ell(v_k)$. For a node v and a path P we use $\|\cdot\|$ to denote its *string length*, that is, $\|v\| = |\ell(v)|$ and

$\|P\| = |\ell(P)|$. We say that a length-κ' substring $Q[x..x + \kappa' - 1]$ *occurs* in G if $Q[x..x + \kappa' - 1]$ occurs in $\ell(P)$ for some path P. In this case, we say that $([x..x + \kappa' - 1], (i, P = v_1 \ldots v_k, j))$ is an *(exact) match* between Q and G, where $Q[x..x + \kappa' - 1] = \ell(v_1)[i..] \cdot \ell(v_2) \cdots \ell(v_{k-1}) \cdot \ell(v_k)[..j]$, with $1 \leq i \leq \|v_1\|$ and $1 \leq j \leq \|v_k\|$. We call the triple (i, P, j) a *substring* of G and we define its *left-extension* $\text{lext}(i, P, j)$ as the singleton $\{\ell(v_1)[i - 1]\}$ if $i > 1$ and $\{\ell(u)[\|u\|] \mid (u, v_1) \in E\}$ otherwise. Analogously, the *right-extension* $\text{rext}(i, P, j)$ is $\{\ell(v_k)[j + 1]\}$ if $j < \|v_k\|$ and $\{\ell(v)[1] \mid (v_k, v) \in E\}$ otherwise. Note that the left (right) extension can be equal to the empty set $\emptyset$, if the start (end) node of P does not have incoming (outgoing) edges. See Fig. 1.

Chaining of Matches. An *asymmetric chain* $A'[1..N']$ is an ordered subset of a set A of N exact matches between a labeled DAG $G = (V, E, \ell)$ and a query string Q, with the ordering $A'[l] < A'[l + 1]$ for $1 \leq l < N'$ defined as $([x'..x' + \kappa'' - 1], (i', P_l, j')) < ([x..x + \kappa' - 1], (i, P_{l+1}, j))$ iff the start of path P_{l+1} is strictly reachable from the end of path P_l and $x' \leq x$. Such ordering is also called *co-linear* as it enforces linear order in two dimensions. The asymmetry comes from the fact that overlaps are not allowed in G, but they are allowed in Q. We are interested in chains that maximize the length of an induced subsequence Q', denoted $Q' = Q \mid A'$, that is obtained by deleting all parts of Q that are not covered by chain A'. For example, consider $Q = \texttt{ACATTCAGTA}$ and $A' = ([2..4], (i_1, P_1, j_1)), ([3..6], (i_2, P_2, j_2)), ([9..10], (i_3, P_3, j_3))$. Then $Q' = Q \mid A' = \texttt{CATTCTA}$; anchors cover the underlined part of $Q = \underline{\texttt{ACATTC}}\texttt{AG}\underline{\texttt{TA}}$.

We could define symmetric chains by considering overlaps of long paths, but for the purposes of this paper it will be sufficient to consider the chaining of exact matches involving length-1 paths in G, along with their possible overlaps: a *symmetric chain* $A'[1..N']$ is an ordered subset of a set A of N exact matches between the nodes of a labeled DAG $G = (V, E, \ell)$ and a query string Q, with the ordering $A'[l] < A'[l + 1]$ for $1 \leq l < N'$ defined as $([x'..x' + \kappa'' - 1], (i', v, j')) < ([x..x + \kappa' - 1], (i, w, j))$ iff (i) w is strictly reachable from v, or $v = w$ and $i' \leq i$ and (ii) $x' \leq x$. We extend the notation $Q' = Q \mid A'$ to cover symmetric chains A' so that Q' is obtained by deleting all parts of Q that are not *mutually* covered by chain A'. We define mutual coverage in Sect. 4.2: informally, Q' is formed by concatenating the prefixes of exact matches until reaching the overlap between the next exact match in the chain. Figure 1 illustrates the concept.

3 Finding MEMs in Labeled DAGs

We now consider the problem of finding all *maximal exact matches* (MEMs) between a labeled graph G and a query string Q for the purpose of chaining.

Definition 1 (MEM between a pattern and a graph [20]). *Let $G = (V, E, \ell)$ be a labeled graph, with $\ell : V \to \Sigma^+$, and $Q \in \Sigma^+$. We say that a match $([x..y], (i, P, j))$ between Q and G is left-maximal (right-maximal) if it cannot be extended to the left (right) in both Q and G, that is,*

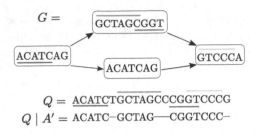

$$Q = \underline{\text{ACATC}}\text{TGCTAGCCCGGTCCCG}$$
$$Q \mid A' = \text{ACATC-GCTAG---CGGTCCC-}$$

Fig. 1. Co-linear chaining setting between a string Q and a labeled graph G. If v is the last node to the right, then $([16..20], (1, v, 5))$ is a match, with $\text{lext}(1, v, 5) = \{\texttt{G}, \texttt{T}\}$ and $\text{rext}(1, v, 5) = \{\texttt{A}\}$. It is a MEM since $|\text{lext}(i, P, j)| \geq 2$ and it cannot be extended to the right (Definition 1). In fact, all exact matches are MEMs and they form a symmetric chain A' (blue-green-red-yellow) inducing the subsequence $Q \mid A'$ (the last $\texttt{C}$ of the green match is omitted due to overlap with the red match).

(LeftMax) $x = 1 \vee \text{lext}(i, P, j) = \emptyset \vee Q[x-1] \notin \text{lext}(i, P, j)$ *and*
(RightMax) $y = |Q| \vee \text{rext}(i, P, j) = \emptyset \vee Q[y+1] \notin \text{rext}(i, P, j).$

The pair $([x..y], (i, P, j))$ is a MEM if it is left-maximal or its left (graph) extension is not a singleton, and right-maximal or its right (graph) extension is not a singleton, that is LeftMax $\vee |\text{lext}(i, P, j)| \geq 2$ *and* RightMax $\vee |\text{rext}(i, P, j)| \geq 2$.

See Fig. 1 for an example. We use this particular extension of MEMs to graphs—with the additional conditions on non-singletons lext and rext—as it captures all MEMs between Q and $\ell(P)$, where P is a source-to-sink path in G. Moreover, we will show that this MEM formulation captures LCS through co-linear chaining, whereas avoiding the additional conditions would fail. Indeed, consider Q, G, and match $([16..20], (1, v, 5))$ from Fig. 1: the match is not left-maximal, since $Q[15] = \texttt{G}$ and $\texttt{G} \in \text{lext}(1, v, 5)$, but extending it would impose any chain using it as an anchor to go through the bottom suboptimal path, that in this case does not capture the LCS between Q and G. Also, it turns out that we can focus on MEMs between the node labels and the query, as chaining will cover longer MEMs implicitly.

To formalize the intuition, we say that a *node MEM* is a match (i, P, j) of $Q[x..y]$ in G such that $P = v$ for some node v, and it is left and right maximal w.r.t. $\ell(P)$ only in the string sense: conditions LextMax$\vee i = 1$ and RightMax$\vee j = \|v\|$ hold. Consider the text $T_{\text{nodes}} = \prod_{v \in V} \mathbf{0} \cdot \ell(v)$, where $\mathbf{0} \notin \Sigma$ is used as a delimiter to prevent MEMs spanning more than a node label. Running the MEM finding algorithm [2] on Q and T_{nodes} will retrieve exactly the node MEMs we are looking for [20] (a more involved problem of finding graph MEMs with a length threshold is studied in [20], but here a simplified result without the threshold is sufficient):

Lemma 1 ([20]). *Given a labeled DAG $G = (V, E, \ell)$, with $\ell : V \rightarrow \Sigma^+$, and a query string Q, we can compute all node MEMs between Q and G in time*

$O(n + m + N)$, where n is the total length of node labels, $m = |Q|$, and N is the number of node MEMs.

Let A be the set of node MEMs found using Lemma 1. In an extended version of this paper [19], we show that any long MEM spanning two or more nodes in G can be formed by concatenating node MEMs into *perfect chains*—chains that have no gap between consecutive matches.

Theorem 1 (Appendix B in [19]). *For every MEM $([x..y], (i, P, j))$ between G and Q, there is a perfect chain $A'[1..p] \subseteq A$ such that $A'[1] \cdots A'[p] = ([x..y], (i, P, j))$.*

Corollary 1. *The set A is a* compact representation *of the set M of MEMs between query Q and a labeled DAG $G = (V, E, \ell)$: it holds $|A| \leq \|M\|$, where $\|M\|$ is the length of the encoding of the paths in MEMs as the explicit sequence of its nodes.*

Our strategy is to use set A as the representation of MEMs: perfect chains are implicitly covered by the chaining algorithms of the next section.

4   Symmetric Co-Linear Chaining in Labeled DAGs

Mäkinen et al. [18, Theorem 6.4] gave an $O(kN \log N + k|V|)$-time algorithm to find an asymmetric chain $A'[1..N']$ of a set A of N anchors[1] between a labeled DAG $G = (V, E, \ell)$ and a query string Q maximizing the length of an induced subsequence $Q' = Q \mid A'$. Here k is the *width* of G, that is, the minimum number of paths covering nodes V of G. The algorithm assumes a minimum path cover as its input, which can be computed in $O(k^2|V| + |E|)$ time [6,7]. A limitation of this chaining algorithm is that anchors in the solution are not allowed to overlap in the graph, which has been partially solved by considering one-node overlaps [16]. However, both of these approaches maximize the length of the sequence induced by the reported chain only on the string Q, which makes the problem formulation asymmetric.

In the case of two strings as input, the asymmetry of the coverage metric was solved by Mäkinen and Sahlin [17] applying the technique by Shibuya and Kurochkin [22]. They provided an $O(N \log N)$-time algorithm to find a symmetric chain $A'[1..N']$ of a set A of N anchors maximizing the length of an *induced common subsequence* $C = Q \mid A' = T \mid A'$ between two input strings Q and T, that is obtained by deleting all parts of Q, or equivalently all parts of T, that are not *mutually covered* by chain A' (to be defined below). Here anchors are assumed to be exact matches (x, i, κ') (not necessarily maximal) such that $Q[x..x + \kappa' - 1] = T[i..i + \kappa' - 1]$, and $A'[j] < A'[j + 1]$ for $1 \leq j < N'$, where the order $<$ between anchors is defined as $(x', i', \kappa'') < (x, i, \kappa')$ iff $x' \leq x$ and $i' \leq i$. For completeness, in an extended version of this paper [19], we include

---

[1] Anchors have the same representation as graph MEMs, $([x..y], (i, P, j))$, but they do not necessarily represent exact matches.

a revised proof that this algorithm computes the length of a longest common subsequence of strings Q and T if it is given all (string) MEMs between Q and T as input [17]. The concept of mutual coverage [17, Problem 1] is defined through the score

$$\text{coverage}(A') = \sum_{j=1}^{N'} \min_{\substack{(i,x,\kappa') := A'[j+1], \\ (i',x',\kappa'') := A'[j]}} \begin{cases} \min(i, i' + \kappa'') - i', \\ \min(x, x' + \kappa'') - x', \end{cases}$$

where $A'[N' + 1] = (\infty, \infty, 0)$. Each part of the sum contributes the corresponding number of character matches from the beginning of the anchors to the induced common subsequence. These form the mutually covered part of the inputs; see Fig. 1 for an illustration of an extension of this concept to graphs.

Consider now the symmetric chaining problem between a DAG and a string:

Problem 1 (Symmetric DAG chaining with overlaps). Find a symmetric chain $A'[1..N']$ of a set A of N anchors between a labeled DAG $G = (V, E, \ell)$ and a query string Q maximizing the length of an induced common subsequence $C = P \mid A' = Q \mid A'$ for some path P of G, where $P \mid A'$ denotes the subsequence obtained by deleting the parts of $\ell(P)$ that are not mutually covered by chain A' and $Q \mid A'$ denotes the subsequence obtained by deleting the parts of Q not mutually covered by chain A'.[2]

In this section, we will solve this problem in the special case where the anchors are all node MEMs between G and Q: thanks to Theorem 1 we know that the algorithm by Mäkinen et al. [18] solves the problem when a longest induced common subsequence C is covered by node MEMs that appear in different nodes. Since in our setting the overlaps can only occur inside node labels, we are left with what essentially is the symmetric string-to-string chaining problem [17, 22]. However, we cannot separate these subproblems and call the respective algorithms as black boxes, but instead we need to carefully interleave the computation of both techniques in one algorithm.

4.1 DAG Chaining with Node MEMs

Algorithm 1 shows the pseudocode of [18, Algorithm 1] simplified to take node MEMs as anchors. The original algorithm uses two arrays to store the start and the end nodes of anchor paths, but in the case of node MEMs one array suffices. We also modified [18, Lemma 3.2] below to explicitly use primary and secondary keys (the original algorithms [17, 18] implicitly assumed distinct keys). We still use primary keys to store MEM ending positions in Q to do range searches, and we use the secondary key to store the MEM identifiers to update the values of the corresponding anchors.

---

[2] Identically, A' maximizes $|C| = \text{coverage}(A')$ when the anchors are interpreted as exact matches between Q and $\ell(P)$, with P some path of G containing the nodes involved by the matches in A', in the order specified by the chain.

ALGORITHM 1: Asymmetric co-linear chaining between a sequence and a DAG using a path cover and node MEMs.

Input: A DAG $G = (V, E, \ell)$, a query string Q, a path cover $P_1, P_2, \ldots, P_k$ of G, and node MEMs $A[1..N]$ of the form $([x..x + \kappa' - 1], (i, v, i + \kappa' - 1))$ where $\ell(v)[i..i + \kappa' - 1] = Q[x..x + \kappa' - 1]$.

Output: Index of a MEM ending at a chain with maximum coverage $\max_j C[j]$ allowing at most one MEM per node of G.

1 Use Lemma 3 to find all forward propagation links;
2 **for** $k' \leftarrow 1$ *to* k **do**
3   Initialize data structures $\mathcal{T}_{k'}^a$ and $\mathcal{T}_{k'}^b$ with keys $(x + \kappa' - 1, j)$ such that $([x..x + \kappa' - 1], (i, v, i + \kappa' - 1)) = A[j]$, $1 \leq j \leq N$, and with key $(0, 0)$, all keys associated with values $-\infty$;
4    $\mathcal{T}_{k'}^a$.update$((0, 0), 0)$;
5    $\mathcal{T}_{k'}^b$.update$((0, 0), 0)$;

   /* Save to anchors[v] all node MEMs of node v.   */
6 **for** $j \leftarrow 1$ *to* N **do**
7    $([x..x + \kappa' - 1], (i, v, i + \kappa' - 1)) = A[j]$;
8   anchors$[v]$.push(j);
9    $C^-[j] \leftarrow 0$;
10    $C[j] \leftarrow \kappa'$;

11 **for** $v \in V$ *in topological order* **do**
12   **for** $j \in$ anchors$[v]$ **do**
      /* Update the data structures for every path that covers v,
         stored in paths[v].   */
13       $([x..x + \kappa' - 1], (i, v, i + \kappa' - 1)) = A[j]$;
14   **for** $k' \in$ paths$[v]$ **do**
15          $\mathcal{T}_{k'}^a$.upgrade$((x + \kappa' - 1, j), C[j])$;
16          $\mathcal{T}_{k'}^b$.upgrade$((x + \kappa' - 1, j), C^-[j] - x)$;

   /* PROPAGATE FORWARD STARTS   */
17   **for** $(w, k') \in$ forward$[v]$ **do**
18   **for** $j \in$ anchors$[w]$ **do**
19          $([x..x + \kappa' - 1], (i, v, i + \kappa' - 1)) = A[j]$;
20          $C^a[j] \leftarrow \mathcal{T}_{k'}^a$.RMaxQ$(0, x - 1)$;
21          $C^b[j] \leftarrow x + \mathcal{T}_{k'}^b$.RMaxQ$(x, x + \kappa' - 1)$;
22          $C^-[j] \leftarrow \max(C^-[j], C^a[j], C^b[j])$;
23          $C[j] = C^-[j] + \kappa'$;

   /* PROPAGATE FORWARD ENDS   */
24 **return** $\operatorname{argmax}_j C[j]$;

Just like the original algorithm, our simplified version fills a table $C[1..N]$ such that $C[j]$ is the maximum coverage of an asymmetric chain that uses the j-th node MEM as its last item. That is, there is an asymmetric chain that induces a subsequence Q' of the query Q of length $C[j]$. In addition, our version is restricted to chains including at most one MEM per node and performs an intermediate step to fill table $C^-[1..N]$ such that $C^-[j] = C[j] - \kappa'$, where κ' is the length of the j-th node MEM. The reason for these modifications will become clear when we integrate the algorithm with the symmetric string-to-string chaining.

To fill tables $C[1..N]$ and $C^-[1..N]$, the algorithm considers a) MEMs from different nodes without overlap in the query and b) MEMs from different nodes with overlap in the query. These cases are illustrated in the left panel of Fig. 2. The algorithm maintains the following data structure for each case and for each path in a given path cover of k paths (see e.g. [3, Chapter 5]):

Lemma 2. *The following four operations can be supported with a balanced binary search tree T in time $O(\log n)$, where n is the number of key-value pairs $((k,j), \mathtt{val})$ stored in the tree. Here k is the primary key, j is the secondary key to break ties, and $k, j, \mathtt{val}$ are integers.*

- value(k, j): *Return the value associated to key (k, j) or $-\infty$ if (k, j) is not a proper key.*
- update$((k, j), \mathtt{val})$: *Associate value* $\mathtt{val}$ *to key (k, j).*
- upgrade$((k, j), \mathtt{val})$: *Associate value* $\max(\mathtt{val}, \mathtt{value}(k, j))$ *to key (k, j).*
- RMaxQ(l, r): *Return* $\max_{l \leq k \leq r, (k,j) \text{ is a key in } T} \mathtt{value}(k, j)$, *or* $-\infty$ *if range $[l..r]$ is empty (Range Maximum Query).*

Moreover, the balanced binary search tree can be constructed in $O(n)$ time, given the n pairs $((k, j), \mathtt{val})$ sorted by component (k, j).

The algorithm processes the nodes in topological order, keeping the invariant that once node v is visited, the final values $C[j]$ and $C^-[j]$ are known for all anchors j included in node v. These values are then stored in the search trees. As a final step in the processing of v, the information stored in the search trees is propagated forward to nodes w, where v is the last node reaching w on some path-cover path, in order to update the intermediate values for MEMs at node w. These forward links are preprocessed with the following lemma:

Lemma 3 (Adaptation of [18, Lemma 3.1]). *Let $G = (V, E)$ be a DAG, and let $P_1, \ldots, P_k$ be a path cover of G. We can compute in $O(k^2|V|)$ time the set of forward propagation links* $\mathtt{forward}[u]$ *defined as follows: for any node v and path k', $(v, k') \in \mathtt{forward}[u]$ if and only if u is the last node on path k' that reaches v such that $u \neq v$.*

Proof. The original DP algorithm [18] runs in $O(k|E|)$ time, but recently it has been shown [14, Algorithms 6 and 7] how to do this in time $O(k|E_{red}|)$, where E_{red} are the edges in the transitive reduction of G. Finally, Cáceres et al. [6,7] showed a transitive sparsification scheme proving that $|E_{red}| \leq k|V|$.

Data structures $T_{k'}^a$ store as primary keys all ending positions of MEMs in Q and as values the corresponding $C[j]$s for node MEMs $A[j]$ processed so far and reaching path $P_{k'}$ (line 15). When a new node MEM is added to a chain at line 20, the range query on $T_{k'}^a$ guarantees that only chains ending before v in G

and before the start of the new node MEM in Q are taken into account. Data structures $T_{k'}^b$ also store as primary keys all ending positions of node MEMs in Q, but as values they store the values $C^-[j]$ with an invariant subtracted (line 16). This invariant is explained by the range query at line 21, that considers chains overlapping (only) in Q with the new node MEM to be added: consider the chain ending at node MEM $A[j'] = ([x'..x' + \kappa'' - 1], (i', v', i' + \kappa'' - 1))$ and the new node MEM $A[j] = ([x..x + \kappa' - 1], (i, v, i + \kappa' - 1)$ is to be added to this chain, where $x \leq x' + \kappa'' - 1 \leq x + \kappa' - 1$. This addition increases the part of Q covered by the chain (excluding the new node MEM) by $x - x'$. This is exactly the value computed at line 21, maximizing over such overlapping node MEMs.

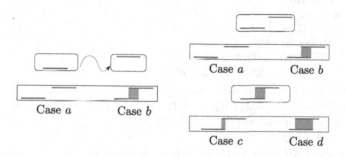

Case a   Case b

Case a   Case b

Case c   Case d

Fig. 2. Precedence of MEMs partitioned to three classes (left, top right, and bottom right subfigures) by occurrence in graph/text (top part of each subfigure) and thereafter to two out of total four cases that require different data structure on the query (bottom part of each subfigure).

4.2   Revisiting Symmetric String-to-string Chaining with MEMs

Before modifying the algorithm to properly consider overlaps of node MEMs in G, let us first modify the symmetric string-to-string chaining algorithm of Mäkinen and Sahlin [17, Algorithm 2] to harmonize the notation and to consider the simplification of [17, Theorem 6] that applies in the case of (string) MEMs. This modification computes the optimal chain given MEMs $A[1..N]$ between strings T and Q and is given as Algorithm 2.

The algorithm uses the same two data structures as before to handle the cases illustrated at the top right of Fig. 2. Moreover, the two additional data structures (balanced binary search trees) in Algorithm 2 handle the overlaps in T by dividing the computation further into cases c) and d) illustrated at the bottom right of Fig. 2): c) if two MEMs overlap more in T than in Q, tree T^c is used for storing the solution; d) otherwise, tree T^d is used for storing the solution. We refer to the original work [17] for the derivation of the invariants and the range queries to handle these cases. The handling of these cases is highlighted with gray background in Algorithm 2.

ALGORITHM 2: Symmetric chaining with two-sided overlaps using MEMs.

Input: An array $A[1..N]$ of (string) MEMs (x, i, κ') between Q and T.

Output: Index of a MEM ending a chain with maximum coverage $\max_j C[j]$.

1 Initialize data structures T^a and T^b with keys $(x + \kappa' - 1, j)$ and data structures T^c and T^d with keys $(x - i, j)$, where $(x, i, \kappa') = A[j]$, $1 \le j \le N$, and all trees with key $(0,0)$. Associate values $-\infty$ to all keys.

2 T^a.upgrade$((0,0), 0)$;

3 $M = \{(x, j) \mid (x, i, \kappa') = A[j], 1 \le j \le N\} \cup \{(x + \kappa' - 1, j) \mid (x, i, \kappa') = A[j], 1 \le j \le N\}$;

4 M.sort();

5 **for** $(x', j) \in M$ **do**

6     $(x, i, \kappa') = A[j]$;

7   **if** $x == x'$ **then**

        /* Start of MEM.   */

8         $C^a[j] = T^a$.RMaxQ$(0, x - 1)$;

9         $C^b[j] = x + T^b$.RMaxQ$(x, x + \kappa' - 1)$;

10        $C^c[j] = i + T^c$.RMaxQ$(-\infty, x - i)$;

11        $C^d[j] = x + T^d$.RMaxQ$(x - i + 1, \infty)$;

12        $C^-[j] = \max(C^a[j], C^b[j], C^c[j], C^d[j])$;

13        $C[j] = C^-[j] + \kappa'$;

14   T^c.upgrade$((x - i, j), C^-[j] - i)$;

15   T^d.upgrade$((x - i, j), C^-[j] - x)$;

16   **else**

        /* End of MEM.   */

17   T^a.upgrade$((x + \kappa' - 1, j), C[j])$;

18   T^b.upgrade$((x + \kappa' - 1, j), C^-[j] - x)$;

19   T^c.update$((x - i, j), -\infty)$;

20   T^d.update$((x - i, j), -\infty)$;

21 **return** $\text{argmax}_j C[j]$;

4.3 Integration of Symmetry to DAG Chaining

We will now merge the two algorithms from previous subsections to solve Problem 1. This algorithm is shown as Algorithm 3; lines highlighted with a dark gray background are from Algorithm 2, whereas lines highlighted with a light gray background are a hybrid of both, and the rest are from Algorithm 1. When visiting node v the algorithm executes the steps of Algorithm 2 on anchors included in v, with $C^a[j]$ and $C^b[j]$ having already been updated with anchors not included in v through forward propagation identical to Algorithm 1. The hybrid parts reflect the required changes to Algorithm 1 in order to visit the MEM anchors twice as in Algorithm 2. This merge covers all three cases of Fig. 2.

Theorem 2. *Given labeled DAG* $G = (V, E, \ell)$ *with path cover* $P_1, \ldots, P_k$, *query string* Q, *and set* $A[1..N]$ *of node MEMs between* Q *and* G, *Algorithm 3* *solves the symmetric DAG chaining with overlaps problem (Problem 1) in time* $O(k^2|V| + kN \log N)$.

Corollary 2. *The length of a longest common subsequence (LCS) between a path in a labeled DAG* $G = (V, E, \ell)$ *and string* Q *can be computed in time* $O(n + m + k^2|V| + |E| + kN \log N)$, *where* $m = |Q|$, n *is the total length of node labels,* k *is the width (minimum number of paths covering the nodes) of* G, *and* N *is the number of node MEMs.*

Proof. The node MEMs can be computed in time $O(n+m+N)$ with Lemma 1. A minimum path cover with k paths can be computed in $O(k^2|V|+|E|)$ time [6,7]. Forward propagation links can be computed in $O(k^2|V|)$ time with Lemma 3. Finally, the term $kN \log N$ comes from Theorem 2. The connection between LCS and solution to symmetric chaining follows with identical arguments as in [19, Appendix B] If P is a path containing an LCS of length c, then Algorithm 3 finds a chain of coverage exactly c as its execution considers the corresponding chain between $\ell(P)$ and Q as done in Algorithm 2. In this case, node MEMs are not necessarily MEMs between $\ell(P)$ and Q, but exact matches supporting the necessary character matches [19, Appendix B].

Note that the LCS connection can be easily adapted for long MEMs spanning two or more nodes of G, but we avoided considering symmetric chains of long MEMs due to the difficulty of handling path overlaps efficiently (see also [18]).

5 Discussion

In this paper, we focused on MEMs with no lower threshold on their length to achieve the connection with LCS. In practical applications, chaining is sped up by using as anchors only MEMs that are of length at least κ, a given threshold. Just finding all such κ-MEMs is a non-trivial problem and solvable in subquadratic time only on some specific graph classes [20]. However, once such κ-MEMs are found, one can split them to node-MEMs and then apply Algorithm 3 to chain them. The resulting chain optimizes the length $|C|$ of a longest common subsequence C between the query Q and a path P such that each match $C[k] = Q[i_k] = \ell(P)[j_k]$ is supported by an exact match of length at least κ, where $1 \leq k \leq |C|$, $i_1 < i_2 < \cdots < i_{|C|}$, and $j_1 < j_2 < \cdots < j_{|C|}$. That is, there is a κ-MEM $([x_k, y_k], [c_k, d_k])$ with respect to Q and $\ell(P)$ s.t. $x_k \leq i_k \leq y_k$ and $c_k \leq j_k \leq d_k$ for each k. Additionally, Ma et al. [16, Appendix C] showed that asymmetric co-linear chaining can be extended to graphs with cycles by considering the graph of the strongly connected components. In the extended version of this paper, we will show how to combine our results to obtain symmetric chaining in general graphs.

ALGORITHM 3: Symmetric co-linear chaining between a sequence and a DAG using a path cover and node MEMs.

Input: Same as in Algorithm 1.
Output: Index of a MEM ending at a chain with maximum coverage $\max_j C[j]$ allowing overlaps in G.

1 Use Lemma 3 to find all forward propagation links.
2 **for** $k' \leftarrow 1$ *to* k **do**
3   Initialize data structures $\mathcal{T}_{k'}^a$ and $\mathcal{T}_{k'}^b$ with keys $(x + \kappa' - 1, j)$ and key $(0, 0)$, and data structures $\mathcal{T}_{k'}^c$ and $\mathcal{T}_{k'}^d$ with keys $(x - i, j)$, where $([x..x + \kappa' - 1], (i, v, i + \kappa' - 1)) = A[j], 1 \leq j \leq N$. Associate values $-\infty$ to all keys.
4    $\mathcal{T}_{k'}^a$.update$((0,0), 0)$;
5    $\mathcal{T}_{k'}^b$.update$((0,0), 0)$;

6 Initialize arrays: anchors, C^- and C as in Algorithm 1;
7 **for** $v \in V$ *in topological order* **do**
8    $M = \{(x, j) \mid ([x..x + \kappa' - 1], (i, v, i + \kappa' - 1)) = A[j], j \in \text{anchors}[v]\} \cup \{(x + \kappa' - 1, j) \mid ([x..x + \kappa' - 1], (i, v, i + \kappa' - 1)) = A[j], j \in \text{anchors}[v]\}$;
9   M.sort();
   `/* Update the data structures for every path that covers v,`   `*/`
      `stored in paths[v].`
10   **for** $k' \in \text{paths}[v]$ **do**
11   **for** $(x', j) \in M$ **do**
12          $(x, i, \kappa') = A[j]$;
13   **if** $x == x'$ **then**
            `/* Start of MEM.`   `*/`
14             $C^a[j] = \mathcal{T}_{k'}^a$.RMaxQ$(0, x - 1)$;
15             $C^b[j] = x + \mathcal{T}_{k'}^b$.RMaxQ$(x, x + \kappa' - 1)$;
16             $C^c[j] = i + \mathcal{T}_{k'}^c$.RMaxQ$(-\infty, x - i)$;
17             $C^d[j] = x + \mathcal{T}_{k'}^d$.RMaxQ$(x - i + 1, \infty)$;
18             $C^-[j] = \max(C^-[j], C^a[j], C^b[j], C^c[j], C^d[j])$;
19             $C[j] = C^-[j] + \kappa'$;
20             $\mathcal{T}_{k'}^c$.upgrade$((x - i, j), C^-[j] - i)$;
21             $\mathcal{T}_{k'}^d$.upgrade$((x - i, j), C^-[j] - x)$;
22   **else**
            `/* End of MEM.`   `*/`
23             $\mathcal{T}_{k'}^a$.upgrade$((x + \kappa' - 1, j), C[j])$;
24             $\mathcal{T}_{k'}^b$.upgrade$((x + \kappa' - 1, j), C^-[j] - x)$;
25             $\mathcal{T}_{k'}^c$.update$((x - i, j), -\infty)$;
26             $\mathcal{T}_{k'}^d$.update$((x - i, j), -\infty)$;

27   Execute **PROPAGATE FORWARD** subroutine of Algorithm 1;
28 **return** $\text{argmax}_j C[j]$;

Acknowledgments. This project has received funding from the Academy of Finland grants No. 352821 and 328877 and the European Union's Horizon 2020 research and innovation programme under the Marie Skłodowska-Curie grant agreement No. 956229.

References

1. Abboud, A., Backurs, A., Williams, V.V.: Tight hardness results for LCS and other sequence similarity measures. In: Guruswami, V. (ed.) IEEE 56th Annual Symposium on Foundations of Computer Science, FOCS 2015, Berkeley, CA, USA, 17–20 October 2015, pp. 59–78. IEEE Computer Society (2015). https://doi.org/10.1109/FOCS.2015.14

2. Belazzougui, D., Cunial, F., Kärkkäinen, J., Mäkinen, V.: Linear-time string indexing and analysis in small space. ACM Trans. Algorithms **16**(2), 17:1–17:54 (2020). https://doi.org/10.1145/3381417

3. de Berg, M., Van Kreveld, M., Overmars, M., Schwarzkopf, O.: Computational Geometry: Algorithms and Applications. Springer Science & Business Media, Berlin, Heidelberg (2000). https://doi.org/10.1007/978-3-540-77974-2

4. Bringmann, K., Künnemann, M.: Quadratic conditional lower bounds for string problems and dynamic time warping. In: Guruswami, V. (ed.) IEEE 56th Annual Symposium on Foundations of Computer Science, FOCS 2015, Berkeley, CA, USA, 17–20 October 2015, pp. 79–97. IEEE Computer Society (2015). https://doi.org/10.1109/FOCS.2015.15

5. Cáceres, M.: Parameterized algorithms for string matching to dags: funnels and beyond. In: Bulteau, L., Lipták, Z. (eds.) 34th Annual Symposium on Combinatorial Pattern Matching, CPM 2023, June 26–28, 2023, Marne-la-Vallée, France, France. LIPIcs, vol. 259, pp. 7:1–7:19. Schloss Dagstuhl - Leibniz-Zentrum für Informatik (2023). https://doi.org/10.4230/LIPIcs.CPM.2023.7

6. Caceres, M., Cairo, M., Mumey, B., Rizzi, R., Tomescu, A.I.: Minimum path cover in parameterized linear time. arXiv preprint arXiv:2211.09659 (2022)

7. Cáceres, M., Cairo, M., Mumey, B., Rizzi, R., Tomescu, A.I.: Sparsifying, shrinking and splicing for minimum path cover in parameterized linear time. In: Naor, J.S., Buchbinder, N. (eds.) Proceedings of the 2022 ACM-SIAM Symposium on Discrete Algorithms, SODA 2022, Virtual Conference/Alexandria, VA, USA, 9–12 January 2022, pp. 359–376. SIAM (2022). https://doi.org/10.1137/1.9781611977073.18

8. Chandra, G., Jain, C.: Sequence to graph alignment using gap-sensitive colinear chaining. In: Tang, H. (eds.) Research in Computational Molecular Biology. RECOMB 2023. LNCS, vol. 13976, pp. 58–73. Springer, Cham (2023). https://doi.org/10.1007/978-3-031-29119-7_4

9. Consortium, T.C.P.G.: Computational pan-genomics: status, promises and challenges. Brief. Bioinform. **19**(1), 118–135 (2016). https://doi.org/10.1093/bib/bbw089

10. Cotumaccio, N.: Graphs can be succinctly indexed for pattern matching in $O(|E|^2 + |V|^{5/2})$ time. In: Bilgin, A., Marcellin, M.W., Serra-Sagristà, J., Storer, J.A. (eds.) Data Compression Conference, DCC 2022, Snowbird, UT, USA, 22–25 March 2022, pp. 272–281. IEEE (2022). https://doi.org/10.1109/DCC52660.2022.00035

11. Cotumaccio, N., Prezza, N.: On indexing and compressing finite automata. In: Proceedings of the 2021 ACM-SIAM Symposium on Discrete Algorithms (SODA), pp. 2585–2599. SIAM (2021)

12. Equi, M., Mäkinen, V., Tomescu, A.I.: Graphs cannot be indexed in polynomial time for sub-quadratic time string matching, unless SETH fails. In: Bureš, T., et al. (eds.) SOFSEM 2021. LNCS, vol. 12607, pp. 608–622. Springer, Cham (2021). https://doi.org/10.1007/978-3-030-67731-2_44

13. Equi, M., Mäkinen, V., Tomescu, A.I., Grossi, R.: On the complexity of string matching for graphs. ACM Trans. Algorithms 19(3), 1–25 (2023)

14. Kritikakis, G., Tollis, I.G.: Fast reachability using DAG decomposition. In: Georgiadis, L. (ed.) 21st International Symposium on Experimental Algorithms, SEA 2023, July 24–26 2023, Barcelona, Spain. LIPIcs, vol. 265, pp. 2:1–2:17. Schloss Dagstuhl - Leibniz-Zentrum für Informatik (2023). https://doi.org/10.4230/LIPIcs.SEA.2023.2

15. Li, H., Feng, X., Chu, C.: The design and construction of reference pangenome graphs with minigraph. Genome Biol. 21, 1–19 (2020)

16. Ma, J., Cáceres, M., Salmela, L., Mäkinen, V., Tomescu, A.I.: Chaining for accurate alignment of erroneous long reads to acyclic variation graphs. bioRxiv (2022). https://doi.org/10.1101/2022.01.07.475257, https://www.biorxiv.org/content/early/2022/05/19/2022.01.07.475257, to appear in Bioinformatics

17. Mäkinen, V., Sahlin, K.: Chaining with overlaps revisited. In: Gørtz, I.L., Weimann, O. (eds.) 31st Annual Symposium on Combinatorial Pattern Matching, CPM 2020, 17–19 June 2020, Copenhagen, Denmark. LIPIcs, vol. 161, pp. 25:1–25:12. Schloss Dagstuhl - Leibniz-Zentrum für Informatik (2020). https://doi.org/10.4230/LIPIcs.CPM.2020.25

18. Mäkinen, V., Tomescu, A.I., Kuosmanen, A., Paavilainen, T., Gagie, T., Chikhi, R.: Sparse dynamic programming on DAGs with small width. ACM Trans. Algorithms 15(2), 29:1–29:21 (2019). https://doi.org/10.1145/3301312

19. Rizzo, N., Cáceres, M., Mäkinen, V.: Chaining of maximal exact matches in graphs. https://doi.org/10.48550/arXiv.2302.01748, preprint of an extended version of SPIRE 2023 paper

20. Rizzo, N., Cáceres, M., Mäkinen, V.: Finding maximal exact matches in graphs. In: Belazzougui, D., Ouangraoua, A. (eds.) 23rd International Workshop on Algorithms in Bioinformatics, WABI 2023, September 4–6 2023, Houston, TX, USA. LIPIcs, vol. 273, pp. 10:1–10:17. Schloss Dagstuhl - Leibniz-Zentrum für Informatik (2023). https://doi.org/10.4230/LIPIcs.WABI.2023.10

21. Rizzo, N., Tomescu, A.I., Policriti, A.: Solving string problems on graphs using the labeled direct product. Algorithmica 84(10), 3008–3033 (2022)

22. Shibuya, T., Kurochkin, I.: Match chaining algorithms for cDNA mapping. In: Benson, G., Page, R.D.M. (eds.) WABI 2003. LNCS, vol. 2812, pp. 462–475. Springer, Heidelberg (2003). https://doi.org/10.1007/978-3-540-39763-2_33

Algorithms and Hardness for the Longest Common Subsequence of Three Strings and Related Problems

Lusheng Wang[1,2] and Binhai Zhu[3](✉)

[1] Department of Computer Science, City University of Hong Kong,
Kowloon, Hong Kong
cswangl@cityu.edu.hk
[2] ShenZhen Research Institution, City University of Hong Kong, Shenzhen, China
[3] Gianforte School of Computing, Montana State University,
Bozeman, MT 59717, USA
bhz@montana.edu

Abstract. A string is called a square (resp. cube) if it is in the form of $XX = X^2$ (resp. $XXX = X^3$). Given a sequence S of length n, a fundamental problem studied in the literature is the problem of computing a longest subsequence of S which is a square or cube (i.e., the longest square/cubic subsequence problem). While the longest square subsequence (LSS) can be computed in $O(n^2)$ time, the longest cubic subsequence (LCubS) is only known to be solvable in $O(n^5)$ time, using the longest common subsequence of three strings (LCS-3) as a subroutine (which was much less studied compared with LCS for two strings, or LCS-2). To improve the running time for LCubS, we look at its complementary version and also investigate LCS-3 for three strings S_1, S_2, S_3, with input lengths $m \le n_1 \le n_2$ respectively. Firstly, we generalize an algorithm by Nakatsu et al. for LCS-2 to have an $O(n_1 n_2 \delta)$ algorithm for computing LCS-3, where δ is the minimum number of letters to be deleted in S_1 to have an LCS-3 solution for S_1, S_2 and S_3. This results in an $O(k^3 n^2)$ algorithm for LCubS, where k is the minimum number of letters deleted in S to have a feasible solution. Then, let $\mathcal{R}$ be the number of triples (i, j, k) that match in the input, i.e., $S_1[i] = S_2[j] = S_3[k]$, we show that LCS-3 can be computed in $O(n + \mathcal{R} \log \log n + \mathcal{R}^2)$ time (n is the maximum length of the three input strings). Finally, we define the t-pseudo-subsequence of S under an integer parameter t, which is a string Z containing a subsequence S' of S such that S' can be obtained from Z by deleting at most t letters. Subsequently, we study the longest majority t-pseudo-subsequence (LMtPS) of $S_i, i = 1..3$, which is a t-pseudo-subsequence $T = t_1 t_2 \cdots t_K$ of $S_i, i = 1..3$, with the maximum length K; moreover, when T is aligned with some subsequence S'_i's of length K in $S_i, i = 1..3$, each t_j matches at least two letters with $S'_i, i = 1..3$. We show that LMtPS of three strings S_1, S_2 and S_3 is polynomially solvable, while if we require additionally that all letters in Σ appear in the solution T then it becomes NP-complete, via a reduction to a new SAT instance called Even-(3,B2)-SAT.

F. M. Nardini et al. (Eds.): SPIRE 2023, LNCS 14240, pp. 367–380, 2023.
https://doi.org/10.1007/978-3-031-43980-3_30

Keywords: Longest common subsequence · Longest cubic
subsequence · NP-completeness · Polynomial-time algorithms

1   Introduction

Computing global patterns in a sequence is a fundamental problem with applications. For example, it is known that plants have undergone up to three rounds of whole genome duplications, resulting in a number of duplicates bounded by 8 [19]. In this case, given an extant plant chromosome C, even if some minor mutations occur in between or even after the whole genome duplications, the longest sequences of C in the form of X^2, X^4 and X^8 would give us quite some information about the original chromosome before any whole genome duplication.

In fact, given a sequence S of length n, in 2004 Kosowski already considered the longest square subsequence (LSS, i.e., longest and also in the form of XX) of S, for which he gave an $O(n^2)$ time solution [12]. About 10 years later, Tiskin improved it slightly to $O(n^2(\log\log n)^2/\log^2 n)$, using a method called semi-local string comparison [17]. Shortly after that, Bringmann and Künnemann proved that LSS cannot be solved in time $O(n^{2-\epsilon})$ unless SETH is false [4]. As a matter of fact, Inoue et al. considered solving the problem by introducing the parameters r^* (the length of the optimal LSS) and $\mathcal{R}$ (the number of matching pairs in S), when $r^* = o(n)$ and $\mathcal{R} = o(n^2)$ the LSS can be computed in $o(n^2)$ time [10].

Recently, Lafond et al. considered the *longest subsequence-repeated subsequence* problem of a given sequence S (which models a singleton genome) [13], aiming at retrieving some tandem duplication history, where they need to use *Longest Cubic Subsequence* (LCubS) as a subroutine. They pointed out that the problem can be trivially solved in $O(n^5)$ time (with two cuts cutting S into three substrings, then use the standard dynamic programming algorithm for three sequences). Surprisingly, this trivial solution is the best solution known for LCubS up to this point. The first motivation of this research is to try to solve LCubS faster by considering the complementary version of the problem — delete a minimum number of k letters in S such that the resulting sequence is a cube. We first give a simple $O(k^2n^3)$ time algorithm using some basic observations.

Then, to improve the running time further to $O(k^3n^2)$, we have to come back to the fundamental problem of computing the longest common subsequence of three strings (LCS-3). It turns out that we could extend the solution by Nakatsu et al. for LCS-2 [15] to have a solution running in $O(n^2\delta)$ time, where n is maximum length of the three input strings, and δ is the number of strings one has to be deleted from the shortest input string — to solve LCS-3. In the past, most of research on LCS has been focused on LCS-2 (see [5] for a comprehensive review), although in 1978 Maier already proved that LCS-p (LCS for p sequences, with p unbounded) is NP-hard [14]. In fact, in 1995 Jiang and Li further proved that LCS-p is as hard to approximate as the Maximum Clique problem [11]. (It was noted that Maier's reduction in fact implies that LCS-p is as hard as the Maximum Independent Set problem [20]. With Hästad's stronger inapproximability result, LCS-p hence cannot be approximated within a factor $n^{1-\epsilon}$ [8]; and this was in fact used in showing that aligning many polygonal chains in 3D, modeling protein backbones, is equally hard to approximate [20].) In 2015, Abboud, Backurs and Williams proved that LCS for d strings (d is bounded) cannot be solved in $O(n^{d-\epsilon})$ time unless SETH is false [1], which gives a conditional lower bound close to $\Omega(n^3)$ for LCS-3. For convenience, we just say loosely that this is a cubic lower bound for LCS-3 henceforth.

With the above discussion, we then present another algorithm for LCS-3 parameterized by $\mathcal{R}$, the number of matching triple (i, j, k)'s in the input, i.e., triples satisfying $S_1[i] = S_2[j] = S_3[k]$. Let the lengths of $S_i, i = 1..3$, be $m \leq n_1 \leq n_2(= n)$ respectively. Our algorithm runs in $O(n + \mathcal{R} \log \log n + \mathcal{R}^2)$ time. This algorithm can be used in a scenario when $\mathcal{R}$ is small, for instance, when each letter (gene) appears a constant number of times (say, in a plant genome). To be more precise, if $\mathcal{R} = o(n^{1.5})$ then the algorithm would run in $o(n^3)$ time, beating the cubic lower bound for LCS-3 by Abboud, Backurs and Williams [1]. The algorithm uses the fundamental data structure by van Emde Boas on maintaining a sorted list of integers in the range $[1..n]$ in $O(\log \log n)$ time per insertion and deletion [18], plus the classic algorithm of computing a longest path in a DAG in linear time.

Finally, we propose a new variation of LCS-3. We first define the t-pseudo-subsequence of S under an integer parameter t, which is a string Z containing a subsequence S' of S and S' can be obtained from Z by deleting at most t letters in Z. S' is called the t-host of Z in S. For example, let $S = 123432451$, then $Z = 4243415$ is a 2-pseudo-subsequence of S with $S' = 24341$ being the 2-host of Z in S — S' is obtained from Z by deleting the first and last letters in Z. (It should be easily verified that S' is a subsequence of both S and Z.)

Subsequently, we study the longest *majority* t-pseudo-subsequence (*LMtPS*) of S_1, S_2 and S_3, which is a t-pseudo-subsequence $T = t_1 t_2 \cdots t_K$ of $S_i, i = 1..3$, with the maximum length K; moreover, when T is aligned with three subsequences S_i' of length K in $S_i, i = 1..3$, each t_j matches at least two letters in $S_i', i = 1..3$. (*Note that such a definition on "majority" is only meaningful when the input contains at least three sequences.*) This LMtPS problem is motivated by the fact that in many applications we should tolerate a small amount of errors. In fact, even in string algorithms this idea was not completely new. For example, in 2022 Bhuiyan et al. studied computing the longest common almost-increasing subsequence [3], where the alphabet is a set of comparable items (e.g., integers).

We show that LMtPS of three strings S_1, S_2 and S_3 is polynomially solvable in $O(n^8)$ time. However, if we require additionally that all letters in Σ must appear in the solution T, which we call the problem *LMtPS+*, then we prove that it is NP-complete. The reduction is indirectly from the NP-complete problem *MAX-(3,B2)-SAT*, in which each clause has exactly 3 literals and each variable occurs exactly twice in its positive and twice in the negative form [2]. We require additionally that the number of variables assigned true and false are the same, for which we define a new SAT instance called Even-(3,B2)-SAT and show that it is NP-complete by a reduction from MAX-(3,B2)-SAT.

This paper is organized as follows. In Sect. 2, we give basic definitions. In Sect. 3, we present improved algorithms for the longest cubic subsequence problem, which uses some algorithm for LCS-3. In Sect. 4, we present another algorithm for LCS-3, parameterized by the number of matching triples in the three input strings. In Sect. 5, we prove that LMtPS+ is NP-complete while LMtPS is polynomially solvable. We conclude the paper in Sect. 6.

2   Preliminaries

We first give some basic definitions. Throughout the paper, Σ is a finite alphabet. We use $[n]$ to denote an integer set $\{1, 2, 3, \cdots, n\}$. Let $S = s_1 s_2 s_3 \cdots s_n$ be a string (or sequence) of length $|S| = n$ over an alphabet Σ. When we partition S into $S = S_1 S_2 ... S_m$, the position between the last letter of S_i and first letter of S_{i+1}, $i = 1..m - 1$, is simply called a *cutting point*. A *subsequence* S' of S is a sequence obtained from S by deleting some letters; and, conversely, S is a *supersequence* of S'. A sequence W is a *square* (resp. *cube*) if it can be written as $XX = X^2$ (resp. $XXX = X^3$). For example, abc $\cdot$ abc is a square while abc $\cdot$ abc $\cdot$ abc is a cube.

The longest square (resp. cubic) subsequence of S is a subsequence of S which is a square (resp. cubic). For instance, let $S = \texttt{ACGTAGCTCAGT}$ then ACGT $\cdot$ ACGT is the longest square subsequence of S, while AGT $\cdot$ AGT $\cdot$ AGT is the longest cubic subsequence of S.

Given m strings of the same length n, say $A_i = a_{i,1} a_{i,2} \cdots a_{i,n}$, $i = 1..m$, an alignment of A_i's is composed of n *vertical multi-sets* $V_j = \{a_{1,j}, a_{2,j}, ..., a_{m,j}\}$, $j = 1..n$. Intuitively V_j is the multi-set of letters at the j-th column among all A_i's. Note that our definition of alignment here is stronger than the traditional alignment — in our case, no blank symbols can be used.

The following definitions are somehow covered in the introduction already, we present them here for the reader's convenience. The t-pseudo-subsequence of S under an integer parameter t is a string Z which is a supersequence of S', where S' is a subsequence of S and S' can be obtained from Z by deleting at most t letters in Z. We say that S' is the *t-host* of Z in S. As another example, let $S = \texttt{CAGCGATG}$, then $Z = \texttt{GCAGCAG}$ is a 1-pseudo-subsequence of S with $S' = \texttt{CAGCAG}$ being the 1-host of Z in S.

The decision version of the longest *majority* t-pseudo-subsequence problem (*LMtPS*) is defined as follows:

INPUT: Three sequences S_1, S_2 and S_3 over Σ, positive integers t and K.

QUESTION: Is there a t-pseudo-subsequence $T = t_1 t_2 \cdots t_K$ of $S_i, i = 1..3$, with length K; moreover, when T is aligned with three subsequences S'_i of length K in $S_i, i = 1..3$, each t_j matches at least two letters in $S'_i, i = 1..3$?

For the optimization version, certainly the goal is to maximize K. Note that when T is aligned with S'_1, S'_2 and S'_3, each t_j matches at least two letters in $S'_i (i = 1..3)$ also means that the vertical multi-set V_j (of size 4) contains at least three letters equal to t_j.

We define the problem *LMtPS+* by adding an additional constraint that T must contain all the letters in Σ. Later in Sect. 5 we would prove that LMtPS+ is NP-complete while LMtPS is polynomially solvable.

In the next section, we would first present some improved algorithms for the longest cubic subsequence (LCubS) problem, given an input string S of length n.

3   Improved Algorithms for Longest Cubic Subsequence

As mentioned earlier, the best known algorithm for LCubS runs in $O(n^5)$ time. Here we would try to improve the running time by looking at the complementary problem of LCubS.

3.1   The Complementary Problem

We define the complementary problem of LCubS formally as follows: given a sequence S of length n, delete a minimum number of k letters from S such that the resulting sequence is a cube.

Lemma 1. *Let $S = s_1 s_2 ... s_n$, and assume that the complementary longest cubic subsequence problem on S has a solution of size at most k, then there are at most $4k$ cutting points in S to decompose S into three substrings.*

Proof. Suppose a cutting point in S is between S_1, S_2 or S_3, i.e., $S = S_1 S_2 S_3$; moreover, the final cubic solution is $S_1' S_2' S_3'$ with $S_1' = S_2' = S_3'$ and S_i' is obtained from S_i ($i = 1..3$) after deleting at most k letters in total. Then $n/3 - k \le |S_i'| \le n/3, i = 1, 2, 3$. Consequently, S_1, S_2 and S_3 can be obtained by adding at most k deleted letters back to S_i''s. Hence $n/3 - k \le |S_i| \le n/3 + k, i = 1, 2, 3$. In other words, there are at most $2k$ cuts needed between S_1, S_2 and S_2, S_3. $\square$

Corollary 1. *One could list $O(k^2)$ number of partitions $\langle S_1'', S_2'', S_3'' \rangle$ such that $S = S_1'' S_2'' S_3''$ and an optimal partition $S = S_1 S_2 S_3$ must be among one of these $O(k^2)$ partitions.*

Proof. Even though we have $2k$ cutting points for each S_i, we could make use of the linearity of S. We could first list the cutting points for S_1 (each corresponds to a candidate S_1), there are at most $2k$ of them. Then we repeat the process to cut for a candidate of S_2, again, there are at most $2k$ of them. After S_1 and S_2 are obtained, S_3 is certainly obtained. $\square$

We then have the following theorem.

Theorem 1. *The longest cubic subsequence (also the complementary longest cubic subsequence) problem can be solved in $O(k^2 n^3)$ time, where k is the minimum number of letters deleted in S to obtain a feasible solution.*

It turns out that we could improve this algorithm further to $O(k^3 n^2)$, which means that when $k = O(n^{1/3})$, the improved algorithm in fact runs in $O(n^3)$ time (very much matching the best lower bound). To achieve this, we need a new algorithm for the longest common subsequence problem on three strings. We achieve that by extending the algorithm by Nakatsu et al. for LCS-2 [15].

3.2   Extending the LCS-2 Algorithm by Nakatsu et al. to LCS-3

We briefly review the algorithm of Nakatsu et al. for LCS-2, i.e., we are given two strings σ and τ with lengths m and n_1 ($m \leq n_1$) respectively. (For reader's convenience, we very much adopt the same notations by Nakatsu et al. [15], i.e., we use σ and τ to represent input strings S_1 and S_2.) Let $\sigma = \sigma(1)\sigma(2)\cdots\sigma(m)$ and $\tau = \tau(1)\tau(2)\cdots\tau(n_1)$. Moreover, let $\sigma(i..m) = \sigma(i)\sigma(i+1)\cdots\sigma(m)$ and $\tau(h..n_1) = \tau(h)\tau(h+1)\cdots\tau(n_1)$. The key idea of Nakatsu et al. is defining a concept $L_i(k)$, which is the largest h such that $\sigma(i..m)$ and $\tau(h..n_1)$ has an LCS of length k. Clearly, we have the observation [15].

Observation 1. $L_i(1) > L_i(2) > L_i(3) > \cdots$, for all $i \in [m]$.

Then a table M is defined as $M[i,j] = L_j(i)$, and the remaining task is to fill the table $M[-,-]$. The tricky part is that we only need a part of the upper triangular region of $M[-,-]$ (otherwise, the algorithm would certainly take $O(mn_1)$ time and space). The filling stops when j equals the length p of an LCS, i.e., $L_1(p) \neq 0$ (while $L_1(p+1) = 0$). The running time is analyzed to run in $O(n_1(m-p))$ time.

The generalization of the above algorithm is quite straightforward. Suppose we are given three strings σ, τ and π with lengths m, n_1 and n_2 respectively. (we can assume $\pi = \pi(1)\pi(2)\cdots\pi(n_2)$ and $m \leq n_1 \leq n_2$). Then we could define $L_{i,j}(k)$ as the largest h such that $\sigma(i..m), \tau(j..n_1)$ and $\pi(h..n_2)$ has an LCS of length k. Again, we have the following observation.

Observation 2. When j is fixed, $L_{i,j}(1) > L_{i,j}(2) > L_{i,j}(3) > \cdots$, for all $i \in [m]$.

Then we could define a similar table $M[i,j,k] = L_{j,k}(i)$, which can be filled by adding an outermost loop on $k \in [n_2]$. Hence we have the following theorem.

Theorem 2. *Given three strings σ, τ and π with lengths $m \leq n_1 \leq n_2$ respectively, a longest common subsequence of length p can be computed in $O(n_1 n_2 (m-p))$ time.*

Following the discussion in the previous subsection, we have $O(k^2)$ pairs of cuts to obtain S_1, S_2 and S_3. After each pair of these cuts is fixed, we run the algorithm in Theorem 2, which takes $O(n^2 k)$ time. Hence the total running time to compute a longest cubic subsequence of S is $O(k^2 \cdot n^2 k) = O(k^3 n^2)$.

Theorem 3. *The longest cubic subsequence (also the complementary longest cubic subsequence) problem can be solved in $O(k^3 n^2)$ time, where k is the minimum number of letters deleted in S to obtain a feasible solution.*

In the next section, we investigate LCS-3 by considering another parameter — the number of matching triples in the input strings σ, τ and π.

4 LCS-3 Parameterized by the Number of Matching Triples

Given three strings σ, τ and π with lengths m, n_1 and n_2 respectively, with $m \le n_1 \le n_2 (= n)$, a *matching triple* (i, j, k) is defined as a triple satisfying that $\sigma(i) = \tau(j) = \pi(k)$. Let $R = \{(i, j, k) | (i, j, k)$ is a matching triple of σ, τ and $\pi\}$, and let $\mathcal{R} = |R|$. Clearly, $\mathcal{R} = \Theta(n^3)$ in the worst case — just make three strings of length n using only one single letter in the alphabet.

However, in many applications $\mathcal{R}$ could be potentially small. For instance, plant genomes go over at most three rounds of whole genome duplications [19], hence it is safe to say that a gene in a plant genome is repeated at most 8 times. Consequently, if we take the chromosome of a plant genome as a sequence over its gene set, three such chromosomes from three different plant genomes would incur a value $\mathcal{R}$ which is linear in the length of these chromosomes. Therefore, it makes sense to compute LCS-3 by looking at this important parameter $\mathcal{R}$.

Our initial idea is similar to that of Illiopoulos and Rahman [9], which is to use the data structure by van Emde Boas [18] to maintain a list of integers in the range of $[1..n]$ with a cost of $O(\log \log n)$ per update (i.e., insertion and deletion); in addition, given any element e in the list, Next(e) (successor of e) can be returned in $O(1)$ time.

For each letter $e \in \Sigma$, we build three lists $L_\sigma(e), L_\tau(e)$ and $L_\pi(e)$. $L_\sigma(e)$ stores the (sorted) positions of e in σ. $L_\tau(e)$ and $L_\pi(e)$ can be similarly defined. With a linear scan, the three lists for all $e \in \Sigma$ can be computed in $O(n)$ time.

With these lists, we can construct a van Emde Boas (VEB) data structure $D(\sigma, \tau, \pi)$ on triple (i, j, k)'s, with $i \in L_\sigma(e), j \in L_\tau(e)$ and $k \in L_\pi(e)$. Since $D(\sigma, \tau, \pi)$ must be linearly ordered, we map (i, j, k) uniquely to a number

$$i + (n + 1)j + (n + 1)^2 k.$$

Each of these mapped values is obviously of value $O(n^3)$. Hence, each insertion and deletion in $D(\sigma, \tau, \pi)$ takes $O(\log \log n^3) = O(\log \log n)$ time. All triples in the set R can be computed with three nested loops, first on the lists $L_\sigma(-)$'s, then the lists $L_\tau(-)$'s, and finally the lists $L_\pi(-)$'s. (Note that we have a total of $3 \times |\Sigma|$ such lists across the three input strings.) Consequently, R can be computed in $O(n + \mathcal{R} \log \log n)$ time.

In [9], the remaining steps to obtain a fast algorithm for LCS-2, to avoid a trivial $O(n^2)$ implementation, is to make use of a fast Range Maxima Query on a linear array of numbers (which could be dynamic). However, in our case such a query is not quite possible for a two-dimensional array (in fact, even for a static two-dimensional array there is a lower bound for Range Maxima Query [6,7]). Hence, we need to use a different method. It turns out that we could make use of the linear time algorithm to compute the longest path in a directed acyclic graph (DAG) [16]. Then, the remaining parts are easy to wrap up. Given the data structure $D(\sigma, \tau, \pi)$, we first compute R. We then build a graph G_R whose vertices are all the triples in R. Then for each pair of triples $t_1 = (i_1, j_1, k_1)$ and $t_2 = (i_2, j_2, k_2)$ we test if t_1 is strictly before (or, *precedes*) t_2, i.e., if $i_1 < i_2, j_1 < j_2$ and $k_1 < k_2$; if so, we have a directed edge from t_1 to t_2. (It is easily seen that this graph is acyclic as the precedence relation is transitive: if t_1 precedes t_2 and t_2 precedes $t3$, then t_1 must precede t_3. Moreover, inserting a source s and a sink t to G_R is a standard practice.) Finally, it is easily seen that the longest path from s to t in G_R gives us the longest common subsequence of σ, τ and π. Note that G_R could have $O(\mathcal{R}^2)$ edges, leaving the total cost at $O(\mathcal{R}^2)$ for this step. We summarize with the following theorem.

Theorem 4. *Given three strings σ, τ and π with lengths m, n_1 and n_2 respectively, and $m \leq n_1 \leq n_2(= n)$, let $\mathcal{M}$ be the number of matching triples in σ, τ and π, a longest common subsequence of σ, τ and π can be computed in $O(n + \mathcal{R} \log \log n + \mathcal{R}^2)$ time.*

When $\mathcal{R}$ is relatively small, this algorithm could potentially run below the cubic lower bound for LCS-3. For instance, it could run in $o(n^3)$ time if $\mathcal{R} = o(n^{1.5})$.

In the next section, we consider the longest majority t-pseudo-subsequence (LMtPS) problem for three strings and its relative, LMtPS+.

5   Algorithm and Hardness Result for Longest Majority t-Pseudo-Subsequence and Its Relative

In this section, we prove that LMtPS+ is NP-complete while LMtPS is in P. We focus on the NP-completeness of LMtPS+ first. Recall that in LMtPS+ we require additionally that all letters in Σ must appear in the solution.

5.1   LMtPS+ is NP-Complete

We first review *MAX-(3,B2)-SAT*, which is a 3SAT instance ϕ_1 with n variables $x_1, x_2, \cdots, x_n$ and m clauses $F_1, F_2, \cdots, F_m$, and with the additional condition that each clause contains exactly three literals and each x_i appears exactly twice and each $\bar{x}_i$ also appears twice in ϕ. Berman et al. proved that MAX-(3,B2)-SAT is NP-complete [2].

Theorem 5. MAX-(3,B2)-SAT *is NP-complete [2].*

To prove the NP-completeness for LMtPS+, we need a new variation of MAX-(3,B2)-SAT of $2n$ variables and $2m+2n$ clauses (n and m are the variables and clauses in an instance of MAX-(3,B2)-SAT respectively), where the truth assignment has the property that exactly n variables are assigned TRUE and exactly n variables are assigned FALSE. Such a truth assignment is called an *even* truth assignment. We call this version Even-(3,B2)-SAT. We first show that Even-(3,B2)-SAT is NP-complete.

Theorem 6. *Even-(3,B2)-SAT is NP-complete.*

Proof. It is obvious that Even-(3,B2)-SAT is in NP. Hence we focus on showing that Even-(3,B2)-SAT is NP-hard next.

As noted earlier, we reduce from MAX-(3,B2)-SAT, where the input ϕ_1 is a conjunction of m disjunctive clauses over n variables and each clause contains exactly three literals; moreover, a variable and its negation both appear twice in ϕ_1. It is obvious that a valid truth assignment for a clause, say $(a_{i,1} \vee a_{i,2} \vee a_{i,3})$, is preserved, if we make another copy of the clause (possibly with some name changing).

Follow this idea, a simple method is to construct a new variable y_i for each variable x_i in ϕ such that they are complementary to each other, i.e., their values satisfy that $y_i = \bar{x}_i$ and $\bar{y}_i = x_i$. There is a standard way for doing this and we show that in Fig. 1.

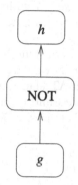

Fig. 1. The negation of g, h, can be enforced by 2SAT clauses $(g \vee h) \wedge (g \vee \bar{h})$.

After introducing a conjunction of $2n$ 2SAT clauses ϕ_3, two for each variable x_i in ϕ (as in Fig. 1), the next step to construct additional m clauses is as follows. For each clause $(a_{i,1} \vee a_{i,2} \vee a_{i,3})$ in ϕ_1, we construct $(\overline{a_{i,1}} \vee \overline{a_{i,2}} \vee \overline{a_{i,3}})$. Then we do a name changing in the second clause with $x_i \rightarrow \bar{y}_i$ and $\bar{x}_i \rightarrow y_i$. Let ϕ_2 be the conjunction of these m new 3SAT clauses involving only with $y_j, \bar{y}_j$. The instance for Even-(3,B2)-SAT is then $\phi = \phi_1 \wedge \phi_2 \wedge \phi_3$. It is clear that this construction takes linear time and ϕ obviously has $2n$ variables and $2m + 2n$ clauses. Moreover, it is easy to see the following relation: ϕ_1 has a truth assignment if and only if ϕ has an even truth assignment. □

Let ϕ be an instance of Even-(3,B2)-SAT, constructed directly from an instance of MAX-(3,B2)-SAT in the above theorem, which is a conjunction of $2m + 2n$ disjunctive clauses over $2n$ variables; moreover, each clause contains at most three literals. Let the $2n$ variables be $x_1, x_2, \cdots, x_n, y_1, y_2, \cdots, y_n$ and let the $2m + 2n$ clauses of ϕ be $F_1, F_2, \cdots, F_{2m+2n}$.

We define $L(i)$ as the list of clauses containing x_i and $\overline{L}(i)$ as the list of clauses containing $\bar{x}_i$ (ordered by the indices of F_k's). Let $g_j, j = 1..2n - 1$, be the peg letters each appearing once in a given string. We define three sequences as follows.

$$S_1 = L(1)\overline{L}(1)g_1 \cdot L(2)\overline{L}(2)g_2 \cdots L(2n-1)\overline{L}(2n-1)g_{2n-1} \cdot L(2n)\overline{L}(2n),$$

$$S_2 = L(1)g_1 \cdot L(2)g_2 \cdots L(2n-1)g_{2n-1} \cdot L(2n),$$

and

$$S_3 = \overline{L}(1)g_1 \cdot \overline{L}(2)g_2 \cdots \overline{L}(2n-1)g_{2n-1} \cdot \overline{L}(2n).$$

We claim that ϕ has an even truth assignment if and only if the LMtPS+ instance has a solution T of length $6n + (2n-1) = 8n - 1$ and T is a majority t-pseudo-subsequence of $S_i, i = 1..3$, with $t = 3n$.

Since the "only-if" part is easy, we only focus on the "if" part. If the LMtPS+ instance has a solution T of length $8n - 1$ which forms a majority $(3n)$-pseudo-subsequence of S_1, S_2 and S_3, the first thing we note is that $|S_1| = 12n + (2n-1) = 14n - 1$ and $|S_2| = |S_3| = 6n + (2n-1) = 8n - 1$ and all the F_i's and g_j's must appear in S. Moreover, by our construction, each $L(i)$ (and $\overline{L}(i)$) must be of length 3. The reason is that each x_i (and $\bar{x}_i$) appears twice in a 3SAT clause in ϕ_1 and each appears once in a 2SAT clause in ϕ_3. Similarly, each y_j (and $\bar{y}_j$) appears twice in a 3SAT clause in ϕ_2 and each appears once in a 2SAT clause in ϕ_3. By the construction of 2SAT clauses, x_i and y_i must have complementary values (or, x_i and $\bar{y}_i$ must have the same T/F value).

Now, it is noted that between g_{i-1} and g_i, we have $L(i)\overline{L}(i)$ in S_1, $L(i)$ in S_2 and $\overline{L}(i)$ in S_3. Since we must put g_{i-1} and g_i in T, for a majority solution we must choose to put either $L(i)$ or $\overline{L}(i)$ in T. Consequently, we set the truth assignment as follows: if $L(i)$ is chosen in T, then set $x_i \leftarrow$ TRUE; if $\overline{L}(i)$ is chosen in T, then set $x_i \leftarrow$ FALSE. Since x_i and $\bar{y}_i$ must have the same T/F value, this assignment is obviously a truth assignment as all F_j's must appear in T — meaning that they are satisfied. Clearly, this truth assignment of x_i's and y_i's form an even truth assignment; moreover, since exactly one of $L(i)$ and $\overline{L}(i)$ is selected for T, T is a $(3n)$-pseudo-subsequence of S_1, S_2 and S_3.

For the membership in NP, it is easily seen that when T and the three subsequences $S_i', i = 1..3$, are given, it is easy to verify whether they form a valid solution in polynomial time. Therefore we have the following theorem.

Theorem 7. *The decision version of LMtPS+ is NP-complete.*

Note that we could make g_i's arbitrarily long, hence t could be arbitrarily small relative to the length of S_1, S_2 and S_3, while the NP-hardness proof still holds. Formally, if $N = \max\{|S_1|, |S_2|, |S_3|\}$ then the NP-hardness result holds even if $\lim_{N \to +\infty} \frac{t}{N} = 0$. (We did not use n here as in this subsection n represents the number of variables in ϕ_1.)

An example of this reduction can be given as follows:

$$\phi_1 = (x_1 \vee x_2 \vee x_3) \wedge (x_1 \vee \bar{x}_2 \vee \bar{x}_3) \wedge (\bar{x}_1 \vee x_2 \vee \bar{x}_3) \wedge (\bar{x}_1 \vee \bar{x}_2 \vee x_3) = F_1 \wedge F_2 \wedge F_3 \wedge F_4,$$

$$\phi_2 = (\bar{y}_1 \vee \bar{y}_2 \vee \bar{y}_3) \wedge (\bar{y}_1 \vee y_2 \vee y_3) \wedge (y_1 \vee \bar{y}_2 \vee y_3) \wedge (y_1 \vee y_2 \vee \bar{y}_3) = F_5 \wedge F_6 \wedge F_7 \wedge F_8,$$

$\phi_3 = (x_1 \vee y_1) \wedge (\bar{x}_1 \vee \bar{y}_1) \wedge (x_2 \vee y_2) \wedge (\bar{x}_2 \vee \bar{y}_2) \wedge (x_3 \vee y_3) \wedge (\bar{x}_3 \vee \bar{y}_3) = F_9 \wedge F_{10} \wedge \cdots \wedge F_{14}.$

Note that we have $m = 4$ and $n = 3$. So ϕ ($= \phi_1 \wedge \phi_2 \wedge \phi_3$) has $2m + 2n = 14$ clauses. Note that they are labelled as they appear in sequence in $\phi_i, i = 1..3$, as $F_1, F_2, \cdots, F_{14}$. Then S_1, S_2 and S_3 can be constructed as follows.

$$S_1 = F_1 F_2 F_5 F_3 F_4 F_6 \cdot g_1 \cdot F_1 F_3 F_7 F_2 F_4 F_8 \cdot g_2 \cdot F_1 F_4 F_9 F_2 F_3 F_{10} \cdot g_3$$

$$F_5 F_{13} F_{14} F_6 F_{11} F_{12} \cdot g_4 \cdot F_7 F_{12} F_{14} F_8 F_{11} F_{13} \cdot g_5 \cdot F_9 F_{12} F_{13} F_{10} F_{11} F_{14}.$$

$$S_2 = F_1 F_2 F_5 \cdot g_1 \cdot F_1 F_3 F_7 \cdot g_2 \cdot F_1 F_4 F_9 \cdot g_3 \cdot F_5 F_{13} F_{14} \cdot g_4 \cdot F_7 F_{12} F_{14} \cdot g_5 \cdot F_9 F_{12} F_{13}.$$

$$S_3 = F_3 F_4 F_6 \cdot g_1 \cdot F_2 F_4 F_8 \cdot g_2 \cdot F_2 F_3 F_{10} \cdot g_3 \cdot F_6 F_{11} F_{12} \cdot g_4 \cdot F_8 F_{11} F_{13} \cdot g_5 \cdot F_{10} F_{11} F_{14}.$$

Corresponding to the even truth assignment $x_1 = $ TRUE, $x_2 = $ FALSE, $x_3 = $ FALSE, $y_1 = $ FALSE, $y_2 = $ TRUE and $y_3 = $ TRUE, the LMtPS+ solution is

$$T = F_1 F_2 F_5 \cdot g_1 \cdot F_2 F_4 F_8 \cdot g_2 \cdot F_2 F_3 F_{10} \cdot g_3 \cdot F_6 F_{11} F_{12} \cdot g_4 \cdot F_7 F_{12} F_{14} \cdot g_5 \cdot F_9 F_{12} F_{13},$$

and

$$S_1' = T, S_2' = S_2, S_3' = S_3.$$

Note that $|S_1| = 12n + (2n - 1) = 41$, $|S_2| = |S_3| = |T| = 6n + (2n - 1) = 23$, but T has 14 matches with both S_2 and S_3. Hence $t = 3n = 23 - 14 = 9$. The t-hosts of T in S_i's are: the subsequence equal to T in S_1, $F_1 F_2 F_5 \cdot g_1 \cdot g_2 \cdot g_3 \cdot g_4 \cdot F_7 F_{12} F_{14} \cdot g_5 \cdot F_9 F_{12} F_{13}$ in S_2, and $g_1 \cdot F_2 F_4 F_8 \cdot g_2 \cdot F_2 F_3 F_{10} \cdot g_3 \cdot F_6 F_{11} F_{12} \cdot g_4 \cdot g_5$ in S_3.

The above reduction breaks for LMtPS in which not all the letters in Σ need to appear in the solution T. It turns out that LMtPS is in fact polynomially solvable.

5.2   LMtPS is Polynomially Solvable

To solve LMtPS, we use a variation of the longest path algorithm for a DAG [16]; namely, deciding if a colored path in a vertex-colored DAG, with at least a certain length and satisfying some constraint on the nodes, exists. The details to solve the problem are as follows.

1. Enumerate all triples (i, j, k) such that $S_1[i], S_2[j]$ and $S_3[k]$ have at least two matches, call them *valid* triples. Construct a DAG G with valid triples as vertices and there is an edge from a valid triple (i_1, j_1, k_1) to another one (i_2, j_2, k_2) if $i_1 < i_2, j_1 < j_2$ and $k_1 < k_2$. Add a source r going to all the other nodes and a sink s coming from all the other nodes (including r), with directed edges.
2. Color the triple nodes sequentially as follows: given (i, j, k), if $S_1[i] = S_2[j] = S_3[k]$, then color it *white*; if $S_1[i] = S_2[j]$ then color it *red*; if $S_2[j] = S_3[k]$ then color it *yellow*; and if $S_1[i] = S_3[k]$ then color it *blue*.

3. Each node v stores a set S_v of 5-vectors in the form $\langle v_K, v_w, v_r, v_y, v_b \rangle$, where v_K represents the length of a path from r to v, v_w (resp. v_r, v_y, v_b) represents the number of white (resp. red, yellow, blue) nodes on this path. Initialize this set S_v by considering the directed path (edge) from r to v. For instance, if v is white, then $S_v \leftarrow \{\langle 1, 1, 0, 0, 0 \rangle\}$.
4. Fix any topological ordering of the DAG G. We can update the 5-vectors of a node v as follows. For each incoming neighbor u (i.e., $(u, v) \in E(G)$), update S_v from S_u as follows: for any 5-vector $(u_K, u_w, u_r, u_y, u_b)$ in S_u, if v is white, then $S_v \leftarrow S_v \cup \{\langle u_K + 1, u_w + 1, u_r, u_y, u_b \rangle\}$ (if v is of a different color, update accordingly). The sink s has no color, hence the 5-vectors in S_s are only updated in the length component.
5. At the end, check if there is a path from r to s with length at least $K+1$ and with at most t nodes in each of the red, yellow and blue colors. If so, return the corresponding solution; otherwise, report no valid solution exists.

Let n be the maximum length of three input strings S_1, S_2 and S_3. The algorithm obviously runs in $O(n^8)$ time as, (1) G could have $O(n^3)$ vertices and $O(n^6)$ edges; and (2) $|S_v| = O(n^5)$ in the worst case (though we could set a bar that once a value in v_r, v_y and v_b, components of a 5-vector in S_v, reaches $t+1$ then we prune the corresponding path from r to v, which could reduce the cost for storing S_v to $O(n^2 t^3)$ — if $t = O(1)$ then the running time of the algorithm can be reduced to $O(n^6)$). It should be noted that we are not storing all the paths from r to v, whose number could certainly be exponential.

Theorem 8. *LMtPS can be solved in $O(n^8)$ time, where n is the maximum length of three input strings S_1, S_2 and S_3.*

6 Concluding Remarks

An interesting question is for the second LCS-3 algorithm, after R is computed, can we compute G_R more carefully so that the number of edges is also linear (or almost linear)? (A similar question can be asked on the algorithm in Sect. 5.2, even though the worst-case running time might not be improved with an affirmative answer.) Obviously, the transitivity property holds in G_R, but the current algorithm does not make use of it to reduce the number of edges in G_R. To be more precise, suppose that there are three directed edges $\langle t_1, t_2 \rangle, \langle t_2, t_3 \rangle$ and $\langle t_1, t_3 \rangle$ in G_R, by transitivity, the third edge between the triples t_1 and t_3 is implied by the first two and hence is redundant. The current algorithm would include such a redundant edge.

Acknowledgments. This research is supported by a grant from National Science Foundation of China (NSFC: 61972328) and GRF grants for Hong Kong Special Administrative Region (CityU 11206120, CityU 11210119).

References

1. Abboud, A., Backurs, A., Williams, V.V.: Tight hardness results for LCS and other sequence similarity measures. In: IEEE 56th Annual Symposium on Foundations of Computer Science (FOCS'15), pp. 59–78. IEEE Computer Society (2015)
2. Berman, P.R., Scott, A.D., Karpinski, M.: Approximation hardness and satisfiability of bounded occurrence instances of SAT. Electron. Colloquium Comput. Complex., TR03-022 (2003)
3. Bhuiyan, M.T.H., Alam, M.R., Rahman, M.S.: Computing the longest common almost-increasing subsequence. Theor. Comput. Sci. **930**, 157–178 (2022)
4. Bringmann, K., Künnemann, M.: Quadratic conditional lower bounds for string problems and dynamic time warping. In: Proceedings of the 56th Annual Symposium on Foundations of Computer Science (FOCS'15), pp. 79–97. IEEE Computer Society (2015)
5. Bringmann, K., Künnemann, M.: Multivariate fine-grained complexity of longest common subsequence. In: Proceedings of the 29th Annual ACM-SIAM Symposium on Discrete Algorithms (SODA'18), pp. 1216–1235. SIAM (2018)
6. Brodal, G.S., Davoodi, P., Rao, S.S.: On space efficient two dimensional range minimum data structures. Algorithmica **63**(4), 815–830 (2012)
7. Demaine, E.D., Landau, G.M., Weimann, O.: On cartesian trees and range minimum queries. In: Albers, S., Marchetti-Spaccamela, A., Matias, Y., Nikoletseas, S., Thomas, W. (eds.) ICALP 2009. LNCS, vol. 5555, pp. 341–353. Springer, Heidelberg (2009). https://doi.org/10.1007/978-3-642-02927-1_29
8. Hästad, J.: Clique is hard to approximate within $n^{1-\epsilon}$. Acta Math. **182**, 105–142 (1999)
9. Iliopoulos, C.S., Rahman, M.S.: A new efficient algorithm for computing the longest common subsequence. Theory Comput. Syst. **45**(2), 355–371 (2009)
10. Inoue, T., Inenaga, S., Bannai, H.: Longest square subsequence problem revisited. In: Boucher, C., Thankachan, S.V. (eds.) SPIRE 2020. LNCS, vol. 12303, pp. 147–154. Springer, Cham (2020). https://doi.org/10.1007/978-3-030-59212-7_11
11. Jiang, T., Li, M.: On the approximation of shortest common supersequences and longest common subsequences. SIAM J. Comput. **24**(5), 1122–1139 (1995)
12. Kosowski, A.: An efficient algorithm for the longest tandem scattered subsequence problem. In: Apostolico, A., Melucci, M. (eds.) SPIRE 2004. LNCS, vol. 3246, pp. 93–100. Springer, Heidelberg (2004). https://doi.org/10.1007/978-3-540-30213-1_13
13. Lafond, M., Lai, W., Liyanage, A., Zhu, B.: The longest subsequence-repeated subsequence problem. arXiv:2302.03797 (2023)
14. Maier, D.: The complexity of some problems on subsequences and supersequences. J. ACM **25**(2), 322–336 (1978)
15. Nakatsu, N., Kambayashi, Y., Yajima, S.: A longest common subsequence algorithm suitable for similar text strings. Acta Informatica **18**, 171–179 (1982)
16. Sedgewick, R., Wayne, K.: Algorithms, 4th edn. Addison-Wesley, Boston (2011)
17. Tiskin, A.: Semi-local string comparison: algorithmic techniques and applications. arXiv:0707.3619 (2013)
18. Peter van Emde Boas: Preserving order in a forest in less than logarithmic time and linear space. Inf. Process. Lett. **6**(3), 80–82 (1977)

19. Zheng, C., Kerr Wall, P., Leebens-Mack, J., de Pamphilis, C., Albert, V.A., Sankoff, D.: Gene loss under neighborhood selection following whole genome duplication and the reconstruction of the ancestral Populus genome. J. Bioinform. Comput. Biol. **7**(03), 499–520 (2009)
20. Zhu, B.: Protein local structure alignment under the discrete Fréchet distance. J. Comput. Biol. **14**(10), 1343–1351 (2007)

Binary Mixed-Digit Data Compression Codes

Igor Zavadskyi(✉) ⓘ and Maksym Kovalchuk ⓘ

Taras Shevchenko National University of Kyiv, 2d Glushkova ave, Kyiv, Ukraine
ihorzavadskyi@knu.ua, max.koval4uk@ukr.net

Abstract. One of the most important trade-offs in data compression is between the compression ratio and decoding speed. The latter can be increased due to a step structure of a codeword length distribution. E.g., in byte-aligned codes (ETDC, SCDC, or RPBC), codewords are composed of whole bytes and thus can be processed easily and quickly. However, this is achieved at the cost of compression ratio. We investigate a new family of data compression codes with codewords composed of digits of different bit lengths. Thus we called them *mixed-digit* codes. Their codeword length distribution is agile, allowing us to outperform the byte-aligned, Fibonacci, and some other recently invented variable-length codes both in compression ratio and decoding speed. We developed and tested the encoding, fast decoding, and optimal code search algorithms.

Keywords: Codes · Compression · Fast decoding · Decoding in parts

1 Introduction

Two main characteristics of lossless data compression methods are the compression ratio and decoding speed. In statistic compression, the compression ratio is theoretically upper-bounded by the Shannon entropy, and codes approaching this bound are well known. They are arithmetic encoding, codes based on Asymmetric Numeration Systems [1], and to some extent, Huffman codes [2]. However, all these codes are not so good in terms of the decoding speed as they require processing the encoded bitstream in a bit-by-bit manner. That is why in recent two decades, special attention was paid to modifying Huffman codes to process the whole bytes of a code or whole codewords. For example, in byte-aligned codes such as Tagged Huffman [3], End Tagged Huffman [4], more advanced (s, c)-dense codes [4] (SCDC) and Byte Codes with Restricted Prefix Properties (RPBC) [5] codewords consist of a whole number of bytes only. This accelerates the decoding significantly at the cost of compression ratio. In word-based text compression (one of the main applications of the above codes), the SCDC and RPBC produce archives 11–16% bigger than Huffman codes.

Another approach to fast decoding relies on using lookup tables consisting of necessary decoding information for all possible values of a bit block. It assumes no limitations regarding an integral number of bytes in a codeword, which improves

F. M. Nardini et al. (Eds.): SPIRE 2023, LNCS 14240, pp. 381–392, 2023.
https://doi.org/10.1007/978-3-031-43980-3_31

the compression ratio compared to byte-aligned codes. An algorithm runs significantly faster if its data fits into the cache memory (preferably L1 cache, which typical size is about 24–128 KB). As a lookup table for n-bit blocks consists of about 2^n elements, several bytes each, blocks should be short enough, preferably at most 12–15 bits. If any codeword fits into a block, the decoding can be extremely simple and fast as, for instance, shown in [6] for no longer than 11 bits Huffman codewords. We need to store the decoded values in one lookup table and the codeword lengths in another and perform only two operations at each iteration of the decoding loop: output the decoded value and shift the encoded bitstream by the codeword length. Of course, this approach is applicable only when the input data dictionary is short, e.g., in character-based text compression. However, in that case, statistic compression often is not efficient enough and gives way to other methods, such as coding of Lempel-Ziv type. For longer alphabets, a problem of efficient decoding of codewords consisting of several bit blocks arises, which is one of the main subjects of this research.

This problem can be solved if a code agrees with the 'decoding in parts' principle formulated in [7]. Let us split the bit representation of a codeword into parts $P_1, \ldots, P_m$. Let $D(P_1), \ldots, D(P_m)$ be the results of independent decoding or other independent transformations of these parts. A code can be decoded in parts if any codeword can be split into parts such that its decoded value is equal to $f(D(P_1), \ldots, D(P_m))$, where f is some easily computed function. Let us note that this is not the case for Huffman codes, as a prefix of a Huffman codeword determines how the following bits should be handled. Thus, every value of a prefix corresponds to a separate decoding table for the suffix, and the total size of the tables grows exponentially depending on the length of a codeword.

Other variable length codes may better correspond to the 'decoding in parts' principle. In [8], the fast decoding method based on lookup tables was developed for the Fibonacci code Fib3, which, applied for natural language texts, under-performs Huffman codes by 4.5–7% in compression ratio. The total size of the Fib3 decoding lookup tables is few megabytes, and the decoding is found to be 2 times slower than for SCDC. The Reverse Multidelimiter Codes (RMD) proposed in [9] outperform the Fib3 both in compression ratio and decoding speed (2–3.5% worse compression than Huffman and up to 30% slower decoding than SCDC/RPBC). The lookup tables for every next part of an RMD codeword depend only on a few rightmost bits of the previous part and the length of a decoded prefix, occupying hundreds of kilobytes of memory in total. The improved version of the RMD fast decoding algorithm given in [10] uses the lookup tables of approximately the same size but performs the decoding 10–15% faster.

The first code fully supporting the 'decoding in parts' principle was proposed in [7]. It is based on the binary-coded ternary number representation (BCT). All codewords have an even bit length as they are composed of trits, i.e., ternary digits 00, 01, and 10, while the pair of bits 11 represents the delimiter. Assume a codeword is split into parts $P_1, \ldots, P_m$ consisting of $t_1, \ldots, t_m$ trits. Then the decoded value can be calculated using a simple recursive formula $D(P_1) + 3^{t_1}(D(P_2) + \cdots + 3^{t_{m-2}}(D(P_{m-1}) + 3^{t_{m-1}}D(P_m)))$, where $D(P_1), \ldots, D(P_m)$ are

Table 1. The first 24 codewords of the code M34

number	codeword	number	codeword	number	codeword	number	codeword
0	111	6	101 1111	12	100 0000 11	18	011 0001 11
1	000 1111	7	110 1111	13	101 0000 11	19	100 0001 11
2	001 1111	8	000 0000 11	14	110 0000 11	20	101 0001 11
3	010 1111	9	001 0000 11	15	000 0001 11	21	110 0001 11
4	011 1111	10	010 0000 11	16	001 0001 11	22	000 0010 11
5	100 1111	11	011 0000 11	17	010 0001 11	23	001 0010 11

the results of parts independent decoding. This principle allows us to construct a high-speed decoding algorithm running more than twice faster than SCDC or RPBC decoding, while the compression ratio of the BCT code is on the level with the Fib3 code. (We assume that the decoding is finding of the index of a codeword in the length-ordered codeword set. Then the decompressed value can be taken as the dictionary element with that index).

Naturally, the question arises whether analogs of the BCT code with different digit bit lengths perform better. Experiments show that the code with 3-bit digits may outperform the BCT code in compression of huge textual databases only, starting from few gigabytes in size. But what if the bit length of a digit depends on its position? Investigating this question, we developed a new, more general code family introduced in Sect. 2, the *binary coded mixed-digit codes* (BCMix). Their codewords are composed of digits, just like in BCT. However, the bitlengths of different digits can differ. Thus, we obtain a flexible code, which can be adjusted to the distribution of source symbol frequencies more tightly by varying the bit lengths of code's digits. We describe the efficient algorithm for finding the optimal code for a given source in Sect. 3 and the fast decoding method in Sect. 4. The results of experiments on compression ratio and decoding speed we discuss in Sect. 5 and make conclusions in Sect. 6.

2   Codes Definition. Straightforward Decoding and Encoding

Let us describe the construction of BCMix codewords. Let b_i be the bit length of the i-th digit and $mask_i = 2^{b_i} - 1$ is the number of possible values of this digit (the b_i-bit value $1 \ldots 1 = mask_i$ is reserved to be the delimiter). We define an n-digit BCMix codeword as $x_0, x_1, \ldots, x_{n-1}(mask_n)$, where $x_0, \ldots, x_{n-1}$ are *digits*, i.e. bit sequences of lengths $b_0, \ldots, b_{n-1}$.

We denote a particular mixed code by the letter M and the bit lengths of several leftmost digits. For example, M3242 denotes a code in which the leftmost digit consists of 3 bits, the second digit of 2 bits, the third digit of 4 bits, and the fourth and all higher digits consist of 2 bits. All digits whose size is not specified in the code name have a length of 2 bits. Thus, M3222 and M3 denote the same code. As an example, the first 24 codewords of code M34 are listed in Table 1.

In a BCMix code, digits have almost the same meaning as in any positional number system. However, the least significant digit is the leftmost digit - this principle allows us to make the left-to-right decoding faster[1].

Let us also define two variables Pow_i and $Pref_i$ denoting the number of i-digit codewords and the number of codewords consisting of less than i digits respectively (excluding the 1-digit delimiter):

$$Pow_i = \begin{cases} 1, & i = 0 \\ Pow_{i-1} \cdot mask_{i-1}, & i > 0 \end{cases}$$

$$Pref_i = \begin{cases} 0, & i = 0 \\ Pref_{i-1} + Pow_{i-1}, & i > 0 \end{cases}$$

To decode the BCMix-codeword $x_0, x_1, \ldots, x_{n-1}(mask_n)$ one can use the Eq. (1). For example, for code M34 $Pow_0 = 1$, $Pow_1 = 7$, $Pref_0 = 0$, $Pref_1 = 1$, $Pref_2 = 8$, and the last codeword in Table 1 can be decoded as $(001)_2 \cdot Pow_0 + (0010)_2 \cdot Pow_1 + Pref_2 = 1 \cdot 1 + 2 \cdot 7 + 8 = 23$.

$$x = \sum_{i=0}^{n-1} (x_i \cdot Pow_i) + Pref_n \tag{1}$$

Quite straightforward Algorithm 1 implements the Eq. (1) to decode a BCMix-encoded bitstream. Let i be the index of the current digit x_i within a codeword. The function $GetDigit(bitstream, i)$ in line 3 reads b_i bits from the bitstream, aligns them to the right edge of a machine word, and shifts the bitstream accordingly. Then we save the obtained value x_i in the variable $digit$ and analyze it. If it is a delimiter, we calculate the $+Pref_n$ part of the Eq. (1) in line 5 and output the decoded number in line 6; otherwise, we add the summand $x_i \cdot Pow_i$ to the result and increase i (lines 9–10).

Algorithm 1: Decoding the BCMix bitstream

---

input : Encoded **bitstream**
output: Sequence of numbers

1 $x, i \leftarrow 0$;
2 **while** *bitstream is not empty* **do**
3 $\quad$ $digit \leftarrow GetDigit(bitstream, i)$;
4 $\quad$ **if** $digit = mask[i]$ **then**
5 $\quad\quad$ $x \leftarrow x + Pref[i]$;
6 $\quad\quad$ **output** x;
7 $\quad\quad$ $x, i \leftarrow 0$;
8 $\quad$ **else**
9 $\quad\quad$ $x \leftarrow x + digit \cdot Pow[i]$;
10 $\quad\quad$ $i \leftarrow i + 1$;

---

[1] If we do not know the number digit length in advance, it is easier to calculate the numerical value from the least significant digit to the most significant. The conventional 'least significant to the right' digit order seems to be borrowed from the right-to-left Arabic script, where it is natural.

The BCMix encoding algorithm is similar to getting digits of a value in any positional number system. The difference is that powers of a number system base are replaced with the array Pow, and the encoded value x is replaced with the position of x-th codeword in the group of codewords of the same digit length (i.e., with $x - Pref_i$, where $Pref_i$ is the largest prefix value that is equal to or less than x). However, to encode a sequence of integers, it would be more time-efficient to store both codewords and their lengths for all numbers up to some value n into two arrays and then get them from those arrays.

Generation of all codewords and their lengths, up to n-th codeword, is given in Algorithm 2. Iterating i from 0 to n, in variable $curNum$, we store the position of the i-th codeword in the group of codewords of digit length $digitN$. In lines 6–13, the i-th codeword is calculated in the i-th element of the array $codes$, and its length - in the i-th element of the array $codeLen$.

Algorithm 2: Generation of all codewords up to some number

---

input : Maximal number **n**
output: Codewords **codes** and their lengths **codeLen**

1 $\quad digitN, curNum \leftarrow 0;$
2 **for** $i \leftarrow 0$ **to** n **do**
3 $\quad\quad$ **if** $curNum = Pow[digitN]$ **then**
4 $\quad\quad\quad digitN \leftarrow digitN + 1;$ $\qquad\qquad\qquad$ // digit length increases
5 $\quad\quad\quad curNum \leftarrow 0;$ $\qquad$ // position among the digitN-integers
6 $\quad\quad codes[i] \leftarrow 0;$
7 $\quad\quad codeLen[i] \leftarrow b[digitN];$
8 $\quad\quad t \leftarrow curNum;$
9 $\quad\quad$ **for** $j \leftarrow 0$ **to** $digitN - 1$ **do**
10 $\quad\quad\quad codes[i] \leftarrow (codes[i] \mid (t \bmod mask[j])) << b[j+1];$
11 $\quad\quad\quad t \leftarrow t \textbf{ div } mask[j];$
12 $\quad\quad\quad codeLen[i] \leftarrow codeLen[i] + b[j];$
13 $\quad\quad codes[i] \leftarrow codes[i] \mid mask[digitN];$
14 $\quad\quad curNum \leftarrow curNum + 1;$

---

3 Searching the Optimal Code

By adjusting the bitlengths of different digits, we can align BCMix codes more tightly to the distribution of source symbol frequencies than BCT or byte-aligned codes. However, we need an efficient algorithm that finds the best code for a given text. The lengths of digits fully determine a BCMix-code. Experiments show that in natural language text compression, it is enough to test 3 possible digit bitlength: 2, 3, and 4 bits. Assuming the biggest codeword consists of 10 digits, there are $3^{10} = 59\,049$ different codes. Taking each of them separately and scanning the whole text to compute the compressed size is impractical. Knowing symbol frequencies is enough to iterate over unique symbols instead of iterating over full text. Nonetheless, the search continues to be relatively slow, taking $O(|\Sigma| \cdot |C|)$ time, where $|\Sigma|$ is the size of an alphabet and $|C|$ is the number of codes. Below we construct the algorithm that finds the optimal code in just $O(|\Sigma| + |C|)$ time, given the ordered frequencies of symbols in the text.

Let us call the *rank* of a symbol its position in the sequence of symbols sorted by descending frequencies. Calculate prefix sums $PSum[i]$ of the array of descending frequencies. Then, in $O(1)$ time, we can answer how many times symbols with ranks from the segment $[l : r]$ occur in the text just by calculating $PSum[r] - PSum[l - 1]$ and assuming $PSum[0] = 0$.

This is the key formula in the optimal code search Algorithm 3 represented as a recursive function BCS with six arguments described in the pseudocode. For a given text, it calculates the compression ratios of all possible BCMix codes, starting with codes consisting of one digit x_0. Apart from arguments and prefix sums array $PSum$, the following constants are used in the function: $|\Sigma|$ is the alphabet size, $minDigitLen$ and $maxDigitLen$ are minimal and maximal bit lengths of code digits, respectively (usually 2 and 4).

At each step of recursion, a new digit is added to a code ($dN+1$ argument in line 7). $fullN$ denotes the number of full codewords consisting of not more than dN digits, including the delimiter, while $incN$ denotes the number of 'incomplete' codewords of length dN, i.e., left parts of longer codewords. A code is constructed when $fullN$ becomes greater or equal to $|\Sigma|$. Then we compare the compressed text size $fullSz$ with the best result among all codes and update this result if needed (line 2). Otherwise, for all possible bit lengths of a new digit, we add it to the code and call the function recursively (lines 4–7).

Let us discuss what happens when the sequence of digits of a code is appended with a new i-bit digit x_{dN}. First, this means adding all $(dN + 1)$-digit full codewords to the set of code's full codewords. Since any incomplete dN-digit codeword can be appended with the delimiter digit x_{dN}, there are $incN$ full codewords ending with x_{dN}. That is why we assign $fullN+incN$ to the argument $fullN$ at the next level of recursion (line 7). Second, the compressed text size is increased by the total length of all new full codewords presented in the text. This value is calculated as the product in line 6: $cLen+i$ is the bit length of a new full codeword, while the function $GetSum(fullN, fullN+incN)$ calculates the total number of words in the text with ranks in the range $[fullN+1; fullN+incN]$. It is equal to $PSum[\min(fullN+incN, |\Sigma|)] - PSum[fullN]$. Third, if we append a dN-digit incomplete codeword with any of 2^i-1 possible values of a non-delimiter digit x_{dN}, we get the $(dN+1)$-digit incomplete codeword, which implies passing the value $incN \cdot (2^i - 1)$ to the argument $incN$ at the next level of recursion. At last, as the bitlength of a code is increased by i, we increase the $cLen$ by i at the next level (line 7).

Initially, the algorithm is invoked as $BCS([\,], 0, 0, 1, 0, 0)$. This means that we start from the digit x_0; there are no full codewords with 0 digits; the length of a 0-digit codeword is 0, as well as the size of a compressed text. However, we assume there is one incomplete codeword of zero length, which will be appended with a delimiter to create a full 1-digit codeword and with all other digits to create incomplete 1-digit codewords.

Obviously, Algorithm 3 has $O(|C|)$ time complexity, where $|C|$ is the number of codes consisting of $|\Sigma|$ codewords. Also, $O(|\Sigma|)$ time is needed to compute the prefix sums, which gives $O(|\Sigma| + |C|)$ time in total. According to experimental results, for all tested texts, it is enough to brute force the lengths of the leftmost four digits. All other digits always consist of 2 bits. Thus, it is enough to test only 81 different codes, which improves the optimal code search time even more. Of course, we also have to calculate and sort the symbol frequencies. However, this is a standard preliminary procedure performed before any statistical encoding.

Algorithm 3: Effective search of the best BCMix code $BCS(b, dN,$ $fullN, incN, cLen, fullSz)$

---

input :
- array of digit lengths, **b**;
- index of the rightmost digit, **dN**;
- number of dN-digit or shorter codewords with a delimiter, **fullN**;
- number of dN-digit incomplete codewords, **incN**;
- bit length of dN-digit codewords, **cLen**;
- size of the compressed text with full codewords only, **fullSz**.

output: Digit lengths of the optimal BCMix code
1 **if** $fullN \geq |\Sigma|$ **then**
2 $\quad$ $updateTheBestCode(b, dN, fullSz)$;
3 $\quad$ **return**;
4 **for** $i \leftarrow minDigitLen$ **to** $maxDigitLen$ **do**
5 $\quad$ $b[dN] \leftarrow i$;
6 $\quad$ $newSz \leftarrow fullSz + (cLen + i) \cdot GetSum(fullN, fullN + incN)$;
7 $\quad$ $BCS(b, dN + 1, fullN + incN, incN \cdot (2^i - 1), cLen + i, newSz)$;

---

4 Fast Decoding

Let us explain how the 'decoding in parts' principle mentioned in the Introduction can be applied to BCMix codes. A BCMix-codeword can be split into parts on digit boundaries, and different lookup tables should be used to decode different parts. Assume the part P of a codeword consists of digits $x_k, \ldots, x_{k+j}$. Consider the function $Dec(P)$ calculating the sum $\sum_{i=k}^{k+j}(x_i \cdot Pow_i)$ (the part of formula 1). Then, if an l-digit codeword is split into parts $P_1, \ldots, P_m$ with $l_1, \ldots, l_m$ digits in each, to decode the whole codeword, we need to calculate the sum

$$Dec(P_1) + \ldots + Dec(P_m) + Pref_l \qquad (2)$$

The main disadvantage of this approach is that we need to use $MaxLen$ instances of lookup tables, where $MaxLen$ is the maximum possible number of digits in a codeword. However, suppose we always start the decoding from the beginning of a codeword, and the lengths of codeword parts are known. In that case, it is enough to store only $Pmax$ instances of lookup tables where

Pmax is the maximum number of codeword parts. This idea is implemented in Algorithm 4.

In the outer loop of Algorithm 4 (lines 4–15), we read an 8-byte word from the bitstream, starting from the byte position *inputPos*, and assign it to the variable *word64*. Then we process part of its 64 bits in the inner loop (lines 9–15), but u leftmost bits remain unprocessed. Thus, at the next iteration of the outer loop, we shift the next 8-byte word by u bits to the left and append it with the unprocessed u bits from the previous iteration (line 5). Therefore, $\lfloor (64 - u)/8 \rfloor$ new full bytes will be processed at the current iteration of the outer loop. This value is calculated in line 6, while values u and *inputPos* are adjusted accordingly in lines 7 and 8.

Algorithm 4: BCMix code fast decoding algorithm

---

input :
- encoded **bitstream**;
- pointer to the lookup tables structure, **tb**;
- byte length of the encoded bitstream, **codeLen**;
- bit length of a codeword part, **blockLen**.

output: Array of decoded numbers **out**

1 $microIters \leftarrow \lfloor 55/blockLen \rfloor$;
2 $blockMask \leftarrow 2^{blockLen} - 1$;
3 $inputPos, outPos, u, word64, out[0..codeLen] \leftarrow 0$;
4 **while** $inputPos < codeLen$ **do**
5      $word64 \leftarrow word64 \mid (\texttt{(*(uint64\_t*)}(inputPos) << u)$;
6      $newBytes \leftarrow \lfloor (64 - u)/8 \rfloor$;
7      $u \leftarrow u + 8 \cdot newBytes$;
8      $inputPos \leftarrow inputPos + newBytes$;
9   **for** $i \leftarrow 0$ **to** $microIters$ **do**
10         $block \leftarrow word64 \mathbin{\&} blockMask$;
11         $word64 \leftarrow word64 >> tb.shift[block]$;
12         $u \leftarrow u - tb.shift[block]$;
13         $out[outPos] \leftarrow out[outPos] + tb.L[block]$;
14         $outPos \leftarrow outPos + tb.n[block]$;
15         $tb \leftarrow tb.nextTable[block]$;

At each iteration of the inner loop, the longest remaining part of the current codeword that consists of the whole number of digits and does not exceed *blockLen* bits is processed. This is done in the following way. In line 10, we assign to variable *block* the rightmost *blockLen* bits of a 64-bit word. All other values we need are taken from lookup tables by index *block*. These tables are combined into a structure referred to by variable *tb*:

- *tb.shift* - number of bits to be processed;
- *tb.n* - number of decoded results (0 or 1);
- *tb.L* - the decoded value of a codeword part;
- *tb.nextTable* - link to the tables for the next part (or the first part if codeword processing is finished at the current iteration).

In lines 11 and 12, we shift the 64-bit word by $tb.shift[block]$ bits to the right and decrease the u respectively. Then in line 13, the 'decoding in parts' principle is implemented. We increase the current output by $tb.L[block]$, which is the decoded value of the current part of a codeword ($Dec(P_i)$ or $Dec(P_m)+Pref_l$ in formula 2). If the processing of a codeword has been finished at the current iteration, the output position is incremented in line 14, and lookup tables for the first part of the next codeword are selected in line 15. Otherwise, $tb.n[block] = 0$, the output position remains the same, and tb is assigned with lookup tables for the next part of the current codeword.

5 Experimental Results

We conduct experiments on compression ratio and decoding time for three texts of different sizes in English. The word-level compression schema is applied. For each text, the dictionary is composed of unique words, ordered by descending frequencies. A word is assumed as a string between two whitespace characters, uppercase and lowercase letters are considered distinct. E.g., 'word', 'word.', and 'Word' are different words. Then we replaced each word with its index in the dictionary and compressed the sequence of indices. Dictionaries are not included in compressed files. Texts parameters are the following:

- *Small* - The Bible, King James version, 4,047,392 bytes, 766,111 words in total, 28,659 unique words, 9.48 bits per word entropy.
- *Middle-sized* - 200 MB English text from Pizza&Chilie corpus, 209,715,200 bytes, 37,003,242 words in total, 836,002 unique words, 11.416 bits per word entropy (http://pizzachili.dcc.uchile.cl/texts/nlang/).
- *Large* - enwik9, English Wikipedia articles collection, 1,000,000,000 bytes, 129,347,859 words in total, 8,859,143 unique words, 13.734 bits per word entropy (https://archive.org/details/enwik9).

The results of text compression are shown in Table 2 together with code parameters and the excess over the entropy in percentage. For every parameterized code family, we choose the code with the best compression ratio (BCMix, SCDC, RPBC, and RMD).

Decoding tests were provided on a PC with an i7-7700HQ processor, 32×4 KB L1 cache, 256×4 KB L2 cache, 6 MB L3 cache, 16 GB RAM, OS Windows 10, Visual Studio 2022 compiler with full optimization. The program code can be found in [11]. The average times of 5000 decodings of the small text, 1000 of the middle-sized, and 200 of the large one are given in Table 3. Only the time of decoding itself, i.e., obtaining the array of word indices in the dictionary, was measured. We did not restore the full text since converting numbers to strings and their concatenation is quite time-consuming and neutralizes the difference between methods' performance.

Codes with parameters giving the best compression ratio (shown in Table 2) were used in decoding benchmark experiments. For BCMix codes, the optimal

Table 2. Empirical comparison of the compression ratio (bytes)

Text\Code	Huffman	BCT	BCMix	SCDC	RPBC	RMD
Small	911,093	956,580	938,709	1,055,686	1,049,003	933,621
	0.35%	5.37%	3.4%	16.28%	15.55%	2.44%
	-	-	M3	$S = 198$	(191,63,1)	$R_{2-\infty}$
Middle-sized	52,919,565	55,459,308	53,994,533	59,857,909	59,259,232	53,879,073
	0.22%	5.03%	2.25%	13.36%	12.22%	2.03%
	-	-	M4	$S = 180$	(151,91,13)	$R_{2,4-\infty}$
Large	222,505,605	235,624,268	226,788,813	249,945,055	246,840,330	228,112,827
	0.19%	6.11%	2.13%	12.56%	11.16%	2.72%
	-	-	M423	$S = 147$	(122,98,35)	$R_{2,4-\infty}$

block length of 10 bits was determined experimentally (the *blockLen* parameter in Algorithm 4). For Huffman codes, we implemented a bit-by-bit decoding algorithm, where the Huffman tree is represented as an array of pairs, each pair indicating 0-bit and 1-bit links to descendant nodes in the tree. Non-leaf and leaf nodes correspond to array elements with non-negative and negative indexes, respectively. If a negative array index is encountered by the decoder, its absolute value is taken as a decoding result. This decoding method appears to be faster than using the canonical Huffman code.

Table 3. Empirical comparison of the decoding time, in milliseconds

Text\Code	Huffman	BCT	BCMix	SCDC	RPBC	RMD
Small	21.02	1.52	3.23	3.71	3.65	4.25
Middle-sized	1684.55	94.43	162.82	212.86	211.71	290.39
Large	13424	372	747.9	801.2	787.1	1292.1

From the information in Tables 2 and 3, also shown in Fig. 1, the following conclusions can be drawn.

1. Digit-based BCMix codes with different digit lengths significantly improve the compression ratio of the BCT code having 2-bit digits.
2. In terms of compression ratio, BCMix codes are outperformed by RMD codes on small and middle-sized texts but compress the large text better. This is because codes with high parameterization need large alphabets to show their advantage. Both BCMix and RMD codes compress any text much better than byte-aligned SCDC or RPBC.
3. BCMix codes can be decoded 25–40% faster than RMD codes and 5–24% faster than byte-aligned codes. In general, BCMix codes occupy a more attractive position on the (compression ratio, decoding speed) plane than both RMD- and byte-aligned codes.

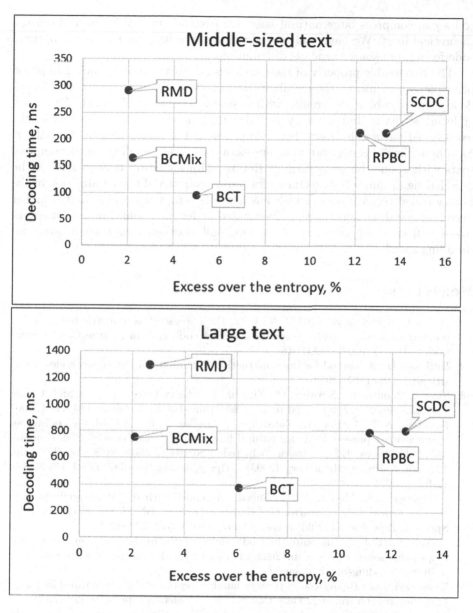

Fig. 1. Compression ratio vs. decoding time.

6 Conclusions

A new class of data compression codes has been investigated. They can be called
digit-aligned or *mixed digit* codes since each codeword consists of a whole number
of digits of different bit lengths. Varying these digit lengths, we can tightly align
a code to the source symbols' distribution. As experiments show, mixed digit

codes can compress large natural language texts about 2–2.5% away from the theoretical limit. We have built the algorithm to efficiently search the optimal code for a given source symbols distribution.

The remarkable property of these codes is supporting the 'decoding in parts' principle, which means the possibility to decode parts of a codeword independently and combine the results with a simple arithmetic formula. This principle allows us to build a very fast decoding algorithm. As a result, in the word-based natural language text compression, the mixed-digit codes outperform byte-aligned codes both in compression ratio and decoding speed, compress texts with about the same ratio as Reverse Multidelimiter codes while can be decoded significantly faster. Our codes can be considered a generalization of the binary-coded ternary code, which has a worse compression ratio but a higher decoding speed. In general, the investigated codes are promising for data compression from the perspective of the trade-off between compression ratio and decoding speed.

References

1. Duda, J., Tahboub, K., Gadgil, N., Delp, E.: The use of asymmetric numeral systems as an accurate replacement for Huffman coding. 2015 Picture Coding Symposium (PCS), pp. 65–69 (2015)
2. Huffman, D.: A method for the construction of minimum-redundancy codes. Proc. IRE **40**, 1098–1101 (1952)
3. Silva de Moura, E., Navarro, G., Ziviani, N., Baeza-Yates, R.: Fast and flexible word searching on compressed text. ACM Trans. Inf. Syst. **18**(2), 113–119 (2000)
4. Brisaboa, N.R., Fariña, A., Navarro, G., Esteller, M.F.: (S,C)-dense coding: an optimized compression code for natural language text databases. In: Nascimento, M.A., de Moura, E.S., Oliveira, A.L. (eds.) SPIRE 2003. LNCS, vol. 2857, pp. 122–136. Springer, Heidelberg (2003). https://doi.org/10.1007/978-3-540-39984-1_10
5. Culpepper, J.S., Moffat, A.: Enhanced byte codes with restricted prefix properties. In: Consens, M., Navarro, G. (eds.) SPIRE 2005. LNCS, vol. 3772, pp. 1–12. Springer, Heidelberg (2005). https://doi.org/10.1007/11575832_1
6. Giesen, F.: Entropy decoding in Oodle data: Huffman decoding on the jaguar. https://fgiesen.wordpress.com/2022/04/04/entropy-decoding-in-oodle-data-huffman-decoding-on-the-jaguar/
7. Zavadskyi, I.O.: Binary-coded ternary number representation in natural language text compression. In: 2022 Data Compression Conference, pp. 419–428 (2022)
8. Klein, S.T., Ben-Nissan, M.: On the usefulness of Fibonacci compression codes. Comput. J. **53**(6), 701–716 (2010)
9. Zavadskyi, I.O., Anisimov, A.V.: Reverse multi-delimiter compression codes. In: 2020 Data Compression Conference, pp. 173–182 (2020)
10. Anisimov, A., Zavadskyi, I., Chudakov, T.: Practical word-based text compression using the reverse multi-delimiter codes. In: Information Technology and Implementation (IT&I-2022), CEUR Workshop Proceedings, pp. 175–183 (2022)
11. Zavadskyi, I., Kovalchuk, M.: The binary mixed digit codes in C programming language. https://github.com/zavadsky/BCMix

Author Index

Printed in the United States
by Baker & Taylor Publisher Services